Oxidative Stress

Canhua Huang • Yuanyuan Zhang
Editors

Oxidative Stress

Human Diseases and Medicine

Springer

Editors
Canhua Huang
Basic Medical Science & Forensic Medicine
West China Hospital of Sichuan University
Chengdu, China

Yuanyuan Zhang
Department of Pharmacology
West China Hospital of Sichuan University
Chengdu, China

ISBN 978-981-16-0521-5 ISBN 978-981-16-0522-2 (eBook)
https://doi.org/10.1007/978-981-16-0522-2

This Springer imprint is published by the registered company Springer Nature Singapore Pte Ltd.
The registered company address is: 152 Beach Road, #21-01/04 Gateway East, Singapore 189721, Singapore

Contents

Oxidative Stress and Antioxidant Strategies in Human Diseases

1

Kefei Yuan and Yuanyuan Zhang

Abstract

The present chapter aims to introduce the reader to its content and indicate the pertinence, relevance, and information of this book. The first section of this chapter, consisting of an overview of oxidative stress, tries to present the basic information of oxidative stress and underlines the importance of oxidative stress in many cellular events. The second part of this chapter elaborates the biological functions and molecular mechanisms of oxidative stress in several common human diseases, including cancer, Alzheimer disease, and cardiovascular disease. More importantly, the second section illustrates the antioxidant strategies to treat the aforementioned human diseases and discusses the achievements, fails, and contradictions of the antioxidant strategies in these diseases.

This chapter that serves as the introduction section provides the basics for the readers to fully appreciate the rest of the book, which consists of the expertise from several excellent scientists in the field of stress medicine.

Keywords

Oxidative stress · Antioxidant strategy · Cancer · Alzheimer disease · Cardiovascular disease

K. Yuan (✉)
Department of Liver Surgery and Liver Transplantation, State Key Laboratory of Biotherapy and Cancer Center, West China Hospital, Sichuan University and Collaborative Innovation Center of Biotherapy, Chengdu, China

Laboratory of Liver Surgery, West China Hospital, Sichuan University, Chengdu, China

Y. Zhang
West China School of Basic Medical Sciences & Forensic Medicine, Sichuan University, Chengdu, PR China

C. Huang, Y. Zhang (eds.), *Oxidative Stress*,
https://doi.org/10.1007/978-981-16-0522-2_1

1.1 What Is Oxidative Stress?

Oxidative stress was first defined as "a disturbance in the prooxidant-antioxidant balance in favor of the former" by a 1985 book entitled *Oxidative Stress* in the introductory chapter [1]. After that, the endogenous and exogenous sources of prooxidants were gradually identified, and the antioxidants, which could be recognized as the metabolic sinks of the prooxidants, were revealed piece by piece [2]. Nowadays, oxidative stress is defined in a more detailed manner as an imbalance between the production of reactive oxygen species (ROS) and reactive nitrogen species (RNS) and antioxidants that favor the former [3]. Likewise, if the production of antioxidants prevails the production of ROS and RNS, it is termed as reductive stress [4].

1.2 Prooxidant and Antioxidant Systems

Prooxidants are mainly consisted of ROS and RNS (Fig. 1.1). According to the chemical nature, ROS include free radicals and non-radical species. Free radicals are molecules or molecular fragments with one or more unpaired electrons, such as superoxide radical anion ($O_2{\bullet}^-$), hydroxyl radical (•OH), peroxyl radicals (ROO•), and alkoxyl radical (RO•). In the meantime, molecules or molecular fragments with paired electrons are termed as non-radicals, for example, hydrogen peroxide (H_2O_2), organic hydroperoxide (ROOH), singlet molecular oxygen ($O_2{}^1\Delta_g$), electronically excited carbonyls (RCO)[a], and ozone (O_3). Likewise, RNS include free radicals (e.g., nitric oxide (NO•), nitrogen dioxide ($NO_2\bullet$)) and non-radicals (nitrite (NO_2^-), nitroxyl anion (NO^-), peroxynitrite ($ONOO^-$), peroxynitrate (O_2NOO^-), and nitrosoperoxycarbonate ($ONOOCO_2^-$)) [5]. The free electron usually makes the free radicals highly reactive. On the other hand, non-radical species are more selective in their reactions.

Prooxidants come from both endogenous and exogenous sources. The endogenous sources of prooxidants are composed of NADPH oxidases, mitochondrial

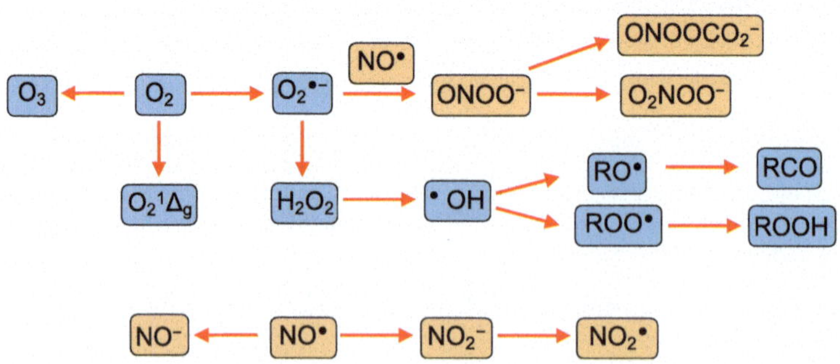

Fig. 1.1 Transformation of ROS and RNS. Blue box represents ROS, and orange box represents RNS

complexes (I, II, and III), xanthine oxidase (XO), and dysfunctional endothelial NOS (eNOS) [6]. Meanwhile, endogenous sources of prooxidants are tobacco smoke, ionizing radiation, organic solvents, certain pollutants, pesticides, and anesthetics [7].

Antioxidants can counteract the effects of prooxidants. The antioxidants in living cells can be divided into two types, low-molecular-weight antioxidant molecules and larger-molecular-weight antioxidant enzymes. The low-molecular-weight antioxidant molecules include glutathione, ascorbate, vitamin E, flavonoids, and carotenoids [8]. Nevertheless, the metabolic sinks of prooxidants, such as superoxide dismutases, catalases, thioredoxin system, glutathione peroxidases, and peroxiredoxins, are recognized as the most important antioxidant enzymes [9].

Oxidation-reduction (redox) homeostasis is critical for life. Under normal physiological conditions, the redox homeostasis is maintained by the balance of prooxidant and antioxidant systems. However, the redox homeostasis is compromised by overproduction of prooxidants that leads to oxidative stress in a series of human diseases, including cancer, Alzheimer disease (AD), and cardiovascular disease (CVD) [10–12].

1.3 Oxidative Stress and Cancer

Oxidative stress caused by uncontrolled formation of ROS has been demonstrated to be a hallmark of cancer [13]. As evidenced by the increase in oxidative stress markers, oxidant generation is implicated in initiation, progression, metastasis, and stem cell and microenvironmental reprogramming of cancer [14]. Cancer cells increase ROS production via turning on the molecular redox switches, which include various signaling pathways and transcription factors [15]. These redox-sensitive transcription factors (e.g., Nrf2 and NF-κB) serve as master regulators of redox homeostasis in cancer cells. They help the cancer cells to rewire their metabolic pathways for adaptation to the nutritional stress and hypoxia during tumor growth and boost cell proliferation, invasion, and migration [16–18].

The transcription factor nuclear factor erythroid 2-related factor 2 (Nrf2) maintains the redox homeostasis in cancer cells by inducing a series of antioxidants and detoxification enzymes [19, 20]. Nrf2 is highly regulated by a cytoplasmic protein, Kelch-like ECH-associated protein 1 (Keap1). Under unstressed conditions, Keap1 binds to Nrf2 and subjects Nrf2 to ubiquitin-dependent protein degradation in combination with an ubiquitin E3 ligase, cullin 3 (Cul3), and thus suppresses the transcriptional activity of Nrf2 [21, 22]. Under oxidative stress, specific cysteinyl residues of Keap1 are oxidized, which causes the disassembly of Nrf2-Keap1 protein complex, blocking the proteasomal degradation of Nrf2 and contributing to the accumulation of Nrf2 in the cytoplasm and its translocation into the nucleus [23]. When localized in the nucleus, Nrf2 is able to promote the transcription of its target genes, such as *Hmox1*, *Nqo1*, *Gclc*, *Gclm*, and *Ggt1*, to enlarge the production of antioxidants for counteracting the oxidative stress [24] (Fig. 1.2).

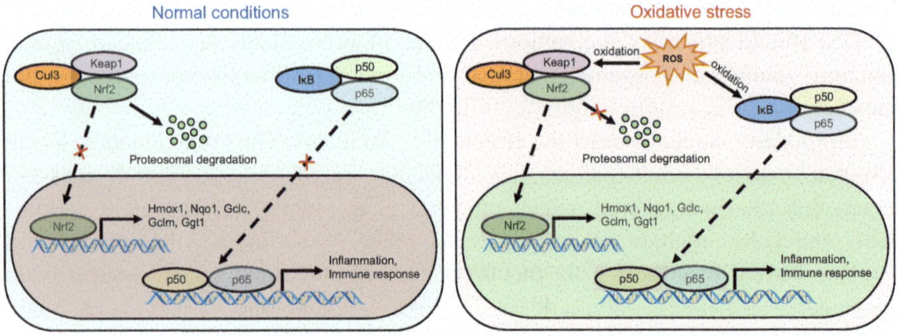

Fig. 1.2 Mechanisms of ROS-induced activation of Nrf2 and NF-κB pathway. Under normal conditions, Nrf2 that binds to Keap1 is degraded by proteasome. NF-κB subunits (p50 and p65) are trapped in the cytoplasm with IκB. Upon oxidative stress, Keap1 and IκB are oxidized by ROS, leading to the nuclear translocation of Nrf2 and NF-κB subunits, therefore contributing to the activation of Nrf2 and NF-κB pathway

Nuclear factor κB (NF-κB) is a widely expressed, pleiotropic transcription factor that can rapidly activate the expression of genes involved in many cellular processes including inflammation and immune responses [25, 26]. In unstimulated cells, NF-κB subunits (p50 and p65) are trapped in the cytoplasm in interaction with the inhibitory subunit of the NF-κB inhibitor (IκB) [27]. Oxidative stress induced by H_2O_2 promotes the phosphorylation of IκB, causing the release of it from NF-κB subunits, thus facilitating the nuclear translocation and activation of NF-κB [28] (Fig. 1.2). On the other hand, NF-κB also possesses redox-sensitive cysteine residues. These cysteine residues locate within the DNA-binding subunits of NF-κB. Oxidation of these cysteine residues significantly repressed the DNA-binding activity of NF-κB [29], while nuclear expression of thioredoxin or redox factor 1 can restore the activity of NF-κB through reduction of the oxidized cysteine residues [30].

1.4 Antioxidant Strategies in Cancer

Since cancer cells rely on the aforementioned signaling pathways mediated by oxidative stress to maintain their survival, it seems rational to target these redox-regulated mechanisms for cancer therapy. Because increased ROS production has been demonstrated to facilitate the development and progression of cancer [31, 32] and the notion that oxidative damage caused by ROS can contribute to tumorigenesis is well established [33], introduction of antioxidants was speculated to be potential anticancer strategy.

Vitamin C (ascorbate or ascorbic acid) is one of the most well-known antioxidant. Vitamin C has already been used to treat cancer in the 1970s and shows some efficacy according to several retrospective studies and case reports [34–37]. However, two subsequent double-blind placebo-controlled prospective trials using the same dose of vitamin C failed to show any benefits [38, 39]. Back then, all those studies used oral administration of high-dose vitamin C only (10 g/day). While it has now been proved that vitamin C levels in human plasma are tightly regulated to maintain physiological concentrations of 50–70 μM, thus, it is impossible to achieve a high plasma vitamin C levels by oral consumption alone [40, 41]. Recent studies have used intravenous (IV) administration of vitamin C at high doses (0.4–1.5 g vitamin C/kg body weight) that can achieve up to 100-fold the concentration of plasma vitamin C compared with oral administration, to treat various solid tumors (e.g., lung cancer, breast cancer, and liver cancer), and gained prominence [42–45]. Nonetheless, the encouraging therapeutic effect of vitamin C on cancer may have nothing to do with its antioxidant nature, because vitamin C exerts its anticancer effect only at high doses, while recent studies have demonstrated that vitamin C functions as an antioxidant at low doses [46], and at higher doses, vitamin C behaves as a prooxidant via targeting GAPDH and contributes to the accumulation of ROS [47].

Vitamin E (α-tocopherol) is another common dietary antioxidant that can protect DNA and cell membranes from oxidative damage [48]. Several studies have demonstrated the tumor-suppressive function of vitamin E in prostate cancer [49, 50]. However, the SELECT trial failed to prove that vitamin E has a beneficial effect on prostate cancer. In fact, this trial showed that the incidence of prostate cancer was actually increased by 17% after vitamin E supplementation [51]. Moreover, a series of other studies showed similar results that vitamin E supplementation did not have any effects on the incidence of colorectal cancer [52] or cause-specific mortality [53].

Nicotinamide (NAM) is a water-soluble form of vitamin B3 (niacin) and the natural precursor for the synthesis of nicotinamide-adenine dinucleotide (NAD^+). Accumulating evidence, including several clinical studies, has demonstrated the vital role of NAM in skin cancer chemoprevention [54–56]. Interestingly, NAM has been shown to exert antioxidative effects in different types of cells [57, 58]. However, whether its chemoprotective effect on skin cancer is related to the antioxidative function remains to be determined.

Bilirubin, the end product of haem catabolism in mammals, is a potent endogenous antioxidant at micromolar concentrations [59]. Bilirubin exerts its antioxidative power through inhibiting the ROS production from mitochondrial respiratory chain [60]. Recent studies have proved that the serum levels of bilirubin are negatively associated with the incidence and mortality of lung cancer and colon cancer [61], suggesting the chemopreventive function of bilirubin in cancer. In addition, phycocyanobilin, a tetrapyrrole analog of bilirubin that can be extracted from the blue-green alga *Spirulina platensis* [62], is also found to function as an antioxidant and is able to repress the tumor growth of pancreatic cancer [63] and the tumorigenesis of breast cancer [64]. Although numerous studies have confirmed the antioxidant roles of bilirubin and its analog, the potential of them to be a clinical antioxidant strategy for cancer therapy remains to be determined because by far there was only one placebo-controlled clinical study with a small sample size showing that consumption of *Spirulina* for a whole year could heal the oral precancerous lesion in tobacco chewers [65].

As summarized above, some studies have proved the efficacy of antioxidants in cancer prevention and treatment. However, most of these reports were based on in vitro studies. The utility of antioxidants in vivo often failed to repress the initiation and growth of cancer. The reasons of this inconsistence might lie in the complex in vivo conditions, such as the type and stage of the tumor, the endogenous antioxidant capacity of the tumor, and the redox state of the tumor niche [66, 67]. Intriguingly, it remains debatable whether antioxidants or prooxidants should be used to treat cancer. In fact, the cytotoxicity of many currently used anticancer drugs is mediated at least partially by their ability to stimulate the production of ROS [68]. The theoretical basis of these prooxidant strategies against cancer is that in cancer cells, the antioxidant capacity is usually heightened to balance the overproduction of ROS driven by hypoxia and metabolic stresses. The heightened antioxidant capacity can contribute to resistance of chemotherapy because many antioxidants, such as GSH, can detoxify the drugs [17]. Therefore, when thinking about preventing or treating cancer via modification of the redox state in cancer cells or tumor microenvironment, it is of great importance to take the particular redox characteristics of the tumor into account.

1.5 Oxidative Stress and Alzheimer Disease

Alzheimer disease (AD) is the leading cause of dementia worldwide and is characterized by diffuse and neuritic extracellular amyloid plaques and intracellular neurofibrillary tangles in the brain, which contribute to dysfunction of synapses [69–71]. Although the etiology and pathogenesis of AD remain unclear, the most prevalent theory of the pathogenesis of AD is that in healthy human brain, the amyloid precursor protein (APP) is cleaved into amyloid-β peptide (Aβ) by β-secretase and γ-secretase. After that, Aβ is cleared by microglial cells to prevent the accumulation of Aβ in the brain, while in AD patient brain, the balance between production and clearance is broken, which facilitates the accumulation of Aβ and the formation of Aβ plaques. Aβ plaques cause progressive synaptic and neuritic injury through inhibition of mitochondrial activity, blockade of proteasome function, activation of inflammatory pathways, and alteration of intracellular calcium levels [72]. Additionally, Aβ interacts with the cholinergic system to suppress the cholinergic transmission. Moreover, Aβ can alter the activities of the phosphorylation signaling pathways to hyperphosphorylate the microtubule-associated protein tau, resulting in the accumulation of neurofibrillary tangles [73]. The accumulation of synaptic and neuritic injury leads to widespread neuronal/neuritic dysfunction and cell death with transmitter deficits, finally contributing to dementia [74].

Oxidative stress has been demonstrated to be closely associated with the pathogenesis and progression of AD [3, 70, 75–85]. Preclinical AD (PCAD) is recognized as the earliest stage of AD. People in this stage have normal cognitive status, while evidence of substantial AD neuropathology can be detected in their brains upon

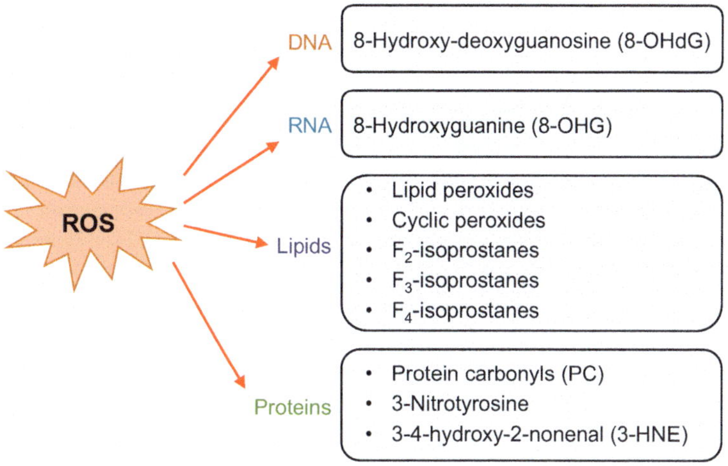

Fig. 1.3 Products generated by oxidative modifications in Alzheimer disease

death and autopsy. After PCAD, patients go through a progressive stage called amnestic mild cognitive impairment (aMCI), which is commonly thought to be a prodromal early stage of AD. People in this stage show certain degree of memory loss that is lesser than patients with AD and can still perform normal activities for daily life. According to estimation of the literature, nearly 15% of persons with aMCI progress into AD every year [86]. Oxidative damage, which is caused by reaction of certain ROS or RNS with biomolecules, is observed during the whole processes of AD from PCAD to AD [3]. For instance, 8-hydroxydeoxyguanosine (8-OHdG), a product of DNA oxidation, and 8-hydroxyguanine (8-OHG), a product of RNA oxidation, can be easily found in the brains of AD patients [87–91]. In addition, the evidence of lipid peroxidation can be found in the brains or cerebrospinal fluid (CSF) of AD patients and aMCI persons [3, 77, 79, 92]. The levels of protein carbonyl (PC), that is, one of the protein oxidation products, are also significantly higher in the brains of AD patients and aMCI persons or cerebrospinal fluid (CSF) compared with normal persons [3, 80, 85] (Fig. 1.3).

Oxidative modification on nuclear DNA has been shown to affect gene transcription via alterations of promoter function and oxidation-driven mutations [88]. On the other hand, oxidative modification on mitochondrial DNA might contribute to AD via impairment of energy metabolism [88, 89]. Likewise, RNA oxidation is able to downregulate protein translation because oxidized RNA is degraded. For example, AUF1 protein has been proved to bind to 8-oxoG and induce selective degradation of oxidized messenger RNA [93]. Interestingly, RNA oxidation can also grant target RNA with novel functions. Oxidized miR-184 associates with the 3′-UTRs of Bcl-xL and Bcl-w that are not its native targets [94]. Moreover, RNA oxidation has been shown to induce translational errors via transcriptional mutation [95]. Recently, accumulating evidence coincidentally identifies that oxidative RNA modification is a hallmark of AD, suggesting that RNA oxidation might contribute to the development and progression of AD [90, 91]. However, the detailed molecular mechanism of oxidative RNA modification in the pathogenesis of AD remains to be determined.

Lipid peroxidation is another well-established feature of AD, as it is easily detected in AD and aMCI brains or CSF and in PCAD hippocampi. Oxidative modification on lipids generates products including lipid peroxides, cyclic peroxides, and isoprostanes (F_2-isoprostanes, F_3-isoprostanes, and F_4-isoprostanes). Most of these oxidized lipids are recognized as biomarkers of AD, especially F_2-isoprostanes and F_4-isoprostanes that are regarded as "gold-standard" biomarkers of lipid peroxidation [79, 96–98]. However, utility of these biomarkers in AD has generated controversial results [99]. Therefore, in order to use lipid peroxidation as biomarkers of AD more wisely, further research is needed to understand the biological functions and mechanisms of oxidative lipid modification in AD.

With the application of redox proteomic techniques, the scientists are able to study the oxidative protein modification specifically [100]. Oxidative modification generates several types of products in AD, including PC-modified, protein-bound 3-nitrotyrosine (NT)-modified, and protein-bound 3-4-hydroxy-2-nonenal (HNE)-modified proteins [100–103]. For instance, oxidative modifications of various glycolytic enzymes are identified in the brain of AD patients, such as α-enolase, triose phosphate isomerase (TPI), aldolase, glyceraldehyde-3-phosphate dehydrogenase (GAPDH), and phosphoglycerate mutase 1 (PGAM1) [78, 100]. In addition, a series of metabolic enzymes located in brain mitochondria can also be oxidized, including ATP synthase, creatine kinase (an enzyme that maintains ATP levels in neurons), and aconitase (a key enzyme that drives synthesis of ATP through OXPHOS) [78, 100, 104]. Oxidative modification of these proteins impairs their enzymatic activities, which results in decreased glucose metabolism and ATP production in the brain of AD patients.

Another mode of oxidative protein modification is called glycoxidation. Glycoxidation is defined as the reaction between the side chain amino groups of lysine residues on proteins and reducing sugars that eventually generates advanced glycation end products (AGEs) [105]. During the pathogenesis of AD, Aβ extracellular fibrillar aggregate, that is, a type of AGEs, can bind to receptors for AGEs (RAGE) in brain endothelial cells and neurons, inducing oxidative stress that leads to vascular dementia and neuronal death in AD [105–109].

More importantly, oxidative protein modification can contribute to the pathogenesis of AD through interference with ubiquitin-proteasome system (UPS). Oxidation-induced dysfunction of ubiquitin carboxy-terminal hydrolase L1 (UCHL1), a key enzyme of UPS, is a typical feature of both AD and aMCI brain [100, 110, 111]. Normally, UCHL1 is in charge of removing ubiquitin from the polymeric ubiquitin chains of damaged proteins to prepare these proteins for entering the 19S cap of the 26S proteasome, while in the brain of AD patients, oxidized UCHL1 failed to play its role, which contributes to the accumulation of ubiquitinylated damaged proteins [112]. Therefore, the dysfunction of UCHL1 by oxidative protein modification leads to neuronal damage and eventually death via disturbing the proteostasis in the human brain.

1.6 Antioxidant Strategies in Alzheimer Disease

Currently, there are very few drugs that have been proved for clinical AD treatment [113]. Most of them are acetylcholinesterase inhibitors, which help to ameliorate the memory deficiency of AD patients by increasing the levels of the neurotransmitter acetylcholine, but do not treat the disease itself [114]. Recently, a novel anti-AD drug is approved in China that is reported to inhibit the progression of AD via remodeling gut microbiota and suppressing gut bacterial amino acids-shaped neuroinflammation [115]. However, none of the approved AD drugs employ antioxidant strategies to treat the disease. In fact, considering that oxidative stress plays such vital roles in the pathogenesis of AD, there have been numerous studies trying to treat AD by inhibition of oxidative stress. Disappointedly, although antioxidant therapies show significant protective effects on AD in studies carried out with neuronal cell cultures or transgenic mice, most of them failed to show similar performance in clinical trials [3, 71, 116–118].

As a common antioxidant, vitamin E has been utilized by a series of studies to treat AD patients. Among these studies, a report using vitamin E (α-tocopherol) at the dosage of 2000 IU a day showed that vitamin E can help to slow the progression of disease in patients with moderately severe impairment from AD [119]. However, the rest of the clinical trials failed to prove that treating AD patients with vitamin E is beneficial [117, 120, 121]. So why did it fail? To answer this question, it is rational to start from the basis of this therapy, that is, whether administration of high-dose vitamin E is able to alleviate oxidative stress in the human brain or not. In mice, vitamin E has been demonstrated to decrease the oxidative modification levels of the brain [122]. While this phenomenon has yet to be confirmed in human, as mentioned above, the human body can restrain the vitamin E in plasma to a certain level. So it is very likely that administration of high-dose vitamin E cannot ensure the transport of extra vitamin E into the brain [123]. Therefore, this treatment is not effective. Another factor that should be considered is the aforementioned studies utilized the most important form of vitamin E, α-tocopherol, which is the major form of vitamin E in the brain. However, there are other three forms of vitamin E (β-, γ-, and ξ-tocopherol) that also hold the potential to function as antioxidants [3]. Interestingly, a recent study has proved that γ-tocopherol is highly reactive in reducing RNS, while α-tocopherol prefers scavenging lipid peroxyl radicals [124]. Therefore, different forms of vitamin E display distinct antioxidative effects. As both ROS and RNS are implicated in the pathogenesis of AD, it might be better for the clinical trials of vitamin E to include both α-tocopherol and γ-tocopherol.

Several epidemiological and intervention studies indicate that "antioxidative" polyphenols (e.g., quercetin and resveratrol), those that are able to enter the brain, may improve memory and cognition in animals and humans [125]. Consistently, one of the "antioxidative" polyphenols, ferulic acid ethyl ester, has shown protective effects of oxidative stress in vivo [116], suggesting its potential to treat AD. The antioxidative function of these polyphenols is probably mediated by their ability to activate Nrf2 and subsequently increase the production of antioxidants, such as GSH [126]. In fact, these polyphenols are actually mild prooxidants that can generate a mild oxidative stress; thus, they are able to induce the activation of Nrf2. According to this theory, it is more prudent to use these polyphenols at relatively low levels to treat AD, as high levels of them are likely to induce oxidative stress and too much activation of Nrf2 may be harmful [127].

Lipid peroxidation in AD is also recognized as a potential drug target. Some preclinical studies have used deuterated lipids to treat AD [128]. The theoretical basis of this therapy is that the C-D bond in deuterated lipids is much stronger than the C-H bond in normal lipids with labile hydrogen at side chains. Therefore, the deuterated lipids are much more resistant to lipid peroxidation. Although some effects have been observed in the preclinical studies, further investigation, especially clinical trial, is needed to clarify the true value of the deuterated lipids in AD therapy.

Combinations of acetylcholinesterase inhibitors with antioxidants are also emerging. As the first approved cholinesterase inhibitor for AD therapy, tacrine (1,2,3,4-tetrahydroacridin-9-amine) has a glorious history in treating AD. However, it was sadly discontinued because of safety issues in the USA (2013) [129], while recently, a series of studies have tried to combine the utilization of tacrine with antioxidants, such as coumarine [130, 131], melatonin [132], and trolox [133]. All these combinations are able to decrease the oxidative stress in AD as they have antioxidative activities and/or metal-chelating abilities. At present, the effect of these combinations on AD in vivo remains to be illustrated.

To sum up, utilization of the antioxidant strategies against AD failed to show significant benefits in clinical trials. There are some valuable experience that we have learned from these studies. First, the optimal dose of antioxidant administration is very important. Second, the stage of AD that is included in clinical trials may affect the outcome of antioxidant therapy, as oxidative stress is an early-stage feature of AD. Third, different antioxidants target different prooxidants, so it is more reasonable to combine the administration of multiple antioxidants to treat AD because various types of prooxidants are implicated in the pathogenesis of AD. Last but not least, there are some novel antioxidant strategies invented for AD therapy, while their value should be tested by clinical trials.

1.7 Oxidative Stress and Cardiovascular Disease

Cardiovascular disease (CVD), which causes the most deaths worldwide with rising morbidity and mortality, is a class of diseases that involve the heart or blood vessels, including atherosclerosis, hypertension, hypercholesterolemia, aortic aneurysm, heart failure, cardiac ischemia-reperfusion injury, myocardial infarction, cardiac arrhythmias, and diabetic vascular complications [134]. Accumulating evidences have demonstrated that sustained oxidative stress is implicated in the pathogenesis of CVD. The main enzymatic sources of ROS in cardiovascular system are NADPH oxidase (NOX), endothelial nitric oxide synthase (eNOS), and mitochondria [135]. Among these ROS sources, NOX plays the leading role in the pathogenesis of CVD, as it is the trigger of other ROS sources. For instance, NOX-derived ROS, such as H_2O_2 and superoxide, can inactivate a vital cofactor of eNOS, tetrahydrobiopterin (H_4B), via oxidative modification, thus leading to the uncoupling of eNOS and switching the production of eNOS from NO to superoxide [136–138]. On the other hand, NOX-derived ROS can also trigger the ROS production of dysfunctional mitochondria through oxidative modification of mitochondrial DNA and components of the membrane permeability transition pore [139–141]. In this section, the detailed biological functions and mechanisms of NOX, eNOS, and dysfunctional mitochondria in the pathogenesis of CVD will be discussed.

NOX is regarded as the predominant source of ROS in CVD [142–144], and its implication in the pathogenesis of CVD has been demonstrated by a series of studies carried out with NOX genetically modified mouse models [145–149]. So far, seven isoforms of NOXs have been identified, including NOX1–NOX5, dual oxidase 1 (DUOX1), and DUOX2 [150]. All NOXs are multi-transmembrane proteins with similar structures. Among them, NOX1–NOX5 are six-transmembrane proteins, while DUOX1 and DUOX2 are seven-transmembrane proteins. As evidenced by structural analysis, the NOXs share a series of common domains, such as six conserved transmembrane domains, four conserved haem-binding histidines, the NADPH-binding domain, and the FAD-binding domain [151]. NOXs exert their function by transferring electron sequentially from NADPH to FAD, FAD to haem, and haem to molecular oxygen, thus contributing to the production of superoxide [152].

NOXs are widely expressed in the cardiovascular system, while their cellular expression patterns and subcellular localizations vary a lot, which endows them with distinct functions in ROS production [153]. The expressions of NOX1, NOX2, NOX4, and NOX5 have been observed in cardiovascular cells [154–156]. Likewise, NOX1 [157], NOX2 [158], NOX4 [157], and NOX5 [159, 160] can also be detected in endothelial cells. Consistently, cardiomyocytes contain NOX1 [161, 162], NOX2 [163, 164], NOX4 [165], and NOX5 [166] as well, while vascular smooth muscle cells only express NOX1 [167], NOX4 [168], and NOX5 [160, 169], with the absence of NOX2.

The subcellular localization of NOXs in the cardiovascular system depends on the cell types where they are expressed. In vascular smooth muscle cells, NOX1 has been reported to localize the caveolae [170]. NOX2 is found in the cytosol and plasma membrane of cardiomyocytes [164] or in the endoplasmic reticulum [171] and perinuclear cytoskeleton [172] of endothelial cells. Likewise, NOX4 is also localized at the endoplasmic reticulum of endothelial cells [171]. Intriguingly, under normal conditions, NOX4 is found in stress fibers, focal adhesions, and nucleus in vascular smooth muscle cells [170, 173, 174], while NOX4 is localized at the endoplasmic reticulum in hypertension patients [175]. In cardiomyocytes, NOX4 has been reported to localize in the nucleus [176] and mitochondria [177]. NOX5 can be found in the endoplasmic reticulum of endothelial cells [159] and the plasma membrane of vascular smooth muscle cells [178]. Most of the NOXs contribute to oxidative stress through production of superoxide, including NOX1, NOX2, and NOX5. Instead of producing superoxide, NOX4 gives rise to the cellular levels of H_2O_2 [179, 180]. The specific biological function of NOX4 may lie in the highly conserved histidine residue of NOX4 [181].

In the pathogenesis process of cardiovascular disease, NOXs are regarded as the driver that initiates the production of ROS, which contribute to sustained oxidative stress via uncoupling eNOS and induction of mitochondrial dysfunction.

eNOS belongs to the nitric oxide synthase (NOS) family. NOS family includes three members, eNOS, neuronal NOS (nNOS), and inducible NOS (iNOS). NOSs are a family of enzymes that catalyze the transition from L-arginine to NO [182]. Under normal conditions, eNOS forms a dimer that is stabilized by H_4B. The dimeric eNOS is able to bind L-arginine (the substrate) and molecular oxygen (co-substrate) at the N-terminal oxygenase domain while binding reduced NADPH (co-substrate that harbors one extra electron) at the C-terminal reductase domain. During the synthesis of NO, eNOS transfers one electron from NADPH to the terminal guanidine group of L-arginine, while under oxidative stress, oxidative modification of H_4B leads to the uncoupling of eNOS dimer, and the monomeric eNOS transfers the electron of reduced NADPH to molecular oxygen to form superoxide rather than NO [183, 184].

Accumulating evidences demonstrate that NOX is an upstream factor of eNOS uncoupling. Angiotensin II, an activator of NOX, causes accumulation of H_2O_2 and subsequent H_4B deficiency and uncoupling of eNOS in bovine endothelial cells [136]. Moreover, utilization of NOX inhibitor, apocynin, can restore the function of H_4B and reverse the uncoupling of eNOS in a hypertension mouse model [185]. Likewise, blocking angiotensin II-induced NOX activation with the angiotensin II receptor type 1 (AT1) inhibitor candesartan can also recouple eNOS [186]. NOX-induced H_4B deficiency and eNOS uncoupling have been demonstrated in various CVDs, including atherosclerosis [187], cardiac IR injury [188], hypertension [189], heart failure [190], and aortic aneurysms [191] (Fig. 1.4).

Mitochondria produce ATP through oxidative phosphorylation [192]. This process relies on the mitochondrial electron transport chain to generate a proton gradient. Under normal conditions, only a few electrons leak from the mitochondrial electron transport chain [193], while upon oxidative stress, superoxide produced by NOX and eNOS can react with NO to generate peroxynitrite, which causes oxidative modifications of proteins within the electron transport chain and membrane lipids of mitochondria [141]. For example, oxidative modified PKC can activate mitoK_{ATP}, which causes the K^+ pours into the mitochondrial matrix and damages the mitochondrial electron transport chain to increase the leak of electrons and the subsequent production of ROS [194, 195]. Mitochondrial dysfunction is commonly observed in aortic aneurysms [196], heart failure [197], atherosclerosis [198], and hypertension [199], suggesting its implication in these CVDs (Fig. 1.4).

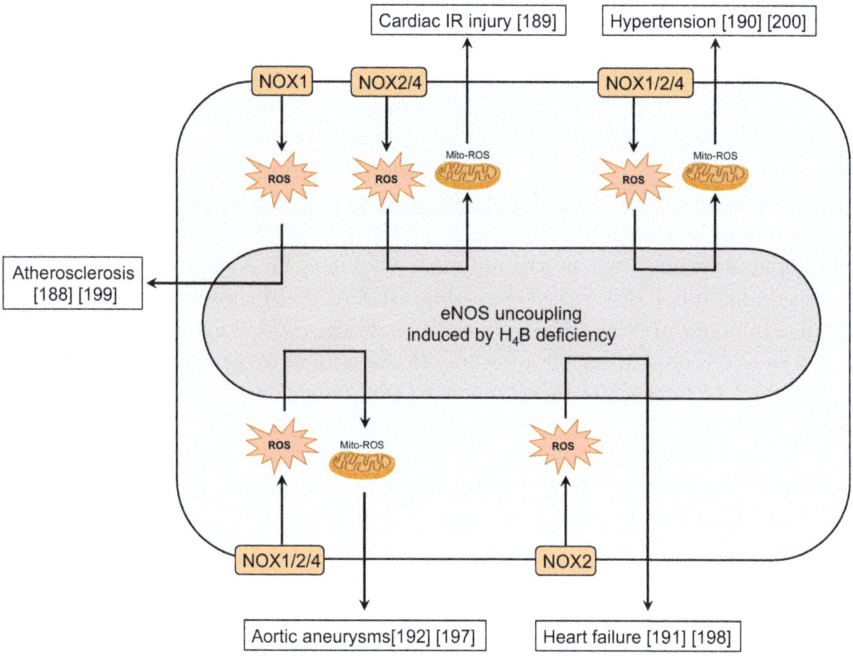

Fig. 1.4 Cross talk of NOXs, eNOS, and mitochondrial dysfunction in the pathogenesis of cardiovascular disease. *NOX* NADPH oxidase, *Mito-ROS* mitochondria-derived reactive oxygen species, *eNOS* endothelial nitric oxide synthase, *IR* ischemia reperfusion, H_4B tetrahydrobiopterin

1.8 Antioxidant Strategies in Cardiovascular Disease

Since NOX, eNOS, and mitochondrial dysfunction are recognized as the key drivers for the pathogenesis of CVD, numerous studies are carried out to invent antioxidant strategies targeting them in order to prevent or treat CVD.

In order to block the function of NOXs, various molecules have been utilized as NOX inhibitors [200]. At first, a series of nonspecific small-molecule inhibitors of NOXs were identified, such as apocynin [201] and diphenyleneiodonium (DPI) [202]. These molecules can inhibit NOXs. Meanwhile, they have strong off-target effects that are able to inhibit other enzymes and systems that produce ROS. For instance, apocynin can decrease the activity of all flavin-containing enzymes [203]. DPI can also inhibit NOS and mitochondrial electron transport chain [204]. These nonspecific NOX inhibitors cannot fulfill the clinical requirement to treat CVD because different isoforms of NOX are implicated in different types of CVD. The ideal NOX inhibitor should be able to selectively target a specific isoform of NOXs without interfering the functions of the rest of the NOXs as they also play important physiological roles in the human body.

Recently, several selective NOX inhibitors are identified. Among them, two pyrazolopyridine dione derivatives, GKT136901 and GKT137831, can selectively inhibit NOX1 and NOX4 [205]. They have been proved to relieve the oxidative stress and are effective in treating CVD in mice, especially GKT137831 [206]. In fact, GKT137831 is regarded as one of the most promising NOX inhibitors at present. It has already been utilized in a phase II clinical trial to treat patients with type 2 diabetes mellitus (T2DM) and albuminuria (NCT02010242).

A series of other specific inhibitors of NOXs, which are identified by high-throughput screening, are newly reported. Of them, GLX481372 is reported to selectively inhibit NOX4 and NOX5, while GLX7013114 is recognized as a highly selective inhibitor of NOX4 [207]. In addition, GSK2795039 is found to selectively inhibit NOX2 [208]. Although according to the present revealed data they all hold the potential to treat CVD by targeting NOXs, their clinical application requires further investigation.

As the downstream ROS production source of NOX, a variety of interventions have been developed to target eNOS. Since uncoupling of eNOS is caused by NOX-induced BH_4 deficiency, numerous studies have been trying to reverse eNOS uncoupling by increasing the levels of BH_4. The supplementation of BH_4 has shown positive results in a set of human studies against CVDs, including coronary artery disease [209], hypertension [210], and heart failure [211]. However, a recent study found that under severe oxidative stress conditions, supplementation of BH_4 could not rescue eNOS uncoupling because BH_4 was rapidly oxidized into BH_2 and lost its biological function [212]. It has been proved that during the pathogenesis of CVD, the supply of L-arginine, the substrate of eNOS, may not be sufficient because of endothelial dysfunction [213]. Therefore, a series of studies were carried out to evaluate the effect of L-arginine supplementation in CVD. However, the results of these studies are contradictory. Some studies showed that L-arginine supplementation was beneficial for CVD patients [214], while others revealed deleterious effects of L-arginine supplementation in treating CVD [215]. The underlying reason for this inconsistence remains to be determined.

Development of mitochondria-targeting antioxidants is another strategy to treat CVD. As the one of most studied mitochondrial antioxidants, MitoQ has been demonstrated to successfully relieve the symptoms of CVD in animal models [216, 217]. Moreover, it has also been used in clinical trials to treat hepatitis C and Parkinson's disease patients (NCT00433108, NCT00329056). The clinical studies proved that administration of MitoQ for 1 year was safe. Therefore, clinical trials of MitoQ to treat CVD are in urgent need to test its efficacy in human. Szeto-Schiller-31 is a small water-soluble tetrapeptide that can scavenge ROS in mitochondria [218]. It has shown protective function against ischemia/reperfusion in animal models and is utilized in three phase II clinical trials to treat heart failure (NCT02914665, NCT02788747, NCT02814097).

1.9 Concluding Remarks

Since oxidative stress was defined in the 1980s, the understanding of its physiological and pathological functions has been improved a lot. At first, oxidative stress was recognized as a merely deleterious factor that can damage almost all biomolecules and impair the function of them, thus leading to the pathogenesis of various diseases. Therefore, antioxidants that can quench oxidative stress have been widely used to treat human diseases that are associated with oxidative stress. However, the therapeutic effects of the antioxidants in human diseases are not satisfactory. In fact, most of the administrations of antioxidants did not show any therapeutic effect on human diseases at all. After that, the physiological functions of oxidative stress have been revealed gradually, and more sophisticated therapeutic methods are proposed. Nowadays, the effects of oxidative stress on cellular events are appreciated as oxidative modifications that regulate various signaling pathways. More importantly, with the understanding of the sources of oxidative stress and its detailed biological functions and mechanisms in the pathogenesis of human diseases, specific antioxidative strategies have been developed that target the key drivers of signaling pathways mediated by oxidative stress. Some of these strategies have already been proved to be beneficial for clinical treatment of a series of human diseases, such as CVD. However, most of the newly developed antioxidative strategies still need to be tested by clinical studies.

References

1. Sies H. Oxidative stress: introductory remarks. London: Elsevier; 1985. p. 1–8.
2. Sies H. Biochemistry of oxidative stress. Angew Chem Int Ed. 1986;25:1058–71.
3. Halliwell B, Gutteridge JM. Free radicals in biology and medicine. Oxford: Oxford University Press; 2015.
4. Rajasekaran NS, Connell P, Christians ES, Yan L-J, Taylor RP, Orosz A, et al. Human αB-crystallin mutation causes oxido-reductive stress and protein aggregation cardiomyopathy in mice. Cell. 2007;130:427–39.
5. Helmut Sies C, Jones D. Oxidative stress. Annu Rev Biochem. 2017;86:715–48.
6. Halliwell B. Free radicals and antioxidants–quo vadis? Trends Pharmacol Sci. 2011;32:125–30.
7. Rao PS, Kalva S, Yerramilli A, Mamidi S. Free radicals and tissue damage: role of antioxidants. Free Radicals Antioxid. 2011;1:2–7.
8. Carrì MT, Ferri A, Cozzolino M, Calabrese L, Rotilio G. Neurodegeneration in amyotrophic lateral sclerosis: the role of oxidative stress and altered homeostasis of metals. Brain Res Bull. 2003;61:365–74.
9. Liu C, Liu Y, Guo K, Fan D, Li G, Zheng Y, et al. Effect of drought on pigments, osmotic adjustment and antioxidant enzymes in six woody plant species in karst habitats of southwestern China. Environ Exp Bot. 2011;71:174–83.
10. Poprac P, Jomova K, Simunkova M, Kollar V, Rhodes CJ, Valko M. Targeting free radicals in oxidative stress-related human diseases. Trends Pharmacol Sci. 2017;38:592–607.
11. Butterfield DA, Halliwell B. Oxidative stress, dysfunctional glucose metabolism and Alzheimer disease. Nat Rev Neurosci. 2019;20:148–60.

12. Daiber A, Chlopicki S. Revisiting pharmacology of oxidative stress and endothelial dysfunction in cardiovascular disease: evidence for redox-based therapies. Free Radic Biol Med. 2020;157:15–37.
13. Hornsveld M, Dansen TB. The hallmarks of cancer from a redox perspective. Antioxid Redox Signal. 2016;25:300–25.
14. Moloney JN, Cotter TG. ROS signalling in the biology of cancer. Semin Cell Dev Biol. 2018;80:50–64.
15. Kalyanaraman B, Cheng G, Hardy M, Ouari O, Bennett B, Zielonka J. Teaching the basics of reactive oxygen species and their relevance to cancer biology: mitochondrial reactive oxygen species detection, redox signaling, and targeted therapies. Redox Biol. 2018;15:347–62.
16. Parascandolo A, Laukkanen MO. Carcinogenesis and reactive oxygen species signaling: interaction of the NADPH oxidase NOX1–5 and superoxide dismutase 1–3 signal transduction pathways. Antioxid Redox Signal. 2019;30:443–86.
17. DeBerardinis RJ, Chandel NS. Fundamentals of cancer metabolism. Sci Adv. 2016;2: e1600200.
18. Kim J, Kim J, Bae J-S. ROS homeostasis and metabolism: a critical liaison for cancer therapy. Exp Mol Med. 2016;48:e269.
19. Kensler TW, Wakabayashi N, Biswal S. Cell survival responses to environmental stresses via the Keap1-Nrf2-ARE pathway. Annu Rev Pharmacol Toxicol. 2007;47:89–116.
20. Malhotra D, Thimmulappa R, Vij N, Navas-Acien A, Sussan T, Merali S, et al. Heightened endoplasmic reticulum stress in the lungs of patients with chronic obstructive pulmonary disease: the role of Nrf2-regulated proteasomal activity. Am J Respir Crit Care Med. 2009;180:1196–207.
21. Eggler AL, Small E, Hannink M, Mesecar AD. Cul3-mediated Nrf2 ubiquitination and antioxidant response element (ARE) activation are dependent on the partial molar volume at position 151 of Keap1. Biochem J. 2009;422:171–80.
22. Yadav SK, Adhikary B, Chand S, Maity B, Bandyopadhyay SK, Chattopadhyay S. Molecular mechanism of indomethacin-induced gastropathy. Free Radic Biol Med. 2012;52:1175–87.
23. Li W, Kong AN. Molecular mechanisms of Nrf2-mediated antioxidant response. Mol Carcinog. 2009;48:91–104.
24. DeNicola GM, Karreth FA, Humpton TJ, Gopinathan A, Wei C, Frese K, et al. Oncogene-induced Nrf2 transcription promotes ROS detoxification and tumorigenesis. Nature. 2011;475:106–9.
25. Baker RG, Hayden MS, Ghosh S. NF-κB, inflammation, and metabolic disease. Cell Metab. 2011;13:11–22.
26. Hayden M, West A, Ghosh S. NF-κ B and the immune response. Oncogene. 2006;25:6758–80.
27. Sen R, Baltimore D. Multiple nuclear factors interact with the immunoglobulin enhancer sequences. Cell. 1986;46:705–16.
28. Schreck R, Rieber P, Baeuerle PA. Reactive oxygen intermediates as apparently widely used messengers in the activation of the NF-kappa B transcription factor and HIV-1. EMBO J. 1991;10:2247–58.
29. Toledano MB, Leonard WJ. Modulation of transcription factor NF-kappa B binding activity by oxidation-reduction in vitro. Proc Natl Acad Sci. 1991;88:4328–32.
30. Hirota K, Murata M, Sachi Y, Nakamura H, Takeuchi J, Mori K, et al. Distinct roles of thioredoxin in the cytoplasm and in the nucleus a two-step mechanism of redox regulation of transcription factor NF-κB. J Biol Chem. 1999;274:27891–7.
31. Myant KB, Cammareri P, McGhee EJ, Ridgway RA, Huels DJ, Cordero JB, et al. ROS production and NF-κB activation triggered by RAC1 facilitate WNT-driven intestinal stem cell proliferation and colorectal cancer initiation. Cell Stem Cell. 2013;12:761–73.
32. Trachootham D, Alexandre J, Huang P. Targeting cancer cells by ROS-mediated mechanisms: a radical therapeutic approach? Nat Rev Drug Discov. 2009;8:579–91.

33. Li M, Fang X, Baker DJ, Guo L, Gao X, Wei Z, et al. The ATM–p53 pathway suppresses aneuploidy-induced tumorigenesis. Proc Natl Acad Sci. 2010;107:14188–93.
34. Cameron E, Campbell A. The orthomolecular treatment of cancer. II. Clinical trial of high-dose ascorbic acid supplements in advanced human cancer. Chem Biol Interact. 1974;9:285–315.
35. Cameron E, Campbell A, Jack T. The orthomolecular treatment of cancer. III. Reticulum cell sarcoma: double complete regression induced by high-dose ascorbic acid therapy. Chem Biol Interact. 1975;11:387–93.
36. Cameron E, Pauling L. Supplemental ascorbate in the supportive treatment of cancer: prolongation of survival times in terminal human cancer. Proc Natl Acad Sci U S A. 1976;73:3685–9.
37. Cameron E, Pauling L. Supplemental ascorbate in the supportive treatment of cancer: reevaluation of prolongation of survival times in terminal human cancer. Proc Natl Acad Sci U S A. 1978;75:4538–42.
38. Creagan ET, Moertel CG, O'Fallon JR, Schutt AJ, O'Connell MJ, Rubin J, et al. Failure of high-dose vitamin C (ascorbic acid) therapy to benefit patients with advanced cancer. A controlled trial. N Engl J Med. 1979;301:687–90.
39. Moertel CG, Fleming TR, Creagan ET, Rubin J, O'Connell MJ, Ames MM. High-dose vitamin C versus placebo in the treatment of patients with advanced cancer who have had no prior chemotherapy. A randomized double-blind comparison. N Engl J Med. 1985;312:137–41.
40. Lindblad M, Tveden-Nyborg P, Lykkesfeldt J. Regulation of vitamin C homeostasis during deficiency. Nutrients. 2013;5:2860–79.
41. Young JI, Zuchner S, Wang G. Regulation of the Epigenome by Vitamin C. Annu Rev Nutr. 2015;35:545–64.
42. Padayatty SJ, Riordan HD, Hewitt SM, Katz A, Hoffer LJ, Levine M. Intravenously administered vitamin C as cancer therapy: three cases. CMAJ. 2006;174:937–42.
43. Stephenson CM, Levin RD, Spector T, Lis CG. Phase I clinical trial to evaluate the safety, tolerability, and pharmacokinetics of high-dose intravenous ascorbic acid in patients with advanced cancer. Cancer Chemother Pharmacol. 2013;72:139–46.
44. Hoffer L, Levine M, Assouline S, Melnychuk D, Padayatty S, Rosadiuk K, et al. Phase I clinical trial of iv ascorbic acid in advanced malignancy. Ann Oncol. 2008;19:1969–74.
45. Raymond YC, Glenda CS, Meng LK. Effects of high doses of vitamin C on cancer patients in Singapore: nine cases. Integr Cancer Ther. 2016;15:197–204.
46. Padayatty SJ, Levine M. Vitamin C: the known and the unknown and Goldilocks. Oral Dis. 2016;22:463–93.
47. Yun J, Mullarky E, Lu C, Bosch KN, Kavalier A, Rivera K, et al. Vitamin C selectively kills KRAS and BRAF mutant colorectal cancer cells by targeting GAPDH. Science. 2015;350:1391–6.
48. Packer L. Protective role of vitamin E in biological systems. Am J Clin Nutr. 1991;53:1050S.
49. Israel K, Sanders BG, Kline K. RRR-alpha-tocopheryl succinate inhibits the proliferation of human prostatic tumor cells with defective cell cycle/differentiation pathways. Nutr Cancer. 1995;24:161–9.
50. Sigounas G, Anagnostou A, Steiner M. Dl-alpha-tocopherol induces apoptosis in erythroleukemia, prostate, and breast cancer cells. Nutr Cancer. 1997;28:30–5.
51. Klein EA, Thompson IM Jr, Tangen CM, Crowley JJ, Lucia MS, Goodman PJ, et al. Vitamin E and the risk of prostate cancer: the selenium and vitamin E cancer prevention trial (SELECT). JAMA. 2011;306:1549–56.
52. Pais R, Dumitrascu DL. Do antioxidants prevent colorectal cancer? A meta-analysis. Rom J Intern Med. 2013;51:152–63.
53. Bjelakovic G, Nikolova D, Gluud LL, Simonetti RG, Gluud C. Antioxidant supplements for prevention of mortality in healthy participants and patients with various diseases. Cochrane Database Syst Rev. 2012;3:CD007176.

54. Chen AC, Martin AJ, Choy B, Fernandez-Penas P, Dalziell RA, McKenzie CA, et al. A phase 3 randomized trial of nicotinamide for skin-cancer chemoprevention. N Engl J Med. 2015;373:1618–26.
55. Surjana D, Halliday GM, Martin AJ, Moloney FJ, Damian DL. Oral nicotinamide reduces actinic keratoses in phase II double-blinded randomized controlled trials. J Invest Dermatol. 2012;132:1497–500.
56. Chen AC, Martin AJ, Dalziell RA, McKenzie CA, Lowe PM, Eris JM, et al. A phase II randomized controlled trial of nicotinamide for skin cancer chemoprevention in renal transplant recipients. Br J Dermatol. 2016;175:1073–5.
57. Choi HJ, Jang SY, Hwang EX. High-dose nicotinamide suppresses ROS generation and augments population expansion during CD8(+) T cell activation. Mol Cells. 2015;38:918–24.
58. Kwang JY, Hwang HJ. Nicotinamide exerts antioxidative effects on senescent cells. Mol Cells. 2015;38:229–35.
59. Stocker R, Yamamoto Y, McDonagh AF, Glazer AN, Ames BN. Bilirubin is an antioxidant of possible physiological importance. Science. 1987;235:1043–6.
60. Zelenka J, Dvorak A, Alan L, Zadinova M, Haluzik M, Vitek L. Hyperbilirubinemia protects against aging-associated inflammation and metabolic deterioration. Oxidative Med Cell Longev. 2016;2016:6190609.
61. Wagner KH, Wallner M, Molzer C, Gazzin S, Bulmer AC, Tiribelli C, et al. Looking to the horizon: the role of bilirubin in the development and prevention of age-related chronic diseases. Clin Sci (Lond). 2015;129:1–25.
62. Vitek L, Majer F, Muchova L, Zelenka J, Jiraskova A, Branny P, et al. Identification of bilirubin reduction products formed by Clostridium perfringens isolated from human neonatal fecal flora. J Chromatogr B Analyt Technol Biomed Life Sci. 2006;833:149–57.
63. Konickova R, Vankova K, Vanikova J, Vanova K, Muchova L, Subhanova I, et al. Anti-cancer effects of blue-green alga Spirulina platensis, a natural source of bilirubin-like tetrapyrrolic compounds. Ann Hepatol. 2014;13:273–83.
64. Ouhtit A, Ismail MF, Othman A, Fernando A, Abdraboh ME, El-Kott AF, et al. Chemoprevention of rat mammary carcinogenesis by spirulina. Am J Pathol. 2014;184:296–303.
65. Mathew B, Sankaranarayanan R, Nair PP, Varghese C, Somanathan T, Amma BP, et al. Evaluation of chemoprevention of oral cancer with Spirulina fusiformis. Nutr Cancer. 1995;24:197–202.
66. Chaiswing L, St Clair WH, St Clair DK. Redox paradox: a novel approach to therapeutics-resistant cancer. Antioxid Redox Signal. 2018;29:1237–72.
67. Panieri E, Santoro MM. ROS homeostasis and metabolism: a dangerous liason in cancer cells. Cell Death Dis. 2016;7:e2253.
68. Yang H, Villani RM, Wang H, Simpson MJ, Roberts MS, Tang M, et al. The role of cellular reactive oxygen species in cancer chemotherapy. J Exp Clin Cancer Res. 2018;37:266.
69. Reitz C, Brayne C, Mayeux R. Epidemiology of Alzheimer disease. Nat Rev Neurol. 2011;7:137–52.
70. Nelson PT, Braak H, Markesbery WR. Neuropathology and cognitive impairment in Alzheimer disease: a complex but coherent relationship. J Neuropathol Exp Neurol. 2009;68:1–14.
71. Martins RN, Villemagne V, Sohrabi HR, Chatterjee P, Shah TM, Verdile G, et al. Alzheimer's disease: a journey from amyloid peptides and oxidative stress, to biomarker technologies and disease prevention strategies-gains from AIBL and DIAN cohort studies. J Alzheimers Dis. 2018;62:965–92.
72. Akundi RS, Huang Z, Eason J, Pandya JD, Zhi L, Cass WA, et al. Increased mitochondrial calcium sensitivity and abnormal expression of innate immunity genes precede dopaminergic defects in Pink1-deficient mice. PLoS One. 2011;6:e16038.
73. Cole SL, Vassar R. The role of amyloid precursor protein processing by BACE1, the beta-secretase, in Alzheimer disease pathophysiology. J Biol Chem. 2008;283:29621–5.

74. Hardy J, Selkoe DJ. The amyloid hypothesis of Alzheimer's disease: progress and problems on the road to therapeutics. Science. 2002;297:353–6.
75. Butterfield DA, Di Domenico F, Barone E. Elevated risk of type 2 diabetes for development of Alzheimer disease: a key role for oxidative stress in brain. Biochim Biophys Acta. 2014;1842:1693–706.
76. Cheignon C, Tomas M, Bonnefont-Rousselot D, Faller P, Hureau C, Collin F. Oxidative stress and the amyloid beta peptide in Alzheimer's disease. Redox Biol. 2018;14:450–64.
77. Halliwell B. Oxidative stress and neurodegeneration: where are we now? J Neurochem. 2006;97:1634–58.
78. Butterfield DA, Boyd-Kimball D. Oxidative stress, amyloid-beta peptide, and altered key molecular pathways in the pathogenesis and progression of Alzheimer's disease. J Alzheimers Dis. 2018;62:1345–67.
79. Nourooz-Zadeh J, Liu EH, Yhlen B, Anggard EE, Halliwell B. F4-isoprostanes as specific marker of docosahexaenoic acid peroxidation in Alzheimer's disease. J Neurochem. 1999;72:734–40.
80. Di Domenico F, Pupo G, Giraldo E, Badia MC, Monllor P, Lloret A, et al. Oxidative signature of cerebrospinal fluid from mild cognitive impairment and Alzheimer disease patients. Free Radic Biol Med. 2016;91:1–9.
81. Di Domenico F, Tramutola A, Butterfield DA. Role of 4-hydroxy-2-nonenal (HNE) in the pathogenesis of Alzheimer disease and other selected age-related neurodegenerative disorders. Free Radic Biol Med. 2017;111:253–61.
82. Lauderback CM, Hackett JM, Huang FF, Keller JN, Szweda LI, Markesbery WR, et al. The glial glutamate transporter, GLT-1, is oxidatively modified by 4-hydroxy-2-nonenal in the Alzheimer's disease brain: the role of Abeta1-42. J Neurochem. 2001;78:413–6.
83. Martins RN, Harper CG, Stokes GB, Masters CL. Increased cerebral glucose-6-phosphate dehydrogenase activity in Alzheimer's disease may reflect oxidative stress. J Neurochem. 1986;46:1042–5.
84. Smith MA, Richey Harris PL, Sayre LM, Beckman JS, Perry G. Widespread peroxynitrite-mediated damage in Alzheimer's disease. J Neurosci. 1997;17:2653–7.
85. Hensley K, Hall N, Subramaniam R, Cole P, Harris M, Aksenov M, et al. Brain regional correspondence between Alzheimer's disease histopathology and biomarkers of protein oxidation. J Neurochem. 1995;65:2146–56.
86. Markesbery WR. Neuropathologic alterations in mild cognitive impairment: a review. J Alzheimers Dis. 2010;19:221–8.
87. Lyras L, Cairns NJ, Jenner A, Jenner P, Halliwell B. An assessment of oxidative damage to proteins, lipids, and DNA in brain from patients with Alzheimer's disease. J Neurochem. 1997;68:2061–9.
88. Santos RX, Correia SC, Zhu X, Lee HG, Petersen RB, Nunomura A, et al. Nuclear and mitochondrial DNA oxidation in Alzheimer's disease. Free Radic Res. 2012;46:565–76.
89. Abolhassani N, Leon J, Sheng Z, Oka S, Hamasaki H, Iwaki T, et al. Molecular pathophysiology of impaired glucose metabolism, mitochondrial dysfunction, and oxidative DNA damage in Alzheimer's disease brain. Mech Ageing Dev. 2017;161:95–104.
90. Shan X, Lin CL. Quantification of oxidized RNAs in Alzheimer's disease. Neurobiol Aging. 2006;27:657–62.
91. Nunomura A, Hofer T, Moreira PI, Castellani RJ, Smith MA, Perry G. RNA oxidation in Alzheimer disease and related neurodegenerative disorders. Acta Neuropathol. 2009;118:151–66.
92. Milne GL, Dai Q, Roberts LJ. The isoprostanes--25 years later. Biochim Biophys Acta. 2015;1851:433–45.
93. Ishii T, Hayakawa H, Igawa T, Sekiguchi T, Sekiguchi M. Specific binding of PCBP1 to heavily oxidized RNA to induce cell death. Proc Natl Acad Sci U S A. 2018;115:6715–20.
94. Wang JX, Gao J, Ding SL, Wang K, Jiao JQ, Wang Y, et al. Oxidative modification of miR-184 enables it to target Bcl-xL and Bcl-w. Mol Cell. 2015;59:50–61.

95. Dai DP, Gan W, Hayakawa H, Zhu JL, Zhang XQ, Hu GX, et al. Transcriptional mutagenesis mediated by 8-oxoG induces translational errors in mammalian cells. Proc Natl Acad Sci U S A. 2018;115:4218–22.
96. Butterfield DA, Reed T, Perluigi M, De Marco C, Coccia R, Cini C, et al. Elevated protein-bound levels of the lipid peroxidation product, 4-hydroxy-2-nonenal, in brain from persons with mild cognitive impairment. Neurosci Lett. 2006;397:170–3.
97. Sultana R, Perluigi M, Butterfield DA. Lipid peroxidation triggers neurodegeneration: a redox proteomics view into the Alzheimer disease brain. Free Radic Biol Med. 2013;62:157–69.
98. Bradley MA, Markesbery WR, Lovell MA. Increased levels of 4-hydroxynonenal and acrolein in the brain in preclinical Alzheimer disease. Free Radic Biol Med. 2010;48:1570–6.
99. Pena-Bautista C, Baquero M, Vento M, Chafer-Pericas C. Free radicals in Alzheimer's disease: lipid peroxidation biomarkers. Clin Chim Acta. 2019;491:85–90.
100. Di Domenico F, Barone E, Perluigi M, Butterfield DA. The triangle of death in Alzheimer's disease brain: the aberrant cross-talk among energy metabolism, mammalian target of rapamycin signaling, and protein homeostasis revealed by redox proteomics. Antioxid Redox Signal. 2017;26:364–87.
101. Butterfield DA, Poon HF, St Clair D, Keller JN, Pierce WM, Klein JB, et al. Redox proteomics identification of oxidatively modified hippocampal proteins in mild cognitive impairment: insights into the development of Alzheimer's disease. Neurobiol Dis. 2006;22:223–32.
102. Reed TT, Pierce WM Jr, Turner DM, Markesbery WR, Butterfield DA. Proteomic identification of nitrated brain proteins in early Alzheimer's disease inferior parietal lobule. J Cell Mol Med. 2009;13:2019–29.
103. Reed T, Perluigi M, Sultana R, Pierce WM, Klein JB, Turner DM, et al. Redox proteomic identification of 4-hydroxy-2-nonenal-modified brain proteins in amnestic mild cognitive impairment: insight into the role of lipid peroxidation in the progression and pathogenesis of Alzheimer's disease. Neurobiol Dis. 2008;30:107–20.
104. Sultana R, Reed T, Perluigi M, Coccia R, Pierce WM, Butterfield DA. Proteomic identification of nitrated brain proteins in amnestic mild cognitive impairment: a regional study. J Cell Mol Med. 2007;11:839–51.
105. Rabbani N, Xue M, Thornalley PJ. Dicarbonyls and glyoxalase in disease mechanisms and clinical therapeutics. Glycoconj J. 2016;33:513–25.
106. Yan SD, Chen X, Fu J, Chen M, Zhu H, Roher A, et al. RAGE and amyloid-beta peptide neurotoxicity in Alzheimer's disease. Nature. 1996;382:685–91.
107. Pugazhenthi S, Qin L, Reddy PH. Common neurodegenerative pathways in obesity, diabetes, and Alzheimer's disease. Biochim Biophys Acta Mol basis Dis. 2017;1863:1037–45.
108. Piras S, Furfaro AL, Piccini A, Passalacqua M, Borghi R, Carminati E, et al. Monomeric Abeta1-42 and RAGE: key players in neuronal differentiation. Neurobiol Aging. 2014;35:1301–8.
109. Emendato A, Milordini G, Zacco E, Sicorello A, Dal Piaz F, Guerrini R, et al. Glycation affects fibril formation of Abeta peptides. J Biol Chem. 2018;293:13100–11.
110. Choi J, Levey AI, Weintraub ST, Rees HD, Gearing M, Chin LS, et al. Oxidative modifications and down-regulation of ubiquitin carboxyl-terminal hydrolase L1 associated with idiopathic Parkinson's and Alzheimer's diseases. J Biol Chem. 2004;279:13256–64.
111. Castegna A, Aksenov M, Aksenova M, Thongboonkerd V, Klein JB, Pierce WM, et al. Proteomic identification of oxidatively modified proteins in Alzheimer's disease brain. Part I: creatine kinase BB, glutamine synthase, and ubiquitin carboxy-terminal hydrolase L-1. Free Radic Biol Med. 2002;33:562–71.
112. Di Domenico F, Head E, Butterfield DA, Perluigi M. Oxidative stress and proteostasis network: culprit and casualty of Alzheimer's-like neurodegeneration. Adv Geriatr. 2014;2014:527518.
113. Cummings J, Aisen PS, DuBois B, Frolich L, Jack CR Jr, Jones RW, et al. Drug development in Alzheimer's disease: the path to 2025. Alzheimers Res Ther. 2016;8:39.

114. Godyn J, Jonczyk J, Panek D, Malawska B. Therapeutic strategies for Alzheimer's disease in clinical trials. Pharmacol Rep. 2016;68:127–38.
115. Wang X, Sun G, Feng T, Zhang J, Huang X, Wang T, et al. Sodium oligomannate therapeutically remodels gut microbiota and suppresses gut bacterial amino acids-shaped neuroinflammation to inhibit Alzheimer's disease progression. Cell Res. 2019;29:787–803.
116. Perluigi M, Joshi G, Sultana R, Calabrese V, De Marco C, Coccia R, et al. In vivo protective effects of ferulic acid ethyl ester against amyloid-beta peptide 1-42-induced oxidative stress. J Neurosci Res. 2006;84:418–26.
117. Polidori MC, Nelles G. Antioxidant clinical trials in mild cognitive impairment and Alzheimer's disease - challenges and perspectives. Curr Pharm Des. 2014;20:3083–92.
118. Butterfield DA, Koppal T, Subramaniam R, Yatin S. Vitamin E as an antioxidant/free radical scavenger against amyloid beta-peptide-induced oxidative stress in neocortical synaptosomal membranes and hippocampal neurons in culture: insights into Alzheimer's disease. Rev Neurosci. 1999;10:141–9.
119. Sano M, Ernesto C, Thomas RG, Klauber MR, Schafer K, Grundman M, et al. A controlled trial of selegiline, alpha-tocopherol, or both as treatment for Alzheimer's disease. The Alzheimer's Disease Cooperative Study. N Engl J Med. 1997;336:1216–22.
120. Winblad B, Amouyel P, Andrieu S, Ballard C, Brayne C, Brodaty H, et al. Defeating Alzheimer's disease and other dementias: a priority for European science and society. Lancet Neurol. 2016;15:455–532.
121. Petersen RC, Thomas RG, Grundman M, Bennett D, Doody R, Ferris S, et al. Vitamin E and donepezil for the treatment of mild cognitive impairment. N Engl J Med. 2005;352:2379–88.
122. Ulatowski L, Parker R, Warrier G, Sultana R, Butterfield DA, Manor D. Vitamin E is essential for Purkinje neuron integrity. Neuroscience. 2014;260:120–9.
123. Spector R, Johanson CE. Vitamin transport and homeostasis in mammalian brain: focus on Vitamins B and E. J Neurochem. 2007;103:425–38.
124. Pei R, Mah E, Leonard SW, Traber MG, Bruno RS. Alpha-tocopherol supplementation reduces 5-nitro-gamma-tocopherol accumulation by decreasing gamma-tocopherol in young adult smokers. Free Radic Res. 2015;49:1114–21.
125. Schaffer S, Halliwell B. Do polyphenols enter the brain and does it matter? Some theoretical and practical considerations. Genes Nutr. 2012;7:99–109.
126. Ma Q, He X. Molecular basis of electrophilic and oxidative defense: promises and perils of Nrf2. Pharmacol Rev. 2012;64:1055–81.
127. Huang Z, Mou Y, Xu X, Zhao D, Lai Y, Xu Y, et al. Novel derivative of bardoxolone methyl improves safety for the treatment of diabetic nephropathy. J Med Chem. 2017;60:8847–57.
128. Raefsky SM, Furman R, Milne G, Pollock E, Axelsen P, Mattson MP, et al. Deuterated polyunsaturated fatty acids reduce brain lipid peroxidation and hippocampal amyloid beta-peptide levels, without discernable behavioral effects in an APP/PS1 mutant transgenic mouse model of Alzheimer's disease. Neurobiol Aging. 2018;66:165–76.
129. Brinkman SD, Gershon S. Measurement of cholinergic drug effects on memory in Alzheimer's disease. Neurobiol Aging. 1983;4:139–45.
130. Sun Q, Peng DY, Yang SG, Zhu XL, Yang WC, Yang GF. Syntheses of coumarin-tacrine hybrids as dual-site acetylcholinesterase inhibitors and their activity against butylcholinesterase, Abeta aggregation, and beta-secretase. Bioorg Med Chem. 2014;22:4784–91.
131. Hamulakova S, Poprac P, Jomova K, Brezova V, Lauro P, Drostinova L, et al. Targeting copper(II)-induced oxidative stress and the acetylcholinesterase system in Alzheimer's disease using multifunctional tacrine-coumarin hybrid molecules. J Inorg Biochem. 2016;161:52–62.
132. Rodriguez-Franco MI, Fernandez-Bachiller MI, Perez C, Hernandez-Ledesma B, Bartolome B. Novel tacrine-melatonin hybrids as dual-acting drugs for Alzheimer disease, with improved acetylcholinesterase inhibitory and antioxidant properties. J Med Chem. 2006;49:459–62.
133. Nepovimova E, Korabecny J, Dolezal R, Babkova K, Ondrejicek A, Jun D, et al. Tacrine-trolox hybrids: a novel class of centrally active, nonhepatotoxic multi-target-directed ligands

exerting anticholinesterase and antioxidant activities with low in vivo toxicity. J Med Chem. 2015;58:8985–9003.
134. Zhang Y, Murugesan P, Huang K, Cai H. NADPH oxidases and oxidase crosstalk in cardiovascular diseases: novel therapeutic targets. Nat Rev Cardiol. 2020;17:170–94.
135. Cai H, Harrison DG. Endothelial dysfunction in cardiovascular diseases: the role of oxidant stress. Circ Res. 2000;87:840–4.
136. Chalupsky K, Cai H. Endothelial dihydrofolate reductase: critical for nitric oxide bioavailability and role in angiotensin II uncoupling of endothelial nitric oxide synthase. Proc Natl Acad Sci U S A. 2005;102:9056–61.
137. Takimoto E, Champion HC, Li M, Ren S, Rodriguez ER, Tavazzi B, et al. Oxidant stress from nitric oxide synthase-3 uncoupling stimulates cardiac pathologic remodeling from chronic pressure load. J Clin Invest. 2005;115:1221–31.
138. Du YH, Guan YY, Alp NJ, Channon KM, Chen AF. Endothelium-specific GTP cyclohydrolase I overexpression attenuates blood pressure progression in salt-sensitive low-renin hypertension. Circulation. 2008;117:1045–54.
139. Zorov DB, Filburn CR, Klotz LO, Zweier JL, Sollott SJ. Reactive oxygen species (ROS)-induced ROS release: a new phenomenon accompanying induction of the mitochondrial permeability transition in cardiac myocytes. J Exp Med. 2000;192:1001–14.
140. Cai H, Griendling KK, Harrison DG. The vascular NAD(P)H oxidases as therapeutic targets in cardiovascular diseases. Trends Pharmacol Sci. 2003;24:471–8.
141. Doughan AK, Harrison DG, Dikalov SI. Molecular mechanisms of angiotensin II-mediated mitochondrial dysfunction: linking mitochondrial oxidative damage and vascular endothelial dysfunction. Circ Res. 2008;102:488–96.
142. Forstermann U, Xia N, Li H. Roles of vascular oxidative stress and nitric oxide in the pathogenesis of atherosclerosis. Circ Res. 2017;120:713–35.
143. Matsushima S, Tsutsui H, Sadoshima J. Physiological and pathological functions of NADPH oxidases during myocardial ischemia-reperfusion. Trends Cardiovasc Med. 2014;24:202–5.
144. Guzik TJ, Harrison DG. Vascular NADPH oxidases as drug targets for novel antioxidant strategies. Drug Discov Today. 2006;11:524–33.
145. Wang HD, Xu S, Johns DG, Du Y, Quinn MT, Cayatte AJ, et al. Role of NADPH oxidase in the vascular hypertrophic and oxidative stress response to angiotensin II in mice. Circ Res. 2001;88:947–53.
146. Barry-Lane PA, Patterson C, van der Merwe M, Hu Z, Holland SM, Yeh ET, et al. p47phox is required for atherosclerotic lesion progression in ApoE(-/-) mice. J Clin Invest. 2001;108:1513–22.
147. Matsushima S, Kuroda J, Ago T, Zhai P, Ikeda Y, Oka S, et al. Broad suppression of NADPH oxidase activity exacerbates ischemia/reperfusion injury through inadvertent downregulation of hypoxia-inducible factor-1alpha and upregulation of peroxisome proliferator-activated receptor-alpha. Circ Res. 2013;112:1135–49.
148. Kuroda J, Ago T, Matsushima S, Zhai P, Schneider MD, Sadoshima J. NADPH oxidase 4 (Nox4) is a major source of oxidative stress in the failing heart. Proc Natl Acad Sci U S A. 2010;107:15565–70.
149. Zhang Y, Shimizu H, Siu KL, Mahajan A, Chen JN, Cai H. NADPH oxidase 4 induces cardiac arrhythmic phenotype in zebrafish. J Biol Chem. 2014;289:23200–8.
150. Bedard K, Krause KH. The NOX family of ROS-generating NADPH oxidases: physiology and pathophysiology. Physiol Rev. 2007;87:245–313.
151. Cheng G, Cao Z, Xu X, van Meir EG, Lambeth JD. Homologs of gp91phox: cloning and tissue expression of Nox3, Nox4, and Nox5. Gene. 2001;269:131–40.
152. Gimenez M, Schickling BM, Lopes LR, Miller FJ Jr. Nox1 in cardiovascular diseases: regulation and pathophysiology. Clin Sci (Lond). 2016;130:151–65.
153. Lassegue B, San Martin A, Griendling KK. Biochemistry, physiology, and pathophysiology of NADPH oxidases in the cardiovascular system. Circ Res. 2012;110:1364–90.

154. Zhang M, Perino A, Ghigo A, Hirsch E, Shah AM. NADPH oxidases in heart failure: poachers or gamekeepers? Antioxid Redox Signal. 2013;18:1024–41.
155. Sahoo S, Meijles DN, Pagano PJ. NADPH oxidases: key modulators in aging and age-related cardiovascular diseases? Clin Sci (Lond). 2016;130:317–35.
156. Brandes RP, Schroder K. Differential vascular functions of Nox family NADPH oxidases. Curr Opin Lipidol. 2008;19:513–8.
157. Ago T, Kitazono T, Kuroda J, Kumai Y, Kamouchi M, Ooboshi H, et al. NAD(P)H oxidases in rat basilar arterial endothelial cells. Stroke. 2005;36:1040–6.
158. Gorlach A, Brandes RP, Nguyen K, Amidi M, Dehghani F, Busse R. A gp91phox containing NADPH oxidase selectively expressed in endothelial cells is a major source of oxygen radical generation in the arterial wall. Circ Res. 2000;87:26–32.
159. BelAiba RS, Djordjevic T, Petry A, Diemer K, Bonello S, Banfi B, et al. NOX5 variants are functionally active in endothelial cells. Free Radic Biol Med. 2007;42:446–59.
160. Guzik TJ, Chen W, Gongora MC, Guzik B, Lob HE, Mangalat D, et al. Calcium-dependent NOX5 nicotinamide adenine dinucleotide phosphate oxidase contributes to vascular oxidative stress in human coronary artery disease. J Am Coll Cardiol. 2008;52:1803–9.
161. Matsuno K, Iwata K, Matsumoto M, Katsuyama M, Cui W, Murata A, et al. NOX1/NADPH oxidase is involved in endotoxin-induced cardiomyocyte apoptosis. Free Radic Biol Med. 2012;53:1718–28.
162. Morawietz H, Bornstein SR. Leptin, endothelin, NADPH oxidase, and heart failure. Hypertension. 2006;47:e20.
163. Heymes C, Bendall JK, Ratajczak P, Cave AC, Samuel JL, Hasenfuss G, et al. Increased myocardial NADPH oxidase activity in human heart failure. J Am Coll Cardiol. 2003;41:2164–71.
164. Krijnen PA, Meischl C, Hack CE, Meijer CJ, Visser CA, Roos D, et al. Increased Nox2 expression in human cardiomyocytes after acute myocardial infarction. J Clin Pathol. 2003;56:194–9.
165. Ago T, Kuroda J, Pain J, Fu C, Li H, Sadoshima J. Upregulation of Nox4 by hypertrophic stimuli promotes apoptosis and mitochondrial dysfunction in cardiac myocytes. Circ Res. 2010;106:1253–64.
166. Hahn NE, Meischl C, Kawahara T, Musters RJ, Verhoef VM, van der Velden J, et al. NOX5 expression is increased in intramyocardial blood vessels and cardiomyocytes after acute myocardial infarction in humans. Am J Pathol. 2012;180:2222–9.
167. Suh YA, Arnold RS, Lassegue B, Shi J, Xu X, Sorescu D, et al. Cell transformation by the superoxide-generating oxidase Mox1. Nature. 1999;401:79–82.
168. Ellmark SH, Dusting GJ, Fui MN, Guzzo-Pernell N, Drummond GR. The contribution of Nox4 to NADPH oxidase activity in mouse vascular smooth muscle. Cardiovasc Res. 2005;65:495–504.
169. Banfi B, Molnar G, Maturana A, Steger K, Hegedus B, Demaurex N, et al. A Ca(2+)-activated NADPH oxidase in testis, spleen, and lymph nodes. J Biol Chem. 2001;276:37594–601.
170. Hilenski LL, Clempus RE, Quinn MT, Lambeth JD, Griendling KK. Distinct subcellular localizations of Nox1 and Nox4 in vascular smooth muscle cells. Arterioscler Thromb Vasc Biol. 2004;24:677–83.
171. Van Buul JD, Fernandez-Borja M, Anthony EC, Hordijk PL. Expression and localization of NOX2 and NOX4 in primary human endothelial cells. Antioxid Redox Signal. 2005;7:308–17.
172. Segal AW, Jones OT. Novel cytochrome B system in phagocytic vacuoles of human granulocytes. Nature. 1978;276:515–7.
173. Clempus RE, Sorescu D, Dikalova AE, Pounkova L, Jo P, Sorescu GP, et al. Nox4 is required for maintenance of the differentiated vascular smooth muscle cell phenotype. Arterioscler Thromb Vasc Biol. 2007;27:42–8.

174. Perrotta I, Sciangula A, Perrotta E, Donato G, Cassese M. Ultrastructural analysis and electron microscopic localization of Nox4 in healthy and atherosclerotic human aorta. Ultrastruct Pathol. 2011;35:1–6.
175. Camargo LL, Harvey AP, Rios FJ, Tsiropoulou S, Da Silva RNO, Cao Z, et al. Vascular Nox (NADPH oxidase) compartmentalization, protein hyperoxidation, and endoplasmic reticulum stress response in hypertension. Hypertension. 2018;72:235–46.
176. Matsushima S, Kuroda J, Ago T, Zhai P, Park JY, Xie LH, et al. Increased oxidative stress in the nucleus caused by Nox4 mediates oxidation of HDAC4 and cardiac hypertrophy. Circ Res. 2013;112:651–63.
177. Ago T, Kitazono T, Ooboshi H, Iyama T, Han YH, Takada J, et al. Nox4 as the major catalytic component of an endothelial NAD(P)H oxidase. Circulation. 2004;109:227–33.
178. Konior A, Schramm A, Czesnikiewicz-Guzik M, Guzik TJ. NADPH oxidases in vascular pathology. Antioxid Redox Signal. 2014;20:2794–814.
179. Dikalov SI, Dikalova AE, Bikineyeva AT, Schmidt HH, Harrison DG, Griendling KK. Distinct roles of Nox1 and Nox4 in basal and angiotensin II-stimulated superoxide and hydrogen peroxide production. Free Radic Biol Med. 2008;45:1340–51.
180. Helmcke I, Heumuller S, Tikkanen R, Schroder K, Brandes RP. Identification of structural elements in Nox1 and Nox4 controlling localization and activity. Antioxid Redox Signal. 2009;11:1279–87.
181. Takac I, Schroder K, Zhang L, Lardy B, Anilkumar N, Lambeth JD, et al. The E-loop is involved in hydrogen peroxide formation by the NADPH oxidase Nox4. J Biol Chem. 2011;286:13304–13.
182. Siu KL, Miao XN, Cai H. Recoupling of eNOS with folic acid prevents abdominal aortic aneurysm formation in angiotensin II-infused apolipoprotein E null mice. PLoS One. 2014;9: e88899.
183. Schulz E, Jansen T, Wenzel P, Daiber A, Munzel T. Nitric oxide, tetrahydrobiopterin, oxidative stress, and endothelial dysfunction in hypertension. Antioxid Redox Signal. 2008;10:1115–26.
184. Forstermann U, Li H. Therapeutic effect of enhancing endothelial nitric oxide synthase (eNOS) expression and preventing eNOS uncoupling. Br J Pharmacol. 2011;164:213–23.
185. Youn JY, Wang T, Blair J, Laude KM, Oak JH, McCann LA, et al. Endothelium-specific sepiapterin reductase deficiency in DOCA-salt hypertension. Am J Physiol Heart Circ Physiol. 2012;302:H2243.
186. Oak JH, Cai H. Attenuation of angiotensin II signaling recouples eNOS and inhibits nonendothelial NOX activity in diabetic mice. Diabetes. 2007;56:118–26.
187. Takaya T, Hirata K, Yamashita T, Shinohara M, Sasaki N, Inoue N, et al. A specific role for eNOS-derived reactive oxygen species in atherosclerosis progression. Arterioscler Thromb Vasc Biol. 2007;27:1632–7.
188. Moens AL, Champion HC, Claeys MJ, Tavazzi B, Kaminski PM, Wolin MS, et al. High-dose folic acid pretreatment blunts cardiac dysfunction during ischemia coupled to maintenance of high-energy phosphates and reduces postreperfusion injury. Circulation. 2008;117:1810–9.
189. Wang S, Xu J, Song P, Wu Y, Zhang J, Chul Choi H, et al. Acute inhibition of guanosine triphosphate cyclohydrolase 1 uncouples endothelial nitric oxide synthase and elevates blood pressure. Hypertension. 2008;52:484–90.
190. Moens AL, Ketner EA, Takimoto E, Schmidt TS, O'Neill CA, Wolin MS, et al. Bi-modal dose-dependent cardiac response to tetrahydrobiopterin in pressure-overload induced hypertrophy and heart failure. J Mol Cell Cardiol. 2011;51:564–9.
191. Gao L, Siu KL, Chalupsky K, Nguyen A, Chen P, Weintraub NL, et al. Role of uncoupled endothelial nitric oxide synthase in abdominal aortic aneurysm formation: treatment with folic acid. Hypertension. 2012;59:158–66.
192. Loperena R, Harrison DG. Oxidative stress and hypertensive diseases. Med Clin North Am. 2017;101:169–93.

193. Chance B, Sies H, Boveris A. Hydroperoxide metabolism in mammalian organs. Physiol Rev. 1979;59:527–605.
194. Trono D, Laus MN, Soccio M, Alfarano M, Pastore D. Modulation of potassium channel activity in the balance of ROS and ATP production by durum wheat mitochondria-an amazing defense tool against hyperosmotic stress. Front Plant Sci. 2015;6:1072.
195. Queliconi BB, Wojtovich AP, Nadtochiy SM, Kowaltowski AJ, Brookes PS. Redox regulation of the mitochondrial K(ATP) channel in cardioprotection. Biochim Biophys Acta. 1813;2011:1309–15.
196. Li Q, Youn JY, Siu KL, Murugesan P, Zhang Y, Cai H. Knockout of dihydrofolate reductase in mice induces hypertension and abdominal aortic aneurysm via mitochondrial dysfunction. Redox Biol. 2019;24:101185.
197. Munzel T, Gori T, Keaney JF Jr, Maack C, Daiber A. Pathophysiological role of oxidative stress in systolic and diastolic heart failure and its therapeutic implications. Eur Heart J. 2015;36:2555–64.
198. Ballinger SW, Patterson C, Knight-Lozano CA, Burow DL, Conklin CA, Hu Z, et al. Mitochondrial integrity and function in atherogenesis. Circulation. 2002;106:544–9.
199. Graham D, Huynh NN, Hamilton CA, Beattie E, Smith RA, Cocheme HM, et al. Mitochondria-targeted antioxidant MitoQ10 improves endothelial function and attenuates cardiac hypertrophy. Hypertension. 2009;54:322–8.
200. Schramm A, Matusik P, Osmenda G, Guzik TJ. Targeting NADPH oxidases in vascular pharmacology. Vasc Pharmacol. 2012;56:216–31.
201. Simons JM, Hart BA, Van Dijk H, Labadie RP. Metabolic activation of natural phenols into selective oxidative burst agonists by activated human neutrophils. Free Radic Biol Med. 1990;8:251–8.
202. Hancock JT, Jones OT. The inhibition by diphenyleneiodonium and its analogues of superoxide generation by macrophages. Biochem J. 1987;242:103–7.
203. Tanriverdi LH, Parlakpinar H, Ozhan O, Ermis N, Polat A, Vardi N, et al. Inhibition of NADPH oxidase by apocynin promotes myocardial antioxidant response and prevents isoproterenol-induced myocardial oxidative stress in rats. Free Radic Res. 2017;51:772–86.
204. O'Donnell VB, Smith GC, Jones OT. Involvement of phenyl radicals in iodonium inhibition of flavoenzymes. Mol Pharmacol. 1994;46:778–85.
205. Laleu B, Gaggini F, Orchard M, Fioraso-Cartier L, Cagnon L, Houngninou-Molango S, et al. First in class, potent, and orally bioavailable NADPH oxidase isoform 4 (Nox4) inhibitors for the treatment of idiopathic pulmonary fibrosis. J Med Chem. 2010;53:7715–30.
206. Di Marco E, Gray SP, Chew P, Koulis C, Ziegler A, Szyndralewiez C, et al. Pharmacological inhibition of NOX reduces atherosclerotic lesions, vascular ROS and immune-inflammatory responses in diabetic Apoe(-/-) mice. Diabetologia. 2014;57:633–42.
207. Wang X, Elksnis A, Wikstrom P, Walum E, Welsh N, Carlsson PO. The novel NADPH oxidase 4 selective inhibitor GLX7013114 counteracts human islet cell death in vitro. PLoS One. 2018;13:e0204271.
208. Pick E. Absolute and relative activity values in assessing the effect of NADPH oxidase inhibitors. Antioxid Redox Signal. 2015;23:1250–1.
209. Maier W, Cosentino F, Lutolf RB, Fleisch M, Seiler C, Hess OM, et al. Tetrahydrobiopterin improves endothelial function in patients with coronary artery disease. J Cardiovasc Pharmacol. 2000;35:173–8.
210. Higashi Y, Sasaki S, Nakagawa K, Fukuda Y, Matsuura H, Oshima T, et al. Tetrahydrobiopterin enhances forearm vascular response to acetylcholine in both normotensive and hypertensive individuals. Am J Hypertens. 2002;15:326–32.
211. Setoguchi S, Hirooka Y, Eshima K, Shimokawa H, Takeshita A. Tetrahydrobiopterin improves impaired endothelium-dependent forearm vasodilation in patients with heart failure. J Cardiovasc Pharmacol. 2002;39:363–8.

212. Cunnington C, Van Assche T, Shirodaria C, Kylintireas I, Lindsay AC, Lee JM, et al. Systemic and vascular oxidation limits the efficacy of oral tetrahydrobiopterin treatment in patients with coronary artery disease. Circulation. 2012;125:1356–66.
213. Boger RH, Bode-Boger SM, Szuba A, Tsao PS, Chan JR, Tangphao O, et al. Asymmetric dimethylarginine (ADMA): a novel risk factor for endothelial dysfunction: its role in hypercholesterolemia. Circulation. 1998;98:1842–7.
214. Monti LD, Casiraghi MC, Setola E, Galluccio E, Pagani MA, Quaglia L, et al. L-arginine enriched biscuits improve endothelial function and glucose metabolism: a pilot study in healthy subjects and a cross-over study in subjects with impaired glucose tolerance and metabolic syndrome. Metabolism. 2013;62:255–64.
215. Schulman SP, Becker LC, Kass DA, Champion HC, Terrin ML, Forman S, et al. L-arginine therapy in acute myocardial infarction: the vascular interaction with age in myocardial infarction (VINTAGE MI) randomized clinical trial. JAMA. 2006;295:58–64.
216. Dare AJ, Bolton EA, Pettigrew GJ, Bradley JA, Saeb-Parsy K, Murphy MP. Protection against renal ischemia-reperfusion injury in vivo by the mitochondria targeted antioxidant MitoQ. Redox Biol. 2015;5:163–8.
217. Dare AJ, Logan A, Prime TA, Rogatti S, Goddard M, Bolton EM, et al. The mitochondria-targeted anti-oxidant MitoQ decreases ischemia-reperfusion injury in a murine syngeneic heart transplant model. J Heart Lung Transplant. 2015;34:1471–80.
218. Szeto HH. Cell-permeable, mitochondrial-targeted, peptide antioxidants. AAPS J. 2006;8:E277–83.

2 Oxidative Stress in the Tumor Immune Microenvironment

Yuexi Wang, Yanan Hu, Yuting Jiang, and Shengtao Zhou

Abstract

Oxidative stress is one of defining features of the tumor microenvironment, which is closely related to the interactions between tumor cells and stromal cells. It affects tumor progression in many aspects. In the tumor microenvironment, primary immune cells have different functions, which together make up the immune defense line of the tumor. This review focuses on the relationship between oxidative stress and tumor-related immune system, specifically the effects and mechanisms of oxidative stress on different cell processes of immune cells in the tumor microenvironment. Then, we discuss the main overall effect of oxidative stress, immunosuppression, and its inspiration for tumor immunotherapy, which provides a theoretical basis for the feasibility of oxidative stress as a new target of tumor immunotherapy.

Keywords

Oxidative stress · Reactive oxygen species (ROS) · Tumor microenvironment · Immune cells · Immunosuppression · Immunotherapy

Y. Wang · Y. Hu
Department of Obstetrics and Gynecology, Key Laboratory of Birth Defects and Related Diseases of Women and Children of MOE, State Key Laboratory of Biotherapy, West China Second University Hospital, Sichuan University and Collaborative Innovation Center, Chengdu, PR China

College of Life Science, Sichuan University, Chengdu, PR China

Y. Jiang · S. Zhou (✉)
Department of Obstetrics and Gynecology, Key Laboratory of Birth Defects and Related Diseases of Women and Children of MOE, State Key Laboratory of Biotherapy, West China Second University Hospital, Sichuan University and Collaborative Innovation Center, Chengdu, PR China
e-mail: shengtaozhou@scu.edu.cn

C. Huang, Y. Zhang (eds.), *Oxidative Stress*,
https://doi.org/10.1007/978-981-16-0522-2_2

2.1 Background

2.1.1 Oxidative Stress in the Tumor Microenvironment

In the process of cancer progression and metastasis, tumor microenvironment (TME) is a niche interacting with the host, which encompasses proliferative tumor cells, infiltrative immune cells, tumor matrix, and related vasculature, given that TME is regulated by tumor cells, in which various cells interact to form a complex environment, with characteristics of limited nutritional supply, hypoxia, pH deregulation, and oxidative stress (OS).

OS is a pathophysiological concept, which refers to a stress signal, reflecting that abundant reactive oxygen species (ROS) produced beyond eliminating oxidizing substance capacity in cells. It not only is a biochemical reaction but also has fundamentally complex cellular and molecular mechanisms, including membrane oxidation, mitochondrial metabolism, endoplasmic reticulum (ER) stress, DNA damage repair, gene transcription and expression, signal transmission, protein folding, and other cellular and molecular changes.

When cancer cell grows, malignant cells adapt to the surrounding matrix and undergo reprogramming, producing ROS and leading to OS status. It is recognized that endogenic or exogenous sources could cause OS in TME. Extrinsic: (1) Some components (such as neutrophils and macrophages) of TME can directly produce ROS. (2) Hypoxia can release the mitochondrial electron transport complex III or the nicotinamide adenine dinucleotide phosphate (NADPH) oxidase to produce ROS. (3) OS, induced by microenvironment or senescence, results in senescent fibroblasts secreting senescent activated secretory pathway (SASP), which facilitates tumor growth through impacting on matrix and tumor cells. Intrinsic: Cancer cells downregulate JunD expression or increase COX-2, Nox-4, and Lox-5 activity to give rise to OS, thus aggravating the oxidative environment [1].

OS in TME has essential research significance. Mainly, the formation of superoxide, initiated by tumor necrosis factor α (TNF-α), platelet-derived growth factor (PDGF), interferon-γ (IFN-γ), transforming growth factor-β (TGF-β), interleukin-1 (IL-1), and epidermal growth factor (EGF), can induce the OS, which is associated with cancer and inflammation. Simultaneously, progressive tumors can induce significant OS and recruit inflammatory cells [2].

The effects of oxygen free radicals on oncology and carcinogenesis are multifaceted and remain unraveled. Herein, we mainly discuss the adverse intervention of OS against tumor immunity in TME. ROS, as soluble immunosuppressive factors, cause immune cell dysfunction in cancer patients [3]. In general, ROS affects proliferating signal regulation, tumor-invasive plasticity, tumor cell metabolism reprogramming, and gene mutation, which is closely related to tumor development [4]. Overall, the effects of OS on tumor progression include (1) the mutagenic potential of ROS, (2) the effect on regulating metabolic pathways to control proliferating and survival of cells, (3) the effect on cell movement and invasiveness, and (4) the effect of OS on stromal cells.

The exact nature of the effects of OS on cancer progression and therapeutic response needs further study. At present, advanced understandings of the processes and effects of OS-related networks are not fully elucidated, which hinders the implementation of new anticancer strategies based on OS.

2.1.2 The Major Tumor-Associated Immune Cells

Similarly, tumor immune microenvironment (TIME), encompassing tumor-infiltrating immune cells and factors, also engages in malignant cells progression and killing through interactions of components. Understanding relationships in TIME could address cancer immunotherapy through modulating various immune cell types.

Dendritic cells (DCs) are characterized as initiating immune reaction and tolerance toward specific antigen, termed antigen-presenting cells (APCs) [5]. Several studies have addressed that conventional DCs induce the antitumor immune response by means of capturing, processing, and then presenting tumor-associated antigens (TAAs) via major histocompatibility complex (MHC) molecules [6]. Concomitantly, soluble and costimulatory factors are secreted by DCs owing to shaping T cell activity. Naïve T cells could also recognize TAAs, cross-presented from DCs, which endows $CD4^+$ T cell antitumor capacity as well as $CD8^+$ T cell cytotoxicity competence [7, 8]. Furthermore, tumor-derived DNA triggers activation of DCs and IFNs production via cyclic GMP-AMP synthase (cGAS), stimulator of interferon genes complex (STING), and interferon regulatory factor 3 (IRF3) pathway in anti-TME [9].

T cells work as a pivotal force in antitumor immunoreaction, which can be divided into regulatory T cells (Tregs) and effector T cells (Teffs), encompassing $CD4^+$ helper T (Th) cells as well as $CD8^+$ cytotoxic T lymphocytes (CTLs) cells. CTLs work as fundamentally effector cells toward tumor damage, while Th cells are necessary for helping CTL-dependent tumor eradication. Firstly, CTLs scan tumor cells to find counterpart peptide-MHC (pMHC) complexes followed by integrin engagement, and migration of T cell slows down [10, 11]. Subsequently, target cells are employed by CTLs and induced killing [12, 13]. Notice that C-X-C motif ligand (CXCL12), a kind of chemokine, could weaken CTL function by restraining migration into the tumor in pancreatic ductal adenocarcinoma in contrast to IL-12 that promotes CTL effector function via facilitating the formation of CTL synapse [14, 15]. Moreover, Th cells could also induce antitumor responses in TME, such as co-stimulation, antigen presentation, and T cell homing [16]. Partial Th cells express tumor necrosis factor-related apoptosis-inducing ligand (TRAIL), perforin, granzymes, or death meditator Fas ligand (FasL) to kill tumor cells, while others exhibit indirect function via enhancing CTL immune responses [17, 18].

Previous studies suggest that Treg-mediating immunosuppression contributes to tumor immune evasion [19]. Moreover, cell-cell contact and soluble factors have been considered as possible suppressive mechanisms [20]. Tregs are capable of crippling antitumor immune cell function, through interacting with DCs, Teffs, and

NK cells, respectively [19–24]. Besides, Tregs, expressing programmed cell death protein 1 (PD-1), obtain intensely immunosuppressive activity by enhancing T cell receptor (TCR) as well as signal of CD28 in the treatment of PD-1 blockade [25].

As significant humoral immune participant, tumor-infiltrating B cells (TIBs) function as killing tumor cells through expressing granzyme B and TRAIL in TME [26]. Nevertheless, B cells also exert antitumor suppression in TME. At first, the tumor-targeted antibody could produce circulating immune complexes (CICs) in tumor tissue, which initiates the complement pathway to promote cancer development [27]. Second, B cells secret pro-tumorigenic factors, such as lymphotoxin. Furthermore, regulatory B cells (Bregs) suppress antitumor immunity via secreting anti-inflammation cytokine, IL-10 [28]. Bregs also induce M2 macrophage polarization and Treg production through secreting TGF-β [29, 30]. Based on the dual roles of B cells in TME, it is necessary to formulate a reasonable patient stratification strategy for B cell-related treatment and take different treatment measures for patients with different tumor-related B cell typing results [31].

Natural killer (NK) cells are characterized as mediating effective cell killing and immunosurveillance in TME. The NK cell activation and inhibition depend on the integrating signals produced by inhibitory and activating NK receptors (iNKRs and aNKRs) and the dynamic balance [32, 33]. Changing of surface markers, for instance, lacking human leukocyte antigen class I molecules and upregulated damage-associated proteins, is required in NK cell recognition and killing in TME [34]. Subsequently, NK cells exert cytotoxicity capacity, depending on producing perforin and granzymes in granules and relying on death receptors, to induce apoptosis [35, 36]. Besides, activated NK cells express cytokines and chemokines, such as IFN-γ and CCL3, to induce indirect killing. Some studies demonstrated that NK cells with memory properties acquired a longer survival time and better cytotoxicity toward tumor cells after the second stimulus [37].

Monocyte/macrophages are pivotal lymphocytes in innate immune responses, the functions of which could be affected by the surrounding microenvironment. Pro-inflammatory cytokines, ROS/RNS, and other cytotoxic mediators are produced by M1 macrophages. Functionally, activated M2 macrophages protect tissue from chronic inflammatory microenvironment through generating a small quantity of IL-12 in contrast to a high amount of IL-10 [38]. In TME, tumor-associated macrophages (TAMs) exhibit $IL\text{-}12^{low}/IL\text{-}10^{high}/TGF\text{-}\beta^{high}$ phenotype like M2 macrophages, which impedes T cell proliferation and cytotoxicity and initiates Tregs [39, 40]. Besides, chemokines, cytokines, ROS, and proteolytic enzymes are also secreted by TAMs to promote tumor progress [39]. TAMs could digest the extracellular matrix (ECM) to promote tumor metastasis. More importantly, due to growth factors and anti-inflammatory cytokines in TME, differentiated M1 macrophages can be "re-educated" to M2 macrophages [39]. Therefore, utilizing the feature of "re-polarization" to convert M2 into M1 macrophages could be a potential strategy for tumor treatment.

2.2 OS and Tumor-Associated Immune Cells

2.2.1 The Effects of OS on Teffs in TME

It has been generally found that cancer-related T cells (T cells refer to Teffs here, if not specified) have low reactivity and impaired function. Based on previous studies, the increase of OS in TME is the leading cause of T cell depletion induced by the tumor cells. The effects of OS on T cells are multifaceted. At present, the studies focus on the consequences of OS, and the mechanism involves multiple pathways, which is complex and needs further research. T cells affected by OS are characterized by activated T cells being more susceptible to OS compared with primary T cells [41]. $CD8^+$ T cells, in particular, are susceptible to apoptosis induced by oxygen free radicals [42]. Besides, different doses of OS have different effects on T cells. ROS can participate in signal transduction and gene expression of T cells as a signal molecule. Incubation of T cells with low-level hydrogen peroxide can promote the activation of NF-κB, induce the expression of IL-2, and stimulate cell proliferation. In contrast to activation, the addition of hydrogen peroxide inhibited cell proliferation to a large extent.

Maintaining the level of redox inside and outside the cell is the premise of T lymphocyte's normal function. Due to various factors, TME becomes a chronic oxidative environment, which has a significant impact on T lymphocytes. OS is an important factor in determining the activation, differentiation, apoptosis, proliferation, and function of T cells. We will discuss the specific effects and central mechanisms of OS on T cells below (Fig. 2.1).

2.2.1.1 T Cell Function

The OS caused by excessive ROS in TME weakened the effect of T cells on tumor resistance, mainly leading to T cell hyporesponsiveness [43]. TCR signaling pathway is vital to T cell activation. Several TCR signaling molecules have been known to be affected by OS, such as the downregulation of p56lck, TCR ζ, and linker of activated T cells (LAT) [44]. In TCR signaling pathway, TCR-induced intracellular signal that originated from tyrosine phosphorylation in ITAM sequence of CD3 molecule and phospholipase C-γ1 (PLC-γ1) was activated through phosphorylation and then mediated the opening of subsequent Ca^{2+} channel and Ca^{2+} influx, and increased intracellular Ca^{2+} concentration catalyzed the regulation of downstream gene expression by related transcription factors [45].

Effects on TCR Signaling Pathway

The molecular mechanism of OS-induced hyporesponsiveness of T cell is achieved by targeting specific molecules in the TCR signaling mechanism. PLC-γ1-dependent tyrosine phosphorylation defect and calcium mobilization damage can cause T cell hyporesponsiveness. Compared with healthy T cells, the pattern of tyrosine phosphorylation protein in low-reactive T cells changed. In the low-reactive T cells, ROS targeted inhibition of PLC-γ1, resulting in TCR connection no longer inducing PLC-γ1 activation, PLC-γ1-dependent tyrosine phosphorylation defect, and Ca^{2+}

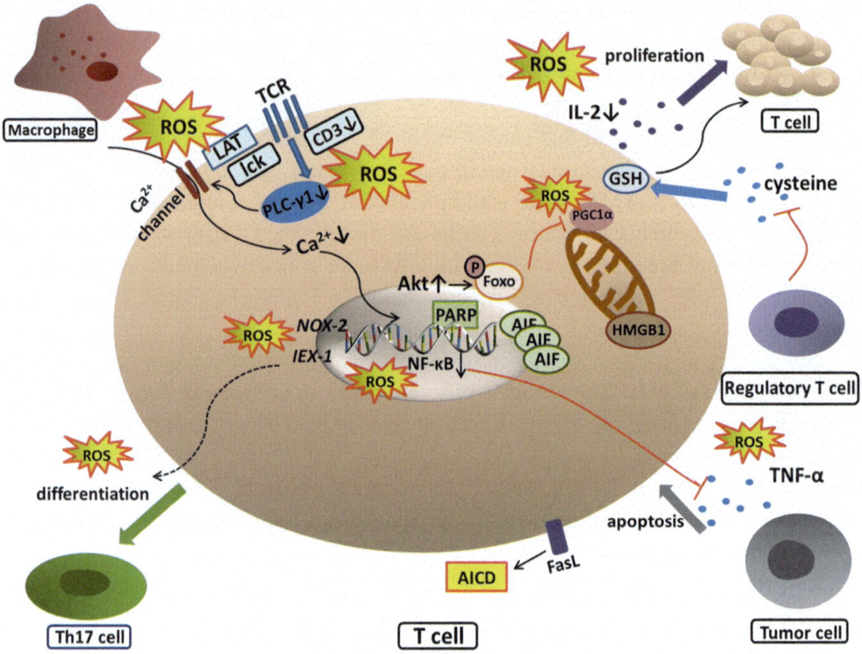

Fig. 2.1 The main effects and central mechanisms of OS on T cells in TME. The molecular mechanism of OS-related T cell hyporesponsiveness can be induced by influencing various components of TCR signaling mechanisms. PGC1α can be a crucial mechanism link between Akt and the inhibited oxidative metabolism of tumor-infiltrating T cells, which impair mitochondrial-related functions of T cells. The ROS-related gene *NOX-2* and *IEX-1* may play a role in Th17 cell differentiation. OS can induce the death of tumor-host T cells through NF-κB-TNF-α interaction. ROS can trigger PARP and apoptosis-inducing factor (AIF)-dependent cell death of T cell. OS can affect proliferation by affecting the production of IL-2 in T cells. Tregs can participate in OS-induced T cell proliferation inhibition in TME through impairing the synthesis of GSH in T cell

flux reduction, that is, downstream TCR signal pathway block [46]. Besides, ROS targets key TCR signal molecules, which have different effects on their structure and stability. The expression of p56lck, TCR ζ, and LAT key TCR near-end signal molecules in low-reactive T cells was impaired. The C-terminus of TCR ζ and the proximal domain of the p56lck membrane were oxidized by ROS, leading to the alteration of protein structure and the impairment of TCR function. The specific manifestations are selective loss of signal molecules in low-reactive T cells, ROS-dependent rapid degradation of p56lck, main structural changes and edge degradation of TCR ζ, and loss of LAT signal. The deterioration of the proteasome is a secondary effect leading to low reactivity [44]. TCR-CD3 complex of T cells in tumor patients showed a structural abnormality, especially the disappearance of the CD3ζ chain [47]. It then results in T cell response inhibition, which is antigen-specific. Studies showed that OS mediates CD3ζ decreasing in the interaction between tumor macrophages and T cells [48]. The OS of tumor-bearing mouse macrophages (TBM-mØs) may lead to the inhibition of the exchange of CD3ζ and newly synthesized CD3ζ on the surface [48].

Effects on Mitochondrial-Related Functions

Oxidative metabolism is the key medium for T cells to exert their functions, and mitochondrial regulation of oxidative metabolism and other functions of T cells are very critical, such as the increase of cytokine production in T cells whose mitochondrial activity is high [49]. PGC1α is a node of multiple cross-linking signal pathways, which can directly interact with Foxo3a to regulate the expression of OS protection genes to control OS response [50]. It also controls mitochondrial biogenesis. PGC1α can be used as the transcription coactivator in the process of mitochondrial replication, and it can be regulated by Akt signaling pathway. Specifically, Foxo family transcription factors can promote the expression of PGC1α [50], while Akt signaling can phosphorylate and deactivate Foxo to play the role of oxidative metabolism inhibition. Continuous and progressive mitochondrial function and quantity loss of tumor-infiltrating T cells have been found, which are caused by the increase of Akt activation, and subsequently cause inhibition of transcription of Foxo-programmed PGC1α. It was demonstrated that PGC1α could link the Akt signaling pathway and the inhibited oxidative metabolism condition of T cells in TME. Moreover, inhibitors of Akt may have immunomodulatory effects, and Akt-targeted therapy may provide new ideas for tumor immunotherapy [51].

2.2.1.2 T Cell Differentiation

In tumor-bearing hosts, OS can cause T cells to tend to polarize to Th2 phenotype [42] and regulate Th1 and Th2 cytokines to varying degrees. These changes of cytokine profile are significant in the process of tumor formation because Th1 cytokine has the antitumor immune function, and the priority transfer to a Th2 profile facilitates tumor progression. This transition from Th1 type to Th2 type also contributes to the reduction of CTL activity [52]. NOX-2 is a ROS-producing enzyme gene. The study showed that in the cells lacking NOX-2 emerged the decrease of IL-4 and the increase of IL-17 production so that the NOX complex may affect Th17 cell differentiation [43]. Besides, the rise of ROS formation in mitochondria after T cell activation can mediate the process of gene IEX-1 deletion, promoting Th17 cell differentiation in early response [53].

2.2.1.3 T Cell Death

NF-κB-Related T Cell Death

NF-κB plays a crucial role in host immunity and lymphoid organ development [54]. NF-κB was downregulated in tumor-bearing mice, indicating that OS inhibited NF-κB [55]. NF-κB can suppress TNF-α expression [56]. In the process of tumor development, if the activity of NF-κB is disturbed, it causes T cell apoptosis induced by TNF-α [57]. On the one hand, TNF-α secreted by the tumor cells has a deadly effect on thymus T cells [58]. It plays a variety of biological activities by activating signal pathways (including IKK, JNK, and caspases) and regulating immune response, inflammation, and apoptosis. On the other hand, ROS itself is a critical factor in TNF-α-induced apoptosis. In T cells, long-term exposure to OS can inhibit the activity of NF-κB translocating into the nucleus [59]. The disorder of NF-κB subcellular distribution induces T cell apoptosis mediated by TNF-α and causes thymus atrophy through activation of TRADD-related caspase-8. The inhibition of NF-κB mediated by ROS and enhancement of TNF-α levels can synergistically lead to T cell death [57]. Moreover, NF-κB can mediate in T cell activation-induced cell death (AICD). The expression of FasL is also closely related to AICD. ROS promotes NF-κB, which increases FasL expression, and then mediates T cell apoptosis. Vitamin E can inhibit the activity of NF-κB, through its function of eliminating free radicals, resulting in FasL expression blockade and T cell AICD development [60, 61].

Oxygen-Free Radicals Induce PARP-Dependent Death of Cytotoxic T Cells
Poly (ADP-ribose) polymerase (PARP), a nick sensor enzyme [62] activated by DNA single-strand breaks, is an essential target in cancer. It was pointed out that phagocyte-derived ROS triggered PARP and apoptosis-inducing factor (AIF)-dependent cell death of cytotoxic T cell. It is accompanied by the decrease of mitochondrial transmembrane potential, AIF nuclear accumulation, and large-scale DNA fragmentation [63].

In addition, OS can regulate the transport of high mobility group box 1 (HMGB1) transport and activity [43]. In the process of T cell apoptosis, mitochondrial ROS oxidation releases HMGB1 in the process of cell apoptosis. HMGB1 can also regulate T cell immune response [64].

2.2.1.4 T Cell Proliferation

The Depletion of GSH in Cells Significantly Inhibited the Proliferation of Lymphocytes to Mitotic Lectin
Glutathione (GSH), as an indicator of OS, is an important intracellular antioxidant. Intracellular GSH level is an essential factor in regulating the redox environment of T cells by interacting with APCs [2]. The results show that the proliferation of lymphocytes to mitosis is directly dependent on the effectiveness of GSH, and limiting the amount of GSH available in cells could inhibit lymphocytes growing during activation [65]. The consumption of GSH in T cells can hamper IL-2 production. IL-2 can stimulate T cell proliferation, and IL-2 secretion regulated by redox regulation is essential in T cell proliferation [66]. Thus, OS can affect the production of IL-2 and thus impair T cell proliferation.

Tregs Participate in OS-Induced T Cell Proliferation Inhibition in TME
In addition to TCR signaling pathway, costimulatory signaling, and related cytokines, the proliferation of T cells also needs a suitable redox microenvironment formed by APCs. Tregs can reshape this environment through a variety of strategies and further inhibit the proliferation of Teffs. On the one hand, Tregs can mediate redox state disturbance and interfere with GSH metabolism of DC and Teffs [67]. On the other hand, cysteine is necessary for T cells for synthesizing GSH, which

provides reduction capacity for synthesis of DNA and cell cycle ([68, 69]). T cells mostly rely on cysteine from DCs to metabolize normally. Tregs can inhibit the accumulation of extracellular cysteine through the competitive mechanism. The signal pathway linking Tregs to redox metabolism may help identify potential new therapeutic targets.

It should be noted that OS not only affects the effect of infiltrating T cells; the T cells in the primary immune chamber can also be affected. For example, the decrease of monocytes in circulation and the spleen, thymic atrophy, and myelosuppression were observed in mice with advanced ascites cancer [57, 70]. Further investigation will help study the role of OS in tumor-induced T cell inhibition. Still, the critical details of the relationship among OS, inflammatory mediators, and T cell function in TME need further study. T cell therapy can combine with ROS scavenger, which can lead to the enhancement of antitumor T cell function, thus improving therapeutic effects [43].

2.2.2 The Effects of OS on NK Cells in TME

The functions of NK cells are increasingly appreciated owing to tumor cell killing without previous sensitization. In TME, various tumor cells and tumor-related cells could produce factors, such as IL-10, IL-6, and TGF-β, which can functionally impair NK cells [71]. As an essential antitumor cell, NK cell cytotoxicity and other functions are strongly suppressed by ROS in TME (Fig. 2.2) [72]. Studies unraveled that immunosuppressive milieu induces NK cell into apoptotic states and losing responses toward activating signals [63, 73].

ROS Activate PARP-Dependent NK Cell Apoptosis

PARP functions as a nick sensor enzyme and signaling molecule. PARP-1 is the majority type of PARP, which could be activated by damaged DNA, OS, and signals in cytoplasmic membrane receptors. In TME, human acute myeloid leukemia (AML) cells generate extracellular ROS by NADPH oxidase complex; thus, PARP-1-dependent apoptosis is initiated in adjacent NK cells [74]. Firstly, overactivated PARP-1 conveys the death signal to mitochondria, and mitochondrial transmembrane undergoes depolarization. Then, high-conductance porosity occurs, and mitochondria release AIF, which is transferred to NK cell nucleus, triggering fragmentation of chromatin into large fragments (50 kbp) [63, 75]. In the downstream of PARP activation, caspases are activated and could modify the fragmentation of DNA. Activating the caspase enzyme system results in partial secondary internucleosomal fragmentation from large fragments into shorter fragments [63].

Therefore, PARP-based immunotherapy is promising in cancer and inflammatory pathologies [76]. Olaparib, a PARP inhibitor, increases the NK killing capacity in breast cancer, non-small-cell lung carcinoma, and chordoma [77]. Moreover, inhibiting ROS production could also be a potential strategy. Studies revealed that

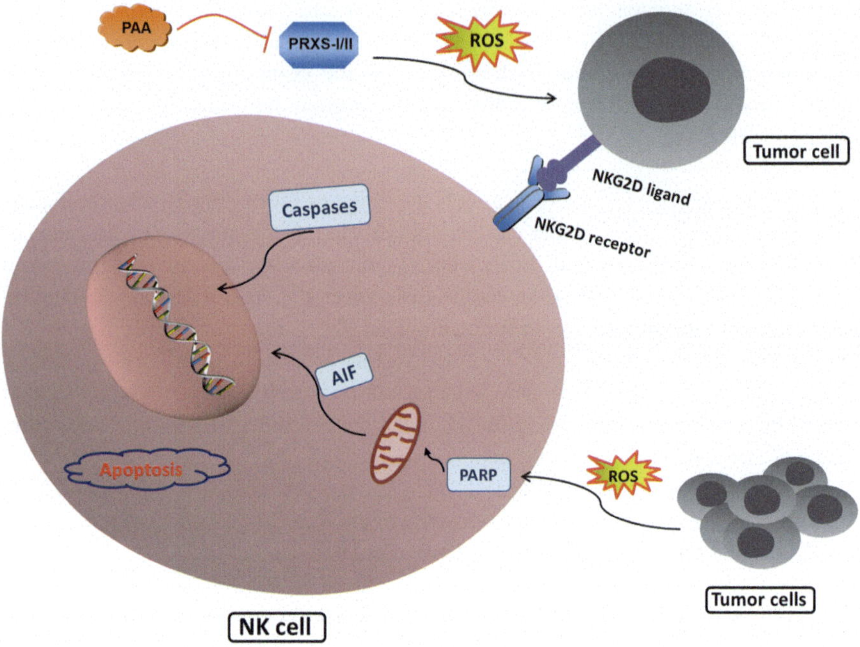

Fig. 2.2 The main effects and central mechanisms of OS on NK cells in TME. ROS impacts on NK cell function and destiny in TME: Tumor cells produce ROS, which results in NK cell PARP activation. Accordingly, mitochondria release AIF, entering cell nucleus. Furthermore, caspases are activated. Then, genomic DNA is degraded into short fragments, and NK cell undergoes apoptosis. New findings demonstrate that parvifoline AA could inhibit peroxiredoxins I/II function, resulting in ROS/ERK production and NKG2D ligands highly expressing in tumor cell. Therefore, NK cell could interact with tumor cells via NKG2D receptors efficiently; thus, NK cell exhibits tumor cell killing

jointly using NADPH oxidase inhibitor histamine dihydrochloride and cytokine IL-2 increases French-American-British M4/M5 AML patients' survival [74]. Besides, destroying the DNA repair system in tumor cells by controlling PARP could enhance chemotherapy and radiotherapy effect due to DNA damaging [78].

However, studies demonstrated that parvifoline AA (PAA) could help tumor killing by NK cells via increasing ROS. PAA is proved to inhibit peroxiredoxins I/II (PrxI/II) function. Accordingly, ROS/ERK was activated, resulting in NK group 2 member D (NKG2D) ligands highly expressing on hepatocellular carcinoma (HCC) cells. Subsequently, NK cells with NKG2D receptors could recognize HCC cells and induce cytotoxicity. Hence, PrxI/II is a potential immunotherapeutic target, and PAA is a promising agent toward HCC treatment [79].

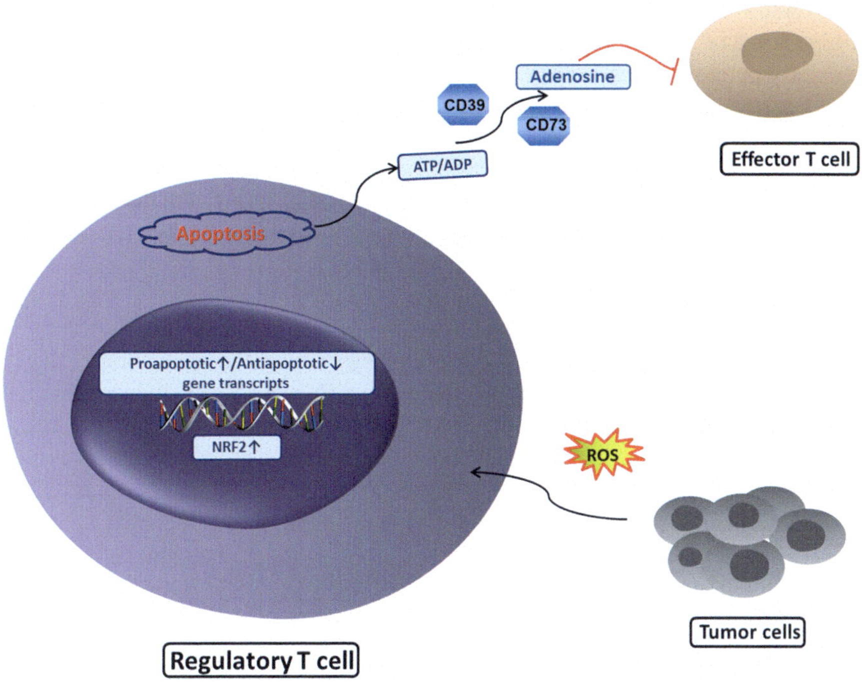

Fig. 2.3 The main effects and central mechanisms of OS on regulatory T cells in TME. ROS impacts on regulatory T cell (Treg) function and destiny in tumor microenvironment: Tumor cells produce ROS, leading to NRF2 and proapoptotic gene transcripts upregulation while antiapoptotic gene transcripts downregulation in Tregs. Then, Tregs endure apoptosis. Furthermore, apoptotic Tregs release ATP/ADP, which convert into adenosine, inhibiting effector T cell function in antitumor responses

2.2.3 The Effects of OS on Tregs in TME

OS has a significant impact on the metabolic behaviors of Tregs (Fig. 2.3). Previous studies demonstrated that Tregs undergo apoptosis in TME, which led to efficient antitumor suppression.

Tregs Endure Apoptosis

In TME, studies have found that vulnerable NRF2-associated antioxidant system resulted in Tregs apoptosis [80, 81]. Transcription factor NRF2 and its relevant genes are controlled by the antioxidative system. In murine and human ovarian-cancer-infiltrating Tregs, the expressions of NRF2, heme oxygenase, and other antioxidant proteins are reduced. As a consequence, Tregs are sensitive to ROS challenge, and high expression of proapoptotic genes and low expression of antiapoptotic genes are detected. Besides, mitochondrial activity is increased in Tregs [81].

Tregs Suppress Teffs by Adenosine
A high level of ATP/ADP is released from apoptotic Tregs and transformed into immunosuppressing adenosine [82]. The relevant energy metabolism, purine metabolism, and pyrimidine metabolism genes are highly expressed [83]. Therefore, apoptotic Tregs not only generate a large number of ATP to self-provide but also release from pannexin-1-dependent channels to the adjacent environment [84]. Subsequently, apoptotic Tregs express ectoenzymes CD39 and CD73 to metabolize ATP to adenosine and suppress Teffs through A_{2A} receptors, which recognize adenosine and induce IL-2 suppression on Teffs [81].

Tregs Promote Immunosuppression
Human ovarian-cancer-infiltrating apoptotic Tregs inhibit the generation of IL-2, TNF, and IFN-γ in Teffs. Moreover, Tregs abolish natural and induced PD-L1 checkpoint blockade Teff antitumor immunity in MC38 colon cancer. Thus, compared with live Tregs, apoptotic Tregs promote tumor growth by suppressing TAA-specific Teffs and inhibiting the expression of cytokines in Teffs more efficiently [81]

In summary, OS enhances the apoptosis of Tregs in TME. Different from Teffs, Tregs are not sensitive to glycolytic restriction but vulnerable to ROS, which leads to antitumor immunosuppression. These studies implied that removing the treatment effect of PD-L1 blockade and undergoing apoptosis in Tregs were novel mechanisms of tumor immune escape.

2.2.4 The Effects of OS on B Cells in TME

It has been demonstrated that the presence and function of B cells play an essential role in tumor prognosis. However, studies about the effects of OS on B cells in TME are relatively few. In immune cells, ROS, as the second messenger, participates in the cell response and function of B cells [85]. ROS can regulate the maturation, activation, proliferation, apoptosis, and function execution of B cells.

There are many investigations on B cell apoptosis among the possible death modes. In the study of chronic lymphocytic leukemia, through the cascade response of cytochrome c and caspases, ROS can cause B cell apoptosis [86]. Besides, ROS can also activate caspase-9, leading to the improvement of X-linked inhibitor of apoptosis-associated factor 1 (XAF1) and follow-up B cell apoptosis. ROS-activated JNK/p38 MAPK signaling pathway can also induce B cell apoptosis, mainly leading to Bax transfer into mitochondria. Then, mitochondrial membrane potential is damaged; caspase-9 and caspase-3 are activated [64]. Thus, OS is a critical inducing factor of B cell apoptosis (Fig. 2.4).

Studies about the ROS in B cell precursor acute lymphoblastic leukemia demonstrated that ROS could contribute to the genetic instability in B cells, which is characterized by double-strand breaks [87]. Besides, no other researches show the influence of ROS in the antitumor immune response of B cells. However, ROS plays

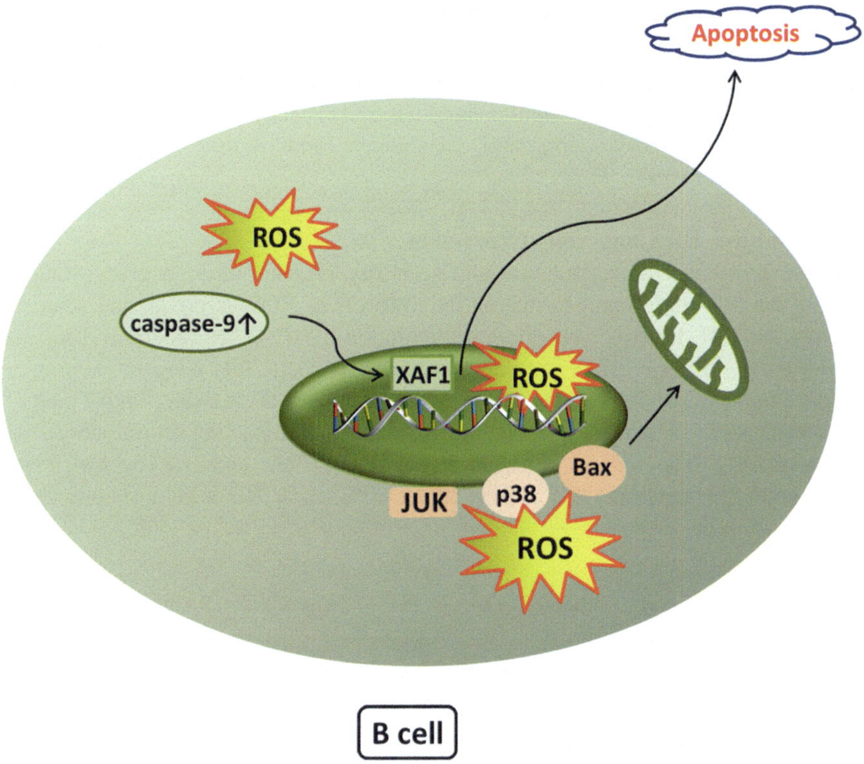

Fig. 2.4 The main effects and central mechanisms of OS on B cells in TME. ROS can activate caspase-9 to increase X-linked inhibitor of apoptosis-associated factor 1 (XAF1) and induce the apoptosis of B cells. JNK/p38 MAPK signal pathway activated by ROS can also cause the apoptosis of B cells, which leads to the translocation of Bax into mitochondria, the destruction of mitochondrial membrane potential, and the activation of caspase-9 and caspase-3. ROS increases the genetic instability and double-strand break of B cells in antitumor immune response

a crucial role in B cell function and its specific subpopulation differentiation, so we can consider that ROS is vital in tumor immune response mediated by B cells [88].

Combined with the above discussion of the dual role of tumor-infiltrating B cells during immune regulation and the specific impact of OS on it, the effects of OS on B cells on tumor immune regulation and the overall process of diseases are also dual. At present, there is no direct study on the effects of OS on B cells in TME. Due to the instability of ROS, its impact on the formation of B cell subsets is not apparent, including regulatory B cell and tumor-related B cell subsets functions. More roles of ROS in B cells are still to be found.

2.2.5 The Effects of OS on DCs in TME

DCs are characterized as specialized APCs. However, intracellular ROS can disrupt tumor-associated DCs (tDCs) lipid metabolism and capacity of antigen presentation, leading to antitumor suppression in human and murine ovarian cancer (Fig. 2.5) [89].

XBP1 Induces ER Chaperone Expression

The ER processes transmembrane and secretory proteins through glycosylation, folding, and disulfide bond formation. High levels of ROS, hypoxia, and acidosis can impair ER function in protein maturation [90]. When aberrant proteins accumulated, "ER stress" or unfolded protein response (UPR) happens, aiming at recovering the metabolic balance of ER. Simultaneously, the activated signal path of UPR engages in increasingly producing ROS through impacting on mitochondrial functions [91]. The sensors of UPR are ATF6, PERK, and IRE1α, which are localized at ER membrane [92]. Among these UPR arms, IRE1α is most conserved.

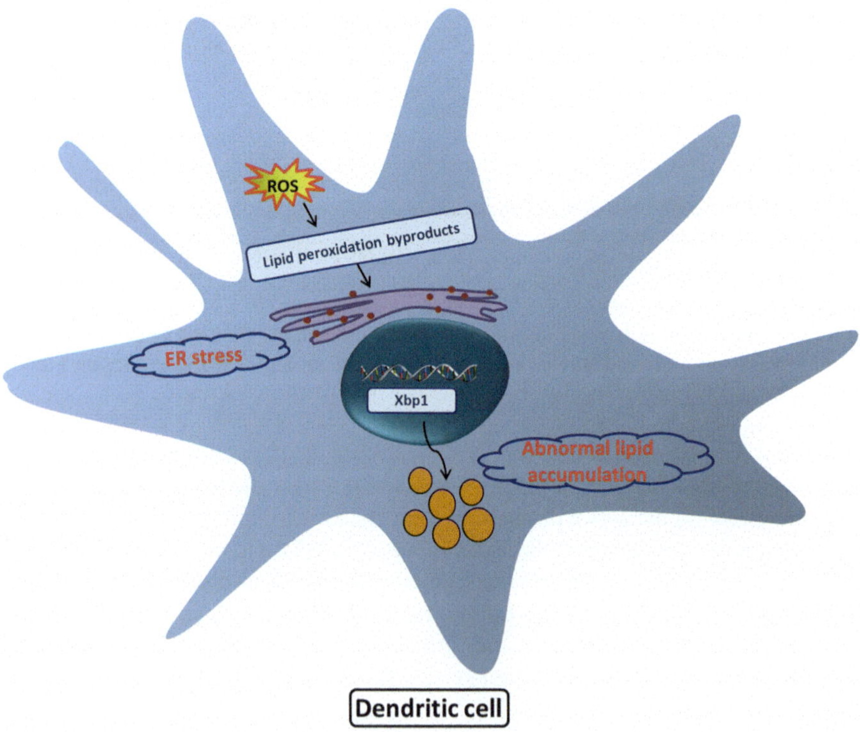

Fig. 2.5 The main effects and central mechanisms of OS on dendritic cells in TME. ROS impacts on dendritic cell (DC) function and destiny in tumor microenvironment: ROS leads to generation of lipid peroxidation byproducts; then, ER stress occurs. Accordingly, transcription factor Xbp1 is produced and facilitates abnormal lipid accumulation. Finally, the lipid metabolic balance is broken, and antitumor immunity capacity is hampered in DCs

The kinase domain oligomerizes and autophosphorylates in ER stress process, and 26 nucleotides are removed from the unspliced *X-box binding protein 1 (Xbp1)* mRNA due to the activated RNase domain, and then *Xbp1* is produced, which encodes transcription factor XBP1 initiating ER chaperone expression ([89, 93]).

ROS Induces ER Stress

In TME, ROS generates lipid peroxidation byproducts, resulting in ER stress and XBP1 activation in tDCs [94]. Moreover, XBP1 promotes lipid accumulation via Agpat6, Fasn, Scd2, and Lpar1, to regulate triglyceride biosynthetic [89]. Apart from 4-HNE, there are some lipid peroxidation byproducts, including acrolein, malondialdehyde, and 4-hydroxyhexenal, that could lead to antitumor suppression via ER stress [95]. Overall, the production of 4-HNE and other lipid peroxidation byproducts caused by ROS makes IRE1α-XBP1 arm overactivated, which destroys lipid metabolism balance in tDCs. Accordingly, the antitumor immunity of T cells is inhibited.

Studies elucidated that targeting IREIα-XBP1 axis could be a new strategy in immunotherapy. Using IRE1α inhibitors could improve anticancer immunity and tumor cell death. Silencing XBP1 by siRNA-encapsulating nanocarriers renders tDCs acquiring immunocompetence to support ovarian-cancer-infiltrating T cells [89].

2.2.6 OS and Macrophages in TME

2.2.6.1 OS Effect of TAMs

A point to be discussed in this review: unlike the relationship between other immune cells and OS conditions in TME, in macrophages, most studies discuss the mechanisms by which they cause OS states in the TME. TAMs infiltration in TME is related mainly to the poor prognosis of cancer, which can mediate a variety of tumor-promoting phenomena: inflammation, vascular regeneration, and OS.

Macrophages are widely recruited in a variety of tumors, and their functions are various: directly produce ROS/RNS, and secrete pro-inflammatory cytokines that promote matrix reaction, both of which can affect the redox state of TME [96]. TAM-induced chronic inflammatory response produces ROS, which plays a vital role in activating the transcription factors that are sensitive to redox conditions and are involved in all processes of tumor development. Therefore, TAMs can promote tumor growth mainly by its contribution to the OS state in TME. They participate in the tumor-promoting physiological effects of downstream immunosuppression, cell proliferation, angiogenesis, and tumor metastasis.

In most tumors, the tumorigenic effect of TAMs is dominant. For example, it has been shown that TAM-regulated OS is the primary process that affects the C26 colon cancer cell proliferation. Through the activity of NADPH oxidase in macrophages, it maintains the physiological range of OS and angiogenesis in the C26 colon cancer environment. It promotes the production of tumor angiogenic proteins [97]. In conclusion, TAMs are one of the main factors in the formation of OS state in the microenvironment, which promotes epithelial-mesenchymal cell transition (EMT), aggressiveness, and the diffusion of metastatic cells [96].

2.2.6.2 The Effect of OS on Macrophages

The Polarization of Macrophages

Reprogramming TAM from the M2 phenotype to M1 is associated with OS in TME and cancer cells, which promotes tumor killing. The switch of TAM polarization from M1 to M2 is parallel to the progressive inhibition of NF-κB in tumor progression [98]. There are studies about mitochondrial Lon that can be induced by OS. The results showed that in TME, the expression of Lon in macrophages and cancer cells increased, which could induce the production of ROS-p38-NF-κB-dependent cytokines, and M2 macrophage polarization was driven by IL-6, IL-13, IL-4, and VEGF-A. The overexpression of Lon induced p38 expression, and p38-NF-κB signal transduction dependent on ROS induces the specific inflammatory cytokines and thus drives the polarization of M2 macrophages [99]. Studies showed that macrophage polarization is related to taking in $KRAS^{G12D}$ protein released from cancer cells that suffer from ferroptosis. ROS can induce cancer cells to release $KRAS^{G12D}$ during OS, and $KRAS^{G12D}$ is packaged into exosomes. After it is absorbed by macrophages through advanced glycosylation end product-specific receptor (AGER)-dependent mechanism, it can promote macrophage polarization through activation of fatty acid oxidation dependent on STAT3. Specifically, in macrophages, $KRAS^{G12D}$ activates STAT3 through AGER, leading to selective upregulation of CPT1A and ACADM. This is conducive to M2 polarization of macrophages and fatty acid oxidation [100].

Phagocytosis

It showed that excessive ROS exposure could lead to the impairment of macrophage function, such as phagocytosis. This macrophage dysfunction is partly due to the oxidation of mannose-binding lectin, and it is necessary for the process of active phagocytosis [101]. Moreover, a new study reveals the role of ROS in macrophage function from a new perspective. CLEC10A is a human sugar receptor expressed in macrophages. CLEC10A on macrophages can mediate the uptake of damaged cells during antigen presentation; CLEC10A ligands (CLEC10AL) exist in human tumor tissues. The increase of ROS induces the expression of CLEC10AL in breast cancer cells. Mechanistically, it is proved by the elevated expression of GalNAc transferase 6 (GalNT6) and GalNT2, and GalNT6 was translocated from cis-Golgi to trans-Golgi compartment. The changes in molecular mechanisms caused by ROS lead to the buildup of truncated glycans on the surface of cells, and they lead to the enhancement of phagocytosis of macrophages [102]. The positive expression of macrophage CLEC10A has been proved to be related to the improvement of prognosis in breast cancer patients (Fig. 2.6).

Overall, the relationship between macrophages and OS has not been comprehensively discussed in cancer. Combined with the interaction between OS and

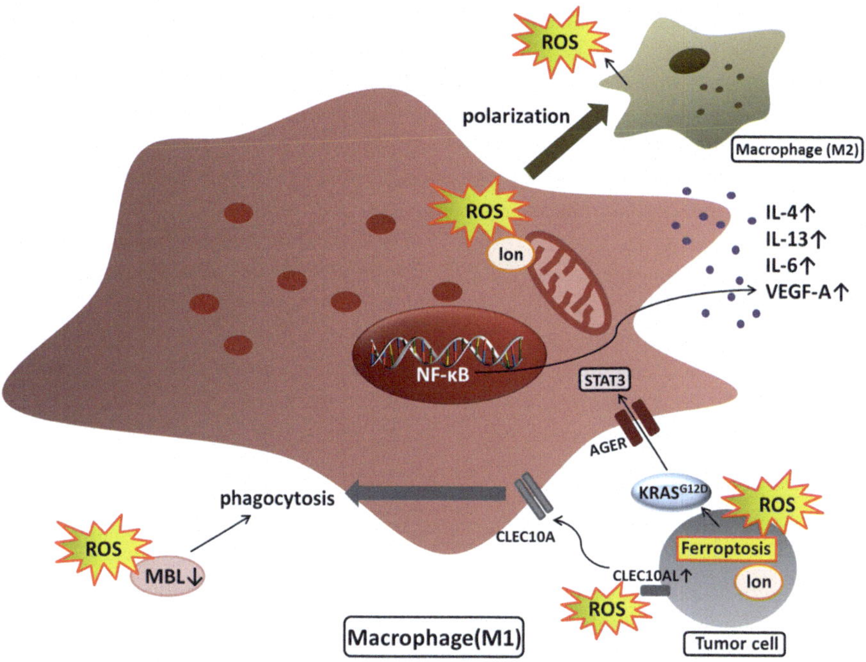

Fig. 2.6 The main effects and central mechanisms of OS on macrophages in TME. The overexpression of Lon triggers ROS-dependent p38, and p38-NF-κB signal transduction induces expression of inflammatory cytokines and thus drives the polarization of M2 macrophages. ROS-induced KRAS^{G12D} releasing from cancer cells during oxidative stress is packaged into exosomes and absorbed by macrophages through AGER-dependent mechanism and then promotes macrophage polarization through STAT3-dependent activation of fatty acid oxidation. The dysfunction of phagocytosis is partly due to the oxidation of the key component necessary for active phagocytosis, mannose-binding lectin (MBL). ROS can also improve the phagocytosis indirectly by inducing the expression of CLEC10AL on the tumor cell surface

macrophages, the effect of OS on the overall process of the tumor is complicated. Therefore, it is imperative to find the right balance in OS-related therapy [98].

2.3 The Significance of OS on TIME

From the above discussion, we can see that the effects of OS on different immune cells are multifaceted. Here, we focus on the information about OS and tumor immunosuppression. Cancer immunosurveillance can be regarded as an inflammatory process, which is vital in the recognition and elimination of early tumors by the immune system. Immunosuppressive TME promotes tumor deterioration. ROS may be the immunosuppressive participants; that is, the impaired immune response of cancer patients is related to OS.

There have been a lot of therapeutic researches based on the role of ROS in tumor cells, such as killing tumor cells by stimulating the increase of ROS production. Besides, a therapeutic strategy targeting TME has been proposed to improve their sensitivity to immune monitoring. There is no study combining OS-related therapy with targeted immunotherapy [4]. However, some studies have shown the crosstalk relationship between OS and tumor immunotherapy. For example, tumor checkpoint inhibitors as immunomodulators can reduce tumor burden by regulating tumor OS. Studies have shown that the response of mice with colorectal cancer to immunotherapy and chemotherapy resistance of ovarian cancer patients can also be mediated by OS [103].

Collectively, recognizing and targeted enhancing the keyway of the antitumor immune response, maintaining a certain level of ROS in tumor cells, and TME can improve the prognosis and the survival of cancer patients.

2.3.1 OS and Immunosuppression in TME

Immunosuppression in tumor development refers to that under the condition of cancer, the immune dysfunction of tumor patients may cause tumor immune escape. The main features of low immune system function are weakened response to antigens, decreased T cell proliferation, decreased signal transduction, and impairment of transcription factor activity in important lymphocytes [55].

ER Stress

The deficient microenvironment of the tumor, such as hypoxia, acidity, and OS, can increase the intracellular ROS content and mediate the ER stress of cells. The function of some immune cells in TME can be regulated by ER stress and further hinder the antitumor immunity. Specifically, ER stress has two tasks for tumor-related immune regulation: one is to mediate the immunosuppression of immune cells through tumor-cell-releasing factors under ER stress; the other is to directly generate ER stress in immune cells, mainly involving myeloid cells (neutrophils, MDSCs, and macrophages), with different mechanisms. For example, the production of 4-HNE and other lipid peroxidation byproducts caused by ROS in DCs makes IRE1a-XBP1 overactivated, which destroys their lipid metabolism balance, impends the antigen presentation of T cells, and thus hinders protective immune response development [104].

Therefore, targeting ER stress could also serve to improve the clinical tumor immune efficacy [104]. Studies demonstrated that silencing of XBP1 in DCs might prolong host survival by siRNA-encapsulating nanocarriers. Then, the antitumor suppression is reversed, and antitumor immune responses are initiated. Targeting XBP1 resulting from ER stress in tDCs provides us a novel therapeutic method [19].

Mitochondrial OS

Apart from the effect of driving the polarization of M2 macrophages, the OS of mitochondria induced by Lon can also promote the metastasis behavior of the tumor and change the cytokine balance in TME to establish immunosuppression. Specifically, cancer cells with high expression of Lon release specific cytokines, which activate the upregulation of endogenous Lon in macrophages and promote macrophages to secrete cytokines supporting tumor. The inflammatory cytokines secreted by cancer cells and macrophages induced by mitochondrial Lon-ROS circulate in the TME, promoting EMT/the plasticity of cancer, angiogenesis, and M2 macrophage polarization, thus triggering immune escape and tumor metastasis [99]. Therefore, regulating the redox balance of mitochondria in specific cells in TME may be a strategy of tumor immunotherapy to improve T cell function [99].

ROS, as an Inflammatory Mediator

ROS, as an inflammatory mediator in TME, is able to inhibit the function of immune cells directly. For example, OS can upregulate IL-8 and TNF in DCs. Additionally, the ROS of MDSCs plays an important role. MDSCs often appear in the TME with OS characteristics, and the ROS production of MDSCs is upregulated after being activated in cancer, while the ROS derived from MDSCs can impair cellular immune response. Important lymphocytes are all the targets of ROS. ROS regulates the metabolism of MDSCs. ROS can maintain the undifferentiated state of MDSCs and exert immunosuppression. The increase of endogenous H_2O_2 levels may be one of the mechanisms of tumor preventing MDSCs from differentiation. HIF-1 and NRF2 participate in the regulation of pathways of MDSCs. Targeted reduction of ROS in MDSCs can enhance the effect of tumor immunotherapy [105].

Therefore, antioxidant programs have been introduced into tumor treatment and prevention. However, the role of ROS in tumor progression is still controversial. Due to its high cytotoxicity, ROS can be used to kill tumor cells. Therefore, the key task of individualized immunotherapy is to identify the inflammatory characteristics of cancer patients and find the balance of ROS-related treatment [106].

Based on the impact of OS on primary tumor-related immune cells, the global immunosuppression consequences are produced. Understanding the molecular events of OS-induced immune cell inactivation may help determine the treatment strategies to alleviate cancer-related immunosuppression.

Taking T cells as an example, OS affects their survival, proliferation, activation, differentiation, and other physiological functions. OS disrupts T cell-mediated immune function, which is one of the crucial reasons for tumor development. The inhibition of T cell response leads to the immunosuppression of the tumor-bearing state. Considering the mechanism of ROS-mediated immunosuppression, it is important to use antioxidants to regulate the reaction of antitumor T cells [43].

In the study of Tregs, OS-mediated apoptotic Tregs can eliminate the anti-PD-L1-mediated antitumor effects, emphasizing that the oxidation pathway can be used as a checkpoint in metabolic process to control Tregs' behavior, which can affect the effect of cancer checkpoint therapy. Thus, the impact of OS on tumor immune cells may be related to the mechanism of immune escape and checkpoint drug resistance [81].

Besides, some immune cells are associated with the formation of OS conditions in TME, that is, the ROS level released by activated macrophages or granulocytes infiltrated by tumor increases, which may become an obstacle for active immunotherapy of cancer. It suggests that in addition to the therapeutic idea of regulating OS to improve the activity of immune cells, targeting tumor immune cells to improve the OS environment specifically is also vital.

2.3.2 Inspiration for Clinical Treatment

All researches above provide a new idea for tumor immunotherapy. The immunotherapy of solid tumors mainly includes the following ways: activating the functions of tumor-associated lymphocytes, enhancing phagocytosis of tumor-related macrophages, targeting inhibition of MDSCs and Tregs, and regulatory therapy of TME. It focuses on the direct enhancement of tumor-related immune cell function or the regulation of microenvironment. Our review provides a theoretical basis for the combination of OS regulation of TME and immune cell function regulation.

According to the above, OS may widely impair the survival and normal function of primary tumor-related immune cells, and treatment aimed at reversing immunosuppression may need to focus on the change of redox state in cancer patients. TME regulation therapy is mediated by interfering with the main feature of tissue hypoxia, OS, metabolic disorder, and chronic inflammation of TME. It can improve the internal environment, provide appropriate environmental signals for the normal physiological activities including the activation and proliferation of killer immune cells, and improve the immune environment by providing a variety of immune regulatory factors [107]. For example, hypoxia can disrupt oxygen metabolism, resulting in a large amount of ROS, making the TME in OS state. It can be restored by adjusting abnormal tumor blood vessels, which can alleviate local hypoxia and promote the function of effector immune cells [107].

Besides, some studies have proposed some drugs that mediate the recovery of immune cell function through OS. It is possible to improve the prognosis of tumor patients by the treatment that can correctly intervene in the OS condition of TME through processes such as improving the T cell-mediated antitumor effects. Traditional Chinese medicine may be used in the treatment of TME regulation [107]. As a component of traditional Chinese medicine, flavonoids are potent antioxidants, which can resist the local OS of tumor cells [108]. Curcumin, as an antioxidant, is used as an antineoplastic drug, which can reduce the systemic toxicity, neutralize the OS in the immune system, and protect the function of T cells. Specifically, it can interfere with the fate of T cells in TME by mediating NF-κB-ROS-TNF-α, neutralize tumor-induced OS, restore the NF-B activity of T cells, inhibit TNF-α, and thus reduce tumor-induced T cell apoptosis [57]. Besides, it can also play a role in Th cell polarization, such as normalizing Th1 cytokine profile, preventing Th2 polarization, and mediating Treg accumulation [109]. All-trans-retinoic acid (ATRA) is a mature promoter of human MDSCs, thus reversing its immunosuppressive function. Specifically, ATRA promotes the accumulation of GSH in MDSCs, which leads to a decreased ROS level, and prevents MDSCs from differentiating into mature myeloid cells [110]. Antioxidant therapy can block the differentiation of TAMs and the occurrence of tumors [111]. For example, caffeic acid (CA) can enhance the cytotoxicity of M1 macrophages and inhibit tumor growth, and its antioxidant activity can mediate the inhibition of TAMs [112]. Specifically, CA can increase the functional capacity of macrophages, which causes an increase in the level of pro-inflammatory cytokines, such as IL-2, IFN-γ, and IL-12, and increase the tumor-killing activity of macrophages. These antitumor treatment measures will cause metabolic changes of TME, thus affecting the function of immune cells [113].

Collectively, immunotherapy and TME regulation should be paid more attention to in the new treatment strategies of cancer patients. Targeted treatment strategies should be designed for different types of immune cells so that they can regulate the innate immune system of the tumor efficiently and establish systematic antitumor immunity. Besides, the use of environmental regulation drugs such as antioxidants can improve TME, restore immune cell function, and prevent tumor growth. Finally, we should pay attention to the combination of universality and individualized therapy.

2.4 Summary and Perspectives

The characteristics of the main effects of OS on tumor-related immune cells can be summarized as multiple aspects and consequences, most of which cause immunosuppression. Among them, the researches on T cells (especially Teffs) and NK cells are more mature, while the existing studies on B cells and macrophages are less. OS is one of the most critical factors that cause the low reactivity of Teffs in the tumor. It mainly affects the TCR signal pathway and causes death, proliferation, and differentiation inhibition of Teffs. The dysfunction of Teffs in TME seriously affects the effectiveness of tumor immune regulation, which leads to the acceleration of tumor progression. In addition to Teffs, NK cells can also kill tumor cells. Under the influence of OS, apoptosis mediated by PARP and AIF occurs. Therefore, PARP can also be used as a potential therapeutic target for tumor and inflammatory pathology. The apoptosis of Tregs in TME is caused by OS and even promotes the tumor immunosuppression, which is a new mechanism of tumor immune escape. Different from T cells, the role of B cells in tumor immunity is less elucidated. Excessive ROS may cause genc instability and double-strand break of B cells or induce apoptosis of B cells in TME. As a typical critical APC, in DCs, intracellular ROS can affect the metabolism and antigen-presenting ability of tumor-related macrophages, mainly through ER stress effects and lipid peroxidation. Finally, unlike other cells, macrophages are an important source of OS in TME, which is the leading cause of the tumorigenic effect of TAMs. Besides, OS can mediate the influence of cell polarization. Macrophages have M1 and M2 types. M1 type mainly plays an inhibitory role in cancer, while M2 type plays a different position. OS can help its polarization to M2 direction. Besides, excessive ROS exposure can lead to macrophage dysfunction.

According to a series of researches, we focus on the consequences of immunosuppression. Tumor immunosuppression is a phenomenon of immune escape caused by the decline of related immune system function in the process of the tumor. Many studies clearly show that many immune cell biological phenomena related to OS, such as ER stress, mitochondrial OS, and ROS as inflammatory mediators, are closely related to tumor immunosuppression. The overall consequences of OS on specific important tumor-related immune cells suggest that we can improve the activity of related immune cells by regulating the level of OS in TME.

Therefore, many related clinical therapies for OS are derived from interfering with the malignant characteristics of TME and improving the conditions for immune cells. Traditional Chinese medicine mediates TME reconstruction, especially the use of antioxidants to improve the immune cell damage caused by OS, such as curcumin. At the same time, we should also pay attention to the rational use of antioxidants. Considering the effect of OS on tumor progression, a high concentration of ROS in cells mediates the death of tumor cells. Still, it may jeopardize tumor immunity, thus increasing the severity of the tumor. Therefore, we should focus on the balance of oxidative regulation in the OS-related treatment of tumors and pay attention to the analysis of the individualized immune status of tumor patients for targeted therapy.

In conclusion, ROS in OS can be used as one of the immunosuppressive factors, which can lead to immune cell dysfunction in tumor patients. OS can mediate the overall downregulation of antitumor immunity. However, the current researches mainly focus on the research results of the operating system, and the relevant mechanisms need to be further studied. These findings provide new targets for tumor immunotherapy and new ideas for clinical treatment.

Conflict of Interest Authors declare no conflict of interest.

References

1. Fiaschi T, Chiarugi P. Oxidative stress, tumor microenvironment, and metabolic reprogramming: a diabolic liaison. Int J Cell Biol. 2012;2012:762825. https://doi.org/10.1155/2012/762825.
2. Bhattacharyya S. Tumour, oxidative stress and host T cell response: cementing the dominance. Scand J Immunol. 2015;6:82.
3. Whiteside TL. Immune suppression in cancer: effects on immune cells, mechanisms and future therapeutic intervention. Semin Cancer Biol. 2006;16(1):3–15. https://doi.org/10.1016/j.semcancer.2005.07.008.
4. Cheng YT, Yang CC, Shyur LF. Phytomedicine-modulating oxidative stress and the tumor microenvironment for cancer therapy. Pharmacol Res. 2016;114:128–43. https://doi.org/10.1016/j.phrs.2016.10.022.
5. Steinman RM. Decisions about dendritic cells: past, present, and future. Annu Rev Immunol. 2012;30:1–22. https://doi.org/10.1146/annurev-immunol-100311-102839.

6. Liu YJ. IPC: professional type 1 interferon-producing cells and plasmacytoid dendritic cell precursors. Annu Rev Immunol. 2005;23:275–306. https://doi.org/10.1146/annurev.immunol.23.021704.115633.
7. Binnewies M, Mujal AM, Pollack JL, Combes AJ, Hardison EA, Barry KC, Tsui J, Ruhland MK, Kersten K, Abushawish MA, Spasic M, Giurintano JP, Chan V, Daud AI, Ha P, Ye CJ, Roberts EW, Krummel MF. Unleashing type-2 dendritic cells to drive protective antitumor CD4(+) T cell immunity. Cell. 2019;177(3):556–71. https://doi.org/10.1016/j.cell.2019.02.005.
8. Broz ML, Binnewies M, Boldajipour B, Nelson AE, Pollack JL, Erle DJ, Barczak A, Rosenblum MD, Daud A, Barber DL, Amigorena S, Veer LJ, Sperling AI, Wolf DM, Krummel MF. Dissecting the tumor myeloid compartment reveals rare activating antigen-presenting cells critical for T cell immunity. Cancer Cell. 2014;26(5):638–52. https://doi.org/10.1016/j.ccell.2014.09.007.
9. Woo SR, Fuertes MB, Corrales L, Spranger S, Furdyna MJ, Leung MY, Duggan R, Wang Y, Barber GN, Fitzgerald KA, Alegre ML, Gajewski TF. STING-dependent cytosolic DNA sensing mediates innate immune recognition of immunogenic tumors. Immunity. 2014;41(5):830–42. https://doi.org/10.1016/j.immuni.2014.10.017.
10. Dustin ML, Bromley SK, Kan Z, Peterson DA, Unanue ER. Antigen receptor engagement delivers a stop signal to migrating T lymphocytes. Proc Natl Acad Sci U S A. 1997;94(8):3909–13. https://doi.org/10.1073/pnas.94.8.3909.
11. Stinchcombe JC, Bossi G, Booth S, Griffiths GM. The immunological synapse of CTL contains a secretory domain and membrane bridges. Immunity. 2001;15(5):751–61. https://doi.org/10.1016/s1074-7613(01)00234-5.
12. Boissonnas A, Fetler L, Zeelenberg IS, Hugues S, Amigorena S. In vivo imaging of cytotoxic T cell infiltration and elimination of a solid tumor. J Exp Med. 2007;204(2):345–56. https://doi.org/10.1084/jem.20061890.
13. Breart B, Lemaitre F, Celli S, Bousso P. Two-photon imaging of intratumoral CD8+ T cell cytotoxic activity during adoptive T cell therapy in mice. J Clin Invest. 2008;118(4):1390–7. https://doi.org/10.1172/JCI34388.
14. Feig C, Jones JO, Kraman M, Wells RJB, Deonarine A, Chan DS, Connell CM, Roberts EW, Zhao Q, Caballero OL, Teichmann SA, Janowitz T, Jodrell DI, Tuveson DA, Fearon DT. Targeting CXCL12 from FAP-expressing carcinoma-associated fibroblasts synergizes with anti-PD-L1 immunotherapy in pancreatic cancer. Proc Natl Acad Sci U S A. 2013;110(50):20212–7. https://doi.org/10.1073/pnas.1320318110.
15. Markiewicz MA, Wise EL, Buchwald ZS, Cheney EE, Hansen TH, Suri A, Cemerski S, Allen PM, Shaw AS. IL-12 enhances CTL synapse formation and induces self-reactivity. J Immunol. 2009;182(3):1351–61. https://doi.org/10.4049/jimmunol.182.3.1351.
16. Melssen M, Slingluff CL Jr. Vaccines targeting helper T cells for cancer immunotherapy. Curr Opin Immunol. 2017;47:85–92. https://doi.org/10.1016/j.coi.2017.07.004.
17. Kennedy R, Celis E. Multiple roles for CD4+ T cells in anti-tumor immune responses. Immunol Rev. 2008;222:129–44. https://doi.org/10.1111/j.1600-065X.2008.00616.x.
18. Tateyama M, Oyaizu N, McCloskey TW, Than S, Pahwa S. CD4 T lymphocytes are primed to express Fas ligand by CD4 cross-linking and to contribute to CD8 T-cell apoptosis via Fas/FasL death signaling pathway. Blood. 2000;96(1):195–202.
19. Kumar P, Bhattacharya P, Prabhakar BS. A comprehensive review on the role of co-signaling receptors and Treg homeostasis in autoimmunity and tumor immunity. J Autoimmun. 2018;95:77–99. https://doi.org/10.1016/j.jaut.2018.08.007.
20. Zou W. Regulatory T cells, tumour immunity and immunotherapy. Nat Rev Immunol. 2006;6(4):295–307. https://doi.org/10.1038/nri1806.
21. Grossman WJ, Verbsky JW, Barchet W, Colonna M, Atkinson JP, Ley TJ. Human T regulatory cells can use the perforin pathway to cause autologous target cell death. Immunity. 2004;21(4):589–601. https://doi.org/10.1016/j.immuni.2004.09.002.

22. Sarhan D, Hippen KL, Lemire A, Hying S, Luo X, Lenvik T, Curtsinger J, Davis Z, Zhang B, Cooley S, Cichocki F, Blazar BR, Miller JS. Adaptive NK cells resist regulatory T-cell suppression driven by IL37. Cancer Immunol Res. 2018;6(7):766–75. https://doi.org/10.1158/2326-6066.cir-17-0498.
23. Shevach EM, Thornton AM. tTregs, pTregs, and iTregs: similarities and differences. Immunol Rev. 2014;259(1):88–102. https://doi.org/10.1111/imr.12160.
24. Wei T, Zhong W, Li Q. Role of heterogeneous regulatory T cells in the tumor microenvironment. Pharmacol Res. 2020;153:104659. https://doi.org/10.1016/j.phrs.2020.104659.
25. Kamada T, Togashi Y, Tay C, Ha D, Sasaki A, Nakamura Y, Sato E, Fukuoka S, Tada Y, Tanaka A, Morikawa H, Kawazoe A, Kinoshita T, Shitara K, Sakaguchi S, Nishikawa H. PD-1 (+) regulatory T cells amplified by PD-1 blockade promote hyperprogression of cancer. Proc Natl Acad Sci U S A. 2019;116(20):9999–10008. https://doi.org/10.1073/pnas.1822001116.
26. Tokunaga R, Naseem M, Lo JH, Battaglin F, Soni S, Puccini A, Berger MD, Zhang W, Baba H, Lenz HJ. B cell and B cell-related pathways for novel cancer treatments. Cancer Treat Rev. 2019;73:10–9. https://doi.org/10.1016/j.ctrv.2018.12.001.
27. Yuen GJ, Demissie E, Pillai S. B lymphocytes and cancer: a love-hate relationship. Trends Cancer. 2016;2(12):747–57. https://doi.org/10.1016/j.trecan.2016.10.010.
28. Zhang CY, Xin H, Zhang W, Yazaki PJ, Zhang ZF, Le K, Li WZ, Lee H, Kwak L, Forman S, Jove R, Yu H. CD5 binds to interleukin-6 and induces a feed-forward loop with the transcription factor STAT3 in B cells to promote cancer. Immunity. 2016;44(4):913–23. https://doi.org/10.1016/j.immuni.2016.04.003.
29. Olkhanud PB, Damdinsuren B, Bodogai M, Gress RE, Sen R, Wejksza K, Malchinkhuu E, Wersto RP, Biragyn A. Tumor-evoked regulatory B cells promote breast cancer metastasis by converting resting CD4(+) T cells to T-regulatory cells. Cancer Res. 2011;71(10):3505–15. https://doi.org/10.1158/0008-5472.CAN-10-4316.
30. Wang SS, Liu W, Ly D, Xu H, Qu LM, Zhang L. Tumor-infiltrating B cells: their role and application in anti-tumor immunity in lung cancer. Cell Mol Immunol. 2019;16(1):6–18. https://doi.org/10.1038/s41423-018-0027-x.
31. Sharonov GV, Serebrovskaya EO, Yuzhakova DV, Britanova OV, Chudakov DM. B cells, plasma cells and antibody repertoires in the tumour microenvironment. Nat Rev Immunol. 2020; https://doi.org/10.1038/s41577-019-0257-x.
32. Di Vito C, Mikulak J, Zaghi E, Pesce S, Marcenaro E, Mavilio D. NK cells to cure cancer. Semin Immunol. 2019;41:101272. https://doi.org/10.1016/j.smim.2019.03.004.
33. Lanier LL. Up on the tightrope: natural killer cell activation and inhibition. Nat Immunol. 2008;9(5):495–502. https://doi.org/10.1038/ni1581.
34. Close HJ, Wurdak H, Short SC, Melcher AA, Stead LF, Wilson EB, Cook GP. Natural killer cell recognition of glioblastoma. Immunology. 2014;143:138.
35. Giorda R, Rudert WA, Vavassori C, Chambers WH, Hiserodt JC, Trucco M. NKR-P1, a signal transduction molecule on natural killer cells. Science. 1990;249(4974):1298–300.
36. Zhu YT, Huang B, Shi J. Fas ligand and lytic granule differentially control cytotoxic dynamics of natural killer cell against cancer target. Oncotarget. 2016;7(30):47163–72. https://doi.org/10.18632/oncotarget.9980.
37. Fehniger TA, Cooper MA. Harnessing NK cell memory for cancer immunotherapy. Trends Immunol. 2016;37(12):877–88. https://doi.org/10.1016/j.it.2016.09.005.
38. Porta C, Riboldi E, Ippolito A, Sica A. Molecular and epigenetic basis of macrophage polarized activation. Semin Immunol. 2015;27(4):237–48. https://doi.org/10.1016/j.smim.2015.10.003.
39. Belgiovine C, D'Incalci M, Allavena P, Frapolli R. Tumor-associated macrophages and anti-tumor therapies: complex links. Cell Mol Life Sci. 2016;73(13):2411–24. https://doi.org/10.1007/s00018-016-2166-5.
40. Chittezhath M, Dhillon MK, Lim JY, Laoui D, Shalova IN, Teo YL, Chen J, Kamaraj R, Raman L, Lum J, Thamboo TP, Chiong E, Zolezzi F, Yang H, Van Ginderachter JA, Poidinger M, Wong AS, Biswas SK. Molecular profiling reveals a tumor-promoting

phenotype of monocytes and macrophages in human cancer progression. Immunity. 2014;41 (5):815–29. https://doi.org/10.1016/j.immuni.2014.09.014.
41. Dimitrios M. Naturally occurring regulatory T cells show reduced sensitivity toward oxidative stress-induced cell death. Blood. 2009;15:113.
42. King MR, Ismail AS, Davis LS, Karp DR. Oxidative stress promotes polarization of human T cell differentiation toward a T helper 2 phenotype. J Immunol. 2006;176(5):2765–72. https://doi.org/10.4049/jimmunol.176.5.2765.
43. Chen X, Song M, Zhang B, Zhang Y. Reactive oxygen species regulate T cell immune response in the tumor microenvironment. Oxidative Med Cell Longev. 2016;2016:1580967. https://doi.org/10.1155/2016/1580967.
44. Cemerski S, van Meerwijk JPM, Romagnoli P. Oxidative-stress-induced T lymphocyte hyporesponsiveness is caused by structural modification rather than proteasomal degradation of crucial TCR signaling molecules. Eur J Immunol. 2003;33(8):2178–85. https://doi.org/10.1002/eji.200323898.
45. Mak TW, Saunders ME. 14 - T cell activation. In: Mak TW, Saunders ME, editors. The immune response. Burlington: Academic; 2006. p. 373–401. https://doi.org/10.1016/B978-012088451-3.50016-8.
46. Cemerski S, Cantagrel A, van Meerwijk JPM, Romagnoli P. Reactive oxygen species differentially affect T cell receptor-signaling pathways. J Biol Chem. 2002;277(22):19585–93. https://doi.org/10.1074/jbc.M111451200.
47. Mizoguchi H. Alterations in signal transduction molecules in T lymphocytes from tumor-bearing mice. Science. 1992;5089:258.
48. Otsuji M. Oxidative stress by tumor-derived macrophages suppresses the expression of CD3 zeta chain of T-cell receptor complex and antigen-specific T-cell responses. Proc Natl Acad Sci U S A. 1996;23:93.
49. Sukumar M, Liu J, Mehta GU, Patel SJ, Roychoudhuri R, Crompton JG, Klebanoff CA, Ji Y, Li P, Yu Z, Whitehill GD, Clever D, Eil RL, Palmer DC, Mitra S, Rao M, Keyvanfar K, Schrump DS, Wang E, Marincola FM, Gattinoni L, Leonard WJ, Muranski P, Finkel T, Restifo NP. Mitochondrial membrane potential identifies cells with enhanced stemness for cellular therapy. Cell Metab. 2016;23(1):63–76. https://doi.org/10.1016/j.cmet.2015.11.002.
50. Olmos Y, Valle I, Borniquel S, Tierrez A, Soria E, Lamas S, Monsalve M. Mutual dependence of Foxo3a and PGC-1alpha in the induction of oxidative stress genes. J Biol Chem. 2009;284 (21):14476–84. https://doi.org/10.1074/jbc.M807397200.
51. Scharping NE, Menk AV, Moreci RS, Whetstone RD, Dadey RE, Watkins SC, Ferris RL, Delgoffe GM. The tumor microenvironment represses T cell mitochondrial biogenesis to drive intratumoral T cell metabolic insufficiency and dysfunction. Immunity. 2016;45(2):374–88. https://doi.org/10.1016/j.immuni.2016.07.009.
52. Shen BC, Huang SC. Predominant Th2/Tc2 polarity of tumor-infiltrating lymphocytes in cervical cancer. Int J Gynecol Obstet. 2000;70:A62. https://doi.org/10.1016/S0020-7292(00)82673-9.
53. Liang Z. Enhanced Th17 differentiation and aggravated arthritis in IEX-1-deficient mice by mitochondrial reactive oxygen species-mediated signaling. J Immunol. 2012;4:189.
54. Jeremias I. Inhibition of nuclear factor kappaB activation attenuates apoptosis resistance in lymphoid cells. Blood. 1998;12:91.
55. Malmberg KJ. Inhibition of activated/memory (CD45RO(+)) T cells by oxidative stress associated with block of NF-kappaB activation. J Immunol. 2001;5:167.
56. Tang G, Minemoto Y, Dibling B, Purcell NH, Li Z, Karin M, Lin A. Inhibition of JNK activation through NF-κB target genes. Nature. 2001;414(6861):313–7. https://doi.org/10.1038/35104568.
57. Sankar B. Tumor-induced oxidative stress perturbs nuclear factor-kappaB activity-augmenting tumor necrosis factor-alpha-mediated T-cell death: protection by curcumin. Cancer Res. 2007;1:67.

58. Stuelten CH, Byfield SD, Arany PR, Karpova TS, Stetler-Stevenson WG, Roberts AB. Breast cancer cells induce stromal fibroblasts to express MMP-9 via secretion of TNF-α and TGF-β. J Cell Sci. 2005;118(10):2143. https://doi.org/10.1242/jcs.02334.
59. Bowie A. Oxidative stress and nuclear factor-kappaB activation: a reassessment of the evidence in the light of recent discoveries. Biochem Pharmacol. 2000;1:59.
60. Belikov AV, Schraven B, Simeoni L. T cells and reactive oxygen species. J Biomed Sci. 2015;22:85. https://doi.org/10.1186/s12929-015-0194-3.
61. Min L-W. Vitamin E inhibits CD95 ligand expression and protects T cells from activation-induced cell death. J Clin Invest. 2002;5:110.
62. Kiessling R, Wasserman K, Horiguchi S, Kono K, Sjöberg J, Pisa P, Petersson M. Tumor-induced immune dysfunction. Cancer Immun. 1999;48(7):353–62. https://doi.org/10.1007/s002620050586.
63. Thor FB. Oxygen radicals induce poly(ADP-ribose) polymerase-dependent cell death in cytotoxic lymphocytes. J Immunol. 2006;176(12):7301–7.
64. Kim EK, Seo HS, Chae MJ, Jeon IS, Song BY, Park YJ, Ahn HM, Yun CO, Kang CY. Enhanced antitumor immunotherapeutic effect of B-cell-based vaccine transduced with modified adenoviral vector containing type 35 fiber structures. Gene Ther. 2014;21(1):106–14. https://doi.org/10.1038/gt.2013.65.
65. Hamilos DL, Zelarney P, Mascali JJ. Lymphocyte proliferation in glutathione-depleted lymphocytes: direct relationship between glutathione availability and the proliferative response. Immunopharmacology. 1989;18(3):223–35. https://doi.org/10.1016/0162-3109(89)90020-9.
66. Tanja H. The role of low molecular weight thiols in T lymphocyte proliferation and IL-2 secretion. J Immunol. 2005;1950(12):175.
67. Yan Z, Garg SK, Banerjee R. Regulatory T cells interfere with glutathione metabolism in dendritic cells and T cells. J Biol Chem. 2010;285(53):41525–32. https://doi.org/10.1074/jbc.M110.189944.
68. Suthanthiran M. Glutathione regulates activation-dependent DNA synthesis in highly purified normal human T lymphocytes stimulated via the CD2 and CD3 antigens. Proc Natl Acad Sci U S A. 1990;9:87.
69. Messina JP. Cell cycle progression of glutathione-depleted human peripheral blood mononuclear cells is inhibited at S phase. J Immunol. 1989;6:143.
70. Debaprasad M. Failure in peripheral immuno-surveillance due to thymic atrophy: importance of thymocyte maturation and apoptosis in adult tumor-bearer. Life Sci. 2005;21:77.
71. Francisco B. NK cell metabolism and tumor microenvironment. Front Immunol. 2019;10:2278.
72. Aydin E, Johansson J, Nazir FH, Hellstrand K, Martner A. Role of NOX2-derived reactive oxygen species in NK cell-mediated control of murine melanoma metastasis. Cancer Immunol Res. 2017;5:9.
73. Mellqvist UH. Natural killer cell dysfunction and apoptosis induced by chronic myelogenous leukemia cells: role of reactive oxygen species and regulation by histamine. Blood. 2000;5:96.
74. Johan A. Monocytic AML cells inactivate antileukemic lymphocytes: role of NADPH oxidase/gp91(phox) expression and the PARP-1/PAR pathway of apoptosis. Blood. 2012;24:119.
75. Seong-Woon Y. Mediation of poly(ADP-ribose) polymerase-1-dependent cell death by apoptosis-inducing factor. Science. 2002;5579:297.
76. Tentori L. Potential clinical applications of poly(ADP-ribose) polymerase (PARP) inhibitors. Pharmacol Res. 2002;2:45.
77. Fenerty KE, Padget M, Wolfson B, Gameiro SR, Su Z, Lee JH, Rabizadeh S, Soon-Shiong P, Hodge JW. Immunotherapy utilizing the combination of natural killer– and antibody dependent cellular cytotoxicity (ADCC)–mediating agents with poly (ADP-ribose) polymerase (PARP) inhibition. J Immunother Cancer. 2018;6:1.
78. Tentori L. Chemopotentiation by PARP inhibitors in cancer therapy. Pharmacol Res. 2005;1:52.

79. Zhu H, Wang B, Kong L, An T, Li Y. Parvifoline AA promotes susceptibility of hepatocarcinoma to natural killer cell-mediated cytolysis by targeting peroxiredoxin. Cell Chem Biol. 2019;26:8.
80. Sporn MB, Liby KT. NRF2 and cancer: the good, the bad and the importance of context. Nat Rev Cancer. 2012;12(8):564–71. https://doi.org/10.1038/nrc3278.
81. Maj T, Wang W, Crespo J, Zhang H, Wang W, Wei S, Zhao L, Vatan L, Shao I, Szeliga W, Lyssiotis C, Liu JR, Kryczek I, Zou W. Oxidative stress controls regulatory T cell apoptosis and suppressor activity and PD-L1-blockade resistance in tumor. Nat Immunol. 2017;18 (12):1332–41. https://doi.org/10.1038/ni.3868.
82. Silvia D. Adenosine generation catalyzed by CD39 and CD73 expressed on regulatory T cells mediates immune suppression. J Exp Med. 2007;6:204.
83. Arvey A, van der Veeken J, Samstein RM, Feng Y, Stamatoyannopoulos JA, Rudensky AY. Inflammation-induced repression of chromatin bound by the transcription factor Foxp3 in regulatory T cells. Nat Immunol. 2014;15(6):580–7. https://doi.org/10.1038/ni.2868.
84. Faraaz BC. Pannexin 1 channels mediate 'find-me' signal release and membrane permeability during apoptosis. Nature. 2010;7317:467.
85. Reth M. Hydrogen peroxide as second messenger in lymphocyte activation. Nat Immunol. 2002;3(12):1129–34. https://doi.org/10.1038/ni1202-1129.
86. Salimi A, Roudkenar MH, Sadeghi L, Mohseni A, Seydi E, Pirahmadi N, Pourahmad J. Ellagic acid, a polyphenolic compound, selectively induces ROS-mediated apoptosis in cancerous B-lymphocytes of CLL patients by directly targeting mitochondria. Redox Biol. 2015;6:461–71. https://doi.org/10.1016/j.redox.2015.08.021.
87. Kantner H-P, Warsch W, Delogu A, Bauer E, Esterbauer H, Casanova E, Sexl V, Stoiber D. ETV6/RUNX1 induces reactive oxygen species and drives the accumulation of DNA damage in B cells. Neoplasia. 2013;15(11):1292. https://doi.org/10.1593/neo.131310.
88. Zhang H, Wang L, Chu Y. Reactive oxygen species: the signal regulator of B cell. Free Radic Biol Med. 2019;142:16–22. https://doi.org/10.1016/j.freeradbiomed.2019.06.004.
89. Cubillos-Ruiz JR, Silberman PC, Rutkowski MR, Chopra S, Perales-Puchalt A, Song M, Zhang S, Bettigole SE, Gupta D, Holcomb K, Ellenson LH, Caputo T, Lee AH, Conejo-Garcia JR, Glimcher LH. ER stress sensor XBP1 controls anti-tumor immunity by disrupting dendritic cell homeostasis. Cell. 2015;161(7):1527–38. https://doi.org/10.1016/j.cell.2015.05.025.
90. Christina P. Crosstalk between advanced glycation and endoplasmic reticulum stress: emerging therapeutic targeting for metabolic diseases. J Clin Endocrinol Metab. 2012;7:97.
91. Sies H, Jones DP. Reactive oxygen species (ROS) as pleiotropic physiological signalling agents. Nat Rev Mol Cell Biol. 2020; https://doi.org/10.1038/s41580-020-0230-3.
92. Claudio H. Targeting the unfolded protein response in disease. Nat Rev Drug Discov. 2013;9:12.
93. Ann-Hwee L. XBP-1 regulates a subset of endoplasmic reticulum resident chaperone genes in the unfolded protein response. Mol Cell Biol. 2003;21:23.
94. Vladykovskaya E, Sithu SD, Haberzettl P, Wickramasinghe NS, Merchant ML, Hill BG, McCracken J, Agarwal A, Dougherty S, Gordon SA, Schuschke DA, Barski OA, O'Toole T, D'Souza SE, Bhatnagar A, Srivastava S. Lipid peroxidation product 4-hydroxy-trans-2-nonenal causes endothelial activation by inducing endoplasmic reticulum stress. J Biol Chem. 2012;287(14):11398–409. https://doi.org/10.1074/jbc.M111.320416.
95. Negre-Salvayre A, Coatrieux C, Ingueneau C, Salvayre R. Advanced lipid peroxidation end products in oxidative damage to proteins. Potential role in diseases and therapeutic prospects for the inhibitors. J Pharmacol. 2008;153(1):6–20. https://doi.org/10.1038/sj.bjp.0707395.
96. Giannoni E, Parri M, Chiarugi P. EMT and oxidative stress: a bidirectional interplay affecting tumor malignancy. Antioxid Redox Signal. 2012;16(11):1248–63. https://doi.org/10.1089/ars.2011.4280.

97. Luput L, Licarete E, Sesarman A, Patras L, Costelalupei M, Banciu M. Tumor-associated macrophages favor C26 murine colon carcinoma cell proliferation in an oxidative stress-dependent manner. Oncol Rep. 2017;37(4):2472–80. https://doi.org/10.3892/or.2017.5466.
98. Mantovani A, Sica A. Macrophages, innate immunity and cancer: balance, tolerance, and diversity. Curr Opin Immunol. 2010;22(2):231–7. https://doi.org/10.1016/j.coi.2010.01.009.
99. Kuo C-L, Chou H-Y, Chiu Y-C, Cheng AN, Fan C-C, Chang Y-N, Chen C-H, Jiang SS, Chen N-J, Lee AY-L. Mitochondrial oxidative stress by Lon-PYCR1 maintains an immunosuppressive tumor microenvironment that promotes cancer progression and metastasis. Cancer Lett. 2020;474:138–50. https://doi.org/10.1016/j.canlet.2020.01.019.
100. Dai E, Han L, Liu J, Xie Y, Kroemer G, Klionsky DJ, Zeh HJ, Kang R, Wang J, Tang D. Autophagy-dependent ferroptosis drives tumor-associated macrophage polarization via release and uptake of oncogenic KRAS protein. Autophagy. 2020; https://doi.org/10.1080/15548627.2020.1714209.
101. Zuo L, He F, Sergakis GG, Koozehchian MS, Stimpfl JN, Rong Y, Diaz PT, Best TM. Interrelated role of cigarette smoking, oxidative stress, and immune response in COPD and corresponding treatments. Am J Phys. 2014;307(3):L205–18. https://doi.org/10.1152/ajplung.00330.2013.
102. Kurze AK, Buhs S, Eggert D, Oliveira-Ferrer L, Muller V, Niendorf A, Wagener C, Nollau P. Immature O-glycans recognized by the macrophage glycoreceptor CLEC10A (MGL) are induced by 4-hydroxy-tamoxifen, oxidative stress and DNA-damage in breast cancer cells. Cell Commun Signal. 2019;17(1):107. https://doi.org/10.1186/s12964-019-0420-9.
103. Tang J, Ramis-Cabrer D, Wang XJ, Barreiro E. Immunotherapy with monoclonal antibodies in lung cancer of mice: oxidative stress and other biological events. Cancer. 2019;11:9.
104. Cubillos-Ruiz JR, Bettigole SE, Glimcher LH. Tumorigenic and immunosuppressive effects of endoplasmic reticulum stress in cancer. Cell. 2017;168(4):692–706. https://doi.org/10.1016/j.cell.2016.12.004.
105. Ohl K, Tenbrock K. Reactive oxygen species as regulators of MDSC-mediated immune suppression. Front Immunol. 2018;9:2499. https://doi.org/10.3389/fimmu.2018.02499.
106. Sheng KC. Inflammatory mediators hold the key to dendritic cell suppression and tumor progression. Curr Med Chem. 2011;36:18.
107. Duan H. Novel therapeutic strategies for solid tumor based on body's intrinsic antitumor immune system. Cell Physiol Biochem. 2018;47(2):441–57. https://doi.org/10.1159/000489979.
108. Panda AK, Chakraborty D, Sarkar I, Khan T, Sa G. New insights into therapeutic activity and anticancer properties of curcumin. J Exp Pharmacol. 2017;9:31–45. https://doi.org/10.2147/JEP.S70568.
109. Bhattacharyya S, Md Sakib Hossain D, Mohanty S, Sankar Sen G, Chattopadhyay S, Banerjee S, Chakraborty J, Das K, Sarkar D, Das T, Sa G. Curcumin reverses T cell-mediated adaptive immune dysfunctions in tumor-bearing hosts. Cell Mol Immunol. 2010;7(4):306–15. https://doi.org/10.1038/cmi.2010.11.
110. Noweeda M. All-trans-retinoic acid improves differentiation of myeloid cells and immune response in cancer patients. Cancer Res. 2006;18:66.
111. Zhang Y, Choksi S, Chen K, Pobezinskaya Y, Linnoila I, Liu Z-G. ROS play a critical role in the differentiation of alternatively activated macrophages and the occurrence of tumor-associated macrophages. Cell Res. 2013;23(7):898–914. https://doi.org/10.1038/cr.2013.75.
112. Orsolic N, Kunstic M, Kukolj M, Gracan R, Nemrava J. Oxidative stress, polarization of macrophages and tumour angiogenesis: efficacy of caffeic acid. Chem Biol Interact. 2016;256:111–24. https://doi.org/10.1016/j.cbi.2016.06.027.
113. Netea-Maier RT, Smit JWA, Netea MG. Metabolic changes in tumor cells and tumor-associated macrophages: a mutual relationship. Cancer Lett. 2018;413:102–9. https://doi.org/10.1016/j.canlet.2017.10.037.

Oxidative Stress-Mediated Stem Cell Aging 3

Zhijie Tian and Xiaozhen Dai

Abstract

Stem cells, which are a class of cells with self-renewal capacity and multipotency of differentiation, play a critical role in tissue maintenance and regeneration in the whole life span of multicellular organism. With the recent researches about stem cell, it is realized that the aging and depletion of stem cells are closely related to organ aging and the occurrence of aging-related diseases, such as cardiovascular and cerebrovascular diseases, autoimmune diseases, and Alzheimer's disease. Thus, understanding the molecular mechanisms of stem cell aging will be important for developing new therapies for aging-related diseases. There are many intrinsic and extrinsic factors promoting stem cell aging. Oxidative stress has been recognized as the major cause of stem cell aging. Oxidative stress is the result of an excessive production of reactive oxygen species (ROS) and an impairment of the antioxidant defense systems. Excessive production of ROS and insufficient cellular antioxidant trigger a variety of aging-related pathways to induce stem cell senescence and aging. In this chapter, we summarized the biologic features of stem cell aging and discussed how oxidative stress affects stem cell aging and the main signal pathways of stem cell aging triggered by oxidative stress. We further explored how stem cells manage ROS accumulation and adapt to oxidative stress. Finally, we discussed the potential strategies against stem cell aging by controlling oxidative stress.

Keywords

Stem cells · Aging · Oxidative stress · ROS

Z. Tian · X. Dai (✉)
School of Bioscience and Technology, Chengdu Medical College, Chengdu, China

C. Huang, Y. Zhang (eds.), *Oxidative Stress*,
https://doi.org/10.1007/978-981-16-0522-2_3

Stem cells are a class of cells with unique ability of self-renewal and multipotent differentiation capacity. In mammals, stem cells can be divided into two basic types: embryonic stem cells (ESCs) and adult stem cells (ASCs) [1]. ESCs originate from the inner cell mass of preimplantation embryos and exhibit both true pluripotency and self-renewal capacity. ASCs reside in variety of adult tissues (e.g., bone marrow, adipose, peripheral blood, liver, pancreas, hair follicle, skeletal muscle) and differentiate into almost all cell types of their originated tissue [2, 3]. Thus, ASCs are critical for maintaining tissue homeostasis and preserving regenerative capacity after injury. Emerging evidence demonstrated that age-associated imbalance of tissue homeostasis is at least partially caused by a loss of the regenerative capacity of ASCs linked to the accumulation of age-associated damage. During aging, there are various stress factors that regulate stem cell activity and promote stem cell aging. Oxidative stress caused by excessive production of reactive oxygen species (ROS) is one of the most significant endogenous toxins and inducer in stem cell senescence and injuries [4]. High-level ROS can trigger stem cell aging through directly inducing DNA, protein, and mitochondrial damage or accelerating telomere shortening. The mechanism underlying oxidative stress-mediated stem cell damage and senescence is a very complex signaling network. Meanwhile, stem cells possess several endogenous antioxidant stress strategies to manage ROS accumulation and adapt to oxidative stress. Therefore, understanding the complex signaling pathways mediated by oxidative stress as well as the antioxidant stress strategies of stem cell is very important to find targets for inhibiting stem cell aging.

3.1 Characteristics of Stem Cell Aging

All multicellular organisms exhibit declines in the function of tissues and organs with age, which is at least partially due to the loss of stem cell number and function over time. Stem cells in many tissues undergo obvious changes with age, exhibiting impaired self-renewal, abnormal proliferative activities, limited differentiation, and declined functional capacities [5]. These changes resulted in stem cell blunted responsiveness to cell replacement and tissue regeneration in old organisms. The mechanisms that underlie cellular aging are composed of intrinsic alterations and extrinsic alterations. Intrinsic alterations include telomere reduction, DNA damage, mitochondrial dysfunction, and loss of proteostasis. Additionally, extrinsic alterations can range from stem cell niche to systemic environment, such as oxidative stress, inflammation, and irradiation.

The characteristics and molecular changes of aged stem cell have been reviewed extensively elsewhere, which demonstrate the biology features of aged stem cell from different tissue are not uniform [5, 6]. For example, the number of functional hematopoietic stem cells (HSCs) decreases with age, and the differentiation potential of aged HSCs is also skewed toward the myeloid lineage at the expense of the lymphoid lineage [7, 8]. Similarly, the number and function of neural stem cells (NSCs) and melanocyte stem cells decline with age. However, the numbers of hair follicle stem cells (HFSCs) and fly intestinal stem cells (ISC) do not change with age, but concomitant with a decline in function [9, 10].

3.2 Oxidative Stress-Mediated Signaling Pathways in Stem Cell Aging

Oxidative stress can be simply defined as excessive production of ROS in the presence of reduced antioxidants. ROS are composed by a diverse range of oxygen-containing entities which possess higher reactivity than molecular oxygen (O_2), such as superoxide radical anion ($O_2\cdot^-$), hydroxyl radical ($\cdot OH$), and hydrogen peroxide (H_2O_2). Mitochondria are primary sources of ROS in most cells, a consequence of its role in energy production via oxidative phosphorylation (OXPHOS). During OXPHOS process, electron can leak prematurely to oxygen from the complexes I and III of mitochondrial electron transport chain [11]. Although mitochondria are primary sources of ROS in most cells, NADPH oxidases (NOX) are another major source of ROS. NOX consumes NADPH to generate $O_2\cdot^-$ and ROS.

Excessive ROS has been viewed as one of the leading causes of stem cell aging, which is attributed to the ROS-mediated oxidation of biologic macromolecules such as DNA, proteins, and lipids, which consequently promote cellular damage and senescence. The mechanism underlying oxidative stress-induced cell aging is involved in complex signaling networks regulated by several of regulators (Fig. 3.1), which are susceptible to redox changes and are recognized as redox

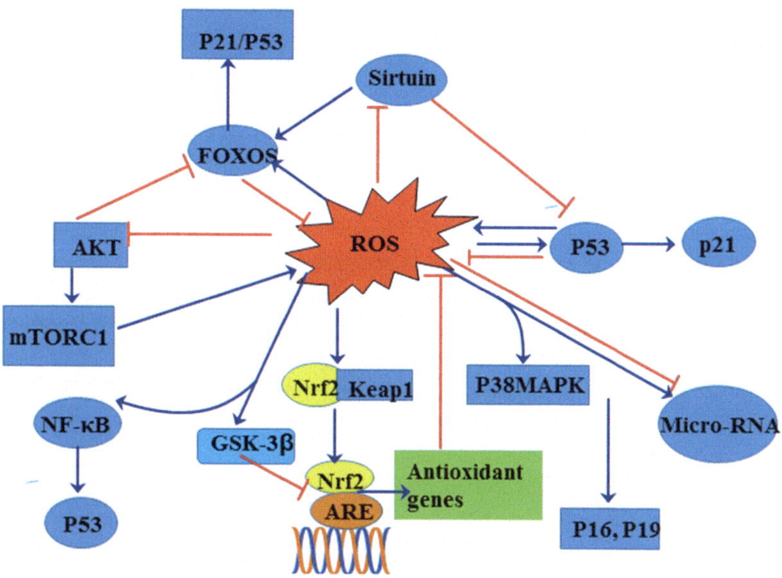

Fig. 3.1 The signaling molecules involved in oxidative stress-induced stem cell aging

sensors. Thus, understanding the role of redox sensors in oxidative stress-induced stem cell aging is essential for development of the molecular basis for improving stem cell therapy efficacy.

3.2.1 p53 Pathway

p53 is a transcription factor that controls the expression of a range of target genes and plays critical roles in regulating cell cycle arrest, apoptosis, or senescence in response to numerous stress that lead to genomic instability, including oxidative stress. The previous studies present evidence that ROS act as an upstream regulator of p53 and p53 can in turn influence cellular ROS production. At low levels of oxidative stress, p53 induces antioxidant gene expression to attenuate oxidative stress and improve cell survival ability; by contrast, under high levels of oxidative stress, p53 shows pro-oxidative activities via upregulating the expression of pro-oxidative genes (e.g., p53-inducible gene 3 and proline oxidase), which further increases oxidative stress levels to induce cell death [12]. Furthermore, p53 exhibits pro-oxidative activities via suppressing the expression of anti-oxidative genes, leading to susceptibility to oxidative stress.

The previous studies reported that a constitutively active form of p53 induced a premature aging like phenotype in mice [13], while impairment of the activity of p53 ortholog CEP-1 extended life span of in *Caenorhabditis elegans* [14]. These results suggest that p53 activity is relative with aging. The aging process of mesenchymal stromal cells (MSCs) is often associated with activation of p53. More recent study reported that excessive ROS upregulated p53 expression, which binds to parkin and inhibits the biologic function of parkin to decrease the level of mitophagy, resulting in senescence of bone marrow MSCs [15]. In human umbilical cord-derived MSCs, it was also found that inhibition of p300 accelerates premature senescence via activation of p53/p21 signal pathway [16]. Wild-type p53-induced phosphatase 1 (Wip1) regulates HSCs aging via directly regulating p53 pathway, since Wip1-dedicicent mice exhibited multifaceted HSC aging phenotypes, while deletion of p53 can rescue the multilineage repopulation defect of Wip1$(^{-/-})$ HSCs [17]. However, in skeletal muscle stem cells (MuSCs), it was discovered that the p53 activity was decreased with age, and pharmacological activation of p53 promotes the expansion of aged MuSCs in vivo [18]. Furthermore, genetically manipulated mice with p53 overexpression decreased aging-associated damage, which indicated that activation of p53 showed protective efforts to aging [19]. In summary, the role of p53 in stem cell aging is dependent on the levels of cellular oxidative stresses and cell type.

3.2.2 Nuclear Factor-E2-Related Factor 2 Pathway

Nuclear factor-erythroid 2-related factor 2 (Nrf2) is a ubiquitously expressed stress-responsive transcription factor, which regulates the expression of cyto-protective genes and antioxidant enzymes in stem cells. The activity of Nrf2 is tightly regulated by Kelch-like ECH-associated protein 1 (Keap1). Under normal condition, Nrf2 is binding to Keap1 in the cytoplasm and subsequently degraded through the ubiquitin-proteasome pathway, which resulted in the low-level activity and expression of Nrf2 [20]. Under oxidative stress, high levels of intracellular ROS promote Nrf2 dissociating from Keap1 either by the oxidization of the reactive cysteine residues (Cys273, Cys288, and Cys151) of Keap1 or by activation of kinase, such as MAPK, PI3Ks, and PKC that phosphorylate Nrf2 [21, 22]. Then, the dissociated Nrf2 is transferred to the nucleus, where it dimerizes with the small Maf proteins and binds to the antioxidant response element (ARE) to transcriptionally upregulate the expression cyto-protective genes to prevent oxidative stress damage. In addition, oxidative stress activates GSK3β, leading to phosphorylation and nuclear import of Src kinase, such as Fyn, Src, Yes, and Fgr, which phosphorylates Nrf2 (Tyr568) and promotes Nrf2 nuclear export and degradation [23, 24]. We previously found that sustained hyperglycemia and hyperlipidemia induced ROS increase, which resulted in the nuclear export of Nrf2 increase by GSK3β/Fyn signaling [25].

We recently reviewed the role of Nrf2 in stem cell state and function, focused on the role of Nrf2 in regulating redox and metabolic of stem cell [1]. The review demonstrated that the roles of Nrf2 in stem cell are unique and type- and lineage-specific. Several studies have confirmed that Nrf2 plays a critical role in stem cell survival and resistance to oxidative stress. Activating Nrf2 protects MSCs, HSCs, NSCs, and endothelial progenitor cells (EPCs) from apoptosis and death induced by oxidative stress through upregulating downstream anti-oxidative genes and growth factors. For example, Nrf2 overexpression protects BM MSCs from death induced by oxidative stress via upregulating HO-1 and SOD [26]. Furthermore, pharmacological activation of Nrf2 with Nrf2 activator, such as ginger oleoresin, could mitigate ionizing radiation-induced oxidative stress and DNA damage of human MSCs. Melatonin suppresses senescence of canine adipose MSCs via activating Nrf2 activation and inhibiting ER stress [27]. Elsewhere, Nrf2 is also crucial for HSCs survival and function under stress condition. Nrf2 deficiency leads to HSCs apoptosis increase and survival decrease under oxidative stress condition. Activating Nrf2 signaling, either by genetic methods or by pharmacological molecules, prevents radiation-induced HSCs injury by scavenging ROS [28, 29].

3.2.3 FoxOs Signal Pathway

Forkhead Box O (FoxO) proteins constitute an evolutionarily conserved family of transcription factors that play a critical role in the regulation of aging and longevity. In mammals, the FoxO family is composed of FoxO1, FoxO3a, FoxO4, and FoxO6, which are highly similar in structure, function, and regulation. FoxO transcription factors are important regulators of oxidative stress in mammals. On one hand, activated FoxOs can counteract oxidative stress via upregulating ROS-scavenging enzymes, such as MnSOD and catalase. On the other hand, ROS can regulate the expression and activity of FoxO proteins at multiple levels, including transcription and post-translation modification. For example, ROS indirectly regulates the expression of FoxO through the upstream regulators of FoxO, such as p53, and miRNA. Furthermore, oxidative stress affects the activation and stabilization of FoxO through phosphorylation or deacetylation.

During the past decades, many evidences have shown that FoxO family members are critical mediators in regulating the self-renewal and maintaining homeostasis of NSCs/NPCs [30] and HSCs [31, 32]. The previous studies suggest that FoxOs play a central role in in keeping the long-term regenerative capability of the HSC compartment; deficiency of FoxOs in HSCs results in increase of ROS and decreased expression of ataxia telangiectasia mutated (ATM) that is key factor in coordinating stem cell cycling with ROS level. FoxO3 is a predominant FoxO isoform; the role FoxO3 in HSC is extensively explored. FoxO3 is a critical factor for protecting HSC from oxidative damage. Similarly, NSCs isolated from adult FoxO3(−/−) mice showed impairment of self-renewal and differentiation abilities with ROS accumulation [33]. The above findings suggest that FoxO plays a critical role in reducing ROS level and keeping redox balance. However, under oxidative stress condition, the expression and activity of FoxOs were altered by ROS. In EPCs, it was found that the expression of FoxO3a was markedly increased in EPCs treated with H_2O_2; increasing FoxO3a resulted in EPCs apoptosis possibly by transcriptional regulation of Bim [30]. FoxO transcription factors are essential for maintaining homeostasis of stem cell, but the mechanism of FoxOs in stem cell aging induced by oxidative stress needs further investigation.

3.2.4 Sirtuin Pathway

Sirtuins are a family of proteins with nicotinamide dinucleotide (NAD^+)-dependent deacylases activity, which play a pivotal role in delaying cellular senescence and extending the organismal life span via regulating various cellular processes. Lee et al. recently reviewed the roles of sirtuins on cellular senescence and life span extension [34]. In mammals, there are seven sirtuins (SIRT1–SIRT7) which are localized in different subcellular compartmentation with different enzymatic activity. SIRT1 prevents premature aging/cellular senescence by regulating FoxOs, p53 and p21 and molecules involved in DNA damage and repair. SIRT1 is a redox-sensitive protein; the level and activity of SIRT1 are regulated by oxidative stress mainly through posttranslational modification, such as phosphorylation, SUMOylation, S-glutathionylation, S-nitrosylation, and carbonylation [35]. In human lung epithelial cells, it was found that the oxidant significantly decreased the protein expression and activity of SIRT1 by carbonyl modifications [36]. Oxidative stress induces rat nucleus pulposus cell (NP) senescence and decreases the expression of SIRT1; specific activation of SIRT1 suppresses oxidative stress-induced senescence of NP cells by Akt-FoxO1 pathway [37]. The other group also found that inhibiting SIRT1 induced premature senescence of human endothelial cells, while overexpression of SIRT1 prevented hydrogen peroxide-induced premature senescence [38]. Besides SIRT1, SIRT6 also plays critical role in cell senescence and aging. SIRT6-deficient mice exhibited genomic instability and several phenotypes of accelerated premature aging. SIRT6 could protect podocytes against apoptosis and inflammation by enhancing autophagy via inhibiting Notch signaling pathway [39].

Besides suppression mitotic cell senescence, sirtuins also regulate the senescence of stem cells. For example, the expression of SIRT1 in EPCs from old subjects was decreased; inhibition of SIRT1 in EPCs from young subjects resulted in marked increase of senescence [40]. Furthermore, cleaving SIRT1 with cathepsin leads to EPC premature senescence [41], while activating SIRT1 with MHY2233 can reduce replicative and oxidative stress-induced senescence in EPCs [42]. Similarly, SIRT1 declines with age and plays an essential role in maintaining MSC's long-term growth without undergoing cellular senescence; overexpression of SIRT1 in MSCs reduced osteogenic cell senescence and enhanced osteogenesis by deacetylating FoxO3a and inhibiting oxidative stress [43]. SIRT1 is also important for the maintenance of HSCs through ROS elimination, FoxO activation, and p53 inhibition [44]. Homma and colleagues have demonstrated that SIRT1 is important in maintaining phenotype and function of induced pluripotent stem (iPS)-derived vascular endothelial cells [45]. Similarly, the previous study demonstrated that SIRT6-deficient hMSCs showed functional decay and are more sensitive to the oxidative stress; further mechanism study uncovered that SIRT6 protects human MSCs against oxidative stress damage via coactiving Nrf2 signal pathway [46]. More recent study demonstrated that the expression of SIRT6 was decreased in BMSCs from the elderly and SIRT6 regulated senescence and osteogenesis of BMSCs via changing autophagy level partly through activating AKT-mTOR pathway [47]. SIRT6 also regulates HSC homeostasis and self-renewal capacity via epigenetic regulation of Wnt signaling [48] or repression NF-κB target genes [49]. The abovementioned findings indicated that SIRT1 or SIRT6 is believed to delay stem cell aging and senescence under the conditions of oxidative stress; activating SIRT1 or SIRT6 in stem cell would be a promising therapeutic strategy for chronic inflammatory diseases associated with aging. A lot of studies have showed that sirtuins activators, such as resveratrol, MHY2233, and melatonin, enhanced cell survival and reduced oxidative stress-mediated senescence in variety of kinds of stem cells [42, 50, 51].

3.2.5 Nuclear Factor Kappa B Pathway

Nuclear factor kappa B (NF-κB) is a family of globally expressed transcription factor involved in a series of cellular process, including proliferation, inflammation, differentiation, senescence, and apoptosis. The family of NF-κB comprises RelA/p65, RelB, c-Rel, p50/p100(NF-κB1), and p52/p100(NF-κB2). NF-κB pathway can be activated by two different pathways, the canonical and noncanonical pathways. The canonical pathway is induced by microbial products, stress factors, pro-inflammatory cytokines, and it depends on the activation of IκB kinase (IKK) complex, leading to phosphorylation and subsequent degradation of IκB molecules via the ubiquitin-proteasome system. Then, released NF-κB translocate into the nucleus to activate target gene expression. By contrast, the noncanonical pathway is induced by B cell-activating factor (BAFF), lymphotoxin β (LTβ), CD40 ligand, CD27 ligand, human T-cell leukemia virus (HTLV), and Epstein-Barr virus (EBV) [39].

Cumulative evidences have indicated that the transcription activity of NF-κB can be directly elevated or inhibited by ROS in a context-dependent manner.

Study found that ROS increased phosphorylation of IκBα, which enables the nuclear translocation of NF-κB to enhance interleukin 8 (IL-8) secretion, and/or increased the stability of p53 protein, leading to cell aging [52]. Oxidative stress can activate NF-κB via triggering HSP90 to induce neural stem cell (NSC) death, while inhibition of HSP90 activity will attenuate the NF-κB/p65 activity under oxidative stress and promote NSCs survival [53]. Lin et al. reported that NF-κB activity was elevated in aged MSCs at basal condition or exposed to inflammatory stimuli; suppression of NF-κB activity in aged MSCs could potentially restore aging-associated bone loss or enhance bone-healing process [54]. Furthermore, the previous studies showed that NF-κB activation inhibits the myogenic differentiation of aged muscle-derived stem/progenitor cells (MDSPCs), while pharmacological or genetic inhibition of NF-κB activation improves myogenic differentiation and reduces oxidative stress-induced damage [55]. Aged HSCs also exhibit increased ground-stage NF-κB activity [56]; constitutive activation of NF-κB pathway leads to loss of HSCs quiescence and functions [57].

3.2.6 PI3K/Akt Signal Pathway

Phosphoinositide 3-kinase (PI3K)/Akt signaling pathway is one of most important prosurvival pathway in cells. Both PI3K and Akt kinases are responsive to cellular redox state. Akt possesses two conserved Cys residues (Cys297 and Cys311), which are susceptible to H_2O_2-mediated oxidation. H_2O_2 induces an intramolecular disulfide bond formation between Cys297 and Cys311of Akt, which leads to Akt dephosphorylation and Akt inactivation [58]. It is well known that PI3K/Akt pathway is a critical mediator in stem cell aging. Inhibition of PI3K-Akt pathway can accelerate cellular senescence in many kinds of stem cell, while activation of PI3K-Akt pathway could slow down the process of senescence. For example, doxorubicin induces MSCs aging via ROS accumulation, which associated with the activity of Akt decrease [59], while activating PI3K-Akt pathway by macrophage migration inhibitory factor (MIF) or high-density lipoprotein can rescue MSCs from oxidative stress-induced senescence or death [59, 60]. The main mechanism of these observations was attributed to the reduction in intracellular oxidative stress and prevents DNA damage. Furthermore, it was found that PI3K-Akt pathway inhibited senescence of human skin-derived precursors (hSKPs); enhancement of the activity of Akt in hSKP senescence could reduce expression of p53, p21, and p16 and mitigate hSKP senescence [61, 62]. Mammalian target of rapamycin (mTOR) is a direct target of Akt in regulating cell growth, metabolism, and autophagy. PI3K/Akt/mTOR signal pathway has been found to associate with cell senescence and aging. Increasing ROS generation correlates with PI3K/Akt/mTOR aberrant activation.

Inhibition of the PI3K/Akt/mTOR pathway prevented aging and prolonged life span, which is accompanied with suppression of ROS, iNOS, Cox-2, NF-κB, SASP, and p53 [63].

3.2.7 p38 MAPK Signal Pathway

The p38 mitogen-activated protein kinase (p38 MAPK), an important member of the MAPK family of signal transduction kinases, is activated in a sequential manner to regulate a variety of cellular process, such as self-renewal, proliferation, differentiation, senescence, and apoptosis. The signal pathway of p38 MAPK is known to be activated in variety of cells in response to oxidative stress. Considering the important role of oxidative stress in stem cell aging, it can be speculated that p38 may be an executor of oxidative stress-induced stem cell aging.

Many researches have showed that p38 signal pathway was activated during stem cell aging induced by elevated ROS. For instance, elevation of ROS levels actives p38 MAPK signal to limit the life span of HSCs; inhibition of p38 MAPK restored the defects of HSC repopulating capacity induced by ROS and extended the life span of HSCs, revealing that the ROS-p38 MAPK pathway contributes to HSC senescence and aging [64]. Connexin-43 deficiency resulted in HSC senescence; the mechanism indicated that connexin-43-deficient HSCs are unable to transfer ROS to the microenvironment, leading to cellular ROS accumulation, which activated p38MAPK/FoxO1 signal [65]. Similarly, thioredoxin-interacting protein also induced HSC aging via elevation of ROS to regulate p38 MAPK activity; inhibiting p38 MAPK activity can rejuvenate aged HSCs [66]. Interestingly, aged satellite cells (SCs) and bona fide muscle stem cells exhibit a cell-autonomous defect in self-renewal due to impaired response to FGF ligands and elevated p38 activity; inhibiting p38 MAPK activity can improve age-associated self-renewal defects [67]. The recent research demonstrated that p38 MAPK activation leads to decreases in the number and activity of ISC; targeting p38 MAPK prevents or rescues ISC aging [68]. Furthermore, there are many studies reporting that antioxidant compounds treatment improved the quantity and function of stem cells via downregulating the activation of p38 MAPK [69, 70].

The abovementioned research suggested that p38 MAPK signaling is a critical regulator in stem cell aging response to oxidative stress. This highlights the importance of uncovering the molecular mechanisms that link ROS production to activation of the p38 MAPK-mediated promotion of aging, longevity, and resistance to oxidative stress. Based on the previous studies on non-stem-cell models, it was known that oxidative stress activated p38 MAPK through two major pathways: one is that oxidative stress-mediated oxidation of apoptosis signal-regulating kinase (ASK1)-thioredoxin (Trx[SH]2) dissociates this complex, releasing activated ASK1-signalosome and leading p38MAPK activation [71–73]; the second is caused by activation of TGF-beta-activated kinase (TAK) 1 and TGF-beta-activated kinase 1-binding protein (TAB) 1 to form TGF-beta/TAK1/TAB1 complex, which activates p38 MAPK pathway either through canonical MAPKKKs or TGF-beta directly through TAB1 by autophosphorylation [74–76]. However, whether oxidative stress activates p38 MAPK via the above pathways in stem cell aging is not clear.

3.2.8 microRNAs

microRNAs (miRNAs) are small single-stranded noncoding RNAs that regulate expression of target genes via modulating the translation or degradation of the targeted transcripts. A number of studies have revealed that miRNAs play a critical role in controlling oxidative stress-induced aging and cellular senescence [77–79]. For example, miR-34a and miR-570-3p are induced by oxidative stress and promote epithelial cell senescence by inhibiting sirtuin-1; inhibition of miR-34a or miR-570-3p can restore sirtuin-1 and suppress markers of cellular senescence [78, 79]. The more recent research reported oxidative stress downregulated expression of miR-20b-5p and miR-106a-5p to suppress the G1/S-phase transition of the cell cycle via promoting p21 expression and suppressing E2F1 in MSC [77]. miR-210 is also an important regulator in controlling stem cell function exposure to oxidative stress. Under oxidative stress condition, miR-210 expression was upregulated by ROS to increase the proliferation and migration of ASCs via targeting protein tyrosine phosphatase and nonreceptor type 2 expression [80]. Furthermore, it was also found that overexpression of miR-210 improved MSC survival under oxidative stress via antioxidation and c-Met pathway activation [81]. Similarly, Ham et al. report that let-7b protects MSCs from ROS damage via targeting caspase 3 [82]. These studies suggest that miRNAs that regulate ROS-induced stem cell are dependent on their distinct targets.

However, the mechanism underlying oxidative stress-regulated miRNAs expression remains to be demonstrated. The early studies showed that oxidative stress regulated aging-related miRNAs expression by either affecting the miRNA processing machinery or regulating the expression of certain specific miRNAs. Mori et al. found that multiple miRNAs were downregulated with age in large part due to a decrease of Dicer, which is one of the components of the miRNA processing machinery and required for synthesis of small interfering RNAs. In additional, they also found that oxidative stress resulted in Dicer expression decrease [83].

3.3 The Antioxidant Strategy of Stem Cell

Given the harmful effects of oxidative stress, stem cells possess several endogenous antioxidant stress strategies to manage ROS accumulation and adapt to oxidative stress. The strategies include resisting oxidative damage, removing excessive ROS, and inhibiting ROS production (Fig. 3.2). Under mild oxidative stress condition, stem cells resist or protect against oxidative damage by regulating antioxidant enzymes and apoptosis-associated gene expression. Conversely, sustained and intensive oxidative stress impairs stem cell self-renewal and accelerates senescence and

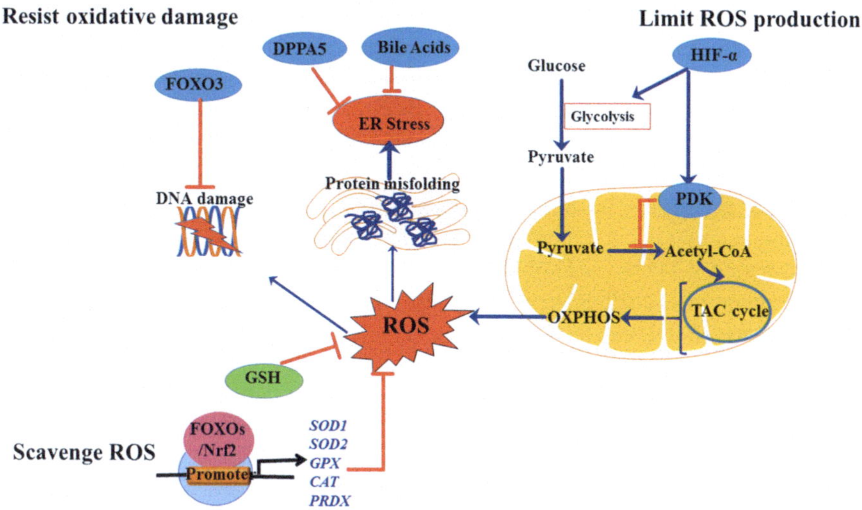

Fig. 3.2 Strategies employed by stem cells to manage ROS accumulation and adapt to oxidative stress

apoptosis of stem cell. Thus, it is critical to understand the defensive anti-oxidative mechanism of stem cells for improving stem cell homeostasis and function.

Firstly, stem cells possess a protective mechanism to resist oxidative damage. For example, oxidative stress results in oxidative protein increase and impairs the protein folding abilities of endoplasmic reticulum (ER), which leads to misfolded proteins accumulation within the ER, resulting in ER stress and ER stress-associated ROS generation. Fetal liver HSCs can suppress oxidative protein aggregation and ER stress via bile acids [84]. Similarly, developmental pluripotency-associated 5 (Dppa5) is also reported to enhance long-term reconstitution of HSCs and improve the activity of HSCs via eliciting ER stress and apoptosis induced by oxidative stress [85]. Furthermore, HSCs can resist against oxidative damage of DNA through FoxO3-mediated upregulation of DNA repair components [31].

Secondly, stem cell prevents ROS accumulation and resists against ROS-induced damage through regulating the expression of endogenous antioxidant enzymes and antioxidants to scavenge excessive ROS. SOD is responsible for transforming $O_2\cdot^-$ into H_2O_2, which is subsequently neutralized to H_2O and O2 by catalases (CAT) or peroxidase, such as glutathione peroxidase (GPx) and peroxiredoxin (PRDX). The expression of antioxidant enzymes in stem cells was regulated by Nrf2 and FoxOs. Under oxidative stress condition, Nrf2 is released from Keap1 and translocated into nucleus, where it binds to ARE and upregulates antioxidant enzymes. In addition, glutathione, an important antioxidant molecule, plays a significant role in cellular ROS neutralization by scavenging ROS directly or serving as a substrate of glutathione peroxidases [86]. The previous studies found that high level of glutathione could maintain stem cell homeostasis and functions through inhibiting cell senescence and preserving stemness and differentiation potential [87, 88]. Therefore, stem cells could manage ROS via ROS-scavenging enzymes and antioxidants.

Finally, stem cells could prevent ROS accumulation via inhibiting ROS generation. Mitochondria are primary sources of ROS in most cells, a consequence of its role in energy production via oxidative phosphorylation (OXPHOS). Stem cell possesses unique metabolic programs to limit ROS production. Stem cells preferentially utilize glycolysis rather than OXPHOS for adenosine-5′-triphosphate (ATP) production in a hypoxic microenvironment to maintain cell quiescence and function. HSCs generate ATP by anaerobic glycolysis through HIF-1α-mediated pyruvate dehydrogenase kinase (PDK) activation [89]. PDK suppress the activity of pyruvate dehydrogenase (PDH) and then block the entry of pyruvate for its catabolism into the tricarboxylic acid (TCA) cycle, thereby inhibiting influx of glycolytic metabolites into the mitochondria to reduce ROS generation. Lactate dehydrogenase A (LDHA) is also required for HSCs to facilitate glycolysis, inhibit OXPHOS, and reduce ROS production [90]. Other study demonstrates an unexpected role of FoxO3 in the maintenance of metabolic homeostasis in NPCs that counteracts oxidative stress and preserves their long-term proliferative potential. In additional, it was proved that FoxO3 is required for NPCs to regulate glucose metabolism via direct transcriptional activation of metabolic enzymes and decrease ROS accumulation [30]. Therefore, metabolism plays a critical role in maintaining redox homeostasis of stem cell.

3.4 Anti-oxidative Stress Strategy in Stem Cell Transplantation Therapy

Stem cell transplantation has emerged as a promising regenerative medicine therapy for a variety of degenerative disorders, such as ischemic stroke, heart attack, wound healing, and kidney failure. However, poor survival of transplanted cell is one of the major problems limiting the therapeutic efficacy. This is partly attributable to insufficient resistance of transplanted stem cell to severe oxidative stress at the injured sites. A recent development in combining the anti-oxidative stress strategy with stem cell therapy improves the efficiency of stem cell transplantation. Work from us and other groups showed that elevating the levels of antioxidants in stem cells with antioxidant molecules or anti-oxidative gene manipulation can significantly influence their survival and regeneration capacity. We summarized the recent researches about improving stem cell therapy efficacy with antioxidants in Table 3.1.

3.5 Conclusion

As stem cells are among the longest-living cells within an organism, stem cell aging may be a major driver of organismal aging. The consequences of stem aging include impaired self-renewal ability, loss of regenerative capacity, increased cell death, and depletion of stem cell pool. The process of stem cell aging is modulated by a range of intrinsic and extrinsic factors. Oxidative stress induces stem cell aging in many

Table 3.1 Anti-oxidative stress strategy in stem cell transplantation therapy

Disease	Model	Tansplanted stem cell	Antioxidant treatment	Benefits	Reference
Myocardial infarction	Lewis male rats with left anterior descending artery ligation	Bone marrow-derived MSCs	HO-1 overexpression	Enhanced anti-apoptotic and anti-oxidative capabilities of engrafted MSCs, and improves myocardial repair after ischemic injury	[91]
	SD male rats with left anterior descending coronary artery ligation	Bone marrow-derived MSCs	NAC	Enhances the adhesion of engrafted MSCs, and improves myocardial repair	[92]
	Male mdx mice with cardiotoxin injury	MDSCs	NAC	Increased the myogenic differentiation of MDSC, improved the survival of MDSCs and cardiac function, decreased scar tissue formation, and increased numbers of CD31 (+) endothelial cell structures	[93]
Stroke	Male SD rats with middle cerebral artery occlusion (MCAO)	Bone marrow-derived MSC	SOD3 overexpression	Reduced the infarct volume, improved neurological function, and promotes angiogenesis	[94]
	MCAO model in male SD rats	Bone marrow-derived MSC	Icariin	Increased MSCs cerebral homing and neuronal differentiation, reduced the infarct volume, improved neurological function, and promotes angiogenesis	[95]
	MCAO model in adult male C57BL/6 mice	NSCs	SOD1 overexpression	Increase the survival of grafted NSCs, enhances angiogenesis, reduced infarct size, and improved behavioral performance	[96]
	Intracerebral hemorrhage (ICH) model in C57BL/6 mice	NSCs	SOD1 overexpression	Protected the grafted NSCs, enhances neuroprotection, and facilitates behavioral recovery	[97]

(continued)

Table 3.1 (continued)

Disease	Model	Tansplanted stem cell	Antioxidant treatment	Benefits	Reference
Hind limb ischemia diseases	Hind limb ischemia model in C57BL/6 mice	ADSCs	IGF-1C-modified hydrogel	Promoted survival and proangiogenic capacity of ADSCs, and ameliorated blood perfusion and muscle regeneration	[98]
	Hind limb ischemia in C57BL/6 mice	Bone marrow-derived MSC	Fucoidan	Enhanced MSC survival and proliferation in ischemic tissues, and improved functional recovery and limb salvage	[99]
Acute liver failure	Male NOD/SCID mice with Gal/LPS challenge	Human umbilical cord MSC	Edaravone	Improved hUCMSCs viability and morphology exposure to oxidative/inflammatory challenge, improves hepatic functions, and promotes host liver regeneration	[70]
Renal ischemia-reperfusion injury	Male SD rats with the release of bilateral renal pedicle clamps following occlusion	Bone marrow-derived MSC	Atorvastatin	Improved survival of implanted stem cells, and inhibited oxidative stress and inflammation in the ischemic kidney	[100]
	FVB mice with the release of renal pedicle clamps following occlusion	ADSCs	Chitosan-IGF-1C hydrogel	Enhanced the retention and survival of transplanted MSC, and protected MSC from oxidative stress	[101]
Acute interstitial cystitis	Male SD rats with cyclophosphamide	ADSCs	Melatonin	Increased the expressions of antioxidants, and attenuated inflammation	[102]
Wound healing	Full-thickness skin defect in STZ-induced diabetic mice	ADSC	HIF-1α overexpression	Promoted wound closure via enhancing angiogenic growth factor expression and suppressing ROS and 8-OHdG levels in ADSCs	[103]

Xenotransplantation	Male NOD/SCID mice irradiated at a dose of 225 cGy	HSC	NAC	NAC-treated mice displayed a significant increase in human HSC engraftment and multilineage hematopoietic differentiation in the mice	[104]

Abbreviations: *MSCs* mesenchymal stem cells, *MDSCs* muscle-derived stem cells, *NSCs* neural stem cells, *HSCs* hematopoietic stem cells, *NAC* N-acetylcysteine, *HO-1* heme oxygenase-1, *SOD* superoxide dismutase, *IGF-1* insulin-like growth factor-1

different signal pathways, which crosstalk with each other to form a complex regulation network. Recent studies have demonstrated the important role of oxidative stress in inducing stem cell aging and emphasized the possibility of maintaining intracellular homeostasis and improving transplantation efficiency with anti-oxidative stress strategy. Furthermore, the signal pathways of oxidative stress-induced stem cell aging have received increasing attention as targets to hinder stem cell aging. However, there are also some problems needed to be clarified in preventing stem cell aging via controlling oxidative stress in the future. Firstly, oxidative stress induces stem cell aging via regulating a complex signaling network, which may vary with different cell types. So, it is essential to define the detailed signaling network of different stem cell aging. Secondly, although emerging evidence has confirmed that antioxidant treatments improve stem cell survival and transplantation efficacy, there is lack of clinical trials to confirm which and whether antioxidants improve the stem cell therapy efficacy. Finally, different levels of ROS play pleiotropic roles in different stem cell populations; it requires the development and application of more advanced technologies and methods for ROS detection and molecular imaging to unveil the mechanisms underlying ROS-induced stem cell aging.

References

1. Dai X, Yan X, Wintergerst KA, Cai L, Keller BB, Tan Y. Nrf2: redox and metabolic regulator of stem cell state and function. Trends Mol Med. 2020;26:185–200.
2. Prentice DA. Adult Stem Cells. Circ Res. 2019;124:837–9.
3. Clevers H. STEM CELLS. What is an adult stem cell? Science. 2015;350:1319–20.
4. Denu RA, Hematti P. Effects of oxidative stress on mesenchymal stem cell biology. Oxidative Med Cell Longev. 2016;2016:2989076.
5. Schultz MB, Sinclair DA. When stem cells grow old: phenotypes and mechanisms of stem cell aging. Development. 2016;143:3–14.
6. Chen F, Liu Y, Wong NK, Xiao J, So KF. Oxidative stress in stem cell aging. Cell Transplant. 2017;26:1483–95.
7. Rossi DJ, Bryder D, Zahn JM, Ahlenius H, Sonu R, Wagers AJ, Weissman IL. Cell intrinsic alterations underlie hematopoietic stem cell aging. Proc Natl Acad Sci U S A. 2005;102:9194–9.
8. Jaiswal S, Fontanillas P, Flannick J, Manning A, Grauman PV, Mar BG, Lindsley RC, Mermel CH, Burtt N, Chavez A, Higgins JM, Moltchanov V, Kuo FC, Kluk MJ, Henderson B, Kinnunen L, Koistinen HA, Ladenvall C, Getz G, Correa A, Banahan BF, Gabriel S, Kathiresan S, Stringham HM, McCarthy MI, Boehnke M, Tuomilehto J, Haiman C, Groop L, Atzmon G, Wilson JG, Neuberg D, Altshuler D, Ebert BL. Age-related clonal hematopoiesis associated with adverse outcomes. N Engl J Med. 2014;371:2488–98.
9. Biteau B, Hochmuth CE, Jasper H. JNK activity in somatic stem cells causes loss of tissue homeostasis in the aging Drosophila gut. Cell Stem Cell. 2008;3:442–55.
10. Rittie L, Stoll SW, Kang S, Voorhees JJ, Fisher GJ. Hedgehog signaling maintains hair follicle stem cell phenotype in young and aged human skin. Aging Cell. 2009;8:738–51.
11. Murphy MP. How mitochondria produce reactive oxygen species. Biochem J. 2009;417:1–13.
12. Liu D, Xu Y. p53, oxidative stress, and aging. Antioxid Redox Signal. 2011;15:1669–78.

13. Tyner SD, Venkatachalam S, Choi J, Jones S, Ghebranious N, Igelmann H, Lu X, Soron G, Cooper B, Brayton C, Park SH, Thompson T, Karsenty G, Bradley A, Donehower LA. p53 mutant mice that display early ageing-associated phenotypes. Nature. 2002;415:45–53.
14. Yanase S, Suda H, Yasuda K, Ishii N. Impaired p53/CEP-1 is associated with lifespan extension through an age-related imbalance in the energy metabolism of C. elegans. Genes Cells. 2017;22:1004–10.
15. Zhang F, Peng W, Zhang J, Dong W, Wu J, Wang T, Xie Z. P53 and Parkin co-regulate mitophagy in bone marrow mesenchymal stem cells to promote the repair of early steroid-induced osteonecrosis of the femoral head. Cell Death Dis. 2020;11:42.
16. Li Y, Zhong H, Wu M, Tan B, Zhao L, Yi Q, Xu X, Pan H, Bi Y, Yang K. Decline of p300 contributes to cell senescence and growth inhibition of hUC-MSCs through p53/p21 signaling pathway. Biochem Biophys Res Commun. 2019;515:24–30.
17. Chen Z, Yi W, Morita Y, Wang H, Cong Y, Liu JP, Xiao Z, Rudolph KL, Cheng T, Ju Z. Wip1 deficiency impairs haematopoietic stem cell function via p53 and mTORC1 pathways. Nat Commun. 2015;6:6808.
18. Liu L, Charville GW, Cheung TH, Yoo B, Santos PJ, Schroeder M, Rando TA. Impaired notch signaling leads to a decrease in p53 activity and mitotic catastrophe in aged muscle stem cells. Cell Stem Cell. 2018;23:544–56.e4.
19. Matheu A, Maraver A, Klatt P, Flores I, Garcia-Cao I, Borras C, Flores JM, Vina J, Blasco MA, Serrano M. Delayed ageing through damage protection by the Arf/p53 pathway. Nature. 2007;448:375–9.
20. Itoh K, Wakabayashi N, Katoh Y, Ishii T, Igarashi K, Engel JD, Yamamoto M. Keap1 represses nuclear activation of antioxidant responsive elements by Nrf2 through binding to the amino-terminal Neh2 domain. Genes Dev. 1999;13:76–86.
21. Kaspar JW, Niture SK, Jaiswal AK. Nrf2:INrf2 (Keap1) signaling in oxidative stress. Free Radic Biol Med. 2009;47:1304–9.
22. Niture SK, Khatri R, Jaiswal AK. Regulation of Nrf2-an update. Free Radic Biol Med. 2014;66:36–44.
23. Jain AK, Jaiswal AK. GSK-3beta acts upstream of Fyn kinase in regulation of nuclear export and degradation of NF-E2 related factor 2. J Biol Chem. 2007;282:16502–10.
24. Niture SK, Jain AK, Shelton PM, Jaiswal AK. Src subfamily kinases regulate nuclear export and degradation of transcription factor Nrf2 to switch off Nrf2-mediated antioxidant activation of cytoprotective gene expression. J Biol Chem. 2011;286:28821–32.
25. Dai X, Yan X, Zeng J, Chen J, Wang Y, Chen J, Li Y, Barati MT, Wintergerst KA, Pan K, Nystoriak MA, Conklin DJ, Rokosh G, Epstein PN, Li X, Tan Y. Elevating CXCR7 improves angiogenic function of EPCs via Akt/GSK-3beta/Fyn-mediated Nrf2 activation in diabetic limb ischemia. Circ Res. 2017;120:e7–e23.
26. Mohammadzadeh M, Halabian R, Gharehbaghian A, Amirizadeh N, Jahanian-Najafabadi A, Roushandeh AM, Roudkenar MH. Nrf-2 overexpression in mesenchymal stem cells reduces oxidative stress-induced apoptosis and cytotoxicity. Cell Stress Chaperones. 2012;17:553–65.
27. Ji K, Fang L, Zhao H, Li Q, Shi Y, Xu C, Wang Y, Du L, Wang J, Liu Q. Ginger oleoresin alleviated gamma-ray irradiation-induced reactive oxygen species via the Nrf2 protective response in human mesenchymal stem cells. Oxidative Med Cell Longev. 2017;2017:1480294.
28. Han X, Zhang J, Xue X, Zhao Y, Lu L, Cui M, Miao W, Fan S. Theaflavin ameliorates ionizing radiation-induced hematopoietic injury via the NRF2 pathway. Free Radic Biol Med. 2017;113:59–70.
29. Kim JH, Thimmulappa RK, Kumar V, Cui W, Kumar S, Kombairaju P, Zhang H, Margolick J, Matsui W, Macvittie T, Malhotra SV, Biswal S. NRF2-mediated notch pathway activation enhances hematopoietic reconstitution following myelosuppressive radiation. J Clin Invest. 2014;124:730–41.

30. Yeo H, Lyssiotis CA, Zhang Y, Ying H, Asara JM, Cantley LC, Paik JH. FoxO3 coordinates metabolic pathways to maintain redox balance in neural stem cells. EMBO J. 2013;32:2589–602.
31. Bigarella CL, Li J, Rimmele P, Liang R, Sobol RW, Ghaffari S. FOXO3 transcription factor is essential for protecting hematopoietic stem and progenitor cells from oxidative DNA damage. J Biol Chem. 2017;292:3005–15.
32. Tothova Z, Kollipara R, Huntly BJ, Lee BH, Castrillon DH, Cullen DE, McDowell EP, Lazo-Kallanian S, Williams IR, Sears C, Armstrong SA, Passegue E, DePinho RA, Gilliland DG. FoxOs are critical mediators of hematopoietic stem cell resistance to physiologic oxidative stress. Cell. 2007;128:325–39.
33. Renault VM, Rafalski VA, Morgan AA, Salih DA, Brett JO, Webb AE, Villeda SA, Thekkat PU, Guillerey C, Denko NC, Palmer TD, Butte AJ, Brunet A. FoxO3 regulates neural stem cell homeostasis. Cell Stem Cell. 2009;5:527–39.
34. Lee SH, Lee JH, Lee HY, Min KJ. Sirtuin signaling in cellular senescence and aging. BMB Rep. 2019;52:24–34.
35. Hwang JW, Yao H, Caito S, Sundar IK, Rahman I. Redox regulation of SIRT1 in inflammation and cellular senescence. Free Radic Biol Med. 2013;61:95–110.
36. Caito S, Rajendrasozhan S, Cook S, Chung S, Yao H, Friedman AE, Brookes PS, Rahman I. SIRT1 is a redox-sensitive deacetylase that is post-translationally modified by oxidants and carbonyl stress. FASEB J. 2010;24:3145–59.
37. He J, Zhang A, Song Z, Guo S, Chen Y, Liu Z, Zhang J, Xu X, Liu J, Chu L. The resistant effect of SIRT1 in oxidative stress-induced senescence of rat nucleus pulposus cell is regulated by Akt-FoxO1 pathway. Biosci Rep. 2019;39:BSR20190112.
38. Ota H, Akishita M, Eto M, Iijima K, Kaneki M, Ouchi Y. Sirt1 modulates premature senescence-like phenotype in human endothelial cells. J Mol Cell Cardiol. 2007;43:571–9.
39. Liu M, Liang K, Zhen J, Zhou M, Wang X, Wang Z, Wei X, Zhang Y, Sun Y, Zhou Z, Su H, Zhang C, Li N, Gao C, Peng J, Yi F. Sirt6 deficiency exacerbates podocyte injury and proteinuria through targeting notch signaling. Nat Commun. 2017;8:413.
40. Kaur I, Rawal P, Rohilla S, Bhat MH, Sharma P, Siddiqui H, Kaur S. Endothelial progenitor cells from aged subjects display decreased expression of sirtuin 1, angiogenic functions, and increased senescence. Cell Biol Int. 2018;42:1212–20.
41. Chen J, Xavier S, Moskowitz-Kassai E, Chen R, Lu CY, Sanduski K, Spes A, Turk B, Goligorsky MS. Cathepsin cleavage of sirtuin 1 in endothelial progenitor cells mediates stress-induced premature senescence. Am J Pathol. 2012;180:973–83.
42. Lamichane S, Baek SH, Kim YJ, Park JH, Dahal Lamichane B, Jang WB, Ji S, Lee NK, Dehua L, Kim DY, Kang S, Seong HJ, Yun J, Lee DH, Moon HR, Chung HY, Kwon SM. MHY2233 attenuates replicative cellular senescence in human endothelial progenitor cells via SIRT1 signaling. Oxidative Med Cell Longev. 2019;2019:6492029.
43. Sun W, Qiao W, Zhou B, Hu Z, Yan Q, Wu J, Wang R, Zhang Q, Miao D. Overexpression of Sirt1 in mesenchymal stem cells protects against bone loss in mice by FOXO3a deacetylation and oxidative stress inhibition. Metab Clin Exp. 2018;88:61–71.
44. Matsui K, Ezoe S, Oritani K, Shibata M, Tokunaga M, Fujita N, Tanimura A, Sudo T, Tanaka H, McBurney MW, Matsumura I, Kanakura Y. NAD-dependent histone deacetylase, SIRT1, plays essential roles in the maintenance of hematopoietic stem cells. Biochem Biophys Res Commun. 2012;418:811–7.
45. Jiang B, Jen M, Perrin L, Wertheim JA, Ameer GA. SIRT1 overexpression maintains cell phenotype and function of endothelial cells derived from induced pluripotent stem cells. Stem Cells Dev. 2015;24:2740–5.
46. Pan H, Guan D, Liu X, Li J, Wang L, Wu J, Zhou J, Zhang W, Ren R, Zhang W, Li Y, Yang J, Hao Y, Yuan T, Yuan G, Wang H, Ju Z, Mao Z, Li J, Qu J, Tang F, Liu GH. SIRT6 safeguards human mesenchymal stem cells from oxidative stress by coactivating NRF2. Cell Res. 2016;26:190–205.

47. Shen X, Chen X, Huang J, Xu R, Cheng J, Jiang H. Age-dependent role of SIRT6 in jawbone via regulating senescence and autophagy of bone marrow stromal cells. J Mol Histol. 2020;51:67–76.
48. Wang H, Diao D, Shi Z, Zhu X, Gao Y, Gao S, Liu X, Wu Y, Rudolph KL, Liu G, Li T, Ju Z. SIRT6 controls hematopoietic stem cell homeostasis through epigenetic regulation of Wnt signaling. Cell Stem Cell. 2016;18:495–507.
49. Li Y, Li X, Cole A, McLaughlin S, Du W. Icariin improves Fanconi anemia hematopoietic stem cell function through SIRT6-mediated NF-kappa B inhibition. Cell Cycle. 2018;17:367–76.
50. Chen W, Chen X, Chen AC, Shi Q, Pan G, Pei M, Yang H, Liu T, He F. Melatonin restores the osteoporosis-impaired osteogenic potential of bone marrow mesenchymal stem cells by preserving SIRT1-mediated intracellular antioxidant properties. Free Radic Biol Med. 2020;146:92–106.
51. Lv YJ, Yang Y, Sui BD, Hu CH, Zhao P, Liao L, Chen J, Zhang LQ, Yang TT, Zhang SF, Jin Y. Resveratrol counteracts bone loss via mitofilin-mediated osteogenic improvement of mesenchymal stem cells in senescence-accelerated mice. Theranostics. 2018;8:2387–406.
52. Bhattacharyya S, Dudeja PK, Tobacman JK. ROS, Hsp27, and IKKbeta mediate dextran sodium sulfate (DSS) activation of IkappaBa, NFkappaB, and IL-8. Inflamm Bowel Dis. 2009;15:673–83.
53. Liu Q, Li Y, Jiang W, Li Y, Zhou L, Song B, Liu X. Inhibition of HSP90 promotes neural stem cell survival from oxidative stress through attenuating NF-kappaB/p65 activation. Oxidative Med Cell Longev. 2016;2016:3507290.
54. Lin TH, Gibon E, Loi F, Pajarinen J, Cordova LA, Nabeshima A, Lu L, Yao Z, Goodman SB. Decreased osteogenesis in mesenchymal stem cells derived from the aged mouse is associated with enhanced NF-kappaB activity. J Orthop Res. 2017;35:281–8.
55. Proto JD, Lu A, Dorronsoro A, Scibetta A, Robbins PD, Niedernhofer LJ, Huard J. Inhibition of NF-kappaB improves the stress resistance and myogenic differentiation of MDSPCs isolated from naturally aged mice. PLoS One. 2017;12:e0179270.
56. Chen Z, Amro EM, Becker F, Holzer M, Rasa SMM, Njeru SN, Han B, Di Sanzo S, Chen Y, Tang D, Tao S, Haenold R, Groth M, Romanov VS, Kirkpatrick JM, Kraus JM, Kestler HA, Marz M, Ori A, Neri F, Morita Y, Rudolph KL. Cohesin-mediated NF-kappaB signaling limits hematopoietic stem cell self-renewal in aging and inflammation. J Exp Med. 2019;216:152–75.
57. Nakagawa MM, Chen H, Rathinam CV. Constitutive activation of NF-kappaB pathway in hematopoietic stem cells causes loss of quiescence and deregulated transcription factor networks. Front Cell Dev Biol. 2018;6:143.
58. Murata H, Ihara Y, Nakamura H, Yodoi J, Sumikawa K, Kondo T. Glutaredoxin exerts an antiapoptotic effect by regulating the redox state of Akt. J Biol Chem. 2003;278:50226–33.
59. Xia W, Hou M. Macrophage migration inhibitory factor rescues mesenchymal stem cells from doxorubicin-induced senescence though the PI3K-Akt signaling pathway. Int J Mol Med. 2018;41:1127–37.
60. Xu J, Qian J, Xie X, Lin L, Zou Y, Fu M, Huang Z, Zhang G, Su Y, Ge J. High density lipoprotein protects mesenchymal stem cells from oxidative stress-induced apoptosis via activation of the PI3K/Akt pathway and suppression of reactive oxygen species. Int J Mol Sci. 2012;13:17104–20.
61. Liu S, Liu S, Wang X, Zhou J, Cao Y, Wang F, Duan E. The PI3K-Akt pathway inhibits senescence and promotes self-renewal of human skin-derived precursors in vitro. Aging Cell. 2011;10:661–74.
62. Liu S, Wang X, Zhao Q, Liu S, Zhang H, Shi J, Li N, Lei X, Zhao H, Deng Z, Cao Y, Ning L, Xia G, Duan E. Senescence of human skin-derived precursors regulated by Akt-FOXO3-p27 (KIP(1))/p15(INK(4)b) signaling. Cell Mol Life Sci. 2015;72:2949–60.
63. Kumar R, Sharma A, Kumari A, Gulati A, Padwad Y, Sharma R. Epigallocatechin gallate suppresses premature senescence of preadipocytes by inhibition of PI3K/Akt/mTOR pathway

and induces senescent cell death by regulation of Bax/Bcl-2 pathway. Biogerontology. 2019;20:171–89.

64. Ito K, Hirao A, Arai F, Takubo K, Matsuoka S, Miyamoto K, Ohmura M, Naka K, Hosokawa K, Ikeda Y, Suda T. Reactive oxygen species act through p38 MAPK to limit the lifespan of hematopoietic stem cells. Nat Med. 2006;12:446–51.
65. Taniguchi Ishikawa E, Gonzalez-Nieto D, Ghiaur G, Dunn SK, Ficker AM, Murali B, Madhu M, Gutstein DE, Fishman GI, Barrio LC, Cancelas JA. Connexin-43 prevents hematopoietic stem cell senescence through transfer of reactive oxygen species to bone marrow stromal cells. Proc Natl Acad Sci U S A. 2012;109:9071–6.
66. Jung H, Kim DO, Byun JE, Kim WS, Kim MJ, Song HY, Kim YK, Kang DK, Park YJ, Kim TD, Yoon SR, Lee HG, Choi EJ, Min SH, Choi I. Thioredoxin-interacting protein regulates haematopoietic stem cell ageing and rejuvenation by inhibiting p38 kinase activity. Nat Commun. 2016;7:13674.
67. Bernet JD, Doles JD, Hall JK, Kelly Tanaka K, Carter TA, Olwin BB. p38 MAPK signaling underlies a cell-autonomous loss of stem cell self-renewal in skeletal muscle of aged mice. Nat Med. 2014;20:265–71.
68. He D, Wu H, Xiang J, Ruan X, Peng P, Ruan Y, Chen YG, Wang Y, Yu Q, Zhang H, Habib SL, De Pinho RA, Liu H, Li B. Gut stem cell aging is driven by mTORC1 via a p38 MAPK-p53 pathway. Nat Commun. 2020;11:37.
69. Kong Y, Shi MM, Zhang YY, Cao XN, Wang Y, Zhang XH, Xu LP, Huang XJ. N-acetyl-L-cysteine improves bone marrow endothelial progenitor cells in prolonged isolated thrombocytopenia patients post allogeneic hematopoietic stem cell transplantation. Am J Hematol. 2018;93:931–42.
70. Zeng W, Xiao J, Zheng G, Xing F, Tipoe GL, Wang X, He C, Chen ZY, Liu Y. Antioxidant treatment enhances human mesenchymal stem cell anti-stress ability and therapeutic efficacy in an acute liver failure model. Sci Rep. 2015;5:11100.
71. Hsieh CC, Kuro-o M, Rosenblatt KP, Brobey R, Papaconstantinou J. The ASK1-signalosome regulates p38 MAPK activity in response to levels of endogenous oxidative stress in the klotho mouse models of aging. Aging. 2010;2:597–611.
72. Hsieh CC, Papaconstantinou J. Thioredoxin-ASK1 complex levels regulate ROS-mediated p38 MAPK pathway activity in livers of aged and long-lived Snell dwarf mice. FASEB J. 2006;20:259–68.
73. Papaconstantinou J. The role of signaling pathways of inflammation and oxidative stress in development of senescence and aging phenotypes in cardiovascular disease. Cell. 2019;8:1383.
74. Ge B, Gram H, Di Padova F, Huang B, New L, Ulevitch RJ, Luo Y, Han J. MAPKK-independent activation of p38alpha mediated by TAB1-dependent autophosphorylation of p38alpha. Science. 2002;295:1291–4.
75. Lu G, Kang YJ, Han J, Herschman HR, Stefani E, Wang Y. TAB-1 modulates intracellular localization of p38 MAP kinase and downstream signaling. J Biol Chem. 2006;281:6087–95.
76. Richardson L, Dixon CL, Aguilera-Aguirre L, Menon R. Oxidative stress-induced TGF-beta/TAB1-mediated p38MAPK activation in human amnion epithelial cells. Biol Reprod. 2018;99:1100–12.
77. Tai L, Huang CJ, Choo KB, Cheong SK, Kamarul T. Oxidative stress Down-regulates MiR-20b-5p, MiR-106a-5p and E2F1 expression to suppress the G1/S transition of the cell cycle in multipotent stromal cells. Int J Med Sci. 2020;17:457–70.
78. Baker JR, Vuppusetty C, Colley T, Hassibi S, Fenwick PS, Donnelly LE, Ito K, Barnes PJ. MicroRNA-570 is a novel regulator of cellular senescence and inflammaging. FASEB J. 2019;33:1605–16.
79. Baker JR, Vuppusetty C, Colley T, Papaioannou AI, Fenwick P, Donnelly L, Ito K, Barnes PJ. Oxidative stress dependent microRNA-34a activation via PI3Kalpha reduces the expression of sirtuin-1 and sirtuin-6 in epithelial cells. Sci Rep. 2016;6:35871.

80. Kim JH, Park SG, Song SY, Kim JK, Sung JH. Reactive oxygen species-responsive miR-210 regulates proliferation and migration of adipose-derived stem cells via PTPN2. Cell Death Dis. 2013;4:e588.
81. Xu J, Huang Z, Lin L, Fu M, Gao Y, Shen Y, Zou Y, Sun A, Qian J, Ge J. miR-210 over-expression enhances mesenchymal stem cell survival in an oxidative stress environment through antioxidation and c-Met pathway activation. Sci China Life Sci. 2014;57:989–97.
82. Ham O, Lee SY, Lee CY, Park JH, Lee J, Seo HH, Cha MJ, Choi E, Kim S, Hwang KC. let-7b suppresses apoptosis and autophagy of human mesenchymal stem cells transplanted into ischemia/reperfusion injured heart 7by targeting caspase-3. Stem Cell Res Ther. 2015;6:147.
83. Mori MA, Raghavan P, Thomou T, Boucher J, Robida-Stubbs S, Macotela Y, Russell SJ, Kirkland JL, Blackwell TK, Kahn CR. Role of microRNA processing in adipose tissue in stress defense and longevity. Cell Metab. 2012;16:336–47.
84. Sigurdsson V, Takei H, Soboleva S, Radulovic V, Galeev R, Siva K, Leeb-Lundberg LM, Iida T, Nittono H, Miharada K. Bile acids protect expanding hematopoietic stem cells from unfolded protein stress in fetal liver. Cell Stem Cell. 2016;18:522–32.
85. Miharada K, Sigurdsson V, Karlsson S. Dppa5 improves hematopoietic stem cell activity by reducing endoplasmic reticulum stress. Cell Rep. 2014;7:1381–92.
86. Wang CK, Yang SC, Hsu SC, Chang FP, Lin YT, Chen SF, Cheng CL, Hsiao M, Lu FL, Lu J. CHAC2 is essential for self-renewal and glutathione maintenance in human embryonic stem cells. Free Radic Biol Med. 2017;113:439–51.
87. Liao N, Shi Y, Zhang C, Zheng Y, Wang Y, Zhao B, Zeng Y, Liu X, Liu J. Antioxidants inhibit cell senescence and preserve stemness of adipose tissue-derived stem cells by reducing ROS generation during long-term in vitro expansion. Stem Cell Res Ther. 2019;10:306.
88. Dannenmann B, Lehle S, Hildebrand DG, Kubler A, Grondona P, Schmid V, Holzer K, Froschl M, Essmann F, Rothfuss O, Schulze-Osthoff K. High glutathione and glutathione peroxidase-2 levels mediate cell-type-specific DNA damage protection in human induced pluripotent stem cells. Stem Cell Rep. 2015;4:886–98.
89. Takubo K, Nagamatsu G, Kobayashi CI, Nakamura-Ishizu A, Kobayashi H, Ikeda E, Goda N, Rahimi Y, Johnson RS, Soga T, Hirao A, Suematsu M, Suda T. Regulation of glycolysis by Pdk functions as a metabolic checkpoint for cell cycle quiescence in hematopoietic stem cells. Cell Stem Cell. 2013;12:49–61.
90. Wang YH, Israelsen WJ, Lee D, Yu VWC, Jeanson NT, Clish CB, Cantley LC, Vander Heiden MG, Scadden DT. Cell-state-specific metabolic dependency in hematopoiesis and leukemogenesis. Cell. 2014;158:1309–23.
91. Tsubokawa T, Yagi K, Nakanishi C, Zuka M, Nohara A, Ino H, Fujino N, Konno T, Kawashiri MA, Ishibashi-Ueda H, Nagaya N, Yamagishi M. Impact of anti-apoptotic and anti-oxidative effects of bone marrow mesenchymal stem cells with transient overexpression of heme oxygenase-1 on myocardial ischemia. Am J Physiol Heart Circ Physiol. 2010;298:H1320–9.
92. Song H, Cha MJ, Song BW, Kim IK, Chang W, Lim S, Choi EJ, Ham O, Lee SY, Chung N, Jang Y, Hwang KC. Reactive oxygen species inhibit adhesion of mesenchymal stem cells implanted into ischemic myocardium via interference of focal adhesion complex. Stem Cells. 2010;28:555–63.
93. Drowley L, Okada M, Beckman S, Vella J, Keller B, Tobita K, Huard J. Cellular antioxidant levels influence muscle stem cell therapy. Mol Ther. 2010;18:1865–73.
94. Sun S, Gao N, Hu X, Luo H, Peng J, Xia Y. SOD3 overexpression alleviates cerebral ischemia-reperfusion injury in rats. Mol Genet Genom Med. 2019;7:e00831.
95. Liu D, Ye Y, Xu L, Yuan W, Zhang Q. Icariin and mesenchymal stem cells synergistically promote angiogenesis and neurogenesis after cerebral ischemia via PI3K and ERK1/2 pathways. Biomed Pharmacother. 2018;108:663–9.
96. Sakata H, Niizuma K, Wakai T, Narasimhan P, Maier CM, Chan PH. Neural stem cells genetically modified to overexpress cu/zn-superoxide dismutase enhance amelioration of ischemic stroke in mice. Stroke. 2012;43:2423–9.

97. Wakai T, Sakata H, Narasimhan P, Yoshioka H, Kinouchi H, Chan PH. Transplantation of neural stem cells that overexpress SOD1 enhances amelioration of intracerebral hemorrhage in mice. J Cereb Blood Flow Metab. 2014;34:441–9.
98. Wang X, Xu L, Gillette TG, Jiang X, Wang ZV. The unfolded protein response in ischemic heart disease. J Mol Cell Cardiol. 2018;117:19–25.
99. Han YS, Lee JH, Jung JS, Noh H, Baek MJ, Ryu JM, Yoon YM, Han HJ, Lee SH. Fucoidan protects mesenchymal stem cells against oxidative stress and enhances vascular regeneration in a murine hindlimb ischemia model. Int J Cardiol. 2015;198:187–95.
100. Cai J, Yu X, Zhang B, Zhang H, Fang Y, Liu S, Liu T, Ding X. Atorvastatin improves survival of implanted stem cells in a rat model of renal ischemia-reperfusion injury. Am J Nephrol. 2014;39:466–75.
101. Feng G, Zhang J, Li Y, Nie Y, Zhu D, Wang R, Liu J, Gao J, Liu N, He N, Du W, Tao H, Che Y, Xu Y, Kong D, Zhao Q, Li Z. IGF-1 C domain-modified hydrogel enhances cell therapy for AKI. J Am Soc Nephrol. 2016;27:2357–69.
102. Chen YT, Chiang HJ, Chen CH, Sung PH, Lee FY, Tsai TH, Chang CL, Chen HH, Sun CK, Leu S, Chang HW, Yang CC, Yip HK. Melatonin treatment further improves adipose-derived mesenchymal stem cell therapy for acute interstitial cystitis in rat. J Pineal Res. 2014;57:248–61.
103. Xu J, Liu X, Zhao F, Zhang Y, Wang Z. HIF1alpha overexpression enhances diabetic wound closure in high glucose and low oxygen conditions by promoting adipose-derived stem cell paracrine function and survival. Stem Cell Res Ther. 2020;11:148.
104. Hu L, Cheng H, Gao Y, Shi M, Liu Y, Hu Z, Xu J, Qiu L, Yuan W, Leung AY, Yang YG, Cheng T. Antioxidant N-acetyl-L-cysteine increases engraftment of human hematopoietic stem cells in immune-deficient mice. Blood. 2014;124:e45–8.

Stress Proteins: Biological Functions, Human Diseases, and Virus Infections

4

Ming-Liang He, Qianya Wan, Dan Song, and Betsy He

Abstract

Stress proteins (SPs) are groups of protein/RNA chaperones that respond fast to intracellular and extracellular stress stimuli to maintain cell homeostasis. Among them, heat-shock proteins (HSPs) play major roles in chaperoning misfolded or unfolded polypeptides, protecting cells from toxic stress, and regulating innate immunity, adaptive immunity, and even inflammatory response; while RNA chaperones are composed of a great number of heterogeneous nuclear ribonucleoproteins (hnRNPs), essential for manipulating both the functions and metabolisms of pre-mRNAs/heterogeneous nuclear RNAs (hnRNAs), including chromatin remodeling, transcription regulation, RNP assembly and stabilization, RNA export, virus replication, histone-like nucleoid structuring, and even intracellular immunity. Dysregulation or dysfunction of stress proteins causes many human diseases, such as human cancer, cardiovascular diseases and neurodegenerative diseases, and infectious diseases. In this chapter, we briefly discuss their biological function in human diseases, particularly the diseases caused by virus infections. Stress proteins have attracted huge interest as targets for combating SARS-CoV-2 (COVID-19) and many other virus infections.

Keywords

Stress proteins · Heat shock proteins · Heterogeneous nuclear ribonucleoproteins

M.-L. He (✉) · Q. Wan · D. Song
Department of Biomedical Sciences, City University of Hong Kong, Kowloon Tong, Hong Kong, China

B. He
Rensselaer Polytechnic Institute, Troy, NY, USA

C. Huang, Y. Zhang (eds.), *Oxidative Stress*,
https://doi.org/10.1007/978-981-16-0522-2_4

4.1 Stress Proteins (SPs)

When an organism (human beings, animals, plants, and even bacteria) meets a stress, they usually find out a way to reduce or eradicate the stress. For example, people would think of a way and execute it, be it tackling the stress source or mediating the harm caused by the stress. Likewise, when cells are exposed to either intracellular or extracellular stressful stimuli, they also have to do similar things. Groups of protein, referred to as stress proteins (SPs), including heat-shock proteins (HSPs), RNA chaperones, and proteins that mainly function in the endoplasmic reticulum (ER), namely, peptidyl-prolyl isomerases, the lectin-binding chaperone system, and protein disulfide isomerases (PDIs), are quickly synthesized to remove or limit the cell damage caused by stresses [1]. This is a common stress response phenomenon for cells to protect their physiological functions and survival by either neutralizing or eradicating stresses from both extracellularly (e.g., UV irradiation, toxic chemicals, hypoxia, pathogen invasion) and intracellularly (e.g., starvation, stimulation by cytokines/chemokines or hormones, oxidative stress, DNA damage). The increase of stress proteins triggers signaling cascades that may promote cell survival, initiate cell death (i.e., apoptosis, necrosis, pyroptosis or autophagic cell death), or modulate innate and adaptive immune response for the elimination of damaged cells to protect a particular organ/tissue under given conditions. Many human diseases, including cardiovascular diseases, stroke, neurodegenerative diseases (e.g., Parkinson's diseases, Alzheimer's disease), cancers, and infectious diseases, are well known to be associated with the dysfunction or dysregulation of stress proteins. In this chapter, we will briefly introduce the functions of stress proteins in human diseases and particularly focus on their functions in diseases caused by viral infections.

4.2 Heat-Shock Proteins (HSPs)

Heat-shock proteins belong to a large chaperone family that is classified as Hsp100, Hsp90, Hsp70, Hsp60, Hsp40, and small HSP (15–40 kDa, sHSP) subfamilies according to their molecule weights. HSPs rapidly increase when a cell is under a physiological or environmental stress such as exposure to UV, irradiation, high temperature, starvation, malnutrition, chemicals, hypoxia or hyperoxia, and pathogen invasion. Heat-shock proteins and their co-chaperones form extensive networks to promote or stabilize the correct folding of newly synthesized peptides and substrate proteins [2]. They participate in many biological processes including cell survival, stemness, transformation and carcinogenesis, differentiation, migration, and death. Accumulating evidence shows that some HSPs participate not only in intracellular innate immunity but also in antigen presentation in adaptive immune response [3, 4]. In this chapter, we mainly focus on the function of Hsp90, Hsp70, and sHSP family members.

Hsp90s are very abundant chaperones located in different cellular compartments of all eukaryotic cells. Hsp90α and Hsp90β are in the cytoplasm, accounting for the greatest proportion in humans. Glucose-related protein GRp94 and TRAP1 are

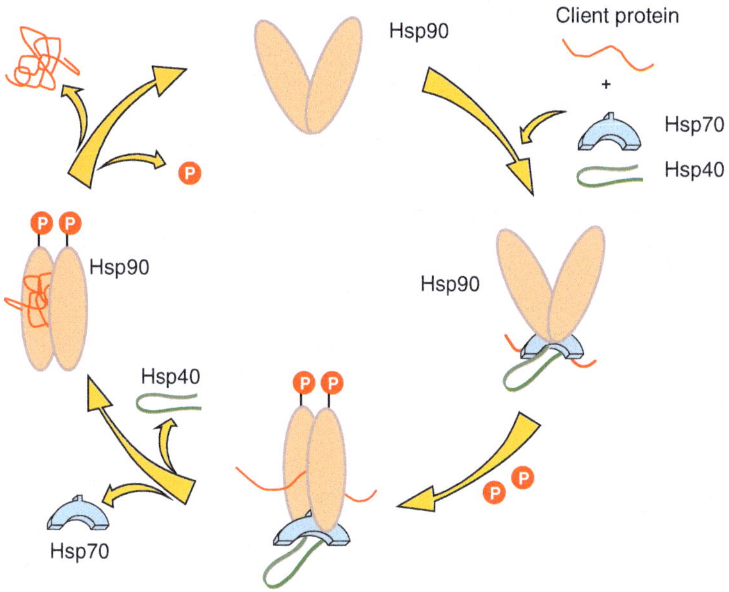

Fig. 4.1 The general chaperone cycle of heat-shock proteins. The unfolded or misfolded client protein binds to Hsp70 and its co-chaperone, Hsp40 chaperones, and then interacts with the free Hsp90 to form a complex. ATP (implicated with a red P) binds Hsp90 and then induces the client protein to be transferred from Hsp70 to Hsp90, followed by Hsp70 and Hsp40 release from the complex. After the hydrolysis of ATP, Hsp90 undergoes additional conformation change, leading to the correctly folded client protein release from Hsp90-client complex

located in the endoplasmic reticulum (ER) and mitochondria, respectively [5]. The functions of Hsp90s are ATP-dependent through three distinct regions: an N-terminal ATPase-dependent hydrolytic domain, a middle linker region, and a C-terminal dimerization domain [6]. Hsp90 binds unfolded peptides or misfolded proteins to prevent their aggregation and degradation. Upon binding with ATP, the N-terminal domain forms a transient dimer to facilitate the substrate binding. Following the hydrolysis of ATP and energy release, a conformational change of the N-terminal domain occurs to correct the folding of the substrate peptides or proteins (Fig. 4.1). In fact, this process is collaborated with or regulated at large by Hsp90 co-chaperones, including CDC37 and CDC37-associated kinase, for inhibiting ATPase activity, E3 ubiquitin ligase carboxyl terminus of Hsp70-interacting protein (CHIP) for protein degradation, and Hsc70/Hsp90-organizing protein (HOP) for inhibiting the N-terminal domain dimerization [7, 8]. Hsp90 also contributes to telomere maintenance, cell survival, growth, cell cycle progression and apoptosis, etc.

Hsp70s account for the majority of molecular chaperones in cells [7], including Hsp72, Hsp70-2, Hsp70B′, and Hsc70 mainly in the cytosol; GRp75 in the mitochondria, and GRp78 mainly in the ER [9]. Some Hsp70 members such as GRp78 can also be presented on the cell surface and even be secreted out. Hsp70 contains a 44-kDa nucleotide-binding domain (NBD) in the N-terminal region and a 28-kDa substrate-binding domain (SBD) in the C-terminal region [10, 11]. Similar to Hsp90, Hsp70s are ATP-dependent molecular chaperones and play key roles in cellular protein surveillance without or under stress. They participate in almost all folding processes, including folding of newly synthesized polypeptides; refolding of misfolded or aggregated proteins, solubilizing or degrading proteins, and transporting proteins; and assembling or disassembling of oligomeric protein complexes [6, 12–14].

The main co-chaperones of Hsp70 are the nucleotide exchange factors (NEFs) and the J-domain proteins (JDPs) of the Hsp40 family. In the process of protein folding, the unfolded peptide substrates first bind JDPs and then form a complex with Hsp70. The Hsp70's ATPase activity is activated for folding and preventing unfolded protein aggregation. NEFs help to release substrates when the correct and active conformation is formed. Therefore, Hsp70s always cooperate with co-chaperones in the process of protein folding or refolding [15, 16].

The small heat-shock protein (sHSP) family contains ubiquitous (HspB5, Hsp20, Hsp22, and Hsp27) and tissue-specific (HspB2, HspB3, HspB4, HSPB7, HspB9, and HspB10) ones [17]. They exist as monomers, dimers, or even large multimeric complexes [18]. The structure of sHSPs is quite different from the other HSPs because of less conserved sequences. The basic structure of sHSPs is a conserved α-crystallin domain (ACD) flanked by a non-conserved N-terminal sequence (NTS) and a C-terminal sequence (CTS). sHSPs are important players and regulators of stress tolerance, apoptosis, aging, and longevity [19–22]. The sHSPs may exhibit different functions when they undergo phosphorylation and oligomerization. The phosphorylated sHSPs normally form monomers or small oligomers that bind F-actin to regulate actin polymerization [23], while the hypophosphorylation of sHSPs promotes large homo- and hetero-oligomer formation crucial for chaperone activity [24].

4.3 Transcriptional Regulation of the HSPs

The expressions of HSPs are mainly regulated by four members of heat-shock factors (HSFs) in vertebrates [25]. Among them, HSF1 plays a major role in regulating HSP expression. Under stress conditions, the monomeric HSF1 trimerizes and translocates from the cytoplasm into the nucleus and binds on the heat-shock elements (HSEs) in the promoter region, therefore activating the *hsp* expression [26]. Experiments show that the fibroblasts without knockout ($hsf1^{-/-}$) undergo apoptosis at a high temperature stress due to no *hsp* transcription [27].

4.4 Protein Disulfide Isomerase (PDI)

Chaperone protein PDIs are multifunctional oxidoreductases mainly locating in the ER. They catalyze the formation, isomerization, and reduction of disulfide bonds of newly synthesized peptides or malformation of disulfide bonds within or among protein subunits. During disulfide bond formations, cysteine residues at the catalytic CGHC active site accept two electrons from the cysteine residues in the polypeptide that causes the PDI reduction and substrate oxidation. Then, the electrons are transferred from the PDI molecule to an acceptor and start a new cycle of disulfide bond formation [28]. Of course, PDIs also keep the molecular chaperone activity to ensure the quality control of glycosylated proteins [29]. ERp57 is one of the most studied PDIs. It shares a similar structure to other PDIs, i.e., four domains (namely, a-b-b′-a′) and two localization sequences (namely, ER retention signal (QDEL) and nucleus localization signal (KKKKKPK) at the C-terminus). The b domains have a high binding affinity for calreticulin and calnexin [30]. When a protein is not correctly folded, UDP-glucose:glycoprotein glucosyltransferase (UGGT) is recruited to ERp57/CRT/CNX complex for reglycosylation [29, 31, 32]. ERp57 is also well known to play crucial roles in antigen presentation during adoptive immune response to infections.

4.5 RNA Chaperones

The 3D structure is the basis of RNA functions. The process of pre-RNA folding into 3D structures, i.e., the multitude of possible RNA base pairings together with the high stability of RNA duplexes that generates alternative secondary and tertiary structures, is highly dependent on RNA chaperones. They promote RNA folding through accelerating the escape from kinetic folding traps and prevention of RNAs from being trapped in nonfunctional conformations. RNA chaperones participate in many biological processes, such as chromatin remodeling, 3'-end formation, mRNA packaging, RNA silencing, RNA transport, transcription regulation, RNP assembly and stabilization, RNA export, translation, DNA repair, telomere biogenesis, histone-like nucleoid structuring, virus replication, and even intracellular immunity. Heterogeneous nuclear ribonucleoproteins (hnRNPs) are essential factors for manipulating both the functions and metabolisms of pre-mRNAs and other heterogeneous nuclear RNAs (hnRNAs) transcribed by RNA polymerase II. An hnRNP molecule normally contains common RNA binding motifs like RGG boxes (arginine glycine boxes), RNA recognition motifs (RRMs), hnRNP K homology (KH)-domains, and zinc finger (ZF)-domains [33]. They are mainly in the nucleus and are redistributed in the cytosol under certain conditions, such as virus infections [34].

4.6 Stress Proteins and Human Diseases

Protein aggregation hallmarks quite a few diseases caused by either loss-of-function or toxic gain-of-function of stress proteins. Cellular homeostasis is mainly maintained by stress proteins. Upon stress stimuli, stress response is rapidly activated to release the pressure through increasing the level of stress proteins. The elevated accumulation of Hsp27, Hsp70, and Hsp90 has been extensively observed in many types of cancer cells [35–37]. When a cell undergoes carcinogenesis, fast growth for escaping the immune response is needed. Under long-term stresses, the

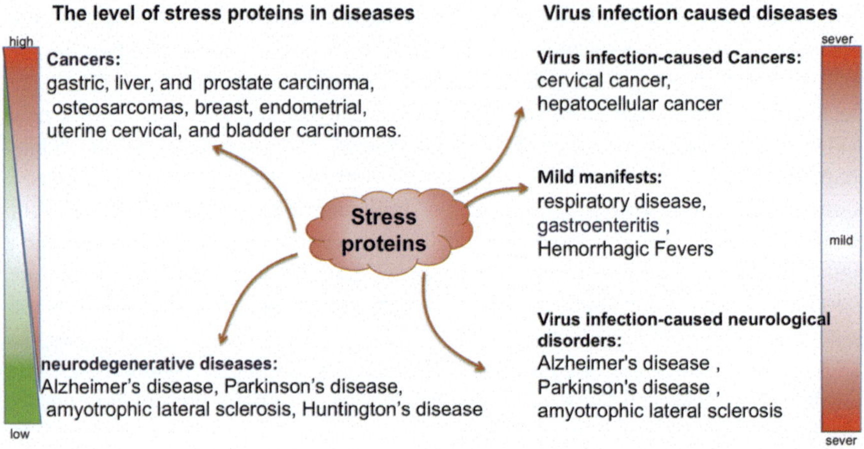

Fig. 4.2 Diseases caused by the dysfunction or dysregulation of stress proteins. The left panel shows the expression level of stress proteins in diverse human diseases, including cancers and neurodegenerative diseases. The expression level of stress proteins is significantly elevated in cancer cells but dramatically decreased in neurodegenerative disease. The right panel shows the virus infection-caused disease from mild to severe

constant high level of HSPs not only helps the mutated cell's survival, but it also participates in or promotes anti-apoptosis, angiogenesis, cell stemness, metastasis, and invasion as well [38]. On the other hand, the expression level of HSPs and RNA chaperones is downregulated in nearly all age-related neurodegenerative diseases, including Parkinson's disease, Alzheimer's disease, amyotrophic lateral sclerosis, several polyglutamine diseases (such as Huntington's disease), and different forms of spinocerebellar ataxias [39, 40] (Fig. 4.2). The decrease of RNA chaperones disturbs and even hinders RNA metabolism [39], while the attenuated HSPs make cells lose protein quality control (PQC) or timely degrade proteins in neurologic disease [40]. The protein aggregations caused by the dysregulation or dysfunction of stress proteins are the hallmarks of age-related neurodegenerative diseases. Stress proteins involved in other diseases are also observed. For examples, cystic fibrosis and Gaucher's disease are associated with loss of functions by either mutation or downregulation of HSPs, whereas diseases (cancer, hepatitis, sever acute respiratory syndrome) caused by pathogen infections may result from the elevated stress proteins' stimulated excessive immune response or enhanced pathogen replication. In the following sections, we mainly discuss the functions of stress proteins in virus infections.

4.7 Stress Proteins and Virus Infections

Disease caused by viruses is a major threat to public health. Throughout human history, pandemic virus infections have caused dozens of million deaths and tremendous economic disasters. In the past, each periodic smallpox virus outbreak killed multiple million people in both Europe and Asia. In 1918, the outbreak of Spanish flu (influenza H1N1) resulted in approximately 50 to 100 million deaths worldwide. In the twenty-first century, we have already experienced several large-scale outbreaks of virus infection which severely damaged the global health and economy, including severe acute respiratory syndrome (SARS) caused by SARS-coronavirus (SARS-CoV) outbreak from Guangzhou-Hong Kong, China, in 2003; swine flu pandemic (2019 H1N1) outbreak from California, United States, in 2009, Middle East respiratory syndrome caused by MERS-CoV outbreak from Saudi Arabia in 2012; avian influenza A virus (H7N9 and H7N10) outbreak in China in 2013; and SARS-CoV-2 outbreak from Wuhan, China, in 2019. Particularly, the SARS-CoV-2 pandemic spread very quickly to almost all countries, infected over 115 million people, and caused over 2.5 million deaths worldwide by February 26, 2021. The quarantine of cities (Wuhan, Beijing, Shanghai, etc.), province (Hubei province), and even countries (e.g., Italy) is used as an emergency measure (China, United States). Almost all public social activities, such as government service, schools, colleges and universities, entertainment, business in companies, international conferences, and travel through airlines or trains, are suspended or arranged online only in many regions and countries (e.g., Hong Kong, China, Italy, Spain, United States). For example, to shut off the virus transmission pathway and better protect citizens' health, all the schools and universities were closed from February 2020 until further notice from government in China, Hong Kong, and United States. Since March 10 of 2020, a great number of prestigious and active research universities have been closed in the United States, including Harvard University, MIT, Stanford University, and the University of California at Berkeley. The reopen date of universities could even not be assured March, 2021.

The virus genome is composed of either DNA or RNA. Based on the composition of virus genome, the virus is classified into seven types (Baltimore classification), i.e., double-stranded DNA virus (dsDNA virus, Type I), single-stranded DNA virus (ssDNA virus, Type II), double-stranded RNA virus (dsRNA virus, Type III), positive single-stranded RNA virus ((+) ssRNA virus, Type IV), negative single-stranded RNA virus ((-) ssRNA virus, Type V), retrovirus (Type VI), and DNA retrovirus (Type VII). The replication of DNA virus occurs in the nucleus, while the RNA virus may replicate its genome in either the cytosol (Type III, Type IV) or in the nucleus (Type V). In the following sections, we will separately discuss the function of stress proteins in virus infection according to the feature of genetic materials.

In the life cycle of viruses, stress proteins play fundamental roles in almost every stage of infections. The dysregulation of stress proteins is often observed in the course of pathogenesis upon virus infections, for example, gastroenteritis [41–44], respiratory symptoms [45, 46], and hemorrhagic fevers [47, 48]. It is well known that at least 20% of human cancers are induced by chronic virus infections such as human papillomaviruses (HPVs) for cervical cancer and several other epithelial malignancies, hepatitis B and C viruses for liver cancer, and Epstein-Barr virus (EBV) for B-cell lymphoma and nasopharyngeal carcinoma. The recently accumulated data indicate that neurodegenerative diseases and neurobehavioral disorders may be caused by chronic and neuropathic viral infections that lead to neural cell deaths or loss of function in the brain followed by ageing [49], such as Alzheimer's disease, Parkinson's disease, amyotrophic lateral sclerosis, multiple sclerosis, and autism spectrum disorders [50, 51].

4.7.1 Hsp90s in Virus Infection

During an RNA virus infection, Hsp90 may be involved in every step of the virus life cycle. At the entry step, Hsp90 functions as a receptor or a co-receptor of enterovirus [52], Japanese encephalitis virus (JEV), and dengue virus (DENV) [53]. The membrane lipid rafts are associated with Hsp90 and Hsp70 in response to DENV infection. Treatment with anti-Hsp90 antibodies or inhibitors blocks the majority of DENV entries into host cells.

At the viral RNA replication step, Hsp90 may exhibit functions in different means. The viral RNA replication is processed by an RNA-dependent RNA polymerase (RdRp). Firstly, Hsp90 stabilizes and extends the RdRp's half-life of influenza virus, vesicular stomatitis virus (VSV), human parainfluenza viruses-2 (HPIV-2), human parainfluenza viruses-3 (HPIV-3), and respiratory syncytial virus (RSV). In addition, Hsp90 also maintains the stability of proteins in the viral replication complex (VRC), e.g., the nonstructural proteins (nsPs) NSP3 of chikungunya virus (CHIKV) and the nonstructural protein NS3 (protease) of HCV. Secondly, Hsp90 promotes viral replication by enhancing the RdRp activity through its chaperone activity. Hsp90 directly binds NS5 (RdRp) to form VRC for HCV replication. In addition, Hsp90 also stabilizes phosphoinositide-dependent kinase 1 (PDK1) that indirectly enhances NS5 activity for HCV replication. Hsp90 inhibitors significantly decrease viral replication by disrupting the Hsp90-NS5 VRC formation [54]. Thirdly, Hsp90 helps redistribute RdRp's location in the host cells. During an influenza virus infection, the viral RdRp subunits (PB1 and PB2) interact with Hsp90 to form a complex and then co-translocate into the nucleus [55, 56]. In the process of nuclear import, Hsp90 maintains the stability of PB1 and PB2. After entry into the nucleus, Hsp90 dissociates from the Hsp90/PB1/PB2 complex and forms a new VRC complex with polymerase acidic protein (PA) [57]. During the viral replication, the function of Hsp90 is often regulated by phosphorylation. Results from our study show that the replication of enterovirus A71 (EV-A71) is promoted by PIM1 kinase [58]. Ectopic expression of PIM1 or Hsp90β promotes viral replication, whereas inhibition of PIM1 or Hsp90β suppresses viral replication. When Hsp90β is knocked out, the sensitivity of viral replication is lost, no matter if PIM1 is overexpressed or knocked down [59]. Hsp90 inhibitors also effectively inhibited SARS-CoV-2 replication in culture cells.

At viral protein maturation, assembly, and release steps, Hsp90 monitors the proper folding of viral proteins. For example, NS2/NS3 is cleaved into two separate proteins right after translation, a key step of HCV NS2/NS3 protein maturation. Hsp90 strictly regulates the proper folding of newly synthesized NS2/3 protein [60]. Hsp90 and its co-chaperone p23 form a complex to assist the proper folding of capsid precursor polyprotein P1 of poliovirus, rhinovirus, and coxsackievirus [61]. Hsp90 binds with neuraminidase (NA) and stabilizes Hsp90-NA complex for facilitating virion release of influenza [62]. The Hsp90-NA complex formation is also necessary for promoting cell survival to generate and release more virions.

During DNA virus infection, the viral DNA replication occurs in the nucleus. Hsp90 may assist the viral nuclear import. This step is dependent on the microtubule (MT)-dependent molecular motor dynein/dynactin complex. Viruses strictly modulate the status of tubulin acetylation that is critical for transportation. In cells infected by herpes simples virus (HSV), Hsp90 forms a complex with the acetylated tubulin and viral capsid protein VP5 [63]. Treatment with Hsp90 inhibitors disrupts the complex, thereby inhibiting the nuclear transport of HSV capsid protein [63]. Similar to that in the RNA virus replication, Hsp90s mainly maintain and promote the function of viral DNA polymerase or reverse-transcriptase (RT). Reverse transcription is the key step for hepatitis B virus (HBV) replication. RT recognizes and interacts with an RNA signal (the packaging signal, ε) on the pregenomic RNA to initiate the reverse transcription. Meanwhile, GRp94 also stabilizes and activates RT by binding to the pregenomic RNA during HBV replication. In addition, Hsp90 also regulates the location of virus DNA polymerase in host cells. When treated with Hsp90 inhibitors, the polymerases of HSV is mis-distributed from the nucleus to the cytoplasm and subsequently degraded in a proteasome-dependent manner [64, 65]. It is observed that Hsp90 is also essential for the nuclear translocation of DNA polymerase of EBV [66, 67]. VP16 is a potent HSV-encoded transcription activator. At viral RNA transcription step, Hsp90 stabilizes and enhances viral protein VP16 activity. Hsp90 also promotes the protein translation of a variety of viruses, including HSV, varicella-zoster virus (VZV), EBV, and Kaposi's sarcoma-associated herpesvirus (KSHV).

During the chronic infection by some DNA viruses, viral proteins together with the long-term elevated Hsp90 family proteins promote cell growth and tumorigenesis. The increased Hsp90 also attenuates apoptosis, induces drug resistance, and causes cancer recurrence. Interestingly, the virus-hijacked Hsp90 may also display antiviral activities by modulating host immunity, instead, under certain conditions. In EBV-infected B cells, the expression of Hsp90 is induced to present in the cell surface at the acute infection stage. The cell surface Hsp90 stimulates the expanding gamma delta T cells (γδ T cells) [68], which exhibit potent abilities in combating infections of HIV, influenza, Sendai, coxsackie, vaccinia, VSV, and HSV-1. It works as both early sentinels of the immune system by providing immediate protection and as bridging elements between innate immunity and adaptive immunity [69].

4.7.2 Hsp70s in Virus Infection

In the course of RNA virus infections, Hsp70s may participate in every step of virus reproduction, i.e., the virus entry, RNA replication, viral protein translation, virion assembly, secretion, and even host immune response to limit virus infection or escape host defense.

At the entry step, many viruses (e.g., families of Flaviviridae, Picornaviridae, and Reoviridae) use Hsp70s to help their entry. Hsp70 functions as a virus receptor to facilitate virus attachment and entry through interacting with the envelope protein of DENV and JEV on the cell surface [70]. Hsp70 participates in multiple steps of Zika virus (ZIKV) infection. Hsp70 inhibitors impair virus entry, RNA replication, and capsid assembly of different ZIKV strains in diverse cell lines [70, 71]. GRp78 is one of the receptors of coxsackievirus A9 (CAV-9) because 50% of the virus trying to bind to the cell membrane is blocked when treated with antibodies against GRp78 [72]. Recently study shows that GRp78 is a reciptor of SARS-CoV-2, sugesting a potential idea drug target for combating COVID-19. GRp78 also interacts with major histocompatibility complex (MHC) I molecules to help with viral internalization into host cells. During the rotavirus infection, Hsc70 helps viral internalization but does not involve in virus attachments. At the replication step, Hsp70s employ different means to facilitate viral RNA replication. Similar to Hsp90s, as a molecule chaperone, Hsp70s facilitate VRC formation and maintain the stability of VRC proteins. Hsc70s can also regulate VRC activity by interacting with other components of the complex. For example, Hsp72 physically interacts with NS5A, NS3, and NS5B (RdRp) of HCV in VRC, while Hsc70 promotes HCV replication by interacting with VRC through attachment on the 3' polyU/UC motif of HCV RNA genome. In addition, Hsp70s may facilitate virus replication through modulating nuclear import of polymerase or nuclear capsid. Influenza virus replicates its genome in the nucleus. The nuclear entry of RdRp subunits PB2 and PB1 depends on the complex formation of PB1/PB2 monomers or PB2/PB1 heterodimers with Hsp70 [73]. At the translation step, Hsp70s employ different mechanisms to promote viral protein translation. The genome of (+) ssRNA virus (such as enterovirus, flavivirus, and coronavirus) encodes a polypeptide. After entry, the first step for initiating the viral life cycle is to translate the polypeptide, followed by autocleavage into functional viral proteins including RNA polymerase. The viral protein translation is often driven by internal ribosome entry site (IRES) for many RNA viruses; in the meantime, the host cellular cap-dependent translation is inhibited by virus-encoded protein(s). The interaction of Hsc70 and 2A protease of EV-A71 enhances the cleavage on eukaryotic translation initiation factor EIF4G for favoring viral protein translation and impairing host cell cap-dependent translation [74]. For some viruses without IRES, plenty of dsRNAs are generated during replication. The dsRNAs activate protein kinase-RNA-activated (PKR) eIF2a signaling to shut down global cellular protein translation for releasing stresses [75, 76]. To avoid the blockage of viral proliferation, influenza A virus downregulates PKR in an Hsp70-dependent way [77–81]. At virion assembly and release steps, Hsp70 participates in the assembly of reovirus trimeric sigma 1 protein, a main viral component for the interaction with the host cell receptor [82]. Hsp70 also interacts with the capsid precursor P1 to form the Hsp70-P1 complex, which is believed to be an assembly intermediate of picornaviruses [83]. Both Hsp70 and Hsc70 interact with NS5A protein to regulate virion release of HCV. Hsp70 only enhances viral protein expression. Hsc70 not only promotes virion release but is also embedded in the viral capsid for ensuring infectivity, although it has no effects on viral protein expression [84, 85].

During DNA virus infection, Hsp70s also play crucial roles in each step. At the step of SV40 entry into the cytosol from the ER, Hsc70, as well as its family member, Hsp105, interacts with and is regulated by Hsc70 co-chaperone SGTA. The Hsc70/Hsp105/SGTA complex facilitates SV40 cytosol entry, a key step in SV40 life cycle. In this process, GRp78 interacts with its co-chaperone Hsp40 family member DNAJB-11 to help SV40 disassembly and enter into the cytoplasm. In the case of adenovirus infection, Hsp70/Hsc70 and co-chaperone Bag3 form a complex with the penton base protein, the viral capsid constituent responsible for the import of viral capsid and DNA into the nucleus [86, 87]. At the viral gene transcription and translation steps, the chaperone activity of Hsp70s guarantees the viral gene expression by stabilizing transcription factors and correct folding of newly synthesized viral proteins. More importantly, Hsp70/Bag1 complex stimulates the general viral gene transcription in an Hsp70-dependent manner upon cells infected by different viruses, such as human polyomavirus and Human cytomegalovirus (HCMV) [88, 89]. Many DNA viruses start the virion assembly in the nucleus. Hsp70 and Hsc70 translocate from the cytosol into the nucleus and form a complex with all three viral capsid proteins (VP1, VP2, and VP3) of polyomavirus for initiating virion assembly.

Cellular transformation, survival, and apoptosis: Some DNA viruses without encoding their own polymerase have to depend on host DNA replication machinery. To ensure replication in quiescent cells, the virus transforms the host cells to keep host cell division by overcoming the cell cycle restriction points. A well-known example is the SV40-induced cell transformation. The large and small T antigens (TAg) are responsible in this process. The J-domain of both types of TAg binds with Hsc70 through its Hsp70-co-chaperone signature motif. Then, the TAg-Hsc70 complex sequesters the retinoblastoma protein (pRb) from the pRb-E2F complex to release free E2F. The free E2F activates transcription of the S-phase genes for DNA replication in an Hsc70-ATP hydrolysis-dependent manner [90, 91]. Alternatively, TAg induces Hsc70 to disassemble an inhibitory silencer complex or to remodel the chromatin. The E7 protein of HPV and E1A protein of adenovirus also exhibit a similar behavior to target the pRb-E2F complex through recruiting Hsc70-interacting protein hTid-1 or directing binding with Hsc70, respectively [92]. In summary, most double-stranded DNA viruses use Hsp70s to reprogram the cell cycle for ensuring virus replication.

In the early stage of infection, the virus has to undergo fast replication to escape host defense, keep host cell alive, and resist cell death. Hsp70s function as chaperones to correct the misfolding of proteins and remove protein aggregations under the stress of fast protein synthesis. In the early stage of EBV infection, the virus-encoded nuclear oncoprotein EBNA3A helps Hsp70 nuclear translocation and Hsp70 chaperone complex formation to immortalize B cells by inhibiting apoptosis [93]. In the late stage of infection, apoptosis is beneficial for virus's spreading. Virions are often observed in apoptotic bodies and subsequently engulfed by phagocytic cells. This makes it possible for the virus to infect neighbor cells without exposure to the host immune system [94, 95]. It has been demonstrated that the decrease of Hsp70 would help induce the timed apoptosis for virion release at the late stage of adenovirus infections.

Innate immunity: It is well known that Hsp70s play an important role in innate immunity and inflammatory responses. Hepatocyte was believed to have a lack of innate immunity for a long time. Instead of promoting virus replication, the elevated GRp78 expression stimulated by HBV replication restricts virus replication through the activation of intracellular innate immunity by inducing the expression of IFNβ and its downstream antiviral proteins [96]. Other studies show that Hsp70s greatly contribute to the innate immunity of host cells either infected by bacteria or virus. Hsp70 interacts with IκB kinase (IKK) and disturbs the function and stability of the NF-κB/IkBα/IKK complex. Hsp70 also impairs IkBα phosphorylation, affecting IkBα proteasomal degradation and nucleus translocation of NF-κB complex [97]. Moreover, Hsp70 also regulates inflammation in an NF-κB independent pathway through impacting the NLRP3/ASC inflammasome formation [98]. More interestingly, Hsp70 can be secreted out, and the extracellular Hsp70 binds to dendritic cells and macrophages before being recognized by innate immune receptors [99, 100]. Upon infection, Toll-like receptors (TLRs) can immediately detect virus and trigger cellular innate immune response. TLR2/TLR4 interact with Hsp70 and participate in the extracellular Hsp70-stimulated pro-inflammatory signaling cascade in a CD14-dependent manner, thereby promoting pro-inflammatory cytokine production through MyD88/IRAK/NF-κB axis signaling cascade [101]. In addition, many other innate immunity-related signal transduction pathways are also activated in the process, such as p38 MAPKs, JNK, and ERK pathways. Then, the downstream transcription factors CREB and AP-1 are activated, resulting in pro-inflammatory cytokine expression, including TNFα, IL-6, IL-1β, and a number of chemokines [102–105].

4.7.3 The Function of Small Heat-Shock Proteins (sHSPs) in Virus Infection

Small heat-shock proteins are the most upregulated proteins responding to stress stimuli such as oxidative stress (elevated reactive oxygen species (ROS) level, hypoxia, etc.), high temperature, UV, starvation, malnutrition, toxic material, or pathogen invasion [106]. sHSPs are capable of recognizing the misfolded proteins and transferring them to other ATP-dependent chaperones for correction or proteasomes/autophagosomes for degradation [107, 108]. Hsp27 is one of the most extensively studied sHSPs. It is ubiquitously expressed at high level in skeletal, smooth, and cardiac muscles. Because of its particular importance in viral infection, in this chapter, we use Hsp27 as a model to address the functions of sHSPs.

During RNA virus infection, Hsp27 may participate in every step of the viral life cycle. Hsp27 is fast elevated to promote viral replication and protein expression during EV-A71 infections. The viral replication is significantly inhibited when

Hsp27 is either silenced by small interfering RNA (siRNA) or knocked out by CRISPR/cas9-mediated gene editing. The suppressed viral activities are recovered after Hsp27 is restored in the host cells. Further studies revealed that Hsp27 promotes $2A^{pro}$-mediated EIF4G cleavage that favors the internal ribosome entry site (IRES)-mediated viral protein translation. In addition, Hsp27 is translocated from the cytosol into the nucleus and reversely correlated with hnRNP A1 redistribution from the nucleus to the cytosol, an important step to initiate viral protein translation [109]. Surprisingly, the knockout of Hsp27 blocks the hnRNP A1 cytosol redistribution, indicating that Hsp27 is an important player for the import/export of some proteins in virus infection. Swine fever virus (CSFV) is a member of the family Flaviviridae. Hsp27 binds with NS5A, an essential protein for viral replication and assembly. Porcine epidemic diarrhoea virus (PEDV) infection causes high fatality in swine. Hsp27 is quickly elevated in response to PEDV infection by triggering antiviral immunity through the activation of NF-κB signaling that stimulates the expression of IFNβ and its downstream interferon stimulated genes (ISGs) [110].

Hsp27 also contributes to DNA virus replication. It may regulate virus replication either in a positive way or a negative way. The increase of phosphorylated Hsp27 is associated with the upregulated virus replication when cells are infected by porcine circovirus type 2 (PCV2). PCV2 is a single-stranded DNA virus responsible for the postweaning multisystemic wasting syndrome (PMWS) in pigs. The phosphorylated Hsp27 is also upregulated in EBV-positive cells via the activation of Akt signaling. However, similar to GRp78, Hsp27 displays antiviral effect against HBV replication through enhancing IFN production in hepatocytes [111]. Hsp27 can alter viral infections through regulating pro-inflammatory cytokine expression via a variety of signaling pathways. The knockdown of Hsp27 causes an increased release of IL-8 from keratinocytes infected by adenovirus and stimulated with UV or TNFα. The increased IL-8 level is correlated with enhanced IκB-α degradation and phosphorylated IκB-α accumulation in Hsp27-depleted cells. It is also found that Hsp27 is associated with the inhibitory protein IκB kinase (IKK) complex. Hsp27 also activates p38 MAPK/MK2 kinase signaling to increase cyclooxygenase-2 and IL-6 expressions. Besides, Hsp27 is of importance in regulating stress caused by reactive oxygen species (ROS) [107]. Hsp27 regulates enzyme activities by upholding glutathione in reduced form, such as glutathione reductase and glucose-6-phosphate dehydrogenase [112]. Hsp27 also greatly contributes to the degradation of oxidized proteins during virus infection, although the underlying mechanisms are not well understood.

4.7.4 The Function of PDIs in Virus Infection

The function of PDIs on virus entry and uncoating: The entry of some viruses into eukaryotic cells is controlled by redox-regulated processes. In the case of the bird newcastle disease virus (NDV) infection, it enters cells through large conformational changes in its F fusion protein via thiol/disulfide exchange because the overexpression of PDIs increases the viral membrane fusion. In endothelial cell surface, PDI decreases β1 and β3 integrins to facilitate DENV entry [113, 114]. The entry of rotavirus into MA104 cells is blocked by thiol blockers and PDI inhibitors [115]. PDIs also participate in HIV entry. The cleavage of two disulfide bonds of gp120 is needed for virion entry into $CD4^+$ cells. PDI inhibitors effectively block the cleavage, leading to loss of entry activity and defect of HIV infection [116].

PDIs regulate viral protein translation: Only little data show that PDIs may participate in viral protein translation. The viral protein synthesis of some (+) ssRNA virus is dependent on IRES-mediated translation. EV-A71 infection is potently inhibited by an ERp57 inhibitor Oblongifolin M (OM) through suppressing the viral IRES activity. This finding is further confirmed because the overexpression of ERp57 enhances the IRES activity and partially rescues the OM-inhibited IRES activity and viral replication [117]. It is generally accepted that PDIs ensure the correct formation and activity of viral glycoproteins.

PDIs regulate viral activities by influencing oxidative stress and ER stress: The rapid and large amount of viral protein synthesis causes disruption of redox balance in the establishment of infection and in the process of pathogenesis. The accumulated ROS in turn modulates viral replication and cellular response to viral pathogenesis that occur during HIV [118], influenza virus [119], HBV [120], HCV [121], encephalomyocarditis virus (EMCV) [122], RSV [123], and JEV [124] infections. ERp57 participates in the mechanisms of cell protection against oxidative stress, regardless of its subcellular location. The expression level of ERp57 and GST is positively correlated with tissue redox status during HPV infection, indicating potential viral-induced oxidative DNA and protein damages [125]. In addition, viral infections may cause the exploitation of the ER membrane, accumulation of misfolded proteins, and imbalance of calcium concentration. Many viruses (influenza virus, hepatitis B virus, Zika virus, etc.) hijack host cell processes to enhance viral pathogenesis, for instances, facilitating viral protein folding, trafficking, and modulating host immune responses [126–128]. As a major effector in ER stress response, ERp57 is critical for viral protein glycosylation and maturation, which in turn affects virus release and infectivity.

4.7.5 The RNA Chaperones in Virus Infection

The function of RNA chaperones in RNA virus infection: The life cycle of most RNA viruses is completed in the cytoplasm. Therefore, the virus is capable of redistributing some nuclear proteins into the cytosol that involves in RNA metabolism and functional regulation in different means. hnRNPs, such as hnRNP A1, C1/C2, D, and I, mainly locate in the nucleus but often shuttle to the cytoplasm to stabilize the viral RNA and initiate the cap-independent translation. For example, VP24 of Ebola virus (EBOV) binds to karyopherin-alpha nuclear import proteins and prevents their interaction with phosphorylated STAT1 (pSTAT1) and hnRNP C1/C2, leading to the cytoplasmic retention of pSTAT1 and hnRNP C1/C2 [129]. The cytosolic retention of pSTAT1 facilitates the virus' escape from the intracellular innate immunity. Rael is an mRNA export factor. VSV infection induces hnRNP A1 cytoplasm redistribution in a Rael-dependent manner [130].

Several hnRNPs are involved in virus replication. hnRNP I/PTB binds UCUAA repeats at IRES and 3'-UTR to modulate coronavirus RNA synthesis [131]. During the HCV RNA replication, hnRNP I/PTB and hnRNP C selectively interact with the 3′ end of the HCV genomic RNA and possibly initiate the (-) strand RNA replication to stabilize the RNA genome and/or to regulate the encapsidation of genomic RNA [132, 133]. Besides, hnRNP C also binds at the 5′- end of negative-strand RNA of poliovirus to facilitate the (+) strand RNA synthesis [134], while miR-555 decreases the expression of hnRNP C, thereby inhibiting poliovirus replication [135, 136].

The main function of hnRNPs is to modulate RNA process and metabolism (splicing, stabilization, transportation, etc.). The genome of influenza A virus is comprised of eight negative-stranded RNA segments. Among them, NS1 and M segments undergo alternative splicing and yield several proteins including nonstructural proteins NS1 and NS2, as well as matrix proteins M1 and ion channel M2. The NS segment encodes NS1 and nuclear export protein (NEP)/NS2; the M segment encodes M1 and M2. hnRNP K bridges the interaction of NS1-BP (binding protein) complex and M mRNA. Lack of neither NS1-BP nor hnRNP K ensures the proper splicing of M mRNA [137, 138]. The mutation of either or both the hnRNP K and NS1-BP-binding sites of viral RNA results in M segment mis-splicing and attenuated virus replication [139].

Most hnRNPs positively regulate viral protein translation. hnRNPs bind on the IRES of different viruses to assist translation. hnRNP I/PTB3, hnRNP L, hnRNP D/AUF1 [140–142], hnRNP A1, and hnRNP K [143] all bind the 5'UTR sequence to regulate HCV protein translation. hnRNP I, one of the polypyrimidine tract-binding proteins (PTBs), binds the IRES of picornaviruses, including cardio-aphthovirus group (including EMCV and EV-A71) and entero-rhinovirus group (including poliovirus and HRV-2). PTBs also bind HCV UTR regions, calicivirus RNA, and coronavirus RNAs. PTBs help IRES to fold into a translation-competent structure [144]. hnRNP D shuttles between the cytosol and the nucleus. It interacts with the stem-loop II of HCV 5' UTR to promote the IRES-dependent translation [140]. PIM1 inhibitors induce hnRNP D relocalization from the nucleus to the cytosol, giving rise to the inhibition of EV-A71 protein translation [58]. hnRNP A1 binds with both 5′- and 3′-UTR of HCV RNA to form a complex with NS5b and septin 6 and assists with viral protein translation. The hnRNP A1 homologs (such as hnRNP A/B, hnRNP A2/B1, and hnRNP A2) may substitute it to modulate the IRES-dependent translation [145]. hnRNP A1 specifically binds on the 5'-UTR of EV-A71 and enhances IRES-dependent translation [146]. Recent studies demonstrate that the cytoplasmic redistribution of hnRNP A1 is strictly regulated by some signaling or stress proteins. P38 inhibitors or Hsp27 knockout dramatically block hnRNP A1 redistribution in the cytosol, leading to the inhibition of virus replication [109, 147].

Apart from binding with IRES and promoting viral protein translation, hnRNPs can directly bind with virus proteins. Upon a CVB3 infection, hnRNP M is cleaved by 3C protease that facilitates virus replication without affecting IRES activity and

RNA stability [148]. The core protein of DENV interacts with hnRNP K, resulting in the release of hnRNP K's inhibitory effects on transcriptional activator C/EBPb [149, 150], while hnRNP C1/C2 interact with viral RNA and NS1 protein to facilitate virus reproduction [151–154]. hnRNP A2/B1 interact with NS1 of influenza virus, leading to the decrease of the viral NS1 RNA/protein levels as well as NS1 RNA nuclear export [155]. The negative regulator of splicing (NRS) element of *gag* gene in Rous sarcoma virus genome is required for 3′ long terminal repeat (LTR) polyadenylation. hnRNP H binds on NRS element and promotes polyadenylation.

The functions of RNA chaperones in DNA virus infections: To initiate viral DNA replication, hnRNP K interacts with DNA replication initiator protein UL84 to promote HCMV replication. hnRNPs regulate DNA virus gene expression at both transcriptional and post-transcriptional levels, either positively or negatively. At the transcription level, hnRNP D0B and hnRNP A/B bind the AAGTATGCA core element of HPV18 late promoter for suppressing the late genes, capsid proteins L1 and L2 [125, 156]. hnRNP K binds enhancer II (EN II) to promote HBV replication [157]. When APOBEC3B and hnRNP K form a complex, it alters the EN II binding activity and suppresses the S gene promoter activity [158]. At the posttranscriptional level, hnRNPs regulate the polyadenylation, splicing, and translation during DNA virus infections. In the process of HPV RNA splicing, hnRNP A1 directly binds with the splicing silencer element to inhibit the splicing of L1 mRNAs; hnRNP I also interferes with the splicing inhibitory elements and activates L1 gene expressions [159]. The genome of HPV consists of an early region and a late region followed by the proximal early (pAE) and the distal late (pAL) polyadenylation signals, respectively. The production of virions is controlled by the expression of L1 and L2 late genes in a differentiation-dependent way. hnRNP H binds the multiple GGG motifs at and promotes polyadenylation at an early polyadenylation signal, leading to inhibiting the L2 transcription. At the late infection stage, hnRNP H is captured by L1 to release the inhibitory effect on L2 [160, 161]. This process is called late gene autoregulation which enables rapid viral capsid protein production [161]. The SM polymerase (pol) mRNA of EBV early protein is cleaved and polyadenylated inefficiently. However, SM can harness hnRNP A1 and C for enhancing SM mRNA process under certain conditions [162]. In the translation process, hnRNP I and hnRNP K inhibit the L2 translation through binding on the 3′ end of L2 mRNA [163].

hnRNPs also exhibit some other functions during virus infections. During EBV infection, AUF1 can bind with the EBER1 noncoding RNA and compete with AUF1-interacting targets. In addition, AUF1/CBF2/EBNA2 forming a complex binds on the latency C promoter (Cp) [164]. Both EBNA2 and EBER1 are important for facilitating viral latency and cell transformation. hnRNPs are also able to regulate viral activities through modulating immune response. In the HSV-1 infection, viral DNA and hnRNP A/B form a complex, followed by homodimerization and demethylation. These events result in translocating the complex to the cytosol and activating natural immunity through type I interferon signaling.

4.8 Antiviral Drug Development and Challenge

Interest in targeting stress proteins in various diseases is increasing. First, targeting stress proteins would provide a wide spectrum of antiviral ability; this in turn increases the chances of curing related diseases. In cultured cells, for example, Hsp90 inhibitors display potent antiviral activity against many DNA and RNA viruses, including enteroviruses (EV-A71, coxsackievirus, poliovirus, rhinovirus, etc.), VSV, RSV, influenza viruses, CHIKV, HCV, HSV, HBV, EBV, HCMV, HTLV and SARS-CoV-2. GRp78 has been show a receptor of SARS-CoV-2. Target GRp78 would not only block viral entry, but aslo it must disrupt viral protein translation. It is also a wonderful strategy to target hnRNPs for antiviral. The depletion of hnRNP A1 inhibits viral reproductions of HPV, HCV, EV-A71, human T-cell lymphotropic virus 1 (HTLV-1), and HIV [165–168]; the depletion of hnRNP C combats viral infections of HCV, DENV, poliovirus, EBV, and HIV. The dysfunction of hnRNP I/PTB also impairs the reproduction of HCV, picornavirus, HPV, and HIV [169, 170]; the deprivation of hnRNP K significantly decreases the reproduction of influenza viruses, EV-A71, DENV, VSV, HCMV, HBV, HSV, HPV, HIV, and KSHV. Therefore, target of stress proteins is very attractive for drug development against viral infections that do not have effective therapies, especially newly emerging viral diseases such as COVID-19. Second, the affinity or dependence of stress proteins is commonly stronger under diseased situations than that in normal conditions. For example, mutant p53 relies much more on Hsp90 function than wild-type p53. Hsp90 inhibitor GM easily disturbs the association of mutant p53 with Hsp90 that causes the mutant p53 degradation only without affecting wild-type p53 [104]. Thirdly, little drug resistance would develop. No drug resistance will occur despite extensive passaging of the virus in the presence of Hsp90 inhibitors both in vitro cell culture and in vivo in animals.

One of the major challenges for drug design against stress proteins is the off-target effects, which is due to structure similarities among different stress proteins. For instance, a drug targeting Hsp90α may affect Hsp90β and even other members of Hsp70s. Over 20 Hsp90 inhibitors for anticancer have been developed; however, clinical studies are often stopped because of strong toxicity and side effects. This challenge would become little importance for treating acute infections such as SARS-CoV infection. However, other challenges appear. Due to the importance of stress proteins in normal cells and virus infections variably dysfunction many biological processes (including stress protein responses) at certain time points, inhibitors targeting stress proteins must be used with great caution in clinical settings because it may cause unexpected side effects and even toxicity. The gone with wind infection manner (e.g., SARS-CoV) limits the opportunity of clinic settings, giving rise to additional challenges for stress protein-based antiviral drug development.

Acknowledgment This chapter was modified from the paper published by our group in *Signal Transduction and Target Therapy* (Wan et al., 5(1): 125, 2020). Related contents are reused with permission.

References

1. Molinari DNHAM. In and out of the ER: protein folding, quality control, degradation, and related human diseases. Physiol Rev. 2006;87:1377–408. https://doi.org/10.1152/physrev.00050.2006.-A.
2. Hendrick JP, F H. Molecular chaperone functions of heat-shock proteins. Annu Rev Biochem. 1993;62:349–84. https://doi.org/10.1146/annurev.bi.62.070193.002025.
3. Murshid A, Gong J, Calderwood S. The role of heat shock proteins in antigen cross presentation. Front Immunol. 2012;3:1–10. https://doi.org/10.3389/fimmu.2012.00063.
4. Asea A, et al. Novel signal transduction pathway utilized by extracellular HSP70: role of toll-like receptor (TLR) 2 and TLR4. J Biol Chem. 2002;277:15028–34.
5. Hoter A, El-Sabban ME, Naim HY. The HSP90 family: structure, regulation, function, and implications in health and disease. Int J Mol Sci. 2018;19:1–33. https://doi.org/10.3390/ijms19092560.
6. Wang X, Chen M, Zhou J, Zhang X. HSP27, 70 and 90, anti-apoptotic proteins, in clinical cancer therapy. Int J Oncol. 2014;45:18–30. https://doi.org/10.3892/ijo.2014.2399.
7. Saibil H. Chaperone machines for protein folding, unfolding and disaggregation. Nat Rev Mol Cell Biol. 2013;14:630–42. https://doi.org/10.1038/nrm3658.
8. Schopf FH, Biebl MM, Buchner J. The HSP90 chaperone machinery. Nat Rev Mol Cell Biol. 2017;18:345–60. https://doi.org/10.1038/nrm.2017.20.
9. Stetler RA, et al. Heat shock proteins: cellular and molecular mechanisms in the central nervous system. Prog Neurobiol. 2010;92:184–211. https://doi.org/10.1016/j.pneurobio.2010.05.002.
10. Zhu X, et al. Structural analysis of substrate binding by the molecular chaperone DnaK. Science. 1996;272:1606.
11. Radons J. The human HSP70 family of chaperones: where do we stand? Cell. 2016;21:379–404. https://doi.org/10.1007/s12192-016-0676-6.
12. Mayer MP. Gymnastics of molecular chaperones. Mol Cell. 2010;39:321–31. https://doi.org/10.1016/j.molcel.2010.07.012.
13. Bukau B, Weissman J, Horwich A. Molecular chaperones and protein quality control. Cell. 2006;125:443–51. https://doi.org/10.1016/j.cell.2006.04.014.
14. Sharma SK, De Los Rios P, Christen P, Lustig A, Goloubinoff P. The kinetic parameters and energy cost of the Hsp70 chaperone as a polypeptide unfoldase. Nat Chem Biol. 2010;6:914–20. https://doi.org/10.1038/nchembio.455.
15. Mayer MP. Hsp70 chaperone dynamics and molecular mechanism. Trends Biochem Sci. 2013;38:507–14. https://doi.org/10.1016/j.tibs.2013.08.001.
16. Kampinga HH, Craig EA. The Hsp70 chaperone machinery: J-proteins as drivers of functional specificity. Nat Rev Mol Cell Biol. 2010;11:579. https://doi.org/10.1038/NRM2941.
17. Garrido C, Paul C, Seigneuric R, Kampinga H. The small heat shock proteins family: the long forgotten chaperones. Int J Biochem Cell Biol. 2012;44:1588–92.
18. van Montfort RL, et al. Crystal structure and assembly of a eukaryotic small heat shock protein. Nat Struct Biol. 2001;8:1025–30.
19. MacRae TH. Structure and function of small heat shock/α-crystallin proteins: established concepts and emerging ideas. Life Sci. 2000;57:899–913.
20. Shashidharamurthy R, Koteiche HA, Dong J, Mchaourab HS. Mechanism of chaperone function in small heat shock proteins. J Biol Chem. 2005;280:5281–9. https://doi.org/10.1074/jbc.M407236200.
21. Hsu A-L. Regulation of aging and age-related disease by DAF-16 and heat-shock factor. Science. 2003;300:1142–5. https://doi.org/10.1126/science.1083701.
22. Bruey J-M, et al. Hsp27 negatively regulates cell death by interacting with cytochrome c. Nat Cell Biol. 2000;2:645–52. https://doi.org/10.1038/35023595.
23. Guay J, et al. Regulation of actin filament dynamics by p38 map kinase-mediated phosphorylation of heat shock protein 27. J Cell Sci. 1997;110(Pt 3):357–68.

24. Parcellier A, et al. Small heat shock proteins HSP27 and αB-crystallin: cytoprotective and oncogenic functions. Antioxid Redox Signal. 2005;7:404–13.
25. Anckar J, Sistonen L. Regulation of HSF 1 function in the heat stress response: implications in aging and disease. Annu Rev Biochem. 2011;80:1089–115. https://doi.org/10.1146/annurev-biochem-060809-095203.
26. Sakurai H, Enoki Y. Novel aspects of heat shock factors: DNA recognition, chromatin modulation and gene expression. FEBS J. 2010;277:4140–9. https://doi.org/10.1111/j.1742-4658.2010.07829.x.
27. McMillan DR, Xiao X, Shao L, Graves K, Benjamin IJ. Targeted disruption of heat shock transcription factor 1 abolishes thermotolerance and protection against heat-inducible apoptosis. J Biol Chem. 1998;273:7523–8. https://doi.org/10.1074/jbc.273.13.7523.
28. Turano C, Coppari S, Altieri F, Ferraro A. Proteins of the PDI family: unpredicted non-ER locations and functions. J Cell Physiol. 2002;193:154–63. https://doi.org/10.1002/jcp.10172.
29. Benham AM. The protein disulfide isomerase family: key players in health and disease. Antioxid Redox Signal. 2012;16:781–9. https://doi.org/10.1089/ars.2011.4439.
30. Oliver JD, Roderick HL, Llewellyn DH, High S. ERp57 functions as a subunit of specific complexes formed with the ER lectins calreticulin and calnexin. Mol Biol Cell. 1999;10:2573–82.
31. Hatahet F, Ruddock LW. Protein disulfide isomerase: a critical evaluation of its function in disulfide bond formation. Antioxid Redox Signal. 2009;11:2807–50. https://doi.org/10.1089/ars.2009.2466.
32. Bedard K, Szabo E, Michalak M, Opas M. Cellular functions of endoplasmic reticulum chaperones calreticulin, calnexin, and ERp57. Int Rev Cytol. 2005;245:91–121. https://doi.org/10.1016/s0074-7696(05)45004-4.
33. Bomsztyk K, Denisenko O, Ostrowski J. hnRNP K: one protein multiple processes. Bioessays. 2004;26:629–38. https://doi.org/10.1002/bies.20048.
34. Zuniga S, Sola I, Cruz JL, Enjuanes L. Role of RNA chaperones in virus replication. Virus Res. 2009;139:253–66. https://doi.org/10.1016/j.virusres.2008.06.015.
35. Ciocca RD, Calderwood SK. Heat shock proteins in cancer: diagnostic, prognostic, predictive, and treatment implications. Cell. 2005;10:86–113.
36. Workman P. Altered states: selectively drugging the Hsp90 cancer chaperone. Trends Mol Med. 2004;10:47–51. https://doi.org/10.1016/j.molmed.2003.12.005.
37. Calderwood SK, Gong J. Heat shock proteins promote cancer: it's a protection racket. Trends Biochem Sci. 2016;41:311–23. https://doi.org/10.1016/j.tibs.2016.01.003.
38. Calderwood SK. Heat shock proteins and cancer: intracellular chaperones or extracellular signalling ligands? Philos Trans R Soc Lond Ser B Biol Sci. 2018;373:524. https://doi.org/10.1098/rstb.2016.0524.
39. Nussbacher JK, Tabet R, Yeo GW, Lagier-Tourenne C. Disruption of RNA metabolism in neurological diseases and emerging therapeutic interventions. Neuron. 2019;102:294–320. https://doi.org/10.1016/j.neuron.2019.03.014.
40. Kampinga HH, Bergink S. Heat shock proteins as potential targets for protective strategies in neurodegeneration. Lancet Neurol. 2016;15:748–59. https://doi.org/10.1016/s1474-4422(16)00099-5.
41. Munnink BBO, Hoek LVD. Viruses causing gastroenteritis: the known, the new and those beyond. Virus. 2016;8:42. https://doi.org/10.3390/V8020042.
42. Siebrasse EA, et al. Identification of MW polyomavirus, a novel polyomavirus in human stool. J Virol. 2012;86:10321. https://doi.org/10.1128/JVI.01210-12.
43. Jones MS, Lukashov VV, Ganac RD, Schnurr DP. Discovery of a novel human picornavirus in a stool sample from a pediatric patient presenting with fever of unknown origin. J Clin Microbiol. 2007;45:2144. https://doi.org/10.1128/JCM.00174-07.
44. Kapoor A, et al. A highly prevalent and genetically diversified Picornaviridae genus in South Asian children. Proc Natl Acad Sci U S A. 2008;105:20482–7. https://doi.org/10.1073/PNAS.0807979105.

45. Walter MJ, Wunderink GR. Severe respiratory viral infections: new evidence and changing paradigms. Infect Dis Clin N Am. 2017;31:455–74.
46. Manjarrez-Zavala ME, Rosete-Olvera DP, Gutiérrez-González LH, Ocadiz-Delgad R, Cabello-Gutiérrez C. Pathogenesis of viral respiratory infection. Resp Dis Infect. 2013; https://doi.org/10.5772/54287.
47. Messaoudi I, Basler CF. Immunological features underlying viral hemorrhagic fevers. Curr Opin Immunol. 2015;36:38. https://doi.org/10.1016/J.COI.2015.06.003.
48. Paessler S, Walker DH. Pathogenesis of the viral hemorrhagic fevers. Annu Rev Pathol Mech. 2013;8:411–40. https://doi.org/10.1146/annurev-pathol-020712-164041.
49. Karim S, et al. The role of viruses in neurodegenerative and neurobehavioral diseases. CNS Neurol Disord. 2014;13:1213–23. https://doi.org/10.2174/187152731307141015122638.
50. Karim S, et al. The role of viruses in neurodegenerative and neurobehavioral diseases. CNS Neurol Disord Drug Targets. 2014;13:1213.
51. Balin BJ, Hudson AP. Herpes viruses and Alzheimer's disease: new evidence in the debate. Lancet Neurol. 2018;17:839–41. https://doi.org/10.1016/s1474-4422(18)30316-8.
52. Tsou Y-L, et al. Heat shock protein 90: role in enterovirus 71 entry and assembly and potential target for therapy. PLoS One. 2013;8:e77133. https://doi.org/10.1371/journal.pone.0077133.
53. Cabrera-Hernandez A, Thepparit C, Suksanpaisan L, Smith DR. Dengue virus entry into liver (HepG2) cells is independent of hsp90 and hsp70. J Med Virol. 2007;79:386–92. https://doi.org/10.1002/jmv.20786.
54. Okamoto T, et al. Hepatitis C virus RNA replication is regulated by FKBP8 and Hsp90. EMBO J. 2006;25:5015–25. https://doi.org/10.1038/sj.emboj.7601367.
55. Momose F, et al. Identification of Hsp90 as a stimulatory host factor involved in influenza virus RNA synthesis. J Biol Chem. 2002;277:45306–14. https://doi.org/10.1074/jbc.M206822200.
56. Naito T, Momose F, Kawaguchi A, Nagata K. Involvement of Hsp90 in assembly and nuclear import of influenza virus RNA polymerase subunits. J Virol. 2007;81:1339–49. https://doi.org/10.1128/JVI.01917-06.
57. Chase G, et al. Hsp90 inhibitors reduce influenza virus replication in cell culture. J Virol. 2008;377:431–9. https://doi.org/10.1016/j.virol.2008.04.040.
58. Zhou F, et al. Pim1 impacts enterovirus A71 replication and represents a potential target in antiviral therapy. Science. 2019;19:715. https://doi.org/10.1016/J.ISCI.2019.08.008.
59. Wan Q, Song D, Li H, He ML. Stress proteins: the biological functions in virus infection, present and challenges for target-based antiviral drug development. Signal Transduct Target Ther. 2020;5(1):125.
60. Waxman L, Whitney M, Pollok BA, Kuo LC, Darke PL. Host cell factor requirement for hepatitis C virus enzyme maturation. Proc Natl Acad Sci U S A. 2001;98:13931–5. https://doi.org/10.1073/pnas.241510898.
61. Geller R, Vignuzzi M, Andino R, Frydman J. Evolutionary constraints on chaperone-mediated folding provide an antiviral approach refractory to development of drug resistance. Genes Dev. 2007;21:195–205. https://doi.org/10.1101/gad.1505307.
62. Kumar P, Gaur P, Kumari R, Lal SK. Influenza A virus neuraminidase protein interacts with Hsp90, to stabilize itself and enhance cell survival. J Cell Biochem. 2019;120:6449–58. https://doi.org/10.1002/jcb.27935.
63. Zhong M, et al. Heat-shock protein 90 promotes nuclear transport of herpes simplex virus 1 capsid protein by interacting with acetylated tubulin. PLoS One. 2014;9:e99425. https://doi.org/10.1371/journal.pone.0099425.
64. Burch AD, Weller SK. Herpes simplex virus type 1 DNA polymerase requires the mammalian chaperone hsp90 for proper localization to the nucleus. J Virol. 2005;79:10740–9. https://doi.org/10.1128/JVI.79.16.10740-10749.2005.
65. Li Y-H, Tao P-Z, Liu Y-Z, Jiang J-D. Geldanamycin, a ligand of heat shock protein 90, inhibits the replication of herpes simplex virus type 1 in vitro. Antimicrob Agents Chemother. 2004;48:867–72. https://doi.org/10.1128/aac.48.3.867-872.2004.

66. Sun X, Kenney SC. Hsp90 inhibitors: a potential treatment for latent EBV infection? Cell Cycle. 2010;9:1665–6. https://doi.org/10.4161/cc.9.9.11594.
67. Kawashima D, et al. Nuclear transport of Epstein-Barr virus DNA polymerase is dependent on the BMRF1 polymerase processivity factor and molecular chaperone Hsp90. J Virol. 2013;87:6482–91. https://doi.org/10.1128/JVI.03428-12.
68. Kotsiopriftis M, Tanner JE, Alfieri C. Heat shock protein 90 expression in Epstein-Barr virus-infected B cells promotes gammadelta T-cell proliferation in vitro. J Virol. 2005;79:7255–61. https://doi.org/10.1128/JVI.79.11.7255-7261.2005.
69. Sciammas R, Bluestone JA. TCRgammadelta cells and viruses. Microbes Infect. 1999;1:203–12.
70. Taguwa S, et al. Zika virus dependence on host hsp70 provides a protective strategy against infection and disease. Cell Rep. 2019;26:906–20. https://doi.org/10.1016/j.celrep.2018.12.095.
71. Pujhari S, et al. Heat shock protein 70 (Hsp70) mediates Zika virus entry, replication, and egress from host cells. Emerg Microbes Infect. 2019;8:8–16. https://doi.org/10.1080/22221751.2018.1557988.
72. Triantafilou K, Fradelizi D, Wilson K, Triantafilou M. GRP78, a coreceptor for coxsackievirus A9, interacts with major histocompatibility complex class I molecules which mediate virus internalization. J Virol. 2002;76:633–43. https://doi.org/10.1128/jvi.76.2.633-643.2002.
73. Manzoor R, et al. Heat shock protein 70 modulates influenza A virus polymerase activity. J Biol Chem. 2014;289:7599–614. https://doi.org/10.1074/jbc.M113.507798.
74. Dong Q, et al. Hsc70 regulates the IRES activity and serves as an antiviral target of enterovirus A71 infection. Antivir Res. 2018;150:39–46. https://doi.org/10.1016/j.antiviral.2017.11.020.
75. Hovanessian AG, Galabru J. The double-stranded RNA-dependent protein kinase is also activated by heparin. Eur J Biochem. 1987;167:467–73. https://doi.org/10.1111/j.1432-1033.1987.tb13360.x.
76. Gale J, Michael J, Korth MJ, Katze MG. Repression of the PKR protein kinase by the hepatitis C virus NS5A protein: a potential mechanism of interferon resistance. Clin Diagn Virol. 1998;10:157–62. https://doi.org/10.1016/S0928-0197(98)00034-8.
77. Lee TG, Tang N, Thompson S, Miller J, Katze MG. The 58,000-dalton cellular inhibitor of the interferon-induced double-stranded RNA-activated protein kinase (PKR) is a member of the tetratricopeptide repeat family of proteins. Mol Cell Biol. 1994;14:2331–42. https://doi.org/10.1128/mcb.14.4.2331.
78. Lee TG, Tomita J, Hovanessian AG, Katze MG. Purification and partial characterization of a cellular inhibitor of the interferon-induced protein kinase of Mr 68,000 from influenza virus-infected cells. Proc Natl Acad Sci U S A. 1990;87:6208–12. https://doi.org/10.1073/pnas.87.16.6208.
79. Lee TG, Tomita J, Hovanessian AG, Katze MG. Characterization and regulation of the 58,000-dalton cellular inhibitor of the interferon-induced, dsRNA-activated protein kinase. J Biol Chem. 1992;267:14238–43.
80. Melville MW, Hansen WJ, Freeman BC, Welch WJ, Katze MG. The molecular chaperone hsp40 regulates the activity of P58IPK, the cellular inhibitor of PKR. Proc Natl Acad Sci U S A. 1997;94:97–102. https://doi.org/10.1073/pnas.94.1.97.
81. Melville MW, et al. The cellular inhibitor of the PKR protein kinase, P58(IPK), is an influenza virus-activated co-chaperone that modulates heat shock protein 70 activity. J Biol Chem. 1999;274:3797–803. https://doi.org/10.1074/jbc.274.6.3797.
82. Leone G, et al. C-terminal trimerization, but not N-terminal trimerization, of the reovirus cell attachment protein Is a posttranslational and Hsp70/ATP-dependent process. J Biol Chem. 1996;271:8466–71. https://doi.org/10.1074/jbc.271.14.8466.
83. Macejak DG, Sarnow P. Association of heat shock protein 70 with enterovirus capsid precursor P1 in infected human cells. J Virol. 1992;66:1520–7.

84. Khachatoorian R, et al. The NS5A-binding heat shock proteins HSC70 and HSP70 play distinct roles in the hepatitis C viral life cycle. J Virol. 2014;454-455:118–27. https://doi.org/10.1016/j.virol.2014.02.016.
85. Khachatoorian R, et al. Structural characterization of the HSP70 interaction domain of the hepatitis C viral protein NS5A. J Virol. 2015;475:46–55. https://doi.org/10.1016/j.virol.2014.10.011.
86. Chroboczek J, Gout E, Favier AL, Galinier R. Novel partner proteins of adenovirus penton. Curr Top Microbiol Immunol. 2003;272:37–55.
87. Gout E, Gutkowska M, Takayama S, Reed JC, Chroboczek J. Co-chaperone BAG3 and adenovirus penton base protein partnership. J Cell Biochem. 2010;111:699–708. https://doi.org/10.1002/jcb.22756.
88. Pater A, Kumar KU, Pater MM, Devireddy LR. BAG-1, a novel Bcl-2-interacting protein, activates expression of human JC virus. J Gen Virol. 2000;81:351–7. https://doi.org/10.1099/0022-1317-81-2-351.
89. Takahashi N, et al. BAG-1M, an isoform of Bcl-2-interacting protein BAG-1, enhances gene expression driven by CMV promoter. Biochem Biophys Res Commun. 2001;286:807–14. https://doi.org/10.1006/bbrc.2001.5473.
90. Sullivan CS, Cantalupo P, Pipas JM. The molecular chaperone activity of simian virus 40 large T antigen is required to disrupt Rb-E2F family complexes by an ATP-dependent mechanism. Mol Cell Biol. 2000;20:6233–43. https://doi.org/10.1128/mcb.20.17.6233-6243.2000.
91. Sullivan CS, Gilbert SP, Pipas JM. ATP-dependent simian virus 40 T-antigen-Hsc70 complex formation. J Virol. 2001;75:1601–10. https://doi.org/10.1128/JVI.75.4.1601-1610.2001.
92. White E, Spector D, Welch W. Differential distribution of the adenovirus E1A proteins and colocalization of E1A with the 70-kilodalton cellular heat shock protein in infected cells. J Virol. 1988;62:4153–66.
93. Young P, Anderton E, Paschos K, White R, Allday MJ. Epstein-Barr virus nuclear antigen (EBNA) 3A induces the expression of and interacts with a subset of chaperones and co-chaperones. J Gen Virol. 2008;89:866–77. https://doi.org/10.1099/vir.0.83414-0.
94. Fazakerley JK. Neurovirology and developmental neurobiology. Adv Virus Res. 2001;56:73–124.
95. Fazakerley JK, Allsopp TE. Programmed cell death in virus infections of the nervous system. Curr Top Microbiol Immunol. 2001;253:95–119.
96. Kuzyk MA, et al. Multiple reaction monitoring-based, multiplexed, absolute quantitation of 45 proteins in human plasma. Mol Cell Probes. 2009;8:1860. https://doi.org/10.1074/mcp.M800540-MCP200.
97. Bromberg Z, Weiss Y. The role of the membrane-initiated heat shock response in cancer. Front Mol Biosci. 2016;3:12. https://doi.org/10.3389/fmolb.2016.00012.
98. Martine P, et al. HSP70 is a negative regulator of NLRP3 inflammasome activation. Cell Death Dis. 2019;10:256. https://doi.org/10.1038/s41419-019-1491-7.
99. Zhang R, et al. Ts-Hsp70 induces protective immunity against Trichinella spiralis infection in mouse by activating dendritic cells through TLR2 and TLR4. PLoS Negl Trop Dis. 2018;12:e0006502. https://doi.org/10.1371/journal.pntd.0006502.
100. Lehner T, et al. Functional domains of HSP70 stimulate generation of cytokines and chemokines, maturation of dendritic cells and adjuvanticity. Biochem Soc Trans. 2004;32:629–32. https://doi.org/10.1042/BST0320629.
101. Fang H, et al. Toll-like receptor 4 (TLR4) is essential for Hsp70-like protein 1 (HSP70L1) to activate dendritic cells and induce Th1 response. J Biol Chem. 2011;286:30393–400. https://doi.org/10.1074/jbc.M111.266528.
102. Al-Huseini LM, et al. Heme oxygenase-1 regulates dendritic cell function through modulation of p38 MAPK-CREB/ATF1 signaling. J Biol Chem. 2014;289:16442–51. https://doi.org/10.1074/jbc.M113.532069.
103. Pathak SK, et al. Activated apoptotic cells induce dendritic cell maturation via engagement of Toll-like receptor 4 (TLR4), dendritic cell-specific intercellular adhesion molecule 3 (ICAM-

3)-grabbing nonintegrin (DC-SIGN), and beta2 integrins. J Biol Chem. 2012;287:13731–42. https://doi.org/10.1074/jbc.M111.336545.
104. Chen Y, et al. IL-1β induction of MUC5AC gene expression is mediated by CREB and NF-κB and repressed by dexamethasone. Lung Cell Mol Phys. 2014;306:L797–807. https://doi.org/10.1152/ajplung.00347.2013.
105. Wong CP, Bray TM, Ho E. Induction of proinflammatory response in prostate cancer epithelial cells by activated macrophages. Cancer Lett. 2009;276:38–46. https://doi.org/10.1016/j.canlet.2008.10.025.
106. Jäättelä M. Heat shock proteins as cellular lifeguards. Ann Med. 1999;31:261–71.
107. Haslbeck M, Franzmann T, Weinfurtner D, Buchner J. Some like it hot: the structure and function of small heat-shock proteins. Nat Struct Mol Biol. 2005;12:842–6. https://doi.org/10.1038/nsmb993.
108. Vos MJ, Hageman J, Carra S, Kampinga HH. Structural and functional diversities between members of the human HSPB, HSPH, HSPA, and DNAJ chaperone families. Biochemist. 2008;47:7001–11.
109. Dan X, et al. Hsp27 responds to and facilitates enterovirus A71 replication by enhancing viral internal ribosome entry site-mediated translation. J Virol. 2019;93:e02322–18. https://doi.org/10.1128/JVI.02322-18.
110. Sun M, et al. Down-regulating heat shock protein 27 is involved in porcine epidemic diarrhea virus escaping from host antiviral mechanism. Vet Microbiol. 2017;205:6–13. https://doi.org/10.1016/j.vetmic.2017.04.031.
111. Tong S-W, et al. HSPB1 is an intracellular antiviral factor against hepatitis B virus. J Cell Biochem. 2013;114:162–73. https://doi.org/10.1002/jcb.24313.
112. Preville X, et al. Mammalian small stress proteins protect against oxidative stress through their ability to increase glucose-6-phosphate dehydrogenase activity and by maintaining optimal cellular detoxifying machinery. Exp Cell Res. 1999;247:61–78. https://doi.org/10.1006/excr.1998.4347.
113. Wan SW, et al. Endothelial cell surface expression of protein disulfide isomerase activates beta1 and beta3 integrins and facilitates dengue virus infection. J Cell Biochem. 2012;113:1681–91. https://doi.org/10.1002/jcb.24037.
114. Diwaker D, Mishra KP, Ganju L, Singh SB. Protein disulfide isomerase mediates dengue virus entry in association with lipid rafts. Viral Immunol. 2015;28:153–60. https://doi.org/10.1089/vim.2014.0095.
115. Calderon MN, Guerrero CA, Acosta O, Lopez S, Arias CF. Inhibiting rotavirus infection by membrane-impermeant thiol/disulfide exchange blockers and antibodies against protein disulfide isomerase. Intervirology. 2012;55:451–64. https://doi.org/10.1159/000335262.
116. Gallina A, et al. Inhibitors of protein-disulfide isomerase prevent cleavage of disulfide bonds in receptor-bound glycoprotein 120 and prevent HIV-1 entry. J Biol Chem. 2002;277:50579–88. https://doi.org/10.1074/jbc.M204547200.
117. Wang M, et al. Oblongifolin M, an active compound isolated from a Chinese medical herb Garcinia oblongifolia, potently inhibits enterovirus 71 reproduction through downregulation of ERp57. Oncotarget. 2015;7:8797–808.
118. Dobmeyer TS, et al. Ex vivo induction of apoptosis in lymphocytes is mediated by oxidative stress: role for lymphocyte loss in HIV infection. Free Radic Biol Med. 1997;22:775–85. https://doi.org/10.1016/s0891-5849(96)00403-0.
119. Knobil K, Choi AM, Weigand GW, Jacoby DB. Role of oxidants in influenza virus-induced gene expression. Am J Phys. 1998;274:L134–42. https://doi.org/10.1152/ajplung.1998.274.1.L134.
120. Acar A, et al. Investigation of oxidative stress and antioxidant defense in patients with hepatitis B virus infection and the effect of interferon-alpha plus lamivudine combination therapy on oxidative stress. Mikrobiyol Bul. 2009;43:411–23.
121. Korenaga M, et al. Hepatitis C virus core protein inhibits mitochondrial electron transport and increases reactive oxygen species (ROS) production. J Biol Chem. 2005;280:37481–8. https://doi.org/10.1074/jbc.M506412200.

122. Ano Y, et al. Oxidative damage to neurons caused by the induction of microglial NADPH oxidase in encephalomyocarditis virus infection. Neurosci Lett. 2010;469:39–43. https://doi.org/10.1016/j.neulet.2009.11.040.
123. Olagnier D, et al. Cellular oxidative stress response controls the antiviral and apoptotic programs in dengue virus-infected dendritic cells. PLoS Pathog. 2014;10:e1004566. https://doi.org/10.1371/journal.ppat.1004566.
124. Liao SL, Raung SL, Chen CJ. Japanese encephalitis virus stimulates superoxide dismutase activity in rat glial cultures. Neurosci Lett. 2002;324:133–6. https://doi.org/10.1016/s0304-3940(02)00236-7.
125. De Marco F, et al. Oxidative stress in HPV-driven viral carcinogenesis: redox proteomics analysis of HPV-16 dysplastic and neoplastic tissues. PLoS One. 2012;7:e34366. https://doi.org/10.1371/journal.pone.0034366.
126. Tatu U, Hammond C, Helenius A. Folding and oligomerization of influenza hemagglutinin in the ER and the intermediate compartment. EMBO J. 1995;14:1340–8.
127. Zai J, et al. N-glycosylation of the premembrane protein of Japanese encephalitis virus is critical for folding of the envelope protein and assembly of virus-like particles. Acta Virol. 2013;57:27–33.
128. Dubuisson J, Rice CM. Hepatitis C virus glycoprotein folding: disulfide bond formation and association with calnexin. J Virol. 1996;70:778–86.
129. Shabman RS, Gulcicek EE, Stone KL, Basler CF. The Ebola virus VP24 protein prevents hnRNP C1/C2 binding to karyopherin α1 and partially alters its nuclear import. J Infect Dis. 2011;204:S904–10. https://doi.org/10.1093/infdis/jir323.
130. Pettit Kneller EL, Connor JH, Lyles DS. hnRNPs Relocalize to the cytoplasm following infection with vesicular stomatitis virus. J Virol. 2009;83:770–80. https://doi.org/10.1128/JVI.01279-08.
131. Shi ST, Lai MM. Viral and cellular proteins involved in coronavirus replication. Curr Top Microbiol Immunol. 2005;287:95–131. https://doi.org/10.1007/3-540-26765-4_4.
132. Clerte C, Hall KB. Characterization of multimeric complexes formed by the human PTB1 protein on RNA. RNA. 2006;12:457–75. https://doi.org/10.1261/rna.2178406.
133. Chung RT, Kaplan LM. Heterogeneous nuclear ribonucleoprotein I (hnRNP-I/PTB) selectively binds the conserved 3′ terminus of Hepatitis C viral RNA. Biochem Biophys Res Commun. 1999;254:351–62. https://doi.org/10.1006/bbrc.1998.9949.
134. Ertel KJ, Brunner JE, Semler BL. Mechanistic consequences of hnRNP C binding to both RNA termini of poliovirus negative-strand RNA intermediates. J Virol. 2010;84:4229–42. https://doi.org/10.1128/JVI.02198-09.
135. Brunner JE, Ertel KJ, Rozovics JM, Semler BL. Delayed kinetics of poliovirus RNA synthesis in a human cell line with reduced levels of hnRNP C proteins. J Virol. 2010;400:240–7. https://doi.org/10.1016/J.VIROL.2010.01.031.
136. Perwitasari O, et al. MicroRNA-555 has potent antiviral properties against poliovirus. J Gen Virol. 2016;97:659–68. https://doi.org/10.1099/jgv.0.000372.
137. Tsai P-L, et al. Cellular RNA binding proteins NS1-BP and hnRNP K regulate influenza A virus RNA splicing. PLoS Pathog. 2013;9:e1003460. https://doi.org/10.1371/journal.ppat.1003460.
138. Wolff T, O'Neill RE, Palese P. NS1-binding protein (NS1-BP): a novel human protein that interacts with the influenza A virus nonstructural NS1 protein is relocalized in the nuclei of infected cells. J Virol. 1998;72:7170–80.
139. Thompson MG, et al. Co-regulatory activity of hnRNP K and NS1-BP in influenza and human mRNA splicing. Nat Commun. 2018;9:2407. https://doi.org/10.1038/s41467-018-04779-4.
140. Paek KY, Kim CS, Park SM, Kim JH, Jang SK. RNA-binding protein hnRNP D modulates internal ribosome entry site-dependent translation of hepatitis C virus RNA. J Virol. 2008;82:12082–93. https://doi.org/10.1128/JVI.01405-08.
141. Bokinsky G, et al. Two distinct binding modes of a protein cofactor with its target RNA. J Mol Biol. 2006;361:771–84. https://doi.org/10.1016/j.jmb.2006.06.048.

142. Williams MC, et al. Mechanism for nucleic acid chaperone activity of HIV-1 nucleocapsid protein revealed by single molecule stretching. PNAS. 2001;98:6121–6. https://doi.org/10.1073/pnas.101033198.
143. Lin J-Y, et al. Heterogeneous nuclear ribonuclear protein K interacts with the enterovirus 71 5' untranslated region and participates in virus replication. J Gen Virol. 2008;89:2540–9. https://doi.org/10.1099/vir.0.2008/003673-0.
144. Anderson EC, Hunt SL, Jackson RJ. Internal initiation of translation from the human rhinovirus-2 internal ribosome entry site requires the binding of Unr to two distinct sites on the 5' untranslated region. J Gen Virol. 2007;88:3043–52. https://doi.org/10.1099/vir.0.82463-0.
145. Kim CS, Seol SK, Song O-K, Park JH, Jang SK. An RNA-binding protein, hnRNP A1, and a scaffold protein, septin 6, facilitate hepatitis C virus replication. J Virol. 2007;81:3852–65. https://doi.org/10.1128/JVI.01311-06.
146. Lin J-Y, et al. hnRNP A1 interacts with the 5' untranslated regions of enterovirus 71 and Sindbis virus RNA and is required for viral replication. J Virol. 2009;83:6106–14. https://doi.org/10.1128/JVI.02476-08.
147. Leong SY, Ong BKT, Chu JJH. The role of Misshapen NCK-related kinase (MINK), a novel Ste20 family kinase, in the IRES-mediated protein translation of human enterovirus 71. PLoS Pathog. 2015;11:e1004686. https://doi.org/10.1371/journal.ppat.1004686.
148. Jagdeo JM, et al. N-terminomics TAILS identifies host cell substrates of poliovirus and Coxsackievirus B3 3C proteinases that modulate virus infection. J Virol. 2018;92:e02211–7. https://doi.org/10.1128/JVI.02211-17.
149. Chang C-J, et al. The heterogeneous nuclear ribonucleoprotein K (hnRNP K) interacts with dengue virus core protein. DNA Cell Biol. 2001;20:569–77. https://doi.org/10.1089/104454901317094981.
150. Miau L-H, Chang C-J, Shen B-J, Tsai W-H, Lee S-C. Identification of heterogeneous nuclear ribonucleoprotein K (hnRNP K) as a repressor of C/EBP-mediated gene activation. J Biol Chem. 1998;273:17789–97.
151. Dechtawewat T, et al. Role of human heterogeneous nuclear ribonucleoprotein C1/C2 in dengue virus replication. J Virol. 2015;12:14. https://doi.org/10.1186/s12985-014-0219-7.
152. Kanlaya R, Pattanakitsakul S-N, Sinchaikul S, Chen S-T, Thongboonkerd V. Vimentin interacts with heterogeneous nuclear ribonucleoproteins and dengue nonstructural protein 1 and is important for viral replication and release. Mol BioSyst. 2010;6:795. https://doi.org/10.1039/b923864f.
153. Noisakran S, et al. Identification of human hnRNP C1/C2 as a dengue virus NS1-interacting protein. Biochem Biophys Res Commun. 2008;372:67–72. https://doi.org/10.1016/J.BBRC.2008.04.165.
154. Dinh PX, Das A, Franco R, Pattnaik AK. Heterogeneous nuclear ribonucleoprotein K supports vesicular stomatitis virus replication by regulating cell survival and cellular gene expression. J Virol. 2013;87:10059–69. https://doi.org/10.1128/JVI.01257-13.
155. Wang Y, Zhou J, Du Y. hnRNP A2/B1 interacts with influenza A viral protein NS1 and inhibits virus replication potentially through suppressing NS1 RNA/protein levels and NS1 mRNA nuclear export. J Virol. 2014;449:53–61. https://doi.org/10.1016/j.virol.2013.11.009.
156. Wang X, et al. Viral DNA replication orientation and hnRNPs regulate transcription of the human papillomavirus 18 late promoter. mBio. 2017;8:e00713–7. https://doi.org/10.1128/mBio.00713-17.
157. Ng LFP, et al. Host heterogeneous ribonucleoprotein K (hnRNP K) as a potential target to suppress Hepatitis B virus replication. PLoS Med. 2005;2:e163. https://doi.org/10.1371/journal.pmed.0020163.
158. Zhang W, et al. Cytidine deaminase APOBEC3B interacts with heterogeneous nuclear ribonucleoprotein K and suppresses hepatitis B virus expression. Cell Microbiol. 2007;10:112–21. https://doi.org/10.1111/j.1462-5822.2007.01020.x.

159. Somberg M, Zhao X, Fröhlich M, Evander M, Schwartz S. Polypyrimidine tract binding protein induces human papillomavirus type 16 late gene expression by interfering with splicing inhibitory elements at the major late 5' splice site, SD3632. J Virol. 2008;82:3665–78. https://doi.org/10.1128/JVI.02140-07.
160. Oberg D, Fay J, Lambkin H, Schwartz S. A downstream polyadenylation element in human papillomavirus type 16 L2 encodes multiple GGG motifs and interacts with hnRNP H. J Virol. 2005;79:9254–69. https://doi.org/10.1128/JVI.79.14.9254-9269.2005.
161. Zheng Z-Z, et al. Specific interaction between hnRNP H and HPV16 L1 proteins: Implications for late gene auto-regulation enabling rapid viral capsid protein production. Biochem Biophys Res Commun. 2013;430:1047–53. https://doi.org/10.1016/J.BBRC.2012.12.042.
162. Catherine Silver S, Yoshizaki T, Pagano JS. The Epstein-Barr virus (EBV) SM protein enhances pre-mRNA processing of the EBV DNA polymerase transcript. J Virol. 1998;72:8485–92.
163. Collier B, Goobar-Larsson L, Sokolowski M, Schwartz S. Translational inhibition in vitro of human papillomavirus type 16 L2 mRNA mediated through interaction with heterogenous ribonucleoprotein K and poly(rC)-binding proteins 1 and 2. J Biol Chem. 1998;273:22648–56. https://doi.org/10.1074/jbc.273.35.22648.
164. Fuentes-Pananá EM, Peng R, Brewer G, Tan J, Ling PD. Regulation of the Epstein-Barr virus C promoter by AUF1 and the cyclic AMP/protein kinase A signaling pathway. J Virol. 2000;74:8166–75. https://doi.org/10.1128/jvi.74.17.8166-8175.2000.
165. Monette A, Ajamian L, López-Lastra M, Mouland AJ. Human immunodeficiency virus type 1 (HIV-1) induces the cytoplasmic retention of heterogeneous nuclear ribonucleoprotein A1 by disrupting nuclear import: implications for HIV-1 gene expression. J Biol Chem. 2009;284:31350–62. https://doi.org/10.1074/jbc.M109.048736.
166. Asai K, Platt C, Cochrane A. Control of HIV-1 env RNA splicing and transport: investigating the role of hnRNP A1 in exon splicing silencer (ESS3a) function. Virology. 2003;314:229–42. https://doi.org/10.1016/S0042-6822(03)00400-8.
167. Tange TO, Damgaard CK, Guth S, Valcárcel J, Kjems J. The hnRNP A1 protein regulates HIV-1 tat splicing via a novel intron silencer element. EMBO J. 2001;20:5748–58. https://doi.org/10.1093/emboj/20.20.5748.
168. Damgaard CK, Tange T s, Kjems J. hnRNP A1 controls HIV-1 mRNA splicing through cooperative binding to intron and exon splicing silencers in the context of a conserved secondary structure. RNA. 2002;8:1401–15.
169. Gordon H, et al. Depletion of hnRNP A2/B1 overrides the nuclear retention of the HIV-1 genomic RNA. RNA Biol. 2013;10:1714–25. https://doi.org/10.4161/rna.26542.
170. Suh D, et al. Mapping of determinants required for the function of the HIV-1 env nuclear retention sequence. J Virol. 2003;310:85–99. https://doi.org/10.1016/S0042-6822(03)00073-4.

Importance of Mitochondrial Quality Control in Parkinson's Disease: The Potential Interplay of Mitochondrial Unfolded Protein Response and Mitophagy

5

Yew Mun Lee, Dongxue Hu, and Yih-Cherng Liou

Abstract

A substantial body of empirical studies has highlighted that the maintenance of a "healthy" mitochondrial network is imperative for the mitigation of neurodegenerative diseases. This maintenance is especially important in the nervous system made up of many neurons that act as important communicators. Neurons are energy-hungry cells that require high amounts of energy to function. A large part of the energy is obtained from the mitochondria, viewed as power plants of mammalian cells. Indeed, mitochondrial dysfunction plays a key role in the pathogenesis of many neurodegenerative diseases such as Parkinson's disease (PD). To prevent this, mitochondria have evolved mitochondrial quality control mechanisms such as the mitochondrial unfolded protein response (UPR^{mt}) and mitophagy to prevent or remove "unhealthy" populations of mitochondria in neurons. Interestingly, numerous studies in the last few decades have shed light on the implications of these two pathways in PD, of which both processes seem to share a close relationship with each other. Henceforth, in this book chapter, we aim to accentuate the mechanisms of both UPR^{mt} and mitophagy in mammalian cells and identify the potential interplay which these two machineries might have in the prevention and potential therapeutic targets of PD.

Keywords

Parkinson's disease · Mitochondrial dynamics · Mitochondrial unfolded protein response (UPR^{mt}) · Mitophagy · Oxidative phosphorylation (OXPHOS) · Reactive oxygen species (ROS) · Parkin · PINK1 · Mitochondrial DNA (mtDNA) · Ubiquitin

Y. M. Lee · D. Hu · Y.-C. Liou (✉)
Department of Biological Sciences, Faculty of Science, National University of Singapore, Singapore, Singapore
e-mail: dbslyc@nus.edu.sg

C. Huang, Y. Zhang (eds.), *Oxidative Stress*,
https://doi.org/10.1007/978-981-16-0522-2_5

5.1 Introduction

A mammalian cell is made up of numerous organelles that work hand in hand to sustain its viability. Although the nucleus holds the most important genetic information that is crucial to the survival of the cell, another organelle is responsible for providing important energy in the form of adenosine triphosphate (ATP) through the process of oxidative phosphorylation (OXPHOS) to sustain its function. This "power-packed" organelle has the most peculiar origin in the mammalian cell, dating back more than 1.5 billion years ago. Then, the prokaryotic cells entered into the eukaryotic cells to become what is commonly known as the mitochondrion or, in plural, the mitochondria [1]. Despite being the power plant of a mammalian cell, its importance cannot be highlighted greater enough than in the brain. The brain is one of the highest energy-consuming organs of the human body, requiring as much as 20% of the basal oxygen consumption to fuel its function [2], of which neurons are estimated to consume as much as 80% of the total energy generated from the brain [3]. The energy produced from the mitochondria fuels crucial neuronal functions, such as the ATP-dependent neurotransmission [4–6]. Therefore, the role of mitochondria is critical for the neurological system to work.

Despite the importance of the mitochondria in sustaining the physiological function of the brain, they bring along an unwanted baggage that can be detrimental to the cell. The main driver of ATP production as highlighted earlier comes from the process of OXPHOS. This exact process would also lead to the generation of reactive oxygen species (ROS). Negatively, an increasing level of accumulated ROS would become a source of cellular toxicity. This observation has led to scientists coining the "free radical theory of aging" [7]. In physiology, besides being a by-product of OXPHOS during ATP production, ROS also have a role as signalling molecules in the brain. As such, ROS are involved in processes such as mounting an immune response, eliciting inflammatory processes, expressing synaptic plasticity, and also playing a role in learning and memory of the brain function [8, 9]. On the flip side, the excessive accumulation of ROS would lead to a buildup of oxidative stress that can damage the proteins and DNA. The accumulative damages incurred would eventually lead to the onset of cellular death as observed in the pathogenesis of many neurodegenerative diseases like Parkinson's disease (PD) [10, 11].

5.1.1 Reactive Oxygen Species

Antioxidative mechanisms are present to alleviate the deleterious accumulation of ROS in the mitochondria. However, there is a tipping point when even such mechanisms are being overwhelmed. This would lead to the onset of mitochondrial dysfunction that is observed in many diseases, including neurodegenerative diseases as well as aging [12]. Furthermore, even with numerous studies demonstrating the prominent role of ROS in the pathogenesis of PD, the use of antioxidants as a treatment for PD remains modest [13]. Therefore, there is a need to improve our understanding of the safeguarding of mitochondrial health beyond that of the ROS in order to better identify new therapeutic interventions for PD. In this chapter, we will emphasize on the importance of mitochondrial quality control that helps to prevent mitochondrial dysfunction in the brain. In particular, we intend to discuss how mitochondrial unfolded protein response (UPR^{mt}) and mitophagy play a responsible role in neurons, as well as the interplay between these two crucial factors. In addition, the content will also cover how the potential loss of mitochondrial quality control will lead to the pathogenesis of PD.

5.2 The Functional Importance of Mitochondria in the Neurons

The brain is a highly complex organ that consists of highly differentiated cells forming different regions of the brain to create a network of intricate signalling processes which governs the functions of the entire body. The major workhorses of the brain are neurons. They are specialized cells which differ in morphology based on their location and function in the brain and the rest of the body. As highlighted previously, neurons are energy-hungry cells, and they consume the majority of the energy produced in the brain to drive their functions. These include propagating electrical signals, maintaining the ionic balance in the cell to generate action potential, and releasing and capturing of neurotransmitters such as dopamine between neurons [14]. As glucose oxidation—which generates less ATP than lipids oxidation—is the major source of energy in the brain to reduce excessive production of ROS, the process of OXPHOS in the mitochondria is crucial for the energy needs of the neurons [15].

Even though mitochondria are found in most of the mammalian cells, a previous research by Reifschneider et al. [16] has shown that cells derived from different organs of the rat exhibit different mitochondrial protein composition, specifically the proteins involved in the process of OXPHOS. Of note, neurons are terminally differentiated cells; their life spans are almost parallel with that of the whole organism. Therefore, damaged neurons are rarely or never regenerated, and they seem to be more susceptible to the accumulation of defective mitochondria during aging [17, 18]. Thus, the mitochondrial quality control in neurons is paramount to their function and survival. In addition, the morphological and structural complexities of neurons such as the presence of synapses require the mitochondria to be distributed to these locations and produce energy required for neuronal function [19, 20]. Indeed, there are functional differences between the synaptic mitochondria and the non-synaptic mitochondria as observed in rats by Borras et al. [21, 22]. Furthermore, the effects of aging have a greater impact on synaptic mitochondria in terms of reduced respiratory rate and increased sensitivity to calcium insult as compared to non-synaptic mitochondria [23, 24]. Collectively, these negative effects highlight the maintenance of a healthy pool of mitochondria as the utmost crucial factor for synaptic neurons' function. The buildup of "unhealthy" mitochondria otherwise would lead to the onset of mitochondrial dysfunction, potentially causing age-related diseases such as PD. Indeed, mitochondrial dysfunction has been observed by Reeve et al. [25] within the synapses of *substantia nigra* (SN) neurons from PD patients. In the next section, we will discuss the role of mitochondrial dysfunction in the manifestation of PD.

5.3 Mitochondrial Dysfunction in Parkinson's Disease

PD is the second most common neurodegenerative disease after Alzheimer's disease (AD) [26]. The established hallmarks of PD include the loss of dopaminergic neurons in the midbrain *SN pars compacta* (SNpc) and the accumulation of α-synuclein (α-syn) containing Lewy bodies. After almost four decades of research, there is a substantial amount of evidence to support the notion that mitochondrial dysfunction is an important trigger for the development of PD (reviewed in [27]). The establishment of mitochondrial dysfunction not only contributes to the loss of OXPHOS and its calcium regulation, but more importantly, the disruption of the mitochondrial quality control processes such as the UPR^{mt} and mitophagy has been implicated to play a pivotal role in the manifestation of PD (Fig. 5.1).

Ever since the application of 1-methyl-4-phenyl-1,2,3,6-tetrahydropyridine (MPTP) that led to the inhibition of complex I (NADH/ubiquinone oxidoreductase) of the mitochondrial electron transport chain (ETC) and the induction of parkinsonism in humans [28], there are ever-increasing evidences that tie mitochondrial dysfunction to PD. More studies have identified the reduction of complex I activity in other areas of the brain, on top of the SNpc in sporadic PD patients [29–35]. These evidences clearly suggest that a perturbation to the normal function of the mitochondria could trigger parkinsonism, highlighting further the relationship between mitochondrial dysfunction and PD.

On top of the perturbation to the ETC, genetic alterations of the mitochondrial DNA (mtDNA) have been identified to contribute to the pathogenesis of PD. Multiple deletions of mtDNA have been observed from postmortem analyses of human brains from both aged people and idiopathic PD patients [36, 37]. The first empirical evidence that altered mtDNA could lead to the pathogenesis of PD came from genetic analysis on the mutation of mtDNA polymerase gamma gene (*POLG*). Mutations of *POLG* lead to the accumulation of deletions to multiple regions of the mtDNA in muscle cells, developing levodopa-responsive parkinsonism with severe SNpc dopaminergic neuronal loss [38, 39]. Interestingly, from the studies on *POLG* mutant mice yielding high levels of mtDNA deletions, the authors did not observe any increase of mitochondrial dysfunction nor degeneration to the SNpc dopaminergic neurons, which are the hallmarks of PD [40, 41]. However, perturbation to the mitochondrial transcription factor A (TFAM), which causes partial depletion of the mtDNA, can disrupt complex VI (cytochrome C oxidase) activity. This decrease in complex VI activity is reminiscence of a progressive parkinsonism phenotype [42]. Collectively, these observations further suggest that mitochondrial dysfunction from the loss of OXPHOS activity contributes greatly to the pathogenesis of PD.

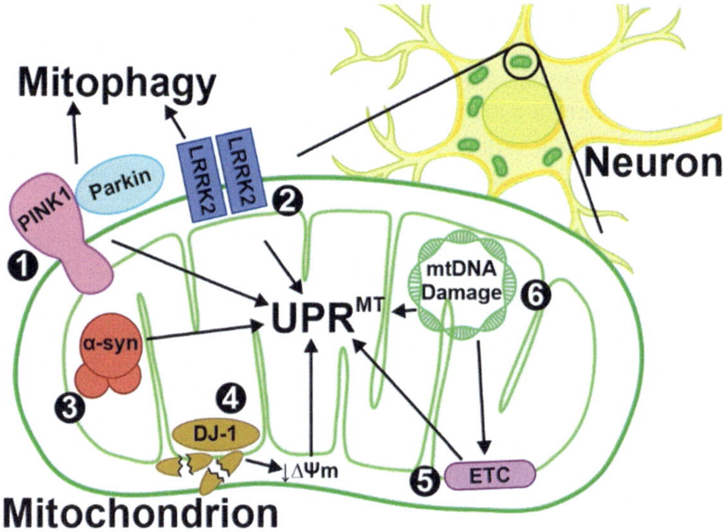

Fig. 5.1 Implications of mitochondrial dysfunction in Parkinson's disease (PD). (1) Phosphatase and tensin homolog (PTEN)-induced putative kinase 1 (PINK1)/parkin RBR E3 ubiquitin protein ligase (Parkin) pathway has been well documented in PD phenotype and has been demonstrated to be involved in both the mitochondrial unfolded protein response (UPR^{mt}) and mitophagy. (2) Leucine-rich repeat kinase 2 (LRRK2) perturbation has also been identified in PD patients and has been known to be involved in mitochondrial dynamics and mitophagy. (3) The accumulation of alpha-synuclein (α-syn) is one of the major hallmarks of PD, and its accumulation in the mitochondria has been suggested to cause mitochondrial dysfunction. (4) The reduced expression of DJ-1 could lead to mitochondrial dysfunction in the form of reduced mitochondrial potential (ΔΨm). (5) Perturbation to the electron transport chain (ETC), particularly complex I, has been demonstrated to induce parkinsonism in humans. (6) The damage and subsequent deletions or mutations to the mitochondrial DNA (mtDNA) have been observed to contribute to mitochondrial dysfunction and lead to the potential pathogenesis of PD. All listed processes have been known to trigger the UPR^{mt} process

Indeed, efforts for identifying important genes have significantly contributed to the understanding of the familial forms of PD. Proteins encoded by these genes include α-syn, leucine-rich repeat kinase 2 (LRRK2), PTEN-induced putative kinase 1 (PINK1), parkin RBR E3 ubiquitin protein ligase (Parkin), DJ-1, and vacuolar protein sorting 35 (VPS35). They play important roles in the function of mitochondria (Fig. 5.1). In addition to the well-documented functions of α-syn and the Pink1/Parkin pathway in PD, mutant LRRK2 has also been identified as a potential pathogenic target of PD as it can disrupt mitochondrial mobility and affect the initiation of mitophagy [43]. Furthermore, the downregulation of DJ-1 was also identified to increase mitochondrial fragmentation and reduce the mitochondrial membrane potential (ΔΨm). These detrimental effects could be mitigated by the overexpression of PINK1 and Parkin [44–46]. The mutation of VPS35 was also found to induce mitochondrial dysfunction by the impairment to the mitochondrial dynamic process [47]. The vital physiological functions of these proteins highly associate with the mitochondrial dysfunction and pathogenesis of PD. They could contribute to important mitochondrial quality control processes such as the UPR^{mt}, mitophagy, and mitochondrial dynamics (reviewed in [27]). These mechanisms are involved in maintaining a pool of "healthy" mitochondria. In the next few sections, we will discuss how the loss of these mechanisms through the age of time could lead to the onset of PD.

5.4 The Mitochondrial Unfolded Protein Response (UPR^{mt})

The process of UPR^{mt}, controlled by the communication between the mitochondria and the nucleus, is a transcriptionally regulated mechanism activated by many types of mitochondrial dysfunction [48–56]. The importance of UPR^{mt} is signified during the decline in mitochondrial function, where it is activated to promote repair and recovery of mitochondrial function from disruptions caused by various insults such as toxins, pathogens [51, 52, 57], and mutations to the respiratory genes [49, 54] in order to maintain cellular functions [49, 51, 54, 58].

The existence of UPR^{mt} was discovered almost 20 years ago by Zhao et al. [56]. The authors overexpressed the mutant form of the mitochondrial matrix protein, ornithine transcarbamylase (OCT), in fibroblasts named COS-7 cells derived from monkey kidney tissue. The mutant OCT was observed to irreversibly misfold and aggregate in the mitochondria, leading to an increased accumulation of nuclear transcripts that encode for several mitochondrial chaperones and proteases. These changes were suggested by the authors for the cell to balance out the amount of protein aggregation (proteostasis) in the mitochondria. Due to its similarity to the UPR mechanism in the endoplasmic reticulum (ER), it was thus coined as UPR^{mt}. During this decade or so, the progress of understanding UPR^{mt} continued mainly in the *Caenorhabditis elegans* model [59]. In recent times, scientists working on UPR^{mt} shifted to the use of mammalian model systems, gaining new knowledge that positioned UPR^{mt} as more than just regulating the proteostasis of the mitochondria [59]. In short, a growing number of evidences have suggested that UPR^{mt} is central in detecting mitochondrial dysfunction and communicates to the nucleus for initiating potential mitochondrial recovery mechanisms.

5.4.1 The UPR^{mt} During Mitochondrial Dysfunction

As highlighted earlier, the onset of mitochondrial dysfunction is triggered by a broad array of mitochondrial stress responses that include the accumulation of misfolded and aggregated proteins in the mitochondrial matrix and the perturbation of mitochondrial OXPHOS activity. These mitochondrial stresses were found to be able to activate UPR^{mt} (Fig. 5.2). Majority of the studies in elucidating the mechanisms of UPR^{mt} were conducted in *C. elegans* (reviewed in [59]). In the later sections of this chapter, we will focus our attention on the current knowledge of UPR^{mt} in the mammalian system.

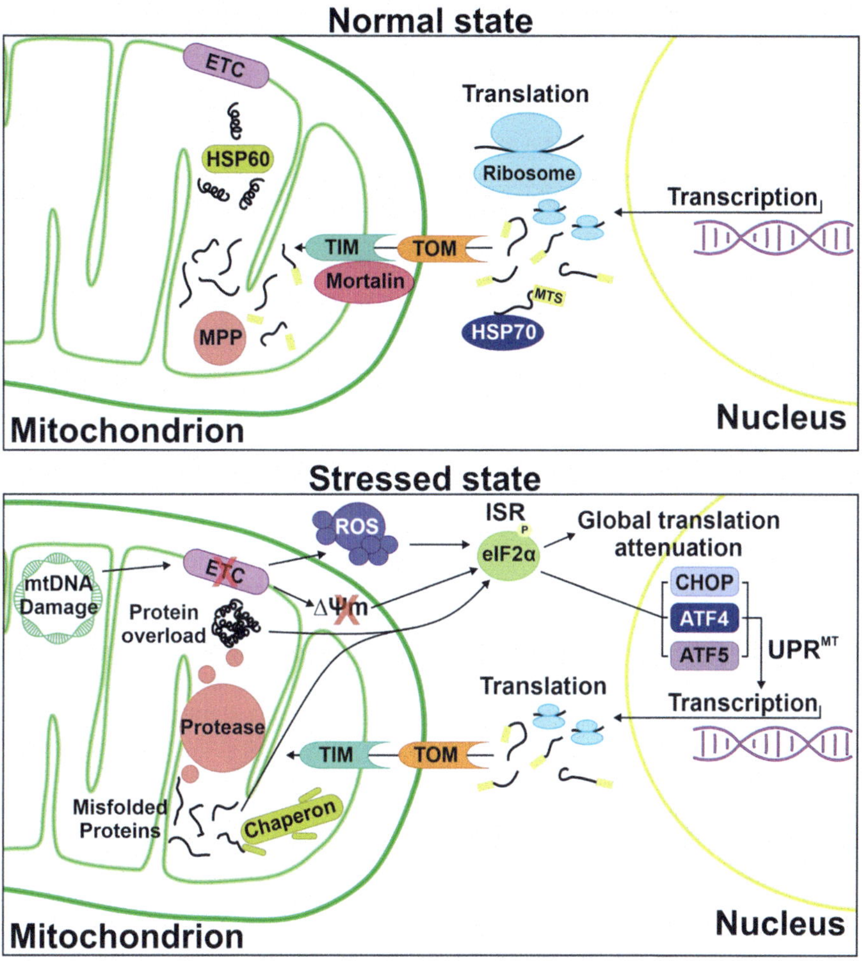

Fig. 5.2 The process of mitochondrial unfolded protein response (UPR^{mt}) and its function during mitochondrial stress. During the normal state of the mitochondria, important protein precursors are synthesized from the nucleolar DNA and imported into the mitochondrial matrix via the translocase of the outer membrane (TOM) and translocase of the inner membrane (TIM) complexes. Each of the protein precursors has a mitochondria-targeting sequence (MTS) that is required for their import. The import process is aided by chaperone proteins such as heat shock protein 70 (HSP70). Mitochondrial chaperone proteins such as mortalin located in the TIM complex are also on hand to help the import of the protein precursors. Once inside the mitochondrial matrix, the protein precursors are processed by first removing the MTS via the protease mitochondrial processing peptidase (MPP) to become mature polypeptides. Subsequently, the mature polypeptides are folded into their proper structure by protein chaperons such as HSP60. Conversely, at the stressed state, mitochondria can be stressed by the presence of mitochondrial DNA (mtDNA) damage, perturbation of the electron transport chain (ETC) leading to reduced membrane potential ($\Delta\Psi m$) and the increase in reactive oxygen species (ROS) production, and the accumulation of excessive mitochondrial proteins or misfolded proteins. The damage to the mtDNA could lead to translation of non-functional proteins of the ETC and cause the ETC to malfunction. The subsequent generation of ROS would lead to damaged proteins and misfolded proteins that would accumulate. In addition, the loss of the $\Delta\Psi m$ will also block the import of the protein precursors. This stressed state would lead to the activation of the integrated stress response (ISR) pathway by the phosphorylation of the

5.4.1.1 The Relationship Between the Mitochondrial Protein Import and UPR

As mentioned previously, many preceding studies of UPR^{mt} have been done using the *C. elegans* model. The most important finding from these studies is that the mitochondrial protein import is a key UPR^{mt} regulatory event [54] and most of the triggers of UPR^{mt} are able to influence the mitochondrial protein import [59]. Of note, the mammalian mitochondria each consists of a mitochondrial genome that encodes for a total of 13 proteins. They are made up of subunits of the respiratory chain and the ATP synthase and the components of the mitochondrial translation system: the ribosomal RNAs and transfer RNAs [60]. As the number of proteins encoded in the mitochondria is small, most of the proteins required for a functional and healthy mitochondrion are transcribed in the nucleus and synthesized in the cytosol, before they are imported into the mitochondrion (Fig. 5.2). However, instead of the fully folded protein, protein precursors that remain unfolded are imported. The association of these protein precursors with ATP-dependent molecular chaperones, heat shock protein 70 (HSP70) and 90 (HSP90), prevents their degradation and aggregation in the cytosol [61]. The protein precursors are subsequently directed and imported into the mitochondria with their N-terminal mitochondria-targeting sequence (MTS) [62]. Once imported, the MTS on the protein precursors is cleaved in the mitochondrial matrix by the mitochondrial processing peptidase (MPP) to allow the precursors to become mature polypeptides.

The import of these protein precursors into the mitochondrial matrix, where most of the mitochondrial proteins are located, involves passing through the two main mitochondrial translocase complexes. First is the translocase of the outer membrane (TOM) complex consists of seven components: Tom5, Tom6, Tom7, Tom20, Tom22, Tom40, and Tom70 [63, 64]. Second is the translocase of the inner membrane (TIM23) complex consists of Tim17, Tim21, Tim23, Tim44, Tim50, Tim14/Pam18, Tim16/Pam16, Pam17, mitochondrial HSP70 (mtHSP70), and Mgel1 [65]. In addition, the import through the TIM23 complex requires an intact mitochondrial $\Delta\Psi m$ along with the hydrolysis of ATP.

Improper mitochondrial protein import, misfolding, and mutations of the mtDNA respiratory genes can affect the proteostasis in the mitochondria, resulting in mitochondrial dysfunction [51]. Henceforth, protein chaperones that avoid protein misfolding or aggregation are crucial for the mitochondrial protein import of the protein precursors (Fig. 5.2) [66]. The importing process is governed by a few important chaperons. They include HSP70 and its interactor HSP60, mitochondrial HSP70 (mtHSP70; also known as mortalin), HSP10, and tumor necrosis factor receptor-associated protein 1 (TRAP1) [67–70]. On top of ensuring proper protein

---

Fig. 5.2 (continued) eukaryotic translation initiation factor 2 subunit 1 (eIF2α). Moreover, the stress response pathway could trigger the inhibition of global protein translation and activate three transcription factors: C/EBP homologous protein (CHOP), activating transcription factor 4 (ATF4), and ATF5. These transcription factors will increase the transcription of important protein chaperons or proteases to help mitigate the mitochondrial stress and prevent mitochondrial dysfunction

folding, the system of removing unfolded or irreversibly misfolded proteins is equally important in maintaining mitochondrial proteostasis. Players in maintaining mitochondrial proteostasis include ATP-dependent proteases iAAA (intermembrane space facing) and mAAA (matrix facing), Lon protease (LONP1), Clp protease proteolytic subunit (CLPP), high-temperature requirement protein A2 (HTRA2), and ATP23 [71–75].

The system of protein chaperons and proteases ensures that the accumulation of misfolded proteins in the mitochondrial matrix and mutations in the mtDNA does not lead to mitochondrial dysfunction. Therefore, the triggering of the UPR^{mt} would lead to the transcriptional activation to produce these chaperons and proteases to mitigate potential mitochondrial dysfunction. Despite the similarity in the transcriptional response of the activation of UPR^{mt} in both *C. elegans* and mammalian models, the regulation of UPR^{mt} is more complicated in the mammalian cells as the research continues to understand this process better.

5.4.1.2 The Regulation of UPR^{mt} in Mammalian Cells

The UPR^{mt} regulator, the activating transcription factor associated with stress-1 (ATFS-1), found in *C. elegans*, belongs to the family of bZIP transcription factor [50, 54]. Similarly, research over the last few years in mammalian models of UPR^{mt} has yielded the identification of three basic leucine zipper (bZIP) transcription factors in C/EBP homologous protein (CHOP), activating transcription factor (ATF4), and ATF5. These transcription factors are associated with the integrated stress response (Fig. 5.2; ISR) [76–79]. The expression of the three transcription factors is governed by the phosphorylation of the eukaryotic translation initiation factor 2 subunit 1 (eIF2α). The phosphorylation of eIF2α is initiated by four kinases as part of the integrated stress response (ISR) pathway originating from a diverse range of cellular stresses such as the ER stress [78].

Studies over the years have revealed that mitochondrial stress of various types is able to induce the expression of CHOP, ATF4, and ATF5, including the genes that are involved in the UPR^{mt} [80–86]. Particularly, studies have demonstrated that CHOP, ATF4, and ATF5 are all required to regulate and induce the expression of UPR^{mt} genes during mitochondrial dysfunction [80, 81, 84, 86]. Although CHOP, ATF4, and ATF5 are involved in mitigating mitochondrial dysfunction, it remains unclear how the three transcription factors together with the ISR pathway are regulated during the whole process of this mitochondrial recovery.

Of interest, among the three transcription factors, ATF5 has been suggested by Florese et al. [84] to be an ortholog to the *C. elegans* ATFS-1. The authors demonstrated that ATF5 could rescue the activity of UPR^{mt} in ATFS-1-deficient *C. elegans*. From the same group, they further highlighted that ATF5 itself is able to respond directly to mitochondrial stress under the mechanism of the mitochondrial import regulation, much like the *C. elegans* ATFS-1. They further demonstrated that ATF5 is required to activate the transcription of several mitochondrial chaperone and protease genes in times of mitochondrial stress. In addition, studies have also demonstrated that ATF4 is able to protect mammalian cells against mitochondrial stress [86, 87] via the expression of ISR metabolic genes [86, 88]. Despite the findings, the coordination among CHOP, ATF4, and ATF5 in the regulation of UPR^{mt} during mitochondrial dysfunction remains unknown. Interestingly, only ATF5 contains the MTS like ATFS-1, suggesting that ATF5 could be the main transcription factor regulating the transcriptional activation of the UPR^{mt} genes. As hypothesized by Shpilka and Haynes [59], CHOP, ATF4, and a phosphorylated eIF2α might induce the expression of ATF5 since the transcription of ATF5 was found to be dependent on the activity of CHOP and ATF4 [77, 89–91]. Much remains to be studied to fully understand the regulation of the UPR^{mt} in mammalian cells.

5.5 Mitophagy

Mitophagy is a mitochondria-specific autophagic process that was first observed under the electron microscope in the 1960s, showing the mitochondria with various extents of degradation in lysosomes [92, 93]. However, it was not until 2005 that Lemasters [94] proposed the term "mitophagy" to describe the selective elimination of mitochondria. Mitophagy mediates the removal of non-functional mitochondria to maintain a pool of "healthy" mitochondria and has been associated with physiological and pathological processes [95]. Mitophagy is involved in various cellular events, such as hypoxic stress, reticulocyte maturation, overt mitochondrial damage, and post-fertilization removal of paternal mtDNA [96]. Furthermore, mitophagy defects have been known to impede mitochondrial function and could lead to the accumulation of abnormal mitochondria, causing impairments to cells and tissues [97].

Upon initiation of mitophagy, isolation membranes (phagophores) are recruited to mitochondria through interactions between receptor proteins and microtubule-associated protein 1 light chain 3 (LC3) anchored on phagophores. The recruited phagophores then expand and encapsulate the targeted mitochondrion, leading to the formation of mitophagosome. The mitophagosome is subsequently fused with a lysosome for the degradation of mitochondria [98]. Multiple mechanisms are involved in the regulation of targeted elimination of mitochondria, and they are largely categorized into two major types: ubiquitin-dependent and ubiquitin-independent mitophagy (Fig. 5.3) [99].

5.5.1 The Mechanism of Ubiquitin-Dependent Mitophagy

PINK1/Parkin-mediated mitophagy is the most well-investigated and the best-described mitophagy pathway [98, 100]. PINK1 contains an MTS, a putative transmembrane (TM) helix, an N-terminal linker region (NT linker), a Ser/Thr kinase domain, and a conserved C-terminal extension (CTE) [101, 102]. Under a healthy physiology state, PINK1 is transported into the mitochondria via its MTS [103] and then cleaved by four proteases: MPP, ClpXP, presenilin-associated rhomboid-like (PARL) protease, and m-AAA [104, 105]. The cleavage yields a 52 kDa

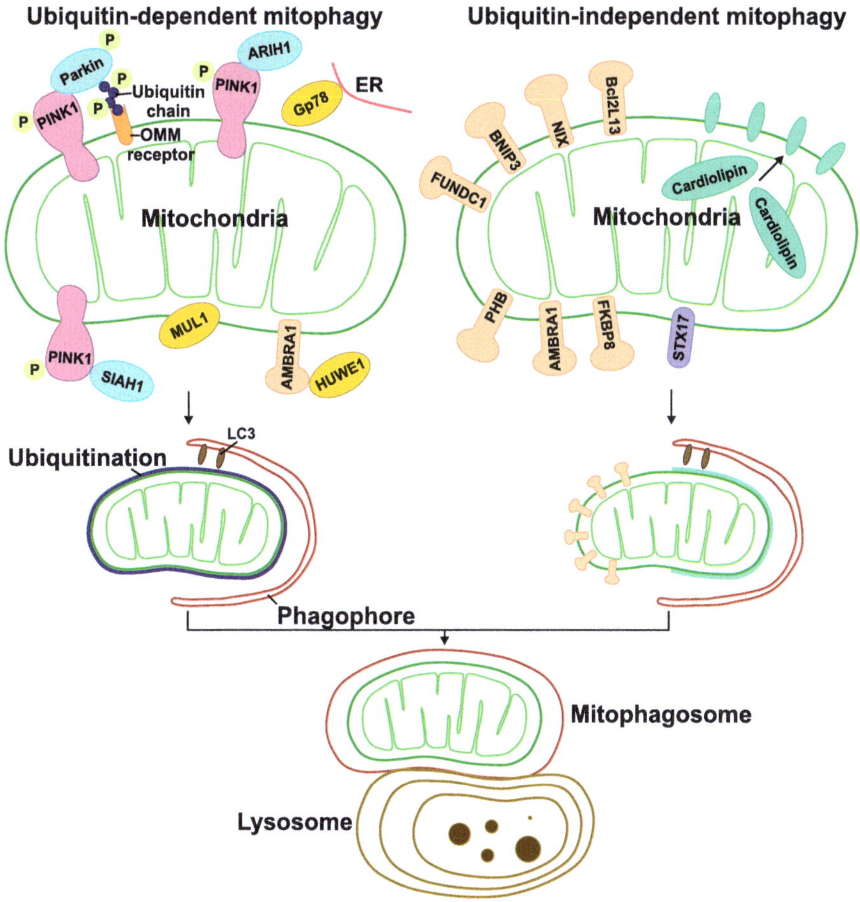

Fig. 5.3 The process of mitophagy. The two main classes of mitophagy lie in how they recruit the phagophore to trigger the process of mitophagy (digestion of the mitochondria). These two main classes are either ubiquitin-dependent or ubiquitin-independent. For the ubiquitin-dependent mitophagy, the main feature is that the outer mitochondrial membrane (OMM) is covered with phospho-ubiquitin chains for phagophore recruitment. The most prominent of this process is initiated by the PTEN-induced putative kinase 1 (PINK1)/parkin RBR E3 ubiquitin protein ligase (Parkin) pathway. The stabilization of PINK1 to the OMM allows it to be autophosphorylated and then phosphorylates ubiquitin. This process allows Parkin to bind to the phospho-ubiquitin and to PINK1 for its phosphorylation and activation. The phosphorylated Parkin will then recruit more phospho-ubiquitin to the OMM proteins to form polyubiquitin chains. Other E3 ubiquitin ligases, such as ariadne RBR E3 Ub protein ligase 1 (ARIH1) and synphilin-1 recruited seven in absentia homolog 1 (SIAH1), have also been identified for this process. Moreover, other E3 ubiquitin ligases that do not require PINK1 to ubiquitinate the OMM have been identified, including mitochondrial ubiquitin ligase 1 (MUL1), WWE domain-containing protein ligase 1 (HUWE1), and glycoprotein 78 (Gp78). HUWE1 works with activating molecule in BECN1-regulated autophagy protein 1 (AMBRA1), an OMM receptor, while Gp78 is localized in the endoplasmic reticulum (ER) membrane. For ubiquitin-independent mitophagy, it mainly functions with OMM receptors that can directly recruit the phagophore, such as Bcl2-like protein 13 (Bcl2L13), Nip3-like protein X (NIX), BCL2-interacting protein 3 (BNIP3), FUN14 domain-containing 1 (FUNDC1), prohibitins (PHB), AMBRA1, and FK506-binding protein 8 (FKBP8). Recently, syntaxin 17 (STX17), not an OMM receptor, has also been identified to be able to initiate mitophagy via a

product, which is retro-translocated into the cytoplasm and further degraded by the proteasome via the N-end rule pathway after being ubiquitinated [106, 107]. These processes account for the low levels of PINK1 under basal conditions.

Upon ΔΨm dissipation, PINK1 is stabilized on the outer mitochondrial membrane (OMM) [108, 109], whereby it is activated by autophosphorylation [110]. Subsequently, PINK1 phosphorylates the E3 ligase Parkin and the neighboring ubiquitin chains located on the OMM. The phosphorylated ubiquitin chains recruit and activate the phosphorylated Parkin. The phosphorylation of both Parkin and ubiquitin chains contributes to the activation of Parkin [111–117]. On the OMM, Parkin next ubiquitinates itself along with other OMM proteins, including mitofusin (Mfn), voltage-dependent anion-selective channel protein 1 (VDAC1), and mitochondrial Rho GTPase (Miro) [112, 118–120]. The increased level of ubiquitin on the OMM would generate a feedforward loop for Parkin recruitment, thereby amplifying the mitophagy signals (Fig. 5.3) [121].

Recently, our group examined the role of ROS as a promoting factor for the process of mitophagy in PD [122]. The idea of ROS as a trigger remains controversial as previous studies have shown that the application of ROS scavengers did not suppress the carbonyl cyanide m-chlorophenylhydrazone (CCCP)-induced Parkin recruitment using HeLa cells [123–125]. We hypothesized that this failure is due to the inadequacy of ROS scavenging, causing the levels of ROS to remain high and initiate mitophagy. Indeed, using a combination of N-acetyl-L-cysteine (NAC) and catalase, we inhibited the increase of ROS levels after CCCP treatment and eventually block the PINK1-dependent Parkin translocation to the mitochondria needed to initiate mitophagy. Moreover, the overexpression of VDAC1 also led to an increase in Parkin translocation to the mitochondria, suggesting that the increased ROS production might have contributed to this phenomenon. Indeed, treatment with NAC and catalase are able to block the Parkin translocation induced by VDAC1 overexpression. Moreover, our recent study also showed that the loss of a mitochondrial fission protein, mitochondrial elongation factor 1 (MIEF1), also known as mitochondrial dynamic protein of 51 kDa (MiD51), can increase mitochondrial ROS levels [126]. The increase in ROS levels can induce PINK1/Parkin-mediated mitophagy, which can be blocked by adding NAC to MIEF1 knockdown (KD) HeLa cells. Taken together, we highlighted that ROS may contribute as a trigger for PINK1/Parkin-dependent mitophagy. Since ROS are also potential triggers for UPR^{mt}, it will be of interest in the future to investigate if this process of ROS-induced mitophagy might be downstream of UPR^{mt} when the ROS level becomes too much for UPR^{mt} to protect the mitochondria

Fig. 5.3 (continued) microtubule-associated protein 1 light chain 3 (LC3)-independent pathway. In addition, cardiolipin, a type of diphosphatidylglycerol lipid in the mitochondria, is also able to recruit the phagophore. Once the phagophore is recruited via interactions with LC3 and surrounds the mitochondrion, a mitophagosome is formed. The mitophagosome will then fuse to a lysosome to degrade the mitochondria and complete the mitophagy process

In addition to Parkin, several alternative ubiquitin E3 ligases, such as mitochondrial ubiquitin ligase 1 (MUL1) [127], synphilin-1 recruited seven in absentia homolog 1 (SIAH1) [128], ariadne RBR E3 Ub protein ligase 1 (ARIH1) [129], glycoprotein 78 (Gp78) [130], and WWE domain-containing protein ligase 1 (HUWE1) [131], have been identified. HUWE1 is recruited to the mitochondria from the cytosol by its cofactor activating molecule in BECN1-regulated autophagy protein 1 (AMBRA1) and works collaboratively to mediate mitophagy [131]. Unique among these E3 ligases, Gp78 is localized on the ER membrane, where it initiates mitophagy at the ER-mitochondria-associated area [130]. Similar with Parkin, these E3 ligases generate polyubiquitin chains on OMM (Fig. 5.3). The polyubiquitin chains can recruit autophagic receptors onto the OMM, which would then connect with the phagophores to damaged mitochondria for cargo engulfment, followed by autophagic degradation. Optineurin (OPTN) [132, 133], calcium-binding and coiled-coil domain-containing protein 2 (NDP52) [133], p62 [118], Tax1-binding protein 1 (TAX1BP1) [133], and next to BRCA1 gene 1 protein (NBR1) [96, 134, 135] are currently known receptors that have been linked to ubiquitin-dependent mitophagy in mammalian cells.

5.5.2 The Mechanism of Ubiquitin-Independent Mitophagy

In the other type of mitophagy, certain OMM receptors or lipid could directly bridge the mitochondria and autophagy machinery to mediate mitophagy, thus activating what is known as the ubiquitin-independent mitophagy [97]. These receptors or lipid includes FUN14 domain-containing 1 (FUNDC1), prohibitins (PHB), BCL2-interacting protein 3 (BNIP3), Nip3-like protein X (NIX), Bcl2-like protein 13 (Bcl2L13), FK506–binding protein 8 (FKBP8), AMBRA1, and the OMM phospholipid cardiolipin (Fig. 5.3) [136, 137]. These proteins mediate mitophagy in response to various cellular stimuli, such as mitochondrial depolarization (e.g., cardiolipin, FKBP8), hypoxia (e.g., NIX, FUNDC1, BNIP3), energetic stress (e.g., PHB), and developmental signals during differentiation (e.g., NIX) [96, 136, 137]. Interestingly, we recently found that the depletion of mitochondrial fission 1 protein (Fis1) leads to aberrant syntaxin 17 (STX17) accumulation on the mitochondria. The translocation of STX17 leads to the interaction with autophagy-related 14 (ATG14) and recruitment of downstream autophagic proteins to induce mitophagy in a PINK1/Parkin-independent manner [138]. These results suggest that STX17 might be a new receptor for mitophagy. Unlike the other receptors mentioned above, STX17 initiates mitophagy by recruiting autophagic proteins other than LC3.

5.5.3 Mitochondrial Dynamics and Mitophagy

Mitochondria are highly dynamic organelles, which continuously undergo the process of fusion and fission. Fusion leads to the formation of interconnected networks, which enables content exchange between mitochondria to maintain the mitochondrial genome and proteome integrity. On the other hand, fission results in the fragmentation of mitochondria and can facilitate the elimination of dysfunctional components [139, 140]. Therefore, mitochondrial dynamics are closely associated with mitophagy. In mammals, several dynamin-related GTPases mediate mitochondrial fusion and fission: mitofusin 1/2 (Mfn1/2), optic atrophy 1 (OPA1), dynamin-related protein 1 (Drp1), fission 1 protein (Fis1), MiD49/51 (MIEF1), and mitochondrial fusion factor (Mff) [141].

Interestingly, we previously found that the loss of death-associated protein 3 (DAP3) can significantly reduce the phosphorylation of Drp1 at Ser637 to promote mitochondrial fission [142]. Physiologically, DAP3 is a mitochondrial ribosomal protein located in the mitochondrial matrix and was initially thought to be solely involved in apoptosis. Depletion of DAP3 leads to defects in the mitochondria-encoded protein synthesis along with the abnormal imbalance of the mitochondrial fission and fusion processes due to the increased fission rate as mentioned above. Furthermore, we observed the inhibition of autophagy by the loss of DAP3. Collectively, our findings provided new insights into the potential interplay between mitochondria-encoded protein synthesis, mitochondrial dynamics, and autophagy. Since the perturbation of mitochondria-encoded protein synthesis could affect mitochondrial proteostasis as mentioned previously, depletion of DAP3 could have triggered UPR^{mt}, further suggesting the link between UPR^{mt} and mitophagy (selective autophagy) via mitochondrial dynamics. In addition, manipulation of the mitochondrial dynamic proteins can also be an indirect method to regulate the mitophagic process. Several lines of evidence have already demonstrated such possibility [143–147]. For example, D'Amico et al. [147] enhanced the process of mitophagy via the upregulation of Mff in both old nematodes and mice. And intriguingly, this intervention brought significant benefits to maintain the pool of "healthy" mitochondria during aging.

5.6 The Relationship of UPR^{mt} and Mitophagy

The processes of UPR^{mt} and mitophagy share a close relationship in that they are both activated by similar mitochondrial stress mechanisms like misfolded proteins [148], mtDNA mutations [53, 149, 150], perturbation to the OXPHOS activity [51], and the decreased levels of the mitochondrial chaperon mortalin (mtHSP70) [151]. Both the activation of UPR^{mt} and mitophagy are dependent on the perturbation to the mitochondrial import. This dependence of the mitochondrial import in both mitochondrial quality control processes suggests that a mitochondrion senses its ability to import proteins as a way to gauge its "health" status. As highlighted in a review by Pickles et al. [96], the authors suggest that if one considers the mitochondria as a large network in a mammalian cell, the global operation of UPR^{mt} would drive the import of chaperons and proteases to improve the general health of the mitochondria. In parallel, those import-compromised mitochondria would be degraded via mitophagy. As for mitochondria with diminished import function, they could potentially be rescued by the activation of UPR^{mt} to avoid mitophagy. Nevertheless, it remains to be further investigated how these two mechanisms are co-regulated. In this section, we will highlight key events that could be considered important triggers of PD in association with both UPR^{mt} and mitophagy.

5.6.1 The Role of UPR^{mt} in Parkinson's Disease

The association of UPR^{mt} and PD was identified in a link between the accumulation of α-syn in the mitochondria and the reduction in mitochondrial protein import. A study by Devi et al. [152] identified a possible MTS within the human α-syn protein sequence, suggesting the potential import of α-syn into the mitochondrial matrix. Furthermore, the authors also observed the increased amount of α-syn in the mitochondrial fractions obtained from the postmortem SNpc region of PD patients. The accumulation of α-syn in the mitochondria could lead to its interaction and inhibition of the function of complex I, resulting in the induction of UPR^{mt} (Fig. 5.4) [51, 55]. Moreover, Ludtmann et al. [153] also demonstrated that α-syn could interact with complex V (ATP synthase) of the mitochondria, leading to changes in the mitochondrial morphology. The interaction between α-syn and complex V will eventually cause cell death by the opening of the mitochondrial permeability transition pore. In addition, a study by Di Maio et al. [154] revealed that posttranslationally modified species of α-syn are able to impair the mitochondrial protein import system, subsequently causing mitochondrial dysfunction. The impairment of the mitochondrial protein import system is mainly caused by the disruption to Tom20-Tom22 interaction due to the binding of α-syn to the MTS receptor site in Tom20 (Fig. 5.4). The interaction of α-syn to the MTS receptor site in Tom20 is further strengthened by the identification of the MTS in α-syn protein sequence highlighted previously. However, more research would need to further understand how α-syn could influence UPR^{mt} and lead to the onset of PD.

Further association of PD to UPR^{mt} lies in the substantial decrease in levels of mortalin observed from the analysis of mitochondria derived from the SN region of PD patients as compared to age-matched healthy patients [155]. A number of studies have identified genetic variations to the gene encoding mortalin in a Spanish and German cohort, although it is only a handful of patients harboring these mortalin variants with pathogenesis to PD [156, 157]. In the German cohort study, Burbulla et al. [157] went on to identify a A476T variant of mortalin as a potential risk factor for PD, where its overexpression would lead to increased ROS levels caused by the induction of proteolytic stress. Moreover, the authors demonstrated that the mitochondrial dysfunction induced by the KD of mortalin could only be rescued by wild-type mortalin, but not PD-associated variants of the mortalin, in human embryonic kidney 293 (HEK293) cells. Furthermore, it was shown that these mortalin variants in yeast could also cause mitochondrial dysfunction, particularly perturbations to mitochondrial proteostasis [158]. This association of mutant mortalin and mitochondrial dysfunction further suggests that mitochondrial dysfunction is a risk factor for

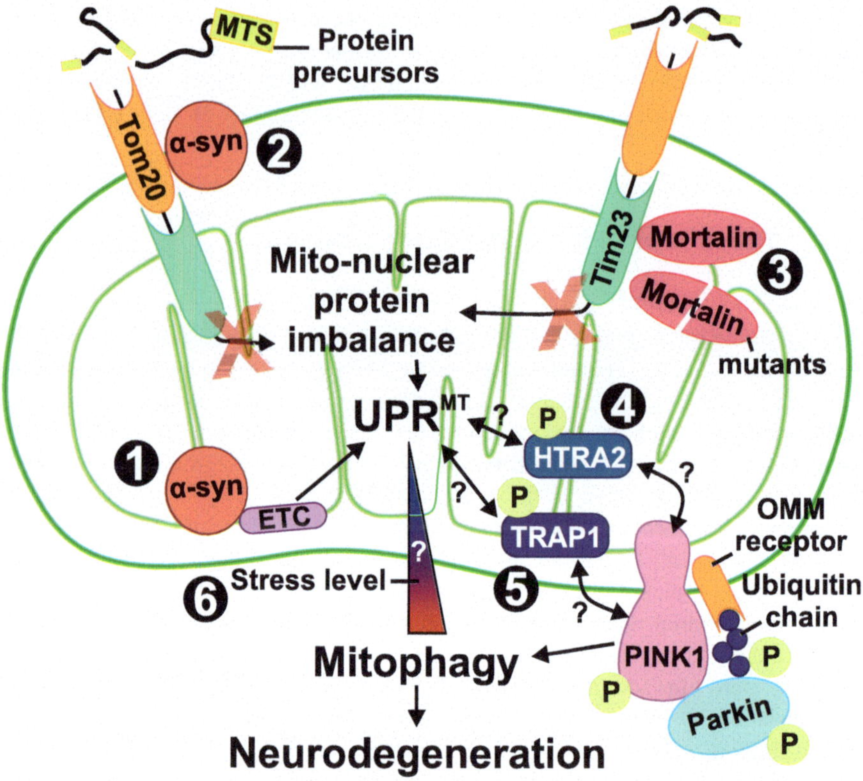

Fig. 5.4 The implications of mitochondrial unfolded protein response (UPR^{mt}) and mitophagy in Parkinson's disease (PD). (1) Alpha-synuclein (α-syn), one of the hallmarks of PD, can bind to components of the electron transport chain (ETC). The binding will perturb the ETC function and induce UPR^{mt}. (2) α-Syn can interact with the translocase of the outer membrane 20 (Tom20) and affect the mitochondrial protein import. The perturbation of the mitochondrial protein import can induce a mito-nuclear protein imbalance and trigger the UPR^{mt}. (3) The mutant mortalin, identified from PD patients, is not able to bind to the translocase of the inner membrane 23 (Tim23) as a chaperon to aid in the import of protein precursors. The absence of mortalin binding to Tim23 can affect mitochondrial import and imbalance on the mito-nuclear protein to trigger UPR^{mt}. (4) High-temperature requirement protein A2 (HTRA2), a protease in the mitochondria, can associate with the protein PINK1. The mutation to HTRA2 can lead to similar mitochondrial dysfunction as PINK1/Parkin deficiency. (5) Tumor necrosis factor receptor-associated protein 1 (TRAP1), another chaperon, can also associate similarly with PINK1/Parkin as with HTRA2. However, the exact mechanism of both TRAP1 and HTRA2 association with PINK1/Parkin and their roles in UPR^{mt} in PD patients remain unknown. (6) It has been suggested that UPR^{mt} is an initial stress response mechanism to protect the function of the mitochondria. However, if the perturbation to the mitochondria reaches a critical point beyond repair, mitophagy will be triggered to remove the "unhealthy" mitochondria and prevent widespread mitochondrial dysfunction. Subsequently, if there are extensive mitochondrial dysfunction throughout the neuron, neuronal death might be triggered, leading to neurodegeneration. MTS: mitochondrial target sequence; OMM: outer mitochondrial membrane

PD (Fig. 5.4). Mortalin has also been found to interact with DJ-1 [157, 159], a protein associated with PD, α-syn [159], and Parkin [160]. Interestingly, studies have shown that mitochondrial dysfunction induced from the depletion of mortalin can be rescued by the overexpression of PINK1 and Parkin [151, 161]. Despite all these findings, the number of PD patients harboring the genetic variants of mortalin remains low [162, 163], suggesting that mortalin might not be a main contributor to the pathogenesis of PD.

The chaperon TRAP1, another member of UPR^{mt}, was found to associate with PINK1 in vivo, potentially linking UPR^{mt} to PD. To this end, Costa et al. [164] generated a TRAP1-deficient *Drosophila* model and identified similar features of mitochondrial dysfunction in PINK1- and Parkin-deficient *Drosophila.* The increased expression of PINK1 and Parkin in the neurons was able to rescue the mitochondrial dysfunction by TRAP1 deficiency. However, TRAP1 expression is only able to partially rescue mitochondrial dysfunction in Parkin mutant flies. Conversely, Zhang et al. [165] showed that TRAP1 is able to rescue the mitochondrial dysfunction in SH-SY5Y neuroblastoma cells after depletion of PINK1, but did not show any effect on the mitochondrial dysfunction induced by depletion of Parkin. Data from these studies seem to point out that TRAP1 has a role in the PINK1-Parkin pathway that seeks to maintain the population of "healthy" mitochondria, and PINK1 alterations affect TRAP1 function [166].

Similar to TRAP1, the phosphorylation of UPR^{mt} serine protease HTRA2 was found to increase in a PINK1-dependent manner [167]. In addition, Tain et al. [168] used a *Drosophila* model and demonstrated that HTRA2 mutant flies have similar locomotor phenotypes that correlated with mitochondrial dysfunction as compared to that in both the PINK1 and Parkin mutant flies. Moreover, HTRA2 expression is able to rescue locomotor deficits induced by PINK1 deficiency. In contrast, Whitworth et al. [169] failed to repeat the rescue effect of HTRA2 in Parkin mutant flies. Collectively, these studies identified the potential association of TRAP1-, HTRA2-, PINK1-, and Parkin-associated stress pathways (Fig. 5.4). It would be interesting to investigate how exactly the UPR^{mt} process could be affected in PD. Moreover, since the PINK1/Parkin pathway is also implicated in the process of mitophagy, the research summarized above also suggests the need to further understand the potential reciprocal regulation of UPR^{mt} and mitophagy in maintaining the health of the mitochondria.

5.6.2 The Role of Mitophagy in Parkinson's Disease

Defects in mitophagy have been identified in many neurodegenerative diseases, including AD, Huntington's disease (HD), amyotrophic lateral sclerosis (ALS), frontotemporal dementia (FTD), and also PD, making the aberrant accumulation of damaged mitochondria a hallmark of major neurological pathologies [98, 136]. Since the accumulation of damaged mitochondria is associated with the development of neuronal health, understanding the pathological relationship between mitophagy defects and neurodegenerative disease is of great importance. Unfortunately, the mechanism of how mitophagy plays a role in PD remains elusive.

As with UPR^{mt}, two PD-associated proteins, PINK1 and Parkin, have been strongly linked to damaged mitochondria tagged for degradation via mitophagy. There have been increasing evidences of a decline in mitophagy during aging, leading to a decline in the clearance of dysfunctional mitochondria and the accumulation of dysfunctional mitochondria during the development of neurodegenerative diseases [170–173]. As highlighted previously, the accumulation of α-syn in PD can make SN neurons vulnerable with the increase in mitochondrial dysfunction. Therefore, the decline in mitophagy in PD would reduce the clearance of "unhealthy" mitochondria, eventually reaching a critical point that would trigger neuronal cell death [174]. Very recently, Moskal et al. designed a high-throughput method to identify small molecules that increase Parkin recruitment to damaged mitochondria upon mitophagy induction. The approach led to the discovery of several rho-associated, coiled-coil-containing protein kinase (ROCK) inhibitors, which promote Parkin-mediated mitophagy via recruiting hexokinase 2 (HK2) to mitochondria. Among them, the molecule SR3677 has been demonstrated to improve the longevity and climbing ability of PD flies challenged with parkinsonian toxin paraquat, suggesting its neuroprotective effects [175].

As highlighted previously, VPS35 deficiency in dopamine (DA) neurons could lead to impaired mitochondrial dynamics. In physiology, VPS35 promotes the degradation of MUL1 which leads to the stabilization of MFN2. The deficiency of VPS35 affects the regulation of the mitochondrial dynamics, leading to mitochondrial dysfunction, potentially underlying the pathogenesis of PD [47]. MUL1 was also reported to be involved in the induction of mitophagy, but its importance in the mitophagy in PD remains to be investigated.

Indeed, the importance of mitophagy in neurons has been highlighted in several studies of mitophagy inducers which have been shown to have promising effects in prolonging health span and protecting neurons in animal models [137, 176–182]. For example, urolithin A (UA) and antibiotic actinonin (AC) are two natural compounds that have been reported to induce mitophagy in *C. elegans* and mouse models [176–178]. Treating either UA or AC on AD mouse or *C. elegans* model showed restored neuronal mitophagy and ameliorated cognitive decline. Of note, UA and AC can elevate the level of mitophagy-related proteins, including PINK1, Parkin, beclin-1, Bcl2L13, AMBRA1, and phospho-unc-51 like autophagy activating kinase 1 (ULK1), indicating a robust induction of mitophagy [178]. UA is a metabolite derived from ellagitannins. Interestingly, in a rotenone-induced PD rat model, the treatment of ellagitannin-rich pomegranate extract was reported to prevent PD-like features and enhance the activity of mitochondrial aldehyde dehydrogenase 2 (ALDH2) in the midbrain homogenate, which can protect cells against oxidative stress. UA was also detected in the brain and plasma after the intake of pomegranate juice [179]. Of note, a recent phase I clinical trial showed UA have a favorable safety profile and induced a molecular signature of improved mitochondrial health [180]. The above results suggest UA is a promising compound as a potential PD therapy. In addition to UA and AC, other mitophagy inducers such as

NAD^+ precursors, including nicotinamide, nicotinamide mononucleotide, and nicotinamide riboside, have also been reported to bring benefits to ameliorate PD phenotypes [137, 178, 181, 182]. The underling mechanism has been discussed by Fang [183].

Collectively, mitophagy has been shown to be an important process in the protection against PD. However, the underlying processes that lead to the loss of mitophagy in PD remain elusive; thus, more studies may help better understand the role of mitophagy in the onset of PD. In addition, further exploration into more mitophagy inducers could lead to potential therapies for PD patients.

5.7 Conclusion and Future Perspectives

Mitochondria are important organelles in mammalian cells. The presence of mitochondrial dysfunction as highlighted in this chapter has been implicated in many diseases including neurodegenerative diseases like PD. Mitochondria have also been associated with the mitochondrial free radical theory of aging. Thus, ensuring that the population of mitochondria remains healthy is an important aspect for mammalian cells to function and survive. Indeed, as summarized in this chapter, we have discussed about the intricate and complex mito-protective pathways that have been evolutionarily placed in the mitochondria for the safekeeping of its function. Of which the UPR^{mt} and mitophagy are present to response to the varying degrees of stress the mitochondria are undergoing. These mitochondrial quality control processes allow the mammalian cells to take precautionary measures to ensure that the accumulation of "unhealthy" mitochondria does not occur. A perturbation to any of these mitochondrial quality control processes would have grave implications such as the triggering of apoptotic death for the cell.

Therefore, it is not a surprise that the loss of these mitochondrial quality control processes would lead to undesirable consequences, such as the onset of neurodegenerative diseases. However, we are still far from understanding how the mitochondrial quality control processes, such as UPR^{mt} and mitophagy, could have contributed to PD. Herein, we have encapsulated that UPR^{mt} and mitophagy do share similar processes such as the PINK/Parkin pathway. However, whether both UPR^{mt} and mitophagy have any crosstalk with each other or how they could co-regulate still requires further investigations. Indeed, despite the decades of research being conducted in understanding the pathogenesis of PD, our increased understanding of PD did not translate to any viable therapeutic approaches. Even when understanding that the end point of PD is the induction of neuronal death, the use of anti-apoptotic drugs is unsuccessful in clinical trials against neurodegenerative diseases [184]. Henceforth, it seems that the use of neuroprotective therapy to slow down or prevent PD is a better alternative. Nevertheless, we are still far from understanding how the process of mitochondrial quality control could be administrated as a neuroprotective treatment.

In this chapter, we have highlighted that the processes of mitophagy and UPR^{mt} are perceived to be closely linked. It has been suggested that the UPR^{mt} is the immediate repair mechanism for the cell to attempt to mitigate the low stress-induced perturbations of the mitochondria, while mitophagy is considered as the later stage of "unhealthy" mitochondrial removal to prevent the onset of cellular death (Fig. 5.4; reviewed in [185]). Therefore, elucidating the co-regulation of these two pathways could improve our knowledge on the mechanism of mitochondrial stress sensors. It could also determine how the neurons track the health status of the mitochondria to determine its fate and to prevent the potential triggering of cell death. With the improved details in the early developments of PD, potential therapeutic targets can be identified.

References

1. Sagan L. On the origin of mitosing cells. J Theor Biol. 1967;14(3):255–74. https://doi.org/10.1016/0022-5193(67)90079-3.
2. Aiello LC, Wheeler P. The expensive-tissue hypothesis: the brain and the digestive system in human and primate evolution. Curr Anthropol. 1995;36(2):199–221.
3. Hyder F, Rothman DL, Bennett MR. Cortical energy demands of signaling and nonsignaling components in brain are conserved across mammalian species and activity levels. Proc Natl Acad Sci. 2013;110(9):3549. https://doi.org/10.1073/pnas.1214912110.
4. Sheng ZH, Cai Q. Mitochondrial transport in neurons: impact on synaptic homeostasis and neurodegeneration. Nat Rev Neurosci. 2012;13(2):77–93. https://doi.org/10.1038/nrn3156.
5. Cai Q, Tammineni P. Mitochondrial aspects of synaptic dysfunction in Alzheimer's disease. J Alzheimers Dis. 2017;57(4):1087–103. https://doi.org/10.3233/jad-160726.
6. Devine MJ, Kittler JT. Mitochondria at the neuronal presynapse in health and disease. Nat Rev Neurosci. 2018;19(2):63–80. https://doi.org/10.1038/nrn.2017.170.
7. Harman D. Aging: a theory based on free radical and radiation chemistry. J Gerontol. 1956;11(3):298–300. https://doi.org/10.1093/geronj/11.3.298.
8. Kishida KT, Klann E. Sources and targets of reactive oxygen species in synaptic plasticity and memory. Antioxid Redox Signal. 2007;9(2):233–44. https://doi.org/10.1089/ars.2007.9.ft-8.
9. Veal EA, Day AM, Morgan BA. Hydrogen peroxide sensing and signaling. Mol Cell. 2007;26(1):1–14. https://doi.org/10.1016/j.molcel.2007.03.016.
10. Casetta I, Govoni V, Granieri E. Oxidative stress, antioxidants and neurodegenerative diseases. Curr Pharm Des. 2005;11(16):2033–52.
11. Weng M, Xie X, Liu C, Lim KL, Zhang CW, Li L. The sources of reactive oxygen species and its possible role in the pathogenesis of Parkinson's disease. Parkinsons Dis. 2018;2018:9163040. https://doi.org/10.1155/2018/9163040.
12. Wu Y, Chen M, Jiang J. Mitochondrial dysfunction in neurodegenerative diseases and drug targets via apoptotic signaling. Mitochondrion. 2019;49:35–45. https://doi.org/10.1016/j.mito.2019.07.003.
13. Filograna R, Beltramini M, Bubacco L, Bisaglia M. Anti-oxidants in Parkinson's disease therapy: a critical point of view. Curr Neuropharmacol. 2016;14(3):260–71. https://doi.org/10.2174/1570159x13666151030102718.
14. Lodish H, Berk A, Zipursky SL, Matsudaira P, Baltimore D, Darnell J. Overview of neuron structure and function. Molecular cell biology. 4th ed. New York: W. H. Freeman; 2000.
15. Schönfeld P, Reiser G. Why does brain metabolism not favor burning of fatty acids to provide energy? Reflections on disadvantages of the use of free fatty acids as fuel for brain. J Cereb Blood Flow Metab. 2013;33(10):1493–9. https://doi.org/10.1038/jcbfm.2013.128.
16. Reifschneider NH, Goto S, Nakamoto H, Takahashi R, Sugawa M, Dencher NA, et al. Defining the mitochondrial proteomes from five rat organs in a physiologically significant

context using 2D blue-native/SDS-PAGE. J Proteome Res. 2006;5(5):1117–32. https://doi.org/10.1021/pr0504440.
17. Kowald A, Kirkwood TB. Accumulation of defective mitochondria through delayed degradation of damaged organelles and its possible role in the ageing of post-mitotic and dividing cells. J Theor Biol. 2000;202(2):145–60. https://doi.org/10.1006/jtbi.1999.1046.
18. Terman A, Kurz T, Navratil M, Arriaga EA, Brunk UT. Mitochondrial turnover and aging of long-lived postmitotic cells: the mitochondrial-lysosomal axis theory of aging. Antioxid Redox Signal. 2010;12(4):503–35. https://doi.org/10.1089/ars.2009.2598.
19. Obashi K, Okabe S. Regulation of mitochondrial dynamics and distribution by synapse position and neuronal activity in the axon. Eur J Neurosci. 2013;38(3):2350–63. https://doi.org/10.1111/ejn.12263.
20. Lin MY, Sheng ZH. Regulation of mitochondrial transport in neurons. Exp Cell Res. 2015;334(1):35–44. https://doi.org/10.1016/j.yexcr.2015.01.004.
21. Borrás C, Sastre J, García-Sala D, Lloret A, Pallardó FV, Viña J. Mitochondria from females exhibit higher antioxidant gene expression and lower oxidative damage than males. Free Radic Biol Med. 2003;34(5):546–52. https://doi.org/10.1016/s0891-5849(02)01356-4.
22. Borrás C, Gambini J, López-Grueso R, Pallardó FV, Viña J. Direct antioxidant and protective effect of estradiol on isolated mitochondria. Biochim Biophys Acta. 2010;1802(1):205–11. https://doi.org/10.1016/j.bbadis.2009.09.007.
23. Lores-Arnaiz S, Bustamante J. Age-related alterations in mitochondrial physiological parameters and nitric oxide production in synaptic and non-synaptic brain cortex mitochondria. Neuroscience. 2011;188:117–24. https://doi.org/10.1016/j.neuroscience.2011.04.060.
24. Lores-Arnaiz S, Lombardi P, Karadayian AG, Orgambide F, Cicerchia D, Bustamante J. Brain cortex mitochondrial bioenergetics in synaptosomes and non-synaptic mitochondria during aging. Neurochem Res. 2016;41(1-2):353–63. https://doi.org/10.1007/s11064-015-1817-5.
25. Reeve AK, Grady JP, Cosgrave EM, Bennison E, Chen C, Hepplewhite PD, et al. Mitochondrial dysfunction within the synapses of substantia nigra neurons in Parkinson's disease. Parkinson's Dis. 2018;4(1):9. https://doi.org/10.1038/s41531-018-0044-6.
26. Lebouvier T, Chaumette T, Paillusson S, Duyckaerts C, Bruley d, Varannes S, Neunlist M, et al. The second brain and Parkinson's disease. Eur J Neurosci. 2009;30(5):735–41. https://doi.org/10.1111/j.1460-9568.2009.06873.x.
27. Chen C, Turnbull DM, Reeve AK. Mitochondrial dysfunction in Parkinson's disease-cause or consequence? Biology. 2019;8(2):38. https://doi.org/10.3390/biology8020038.
28. Langston JW, Ballard P, Tetrud JW, Irwin I. Chronic Parkinsonism in humans due to a product of meperidine-analog synthesis. Science. 1983;219(4587):979–80.
29. Schapira A, Cooper J, Dexter D, Clark J, Jenner P, Marsden C. Mitochondrial complex I deficiency in Parkinson's disease. J Neurochem. 1990;54(3):823–7.
30. Parker WD Jr, Boyson SJ, Parks JK. Abnormalities of the electron transport chain in idiopathic Parkinson's disease. Ann Neurol. 1989;26(6):719–23.
31. Parker WD Jr, Parks JK, Swerdlow RH. Complex I deficiency in Parkinson's disease frontal cortex. Brain Res. 2008;1189:215–8.
32. Krige D, Carroll MT, Cooper JM, Marsden CD, Schapira AH. RKQPDR. Platelet mitochondria function in Parkinson's disease. Ann Neurol. 1992;32(6):782–8.
33. Blin O, Desnuelle C, Rascol O, Borg M, Saint Paul HP, Azulay J, et al. Mitochondrial respiratory failure in skeletal muscle from patients with Parkinson's disease and multiple system atrophy. J Neurol Sci. 1994;125(1):95–101.
34. Haas RH, Nasirian F, Nakano K, Ward D, Pay M, Hill R, et al. Low platelet mitochondrial complex I and complex II/III activity in early untreated Parkinson's disease. Ann Neurol. 1995;37(6):714–22.
35. Swerdlow RH, Parks JK, Miller SW, Davis RE, Tuttle JB, Trimmer PA, et al. Origin and functional consequences of the complex I defect in Parkinson's disease. Ann Neurol. 1996;40(4):663–71.

36. Bender A, Krishnan KJ, Morris CM, Taylor GA, Reeve AK, Perry RH, et al. High levels of mitochondrial DNA deletions in substantia nigra neurons in aging and Parkinson disease. Nat Genet. 2006;38(5):515–7.
37. Kraytsberg Y, Kudryavtseva E, McKee AC, Geula C, Kowall NW, Khrapko K. Mitochondrial DNA deletions are abundant and cause functional impairment in aged human substantia nigra neurons. Nat Genet. 2006;38(5):518–20.
38. Luoma P, Melberg A, Rinne JO, Kaukonen JA, Nupponen NN, Chalmers RM, et al. Parkinsonism, premature menopause, and mitochondrial DNA polymerase γ mutations: clinical and molecular genetic study. Lancet. 2004;364(9437):875–82.
39. Davidzon G, Greene P, Mancuso M, Klos KJ, Ahlskog JE, Hirano M, et al. Early-onset familial parkinsonism due to POLG mutations. Ann Neurol. 2006;59(5):859–62.
40. Perier C, Bender A, García-Arumí E, Melia MJ, Bove J, Laub C, et al. Accumulation of mitochondrial DNA deletions within dopaminergic neurons triggers neuroprotective mechanisms. Brain. 2013;136(8):2369–78.
41. Dai Y, Clark J, Zheng K, Kujoth GC, Prolla TA, Simon DK. Somatic mitochondrial DNA mutations do not increase neuronal vulnerability to MPTP in young POLG mutator mice. Neurotoxicol Teratol. 2014;46:62–7.
42. Ekstrand MI, Terzioglu M, Galter D, Zhu S, Hofstetter C, Lindqvist E, et al. Progressive parkinsonism in mice with respiratory-chain-deficient dopamine neurons. Proc Natl Acad Sci. 2007;104(4):1325–30.
43. Hsieh CH, Shaltouki A, Gonzalez AE, Bettencourt da Cruz A, Burbulla LF, St Lawrence E, et al. Functional impairment in miro degradation and mitophagy is a shared feature in familial and sporadic Parkinson's disease. Cell Stem Cell. 2016;19(6):709–24. https://doi.org/10.1016/j.stem.2016.08.002.
44. Thomas KJ, McCoy MK, Blackinton J, Beilina A, van der Brug M, Sandebring A, et al. DJ-1 acts in parallel to the PINK1/parkin pathway to control mitochondrial function and autophagy. Hum Mol Genet. 2011;20(1):40–50. https://doi.org/10.1093/hmg/ddq430.
45. Wang X, Petrie TG, Liu Y, Liu J, Fujioka H, Zhu X. Parkinson's disease-associated DJ-1 mutations impair mitochondrial dynamics and cause mitochondrial dysfunction. J Neurochem. 2012;121(5):830–9. https://doi.org/10.1111/j.1471-4159.2012.07734.x.
46. McCoy MK, Cookson MR. DJ-1 regulation of mitochondrial function and autophagy through oxidative stress. Autophagy. 2011;7(5):531–2. https://doi.org/10.4161/auto.7.5.14684.
47. Tang FL, Liu W, Hu JX, Erion JR, Ye J, Mei L, et al. VPS35 deficiency or mutation causes dopaminergic neuronal loss by impairing mitochondrial fusion and function. Cell Rep. 2015;12(10):1631–43. https://doi.org/10.1016/j.celrep.2015.08.001.
48. Schulz AM, Haynes CM. UPRmt-mediated cytoprotection and organismal aging. Biochim Biophys Acta. 2015;1847(11):1448–56.
49. Durieux J, Wolff S, Dillin A. The cell-non-autonomous nature of electron transport chain-mediated longevity. Cell. 2011;144(1):79–91.
50. Haynes CM, Yang Y, Blais SP, Neubert TA, Ron D. The matrix peptide exporter HAF-1 signals a mitochondrial UPR by activating the transcription factor ZC376. 7 in C. elegans. Mol Cell. 2010;37(4):529–40.
51. Houtkooper RH, Mouchiroud L, Ryu D, Moullan N, Katsyuba E, Knott G, et al. Mitonuclear protein imbalance as a conserved longevity mechanism. Nature. 2013;497(7450):451–7.
52. Liu Y, Samuel BS, Breen PC, Ruvkun G. Caenorhabditis elegans pathways that surveil and defend mitochondria. Nature. 2014;508(7496):406–10.
53. Martinus RD, Garth GP, Webster TL, Cartwright P, Naylor DJ, Høj PB, et al. Selective induction of mitochondrial chaperones in response to loss of the mitochondrial genome. Eur J Biochem. 1996;240(1):98–103.
54. Nargund AM, Pellegrino MW, Fiorese CJ, Baker BM, Haynes CM. Mitochondrial import efficiency of ATFS-1 regulates mitochondrial UPR activation. Science. 2012;337(6094):587–90.

55. Yoneda T, Benedetti C, Urano F, Clark SG, Harding HP, Ron D. Compartment-specific perturbation of protein handling activates genes encoding mitochondrial chaperones. J Cell Sci. 2004;117(18):4055–66.
56. Zhao Q, Wang J, Levichkin IV, Stasinopoulos S, Ryan MT, Hoogenraad NJ. A mitochondrial specific stress response in mammalian cells. EMBO J. 2002;21(17):4411–9.
57. Pellegrino MW, Nargund AM, Kirienko NV, Gillis R, Fiorese CJ, Haynes CM. Mitochondrial UPR-regulated innate immunity provides resistance to pathogen infection. Nature. 2014;516 (7531):414–7.
58. Lin Y-F, Haynes CM. Metabolism and the UPRmt. Mol Cell. 2016;61(5):677–82.
59. Shpilka T, Haynes CM. The mitochondrial UPR: mechanisms, physiological functions and implications in ageing. Nat Rev Mol Cell Biol. 2018;19(2):109–20. https://doi.org/10.1038/nrm.2017.110.
60. Anderson S, Bankier AT, Barrell BG, de Bruijn MH, Coulson AR, Drouin J, et al. Sequence and organization of the human mitochondrial genome. Nature. 1981;290(5806):457–65. https://doi.org/10.1038/290457a0.
61. Young JC, Hoogenraad NJ, Hartl FU. Molecular chaperones Hsp90 and Hsp70 deliver preproteins to the mitochondrial import receptor Tom70. Cell. 2003;112(1):41–50.
62. Pfanner N. Protein sorting: recognizing mitochondrial presequences. Curr Biol. 2000;10(11): R412–5.
63. Model K, Meisinger C, Kühlbrandt W. Cryo-electron microscopy structure of a yeast mitochondrial preprotein translocase. J Mol Biol. 2008;383(5):1049–57.
64. Bausewein T, Mills DJ, Langer JD, Nitschke B, Nussberger S, Kühlbrandt W. Cryo-EM structure of the TOM core complex from Neurospora crassa. Cell. 2017;170(4):693–700.
65. Mokranjac D, Neupert W. The many faces of the mitochondrial TIM23 complex. Biochim Biophys Acta. 2010;1797(6-7):1045–54.
66. Kim YE, Hipp MS, Bracher A, Hayer-Hartl M, Ulrich Hartl F. Molecular chaperone functions in protein folding and proteostasis. Annu Rev Biochem. 2013;82:323–55.
67. Deocaris CC, Kaul SC, Wadhwa R. On the brotherhood of the mitochondrial chaperones mortalin and heat shock protein 60. Cell Stress Chaperones. 2006;11(2):116.
68. Walter S. Structure and function of the GroE chaperone. Cell Mol Life Sci. 2002;59 (10):1589–97.
69. Okamoto T, Yamamoto H, Kudo I, Matsumoto K, Odaka M, Grave E, et al. HSP60 possesses a GTPase activity and mediates protein folding with HSP10. Sci Rep. 2017;7(1):1–11.
70. Song HY, Dunbar JD, Zhang YX, Guo D, Donner DB. Identification of a protein with homology to hsp90 that binds the type 1 tumor necrosis factor receptor. J Biol Chem. 1995;270(8):3574–81.
71. Gerdes F, Tatsuta T, Langer T. Mitochondrial AAA proteases—towards a molecular understanding of membrane-bound proteolytic machines. Cell Res. 2012;1823(1):49–55.
72. Haynes CM, Petrova K, Benedetti C, Yang Y, Ron D. ClpP mediates activation of a mitochondrial unfolded protein response in C. elegans. Dev Cell. 2007;13(4):467–80.
73. Quirós PM, Español Y, Acín-Pérez R, Rodríguez F, Bárcena C, Watanabe K, et al. ATP-dependent Lon protease controls tumor bioenergetics by reprogramming mitochondrial activity. Cell Rep. 2014;8(2):542–56.
74. Clausen T, Kaiser M, Huber R, Ehrmann M. HTRA proteases: regulated proteolysis in protein quality control. Nat Rev Mol Cell Biol. 2011;12(3):152–62.
75. Quirós PM, Langer T, Lopez-Otin C. New roles for mitochondrial proteases in health, ageing and disease. Nat Rev. 2015;16(6):345–59.
76. Harding HP, Zhang Y, Zeng H, Novoa I, Lu PD, Calfon M, et al. An integrated stress response regulates amino acid metabolism and resistance to oxidative stress. Mol Cell. 2003;11 (3):619–33.
77. Teske BF, Fusakio ME, Zhou D, Shan J, McClintick JN, Kilberg MS, et al. CHOP induces activating transcription factor 5 (ATF5) to trigger apoptosis in response to perturbations in protein homeostasis. Mol Biol Cell. 2013;24(15):2477–90.

78. Pakos-Zebrucka K, Koryga I, Mnich K, Ljujic M, Samali A, Gorman AM. The integrated stress response. EMBO Rep. 2016;17(10):1374–95.
79. Dey S, Baird TD, Zhou D, Palam LR, Spandau DF, Wek RC. Both transcriptional regulation and translational control of ATF4 are central to the integrated stress response. J Biol Chem. 2010;285(43):33165–74.
80. Kim KH, Jeong YT, Oh H, Kim SH, Cho JM, Kim Y-N, et al. Autophagy deficiency leads to protection from obesity and insulin resistance by inducing Fgf21 as a mitokine. Nat Med. 2013;19(1):83.
81. Chung HK, Ryu D, Kim KS, Chang JY, Kim YK, Yi H-S, et al. Growth differentiation factor 15 is a myomitokine governing systemic energy homeostasis. J Cell Biol. 2017;216 (1):149–65.
82. Münch C, Harper JW. Mitochondrial unfolded protein response controls matrix pre-RNA processing and translation. Nature. 2016;534(7609):710–3.
83. Aldridge JE, Horibe T, Hoogenraad NJ. Discovery of genes activated by the mitochondrial unfolded protein response (mtUPR) and cognate promoter elements. PLoS One. 2007;2:9.
84. Fiorese CJ, Schulz AM, Lin Y-F, Rosin N, Pellegrino MW, Haynes CM. The transcription factor ATF5 mediates a mammalian mitochondrial UPR. Curr Biol. 2016;26(15):2037–43.
85. Horibe T, Hoogenraad NJ. The chop gene contains an element for the positive regulation of the mitochondrial unfolded protein response. PLoS One. 2007;2:9.
86. Quirós PM, Prado MA, Zamboni N, D'Amico D, Williams RW, Finley D, et al. Multi-omics analysis identifies ATF4 as a key regulator of the mitochondrial stress response in mammals. J Cell Biol. 2017;216(7):2027–45.
87. Bao XR, Ong S-E, Goldberger O, Peng J, Sharma R, Thompson DA, et al. Mitochondrial dysfunction remodels one-carbon metabolism in human cells. elife. 2016;5:e10575.
88. Mishra P, Chan DC. Mitochondrial dynamics and inheritance during cell division, development and disease. Nat Rev. 2014;15(10):634–46.
89. Zhou D, Palam LR, Jiang L, Narasimhan J, Staschke KA, Wek RC. Phosphorylation of eIF2 directs ATF5 translational control in response to diverse stress conditions. J Biol Chem. 2008;283(11):7064–73.
90. Fusakio ME, Willy JA, Wang Y, Mirek ET, Al Baghdadi RJ, Adams CM, et al. Transcription factor ATF4 directs basal and stress-induced gene expression in the unfolded protein response and cholesterol metabolism in the liver. Mol Biol Cell. 2016;27(9):1536–51.
91. Marciniak SJ, Yun CY, Oyadomari S, Novoa I, Zhang Y, Jungreis R, et al. CHOP induces death by promoting protein synthesis and oxidation in the stressed endoplasmic reticulum. Genes Dev. 2004;18(24):3066–77.
92. Ashford TP, Porter KR. Cytoplasmic components in hepatic cell lysosomes. J Cell Biol. 1962;12(1):198–202. https://doi.org/10.1083/jcb.12.1.198.
93. De Duve C, Wattiaux R. Functions of lysosomes. Annu Rev Physiol. 1966;28:435–92.
94. Lemasters JJ. Selective mitochondrial autophagy, or mitophagy, as a targeted defense against oxidative stress, mitochondrial dysfunction, and aging. Rejuvenation Res. 2005;8(1):3–5.
95. Babbar M, Basu S, Yang B, Croteau DL, Bohr VA. Mitophagy and DNA damage signaling in human aging. Mech Ageing Dev. 2020;186:111207. https://doi.org/10.1016/j.mad.2020.111207.
96. Pickles S, Vigié P, Youle RJ. Mitophagy and quality control mechanisms in mitochondrial maintenance. Curr Biol. 2018;28(4):R170–R85. https://doi.org/10.1016/j.cub.2018.01.004.
97. Palikaras K, Lionaki E, Tavernarakis N. Mechanisms of mitophagy in cellular homeostasis, physiology and pathology. Nat Cell Biol. 2018;20(9):1013–22. https://doi.org/10.1038/s41556-018-0176-2.
98. Cai Q, Jeong YY. Mitophagy in Alzheimer's disease and other age-related neurodegenerative diseases. Cell. 2020;9(1):150. https://doi.org/10.3390/cells9010150.
99. Dombi E, Mortiboys H, Poulton J. Modulating mitophagy in mitochondrial disease. Curr Med Chem. 2018;25(40):5597–612. https://doi.org/10.2174/0929867324666170616101741.

100. Evans CS, Holzbaur ELF. Quality control in neurons: mitophagy and other selective autophagy mechanisms. J Mol Biol. 2020;432(1):240–60. https://doi.org/10.1016/j.jmb.2019.06.031.
101. Rasool S, Trempe J-F. New insights into the structure of PINK1 and the mechanism of ubiquitin phosphorylation. Crit Rev Biochem Mol Biol. 2018;53(5):515–34. https://doi.org/10.1080/10409238.2018.1491525.
102. Schubert AF, Gladkova C, Pardon E, Wagstaff JL, Freund SMV, Steyaert J, et al. Structure of PINK1 in complex with its substrate ubiquitin. Nature. 2017;552(7683):51–6. https://doi.org/10.1038/nature24645.
103. Sekine S. PINK1 import regulation at a crossroad of mitochondrial fate: the molecular mechanisms of PINK1 import. J Biochem. 2019; https://doi.org/10.1093/jb/mvz069.
104. Jin SM, Lazarou M, Wang C, Kane LA, Narendra DP, Youle RJ. Mitochondrial membrane potential regulates PINK1 import and proteolytic destabilization by PARL. J Cell Biol. 2010;191(5):933–42. https://doi.org/10.1083/jcb.201008084.
105. Greene AW, Grenier K, Aguileta MA, Muise S, Farazifard R, Haque ME, et al. Mitochondrial processing peptidase regulates PINK1 processing, import and Parkin recruitment. EMBO Rep. 2012;13(4):378–85. https://doi.org/10.1038/embor.2012.14.
106. Sekine S, Youle RJ. PINK1 import regulation; a fine system to convey mitochondrial stress to the cytosol. BMC Biol. 2018;16(1):2. https://doi.org/10.1186/s12915-017-0470-7.
107. Yamano K, Youle RJ. PINK1 is degraded through the N-end rule pathway. Autophagy. 2013;9(11):1758–69. https://doi.org/10.4161/auto.24633.
108. Matsuda N, Sato S, Shiba K, Okatsu K, Saisho K, Gautier CA, et al. PINK1 stabilized by mitochondrial depolarization recruits Parkin to damaged mitochondria and activates latent Parkin for mitophagy. J Cell Biol. 2010;189(2):211–21. https://doi.org/10.1083/jcb.200910140.
109. Narendra DP, Jin SM, Tanaka A, Suen D-F, Gautier CA, Shen J, et al. PINK1 is selectively stabilized on impaired mitochondria to activate Parkin. PLoS Biol. 2010;8(1):e1000298. https://doi.org/10.1371/journal.pbio.1000298.
110. Rasool S, Soya N, Truong L, Croteau N, Lukacs GL, Trempe J-F. PINK1 autophosphorylation is required for ubiquitin recognition. EMBO Rep. 2018;19:4. https://doi.org/10.15252/embr.201744981.
111. Shiba-Fukushima K, Imai Y, Yoshida S, Ishihama Y, Kanao T, Sato S, et al. PINK1-mediated phosphorylation of the Parkin ubiquitin-like domain primes mitochondrial translocation of Parkin and regulates mitophagy. Sci Rep. 2012;2:1002. https://doi.org/10.1038/srep01002.
112. Durcan TM, Fon EA. The three P's of mitophagy: PARKIN, PINK1, and post-translational modifications. Genes Dev. 2015;29(10):989–99. https://doi.org/10.1101/gad.262758.115.
113. Kane LA, Lazarou M, Fogel AI, Li Y, Yamano K, Sarraf SA, et al. PINK1 phosphorylates ubiquitin to activate Parkin E3 ubiquitin ligase activity. J Cell Biol. 2014;205(2):143–53. https://doi.org/10.1083/jcb.201402104.
114. Kazlauskaite A, Kondapalli C, Gourlay R, Campbell David G, Ritorto Maria S, Hofmann K, et al. Parkin is activated by PINK1-dependent phosphorylation of ubiquitin at Ser65. Biochem J. 2014;460(1):127–41. https://doi.org/10.1042/bj20140334.
115. Kazlauskaite A, Kelly V, Johnson C, Baillie C, Hastie CJ, Peggie M, et al. Phosphorylation of Parkin at Serine65 is essential for activation: elaboration of a Miro1 substrate-based assay of Parkin E3 ligase activity. Open Biol. 2014;4:130213. https://doi.org/10.1098/rsob.130213.
116. Koyano F, Okatsu K, Kosako H, Tamura Y, Go E, Kimura M, et al. Ubiquitin is phosphorylated by PINK1 to activate parkin. Nature. 2014;510(7503):162–6. https://doi.org/10.1038/nature13392.
117. Okatsu K, Koyano F, Kimura M, Kosako H, Saeki Y, Tanaka K, et al. Phosphorylated ubiquitin chain is the genuine Parkin receptor. J Cell Biol. 2015;209(1):111–28. https://doi.org/10.1083/jcb.201410050.

118. Geisler S, Holmström KM, Skujat D, Fiesel FC, Rothfuss OC, Kahle PJ, et al. PINK1/Parkin-mediated mitophagy is dependent on VDAC1 and p62/SQSTM1. Nat Cell Biol. 2010;12(2):119–31. https://doi.org/10.1038/ncb2012.
119. Ziviani E, Tao RN, Whitworth AJ. Drosophila Parkin requires PINK1 for mitochondrial translocation and ubiquitinates. Mitofusin. 2010;107(11):5018–23. https://doi.org/10.1073/pnas.0913485107.
120. Wang X, Winter D, Ashrafi G, Schlehe J, Selkoe D, et al. PINK1 and Parkin target Miro for phosphorylation and degradation to arrest mitochondrial motility. Cell. 2011;147(4):893–906. https://doi.org/10.1016/j.cell.2011.10.018.
121. Ordureau A, Sarraf Shireen A, Duda David M, Heo J-M, Jedrychowski Mark P, Sviderskiy Vladislav O, et al. Quantitative proteomics reveal a feedforward mechanism for mitochondrial parkin translocation and ubiquitin chain synthesis. Mol Cell. 2014;56(3):360–75. https://doi.org/10.1016/j.molcel.2014.09.007.
122. Xiao B, Goh JY, Xiao L, Xian H, Lim KL, Liou YC. Reactive oxygen species trigger Parkin/PINK1 pathway-dependent mitophagy by inducing mitochondrial recruitment of Parkin. J Biol Chem. 2017;292(40):16697–708. https://doi.org/10.1074/jbc.M117.787739.
123. Narendra D, Tanaka A, Suen D-F, Youle RJ. Parkin is recruited selectively to impaired mitochondria and promotes their autophagy. J Cell Biol. 2008;183(5):795–803. https://doi.org/10.1083/jcb.200809125.
124. Ding WX, Ni HM, Li M, Liao Y, Chen X, Stolz DB, et al. Nix is critical to two distinct phases of mitophagy, reactive oxygen species-mediated autophagy induction and Parkin-ubiquitin-p62-mediated mitochondrial priming. J Biol Chem. 2010;285(36):27879–90. https://doi.org/10.1074/jbc.M110.119537.
125. Joselin AP, Hewitt SJ, Callaghan SM, Kim RH, Chung Y-H, Mak TW, et al. ROS-dependent regulation of Parkin and DJ-1 localization during oxidative stress in neurons. Hum Mol Genet. 2012;21(22):4888–903.
126. Xian H, Liou Y-C. Loss of MIEF1/MiD51 confers susceptibility to BAX-mediated cell death and PINK1-PRKN-dependent mitophagy. Autophagy. 2019;15(12):2107–25. https://doi.org/10.1080/15548627.2019.1596494.
127. Rojansky R, Cha M-Y, Chan DC. Elimination of paternal mitochondria in mouse embryos occurs through autophagic degradation dependent on PARKIN and MUL1. elife. 2016;5:e17896. https://doi.org/10.7554/eLife.17896.
128. Szargel R, Shani V, Abd Elghani F, Mekies LN, Liani E, Rott R, et al. The PINK1, synphilin-1 and SIAH-1 complex constitutes a novel mitophagy pathway. Hum Mol Genet. 2016;25(16):3476–90. https://doi.org/10.1093/hmg/ddw189.
129. Villa E, Proïcs E, Rubio-Patiño C, Obba S, Zunino B, Bossowski JP, et al. Parkin-independent mitophagy controls chemotherapeutic response in cancer cells. Cell Rep. 2017;20(12):2846–59. https://doi.org/10.1016/j.celrep.2017.08.087.
130. Fu M, St-Pierre P, Shankar J, Wang PT, Joshi B, Nabi IR. Regulation of mitophagy by the Gp78 E3 ubiquitin ligase. Mol Biol Cell. 2013;24(8):1153–62. https://doi.org/10.1091/mbc.E12-08-0607.
131. Di Rita A, Peschiaroli A, Acunzo D, Strobbe D, Hu Z, Gruber J, et al. HUWE1 E3 ligase promotes PINK1/PARKIN-independent mitophagy by regulating AMBRA1 activation via IKKα. Nat Commun. 2018;9(1):3755. https://doi.org/10.1038/s41467-018-05722-3.
132. Richter B, Sliter DA, Herhaus L, Stolz A, Wang C, Beli P, et al. Phosphorylation of OPTN by TBK1 enhances its binding to Ub chains and promotes selective autophagy of damaged mitochondria. Proc Natl Acad Sci. 2016;113(15):4039. https://doi.org/10.1073/pnas.1523926113.
133. Lazarou M, Sliter DA, Kane LA, Sarraf SA, Wang C, Burman JL, et al. The ubiquitin kinase PINK1 recruits autophagy receptors to induce mitophagy. Nature. 2015;524(7565):309–14. https://doi.org/10.1038/nature14893.

134. Chan NC, Salazar AM, Pham AH, Sweredoski MJ, Kolawa NJ, Graham RLJ, et al. Broad activation of the ubiquitin-proteasome system by Parkin is critical for mitophagy. Hum Mol Genet. 2011;20(9):1726–37. https://doi.org/10.1093/hmg/ddr048.
135. Kirkin V, Lamark T, Sou Y-S, Bjørkøy G, Nunn JL, Bruun J-A, et al. A role for NBR1 in autophagosomal degradation of ubiquitinated substrates. Mol Cell. 2009;33(4):505–16. https://doi.org/10.1016/j.molcel.2009.01.020.
136. Wang Y, Liu N, Lu B. Mechanisms and roles of mitophagy in neurodegenerative diseases. CNS Neurosci Ther. 2019;25(7):859–75. https://doi.org/10.1111/cns.13140.
137. Chakravorty A, Jetto CT, Manjithaya R. Dysfunctional mitochondria and mitophagy as drivers of Alzheimer's disease pathogenesis. Front Aging Neurosci. 2019;11:311. https://doi.org/10.3389/fnagi.2019.00311.
138. Xian H, Yang Q, Xiao L, Shen H-M, Liou Y-C. STX17 dynamically regulated by Fis1 induces mitophagy via hierarchical macroautophagic mechanism. Nat Commun. 2019;10(1):2059. https://doi.org/10.1038/s41467-019-10096-1.
139. Sabouny R, Shutt TE. Reciprocal regulation of mitochondrial fission and fusion. Trends Biochem Sci. 2020; https://doi.org/10.1016/j.tibs.2020.03.009.
140. Held NM, Houtkooper RH. Mitochondrial quality control pathways as determinants of metabolic health. Bioessays. 2015;37(8):867–76. https://doi.org/10.1002/bies.201500013.
141. Wu NN, Zhang Y, Ren J. Mitophagy, mitochondrial dynamics, and homeostasis in cardiovascular aging. Oxidative Med Cell Longev. 2019;2019:9825061. https://doi.org/10.1155/2019/9825061.
142. Xiao L, Xian H, Lee KY, Xiao B, Wang H, Yu F, et al. Death-associated protein 3 regulates mitochondrial-encoded protein synthesis and mitochondrial dynamics. J Biol Chem. 2015;290(41):24961–74. https://doi.org/10.1074/jbc.M115.673343.
143. Burman JL, Pickles S, Wang C, Sekine S, Vargas JNS, Zhang Z, et al. Mitochondrial fission facilitates the selective mitophagy of protein aggregates. J Cell Biol. 2017;216(10):3231–47. https://doi.org/10.1083/jcb.201612106.
144. Lang A, Anand R, Altinoluk-Hambüchen S, Ezzahoini H, Stefanski A, Iram A, et al. SIRT4 interacts with OPA1 and regulates mitochondrial quality control and mitophagy. Aging. 2017;9(10):2163–89. https://doi.org/10.18632/aging.101307.
145. Rana A, Oliveira MP, Khamoui AV, Aparicio R, Rera M, Rossiter HB, et al. Promoting Drp1-mediated mitochondrial fission in midlife prolongs healthy lifespan of Drosophila melanogaster. Nat Commun. 2017;8(1):448. https://doi.org/10.1038/s41467-017-00525-4.
146. Niu Y-J, Nie Z-W, Shin K-T, Zhou W, Cui X-S. PINK1 regulates mitochondrial morphology promoting mitochondrial fission in porcine preimplantation embryos. FASEB J. 2019;33(7):7882–95. https://doi.org/10.1096/fj.201802473R.
147. D'Amico D, Mottis A, Potenza F, Sorrentino V, Li H, Romani M, et al. The RNA-binding protein PUM2 impairs mitochondrial dynamics and mitophagy during aging. Mol Cell. 2019;73(4):775–87. https://doi.org/10.1016/j.molcel.2018.11.034.
148. Jin SM, Youle RJ. The accumulation of misfolded proteins in the mitochondrial matrix is sensed by PINK1 to induce PARK2/Parkin-mediated mitophagy of polarized mitochondria. Autophagy. 2013;9(11):1750–7.
149. Hämäläinen RH, Manninen T, Koivumäki H, Kislin M, Otonkoski T, Suomalainen A. Tissue- and cell-type–specific manifestations of heteroplasmic mtDNA 3243A> G mutation in human induced pluripotent stem cell-derived disease model. Proc Natl Acad Sci. 2013;110(38):E3622–E30.
150. Suen D-F, Narendra DP, Tanaka A, Manfredi G, Youle RJ. Parkin overexpression selects against a deleterious mtDNA mutation in heteroplasmic cybrid cells. Proc Natl Acad Sci. 2010;107(26):11835–40.
151. Burbulla LF, Fitzgerald J, Stegen K, Westermeier J, Thost A, Kato H, et al. Mitochondrial proteolytic stress induced by loss of mortalin function is rescued by Parkin and PINK1. Cell Death Dis. 2014;5(4):e1180.

152. Devi L, Raghavendran V, Prabhu BM, Avadhani NG, Anandatheerthavarada HK. Mitochondrial import and accumulation of α-synuclein impair complex I in human dopaminergic neuronal cultures and Parkinson disease brain. J Biol Chem. 2008;283 (14):9089–100.
153. Ludtmann MHR, Angelova PR, Horrocks MH, Choi ML, Rodrigues M, Baev AY, et al. α-synuclein oligomers interact with ATP synthase and open the permeability transition pore in Parkinson's disease. Nat Commun. 2018;9(1):2293. https://doi.org/10.1038/s41467-018-04422-2.
154. Di Maio R, Barrett PJ, Hoffman EK, Barrett CW, Zharikov A, Borah A, et al. α-Synuclein binds to TOM20 and inhibits mitochondrial protein import in Parkinson's disease. Sci Transl Med. 2016;8(342):342. https://doi.org/10.1126/scitranslmed.aaf3634.
155. Jin J, Hulette C, Wang Y, Zhang T, Pan C, Wadhwa R, et al. Proteomic identification of a stress protein, mortalin/mthsp70/GRP75: relevance to Parkinson disease. Mol Cell Proteomics. 2006;5(7):1193–204.
156. De Mena L, Coto E, Sánchez-Ferrero E, Ribacoba R, Guisasola LM, Salvador C, et al. Mutational screening of the mortalin gene (HSPA9) in Parkinson's disease. J Neural Transm. 2009;116(10):1289–93.
157. Burbulla LF, Schelling C, Kato H, Rapaport D, Woitalla D, Schiesling C, et al. Dissecting the role of the mitochondrial chaperone mortalin in Parkinson's disease: functional impact of disease-related variants on mitochondrial homeostasis. Hum Mol Genet. 2010;19 (22):4437–52.
158. Goswami AV, Samaddar M, Sinha D, Purushotham J, D'Silva P. Enhanced J-protein interaction and compromised protein stability of mtHsp70 variants lead to mitochondrial dysfunction in Parkinson's disease. Hum Mol Genet. 2012;21(15):3317–32.
159. Jin J, Li GJ, Davis J, Zhu D, Wang Y, Pan C, et al. Identification of novel proteins associated with both α-synuclein and DJ-1. Mol Cell Proteomics. 2007;6(5):845–59.
160. Davison EJ, Pennington K, Hung CC, Peng J, Rafiq R, Ostareck-Lederer A, et al. Proteomic analysis of increased Parkin expression and its interactants provides evidence for a role in modulation of mitochondrial function. Proteomics. 2009;9(18):4284–97.
161. Yang H, Zhou X, Liu X, Yang L, Chen Q, Zhao D, et al. Mitochondrial dysfunction induced by knockdown of mortalin is rescued by Parkin. Biochem Biophys Res Commun. 2011;410 (1):114–20.
162. Freimann K, Zschiedrich K, Brüggemann N, Grünewald A, Pawlack H, Hagenah J, et al. Mortalin mutations are not a frequent cause of early-onset Parkinson disease. Neurobiol Aging. 2013;34(11):2694.
163. Chung SJ, Kim M-J, Ryu H-S, Kim J, Kim YJ, Kim K, et al. Lack of association of mortalin (HSPA9) and other mitochondria-related genes with risk of Parkinson's and Alzheimer's diseases. Neurobiol Aging. 2017;49:215.
164. Costa A, Loh S, Martins LM. Drosophila Trap1 protects against mitochondrial dysfunction in a PINK1/parkin model of Parkinson's disease. Cell Death Dis. 2013;4(1):e467.
165. Zhang L, Karsten P, Hamm S, Pogson JH, Müller-Rischart AK, Exner N, et al. TRAP1 rescues PINK1 loss-of-function phenotypes. Hum Mol Genet. 2013;22(14):2829–41.
166. Pridgeon JW, Olzmann JA, Chin L-S, Li L. PINK1 protects against oxidative stress by phosphorylating mitochondrial chaperone TRAP1. PLoS Biol. 2007;5(7):e172.
167. Plun-Favreau H, Klupsch K, Moisoi N, Gandhi S, Kjaer S, Frith D, et al. The mitochondrial protease HtrA2 is regulated by Parkinson's disease-associated kinase PINK1. Nat Cell Biol. 2007;9(11):1243–52.
168. Tain LS, Chowdhury RB, Tao RN, Plun-Favreau H, Moisoi N, Martins LM, et al. Drosophila HtrA2 is dispensable for apoptosis but acts downstream of PINK1 independently from Parkin. Cell Death Differ. 2009;16(8):1118–25.
169. Whitworth AJ, Lee JR, Ho VM-W, Flick R, Chowdhury R, McQuibban GA. Rhomboid-7 and HtrA2/Omi act in a common pathway with the Parkinson's disease factors Pink1 and Parkin. Mechanisms. 2008;1(2-3):168–74. https://doi.org/10.1242/dmm.000109.

170. Diot A, Morten K, Poulton J. Mitophagy plays a central role in mitochondrial ageing. Mamm Genome. 2016;27(7-8):381–95. https://doi.org/10.1007/s00335-016-9651-x.
171. Fivenson EM, Lautrup S, Sun N, Scheibye-Knudsen M, Stevnsner T, Nilsen H, et al. Mitophagy in neurodegeneration and aging. Neurochem Int. 2017;109:202–9. https://doi.org/10.1016/j.neuint.2017.02.007.
172. Shi R, Guberman M, Kirshenbaum LA. Mitochondrial quality control: the role of mitophagy in aging. Trends Cardiovasc Med. 2018;28(4):246–60. https://doi.org/10.1016/j.tcm.2017.11.008.
173. Moreira OC, Estébanez B, Martínez-Florez S, Paz JA, Cuevas MJ, González-Gallego J. Mitochondrial function and mitophagy in the elderly: effects of exercise. Oxidative Med Cell Longev. 2017;2017:2012798. https://doi.org/10.1155/2017/2012798.
174. Reeve A, Simcox E, Turnbull D. Ageing and Parkinson's disease: why is advancing age the biggest risk factor? Ageing Res Rev. 2014;14(100):19–30. https://doi.org/10.1016/j.arr.2014.01.004.
175. Moskal N, Riccio V, Bashkurov M, Taddese R, Datti A, Lewis PN, et al. ROCK inhibitors upregulate the neuroprotective Parkin-mediated mitophagy pathway. Nat Commun. 2020;11(1):88. https://doi.org/10.1038/s41467-019-13781-3.
176. Ryu D, Mouchiroud L, Andreux PA, Katsyuba E, Moullan N, Nicolet-Dit-Félix AA, et al. Urolithin A induces mitophagy and prolongs lifespan in C. elegans and increases muscle function in rodents. Nat Med. 2016;22(8):879–88. https://doi.org/10.1038/nm.4132.
177. Sun N, Yun J, Liu J, Malide D, Liu C, Rovira II, et al. Measuring in vivo mitophagy. Mol Cell. 2015;60(4):685–96. https://doi.org/10.1016/j.molcel.2015.10.009.
178. Fang EF, Hou Y, Palikaras K, Adriaanse BA, Kerr JS, Yang B, et al. Mitophagy inhibits amyloid-β and tau pathology and reverses cognitive deficits in models of Alzheimer's disease. Nat Neurosci. 2019;22(3):401–12. https://doi.org/10.1038/s41593-018-0332-9.
179. Kujawska M, Jourdes M, Kurpik M, Szulc M, Szaefer H, Chmielarz P, et al. Neuroprotective effects of pomegranate juice against Parkinson's disease and presence of ellagitannins-derived metabolite-urolithin A-in the brain. Int J Mol Sci. 2019;21(1):202. https://doi.org/10.3390/ijms21010202.
180. Andreux PA, Blanco-Bose W, Ryu D, Burdet F, Ibberson M, Aebischer P, et al. The mitophagy activator urolithin A is safe and induces a molecular signature of improved mitochondrial and cellular health in humans. Nat Metab. 2019;1(6):595–603. https://doi.org/10.1038/s42255-019-0073-4.
181. Lu L, Tang LE, Wei W, Hong Y, Chen H, Ying W, et al. Nicotinamide mononucleotide improves energy activity and survival rate in an model of Parkinson's disease. Exp Ther Med. 2014;8(3):943–50.
182. Lehmann S, Loh SHY, Martins LM. Enhancing NAD salvage metabolism is neuroprotective in a PINK1 model of Parkinson's disease. Biol Open. 2017;6(2):141–7. https://doi.org/10.1242/bio.022186.
183. Fang EF. Mitophagy and NAD inhibit Alzheimer disease. Autophagy. 2019;15(6):1112–4. https://doi.org/10.1080/15548627.2019.1596497.
184. Investigators TPSGP. Mixed lineage kinase inhibitor CEP-1347 fails to delay disability in early Parkinson disease. Neurology. 2007;69(15):1480–90. https://doi.org/10.1212/01.wnl.0000277648.63931.c0.
185. Jang JY, Blum A, Liu J, Finkel T. The role of mitochondria in aging. J Clin Invest. 2018;128(9):3662–70. https://doi.org/10.1172/jci120842.

Autophagy in Cellular Stress Responses

6

Jingyi Li and Rui Liu

Abstract

Autophagy is an essential cellular mechanism involving the lysosomal degradation of cytoplasmic organelles or cytosolic components. Autophagy can be induced by various stress stimuli including oxidative stress, hypoxia, pathogen infection, and osmotic pressure, which in turn affects a variety of cellular biological functions and plays a role of double-edged sword in determining the fate of cells depending on different stress conditions, degrees of damage, and cell model. We have summarized the molecular events involved in the process of autophagy induced by cellular stress in this chapter.

Keywords

Autophagy · Cellular stress · Endoplasmic reticulum stress

Abbreviation

3-MA	3-Methyladenine
AAC	Abdominal aortic constriction
AEG-1	Astrocyte elevated gene-1

J. Li
The Second Affiliated Hospital of Chengdu Medical College, China National Nuclear Corporation 416 Hospital, Chengdu, Sichuan, China

School of Biological Sciences and Technology, Chengdu Medical College, Chengdu, China

R. Liu (✉)
State Key Laboratory of Oral Diseases, National Clinical Research Center for Oral Diseases, Chinese Academy of Medical Sciences Research Unit of Oral Carcinogenesis and Management, West China Hospital of Stomatology, Sichuan University, Chengdu, China

C. Huang, Y. Zhang (eds.), *Oxidative Stress*,
https://doi.org/10.1007/978-981-16-0522-2_6

AMP	Adenosine monophosphate
AMPK	AMP-activated protein kinase
ATF	Activating transcription factor
ATF4	Activating transcription factor 4
ATG101	Autophagy-related gene 101
ATG12	Autophagy-related gene 12
ATG13	Autophagy-related gene 13
ATG14L	Atg14-like protein
ATG5	Autophagy-related gene 5
ATG9A	Autophagy-related gene 9A
ATP	Adenosine triphosphate
Bcl-2	B cell lymphoma-2
BH3	B cell lymphoma-2 homology domain 3
BNIP3	BCL2/adenovirus E1B-interacting protein 3
cGAS	Cyclic GMP-AMP synthase
CHOP	C/EBP homologous protein
CMA	Chaperone-mediated autophagy
CQ	Chloroquine
CRACs	Ca^{2+} release-activated channels
CXCR4	C-X-C chemokine receptor type 4
E1	Ubiquitin activating
E2	Ubiquitin conjugating
E3	Ubiquitin ligase
eIF2α	Eukaryotic translation initiation factor 2 alpha
ER stress	Endoplasmic reticulum stress
ER	Endoplasmic reticulum
F6P	Fructose-6-phosphate
FIP200	FAK family kinase interacting protein of 200 kDa
FYCO1	FYVE and coiled-coil domain-containing 1
G6P	Glucose-6-phosphate
G6PDH	Glucose-6-phosphate dehydrogenase
GRP78/BiP	78 kDa glucose-regulated protein/binding immunoglobulin protein
HCV	Hepatitis C Virus
HIF-1	Hypoxia-inducible factor-1
HK II	Hexokinase II
HSP	Heat shock protein
HSPA8/HSC70	Heat shock 70 kDa protein 8
HSV-1	Herpes simplex virus type 1
ICS model	Intermittent cold stress in C57BL/6 J mouse model
ICS	Intermittent cold stress
IH	Intermittent hypoxia
IRE1	Inositol-requiring enzyme 1
IRE1α	Inositol-requiring enzyme 1α
IRF3	Interferon regulator 3

IRGM	Immunity-related GTPase family M
K-48	Lysine residue 48
K-63	Lysine residue 63
KFERQ	Recognizes the signal peptide
LAMP2A	Lysosomal-associated membrane protein 2A
LC3	Microtubule-associated protein light chain 3
LC3-II	Microtubule-associated protein light chain 3-II
LIR	LC3 interacts
LUBAC	Linear ubiquitin chain assembly complex
MAP 1LC3B	Microtubule-associated protein 1 light chain 3B
MAVS	Mitochondrial antiviral signaling proteins
MeV	Measles virus
mTOR	Mammalian target of rapamycin
mTORC1	Mammalian target of rapamycin complex 1
NADPH	Nicotinamide adenine dinucleotide phosphate
NDP52	Nuclear dot 10 protein 52
NLRP3	Nucleotide-binding oligomerization domain receptor, pyrin domain-containing 3
NS4B	Nonstructural protein 4B
NS5ATP9	Nonstructural protein 5A upregulation gene
OPTN	Optineurin
OXPHOS	Oxidative phosphorylation
p62	Protein sequestosome 1/p62
PAMPs	Pathogen-associated molecular patterns
PE	Phosphatidylethanolamine
PERK	Protein kinase RNA-like ER kinase
PI3K	Phosphatidylinositol 3-kinase
PI3P	Phosphatidylinositol 3-phosphate
PINK1	Putative kinase protein 1
PKR	Double-stranded RNA-dependent protein kinase
PLEKHM1	Pleckstrin homology and RUN domain containing M1
PRRS	Pattern recognition receptors
R5P	Ribulose-5-phosphate
RIG-I	Retinoic acid-inducible gene I
ROS	Reactive oxygen species
SINV	Sindbis virus
SnRK1	Snf1-related protein kinase 1
SQSTM1	Sequestosome 1
TANK1	TANK-binding kinase 1
TAX1BP1	3'-UTR of Tax1-binding protein 1
TAX1BP1	Tax1-binding protein 1
TBC1D5	Tre-2/Bub2/Cdc16 domain family member 5
TIAM1	T-lymphoma invasion and metastasis-inducing protein 1
TLR	Toll-like receptors
TOR	Target of rapamycin

TRIF	Toll/IL-1R domain-containing adaptor-inducing IFNβ
TRPV1	Transient receptor potential vanilloid 1
TSC1	Tuberous sclerosis complex subunit 1
TSC2	Tuberous sclerosis complex subunit 2
ULK1	UNC-51-like kinase 1
UPR	Unfolded protein reaction
UPR	Unfolded protein response
UPS	Ubiquitin-proteasome system
UVRAG	UV irradiation resistance-associated gene
Vps34	Vacuolar protein sorting 34

6.1 Introduction

Autophagy is a process of dynamic changes of subcellular membrane structure in eukaryotes and lysosome-mediated degradation of intracellular proteins and organelles. There are three types of autophagic degradation, including macroautophagy, microautophagy, and chaperone-mediated autophagy (CMA), mainly differing in the mechanism of cargo delivery to the lysosome, which have been elaborated in these reviews [1, 2]. During macroautophagy, usually referred to as "autophagy", autophagosomes encapsulate part of the cytoplasm and organelles, proteins, and other components that need to be degraded in the cell and fuse with the lysosome to form an autophagolysosome, which degrades the contents of the lysosome under the action of lysosomal enzymes and achieves the energy metabolism needs of the cell itself as well as the renewal of certain organelles.

The autophagosome is a vacuole structure with a diameter of about 100 nm composed of a double-layer phospholipid molecular membrane, the formation of which marks the initiation of autophagy. The processing of phosphatidylethanolamine (PE), phosphatidylinositol 3-phosphate (PI3P), and other phospholipid molecules is the key molecular event in autophagosome formation [3], which is achieved by high-dimensional protein complexes due to large volume of phospholipid molecules containing two 14–18 carbon fatty acid tails. The formation of autophagosomes is a process precisely regulated by multiple protein complexes such as ULK1, ATG14L, and ATG16L1, which can be divided into three stages, including initial signaling stage, functional protein recruitment stage, and membrane extension stage. During the initial signaling stage, ULK1 (autophagy-related gene 1, Atg1) stimulated by upstream signals forms a "ULK1 complex" with FIP200, ATG13, and ATG101, which in turn significantly activates ULK1 and subsequently triggers the functional protein recruitment stage. The activation of ULK1 can further activate beclin-1 (Atg6), Vps34, ATG14L, UVRAG, and other proteins to form "class III PI3K complex." As a PI3P synthase, Vps34 is activated by formation of the complex and thereby synthesizes a large amount of PI3P to recruit more autophagy-related functional proteins, thus laying the foundation for the formation and extension of autophagosome membranes. In response to activated PI3P synthesis, ATG16L1, ATG5, and ATG12 form a three-element "ATG16L1 complex" that further transforms into a higher-dimensional 12-element complex. The complex subsequently promotes the lipidation reaction of LC3 (Atg8) via connecting it to a PE molecule, thus leading to formation of LC3-II anchored to the autophagosome membrane. As a linker protein, LC3-II can assist in transporting the substances that need to be degraded into the autophagosomes, which is also considered as a distinguishable feature of autophagosomes.

When exposed to external stress stimuli, cells initiate a series of reactions in response to cell pressure and protect themselves from potential damage. Autophagy is one of the core pathways that mediates cellular stress response. In 1992, Ohsumi's team reported for the first time that nutrient-deficient conditions stimulated extensive autophagic deg radation of cytosolic components in the vacuoles of yeast cells [4]. Subsequent studies have shown that as a cellular stress response, a utophagy can be induced by various stimuli, including oxidative stress, hypoxia, pathogen infection, and osmotic pressure. In this chapter, we will discuss molecular events involved in the process of autophagy induced by cellular stress.

6.2 Oxidative Stress and Autophagy

Oxidative stress refers to the imbalance between oxidation and antioxidation states in the body, and the production of a large number of reactive oxygen species (ROS) as well as the tendency to oxidation will damage organelles, especially mitochondria [5]. So far, the only known mechanism for mitochondrial renewal is autophagy. Current studies have shown that cellular ROS levels increase under stress conditions, and H_2O_2 and $O_2^{\bullet -}$ in ROS are early inducers of autophagy [6–12].

The mammalian target of rapamycin complex 1 (mTORC1) is involved in the synthesis of proteins [13]. Under nutrient-rich conditions, mTORC1 is activated and phosphorylates Atg13 (Ser258) and ULK1 (Ser757) to inhibit catalytic activation of ULK1 complex (containing ULK1, Atg13, FIP200, and Atg101), thereby inhibiting autophagy [7]. When glucose deficiency and glucose-6-phosphate (G6P) levels are reduced, the glycolytic pathway and pentose phosphate pathway are inhibited, resulting in reduced NADPH and ATP levels. Since NADPH is the common substrate of glutathione reductase and thiol regeneration-related reductase and provides the electrons needed for thiol redox homeostasis, the cellular antioxidant capacity is reduced, and ROS production accordingly increases (Fig. 6.1) [6, 14, 15].

The exposed H_2O_2 and $O_2^{\bullet -}$ actively regulate autophagy through at least three different pathways. Excessive exposure of H_2O_2 and O_2• – can induce the activation of AMP-dependent protein kinase (AMPK) [16, 17]. AMPK stabilizes TSC1–TSC2 complex and shuts down mTORC1 signaling by phosphorylating TSC2, which removes the inhibition of mTORC1 on autophagy induction [18] (1). AMPK also directly phosphorylates ULK1 (Ser777 and Ser317) and activates the catalytic function of ULK1, which subsequently phosphorylates its interaction factors (Atg13 and FIP200) at other residues different from mTORC1. Activated ULK1

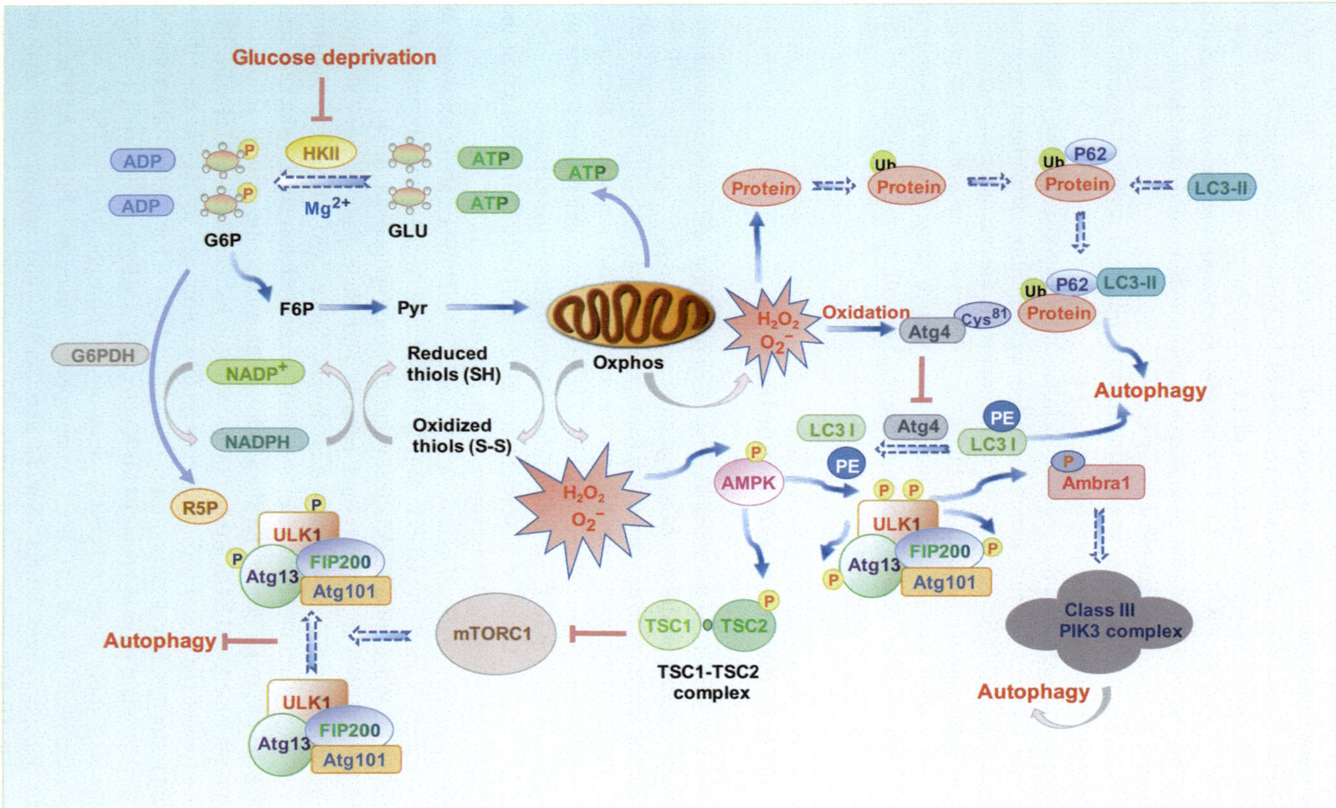

Fig. 6.1 Overview of the oxidative stress and autophagy. Upon glucose deficiency, NADPH and ATP levels are reduced, which results in ROS production. ROS induces autophagy by acting on AMPK, Atg4, and other autophagy pathway-related proteins. *F6P* fructose-6-phosphate, *Pyr* pyruvate, *OXPHOS* oxidative phosphorylation, *G6PDH* glucose-6-phosphate dehydrogenase, *R5P* ribulose-5-phosphate, *HK II* hexokinase II, mitochondrial isoform

complex recruits class III PI3K complex through activating molecules in beclin-1-regulated autophagy Ambra1, which contributes to the extension of autophagic bubble membrane [19, 20] (2). Adaptor molecule p62 has been shown to bind to ubiquitinated proteins and target them to LC3-II to induce autophagy [21, 22], and the study has indicated that p62 may be regulated by redox signaling [23]. In addition, ROS can also promote ubiquitination of the substances to be degraded, which promotes their recognition by p62 and targeted transport to the autophagosomes via binding to LC3-II [7] (3). ROS also causes oxidation of Cys81 on Atg4, resulting in inactivation of its "delipidating" activity on LC3, which facilitates accumulation of LC3-II and helps to extend autophagy (Fig. 6.1) [6, 24].

6.3 Endoplasmic Reticulum Stress and Autophagy

Endoplasmic reticulum (ER) stress involves accumulation of misfolded and unfolded proteins in the endoplasmic reticulum and disturbance of calcium ion balance [25]. The unfolded protein reaction (UPR) is the main pathway of ER stress response, which can help misfolded and unfolded proteins return to normal structure [26, 27]. UPR is mainly regulated by three ER membrane-associated sensor proteins, namely, inositol-requiring enzyme 1α (IRE1α), protein kinase RNA-like ER kinase (PERK), and activating transcription factor (ATF) 6 [28]. Under physiological conditions, the ER chaperone GRP78/BiP binds to ER membrane-located ATF6, IRE1, and PERK proteins. When misfolded proteins accumulate in the ER lumen, BiP/GRP78 dissociates from the three proteins and binds to unfolded proteins, thus activating these sensors and triggering UPR system to clear the accumulated proteins [25]. However, when the accumulation exceeds the UPR clearance capacity, cells may initiate autophagy as a scavenger to the point of protecting cells from damage [29]. Emerging evidence has shown that UPR sensors, PERK and IRE1, are implicated in mediating ER stress-induced autophagy [30, 31]. On the other hand, Ca^{2+} plays a vital role in ER stress-induced autophagy. During ER stress, Ca^{2+} is released from the ER lumen to the cytosol, thereby leading to the activation of AMPK and inducing autophagosome formation via CaMKKβ [32]. ER stress can also mediate autophagy by enhancing the interaction between Ca^{2+} binding protein caldendrin and LC3, which in turn reduces ER stress [33]. Taken together, autophagy can be induced through Ca^{2+}- and UPR-dependent mechanism under ER stress (Fig. 6.2).

6.4 Hypoxia-Induced Autophagy

Oxygen plays an important role in cellular respiration and energy metabolism. It was found that the number of autophagosomes increased under the condition of hypoxic culture [34]. HIF-1 is a hypoxia-inducible factor closely related to hypoxia-induced autophagy [35]. HIF-1 enhances the activity of BNIP3 by acting on BH3 site, so as

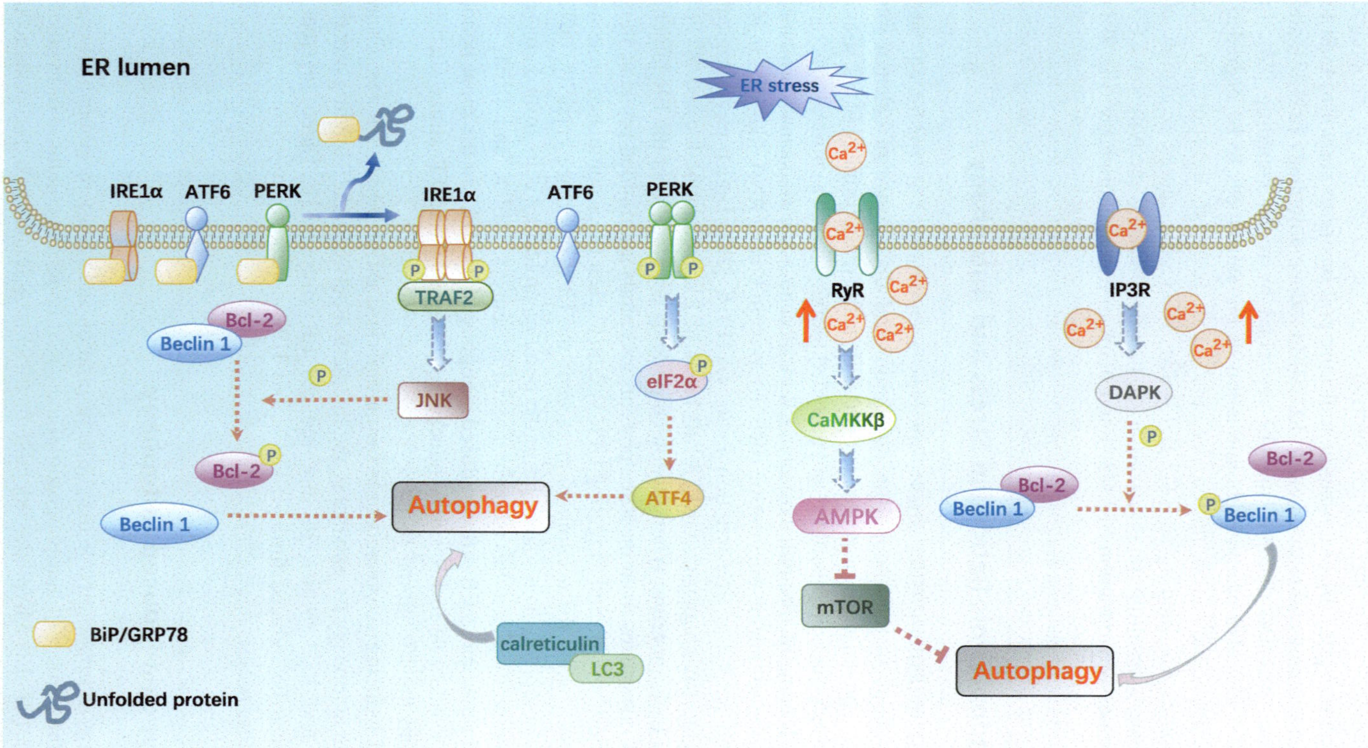

Fig. 6.2 Schematic presentation of ER stress-induced autophagy. IP3R and RyR receptors are ER membrane located in Ca^{2+} release-activated Ca^{2+} channels (CRACs). The PERK-eIF2α or the IRE-TRAF2-JNK pathway is implicated in ER stress-induced autophagy. In addition, the ER stress-induced increase in cytosolic Ca^{2+} can stimulate autophagy via CaMKKβ- or DAPK-mediated pathway

to regulate the interaction between Bcl-2 and beclin-1 as well as inhibit the activity of mTOR, thus stimulating the occurrence of autophagy [36, 37]. Moreover, hypoxia can also cause an increase in intracellular AMP/ATP ratio, thus stimulating AMPK that can promote autophagy by inhibiting the activity of mTOR [38]. Other studies have shown that mRNA levels of autophagy genes *LC3* and *ATG5* increase under hypoxia, which are regulated by PERK-dependent ATF4 and CHOP [39]. In addition, intermittent hypoxia (IH) activates endoplasmic reticulum stress and induces autophagy through the PERK/eIF2/ATF4 signaling pathway, thereby reducing IH-induced apoptosis [40] (Fig. 6.3).

6.5 Osmotic Stress and Autophagy

Normal osmotic pressure is crucial for maintaining cellular homeostasis. However, osmotic stress may occur under both physiological and pathological conditions [41]. As an important stress response, how does autophagy respond to cellular osmotic pressure? Hypertonic stress has been shown to induce autophagy in some organisms [42, 43]. Jiang et al. found that hyperosmotic stress could activate autophagy of nucleus pulposus cells through Ca^{2+}-dependent AMPK/mTOR pathway, thereby promoting the degradation of SQSTM1/p62 and inhibiting apoptosis under hyperosmotic condition, so as to ensure the survival of nucleus pulposus cells [44]. Similarly, Daniel et al. found that polycystin-2, a membrane protein of the polycystin family, downregulated mTOR pathway and thus promoted autophagy to act as an adaptive mechanism of cell survival under hyperosmotic stress [41]. In plants, osmotic stress also inhibits the TOR complex by activating an ortholog of mammalian AMPK and Snf1-related protein kinase 1 (SnRK1) and hinders the inhibitory effect of TOR on Atg1 or other unknown downstream regulatory factors and finally induces autophagy [45, 46]. The above findings indicate that autophagy tends to be activated as a protective mechanism response to osmotic stress.

6.6 Infection-Induced Autophagy

6.6.1 Bacterial Pathogen Infection

Selective autophagy is an important pathway to remove intracellular pathogens (e.g., *Mycobacterium*, *Salmonella*, *Escherichia coli*) [47–49]. In the case of pathogen invasion, host cells respond to the warning signals and damaged cell membranes through the family of galectins [50]. This will help the specific pathogens coated in ubiquitin by E3 ligases to generate a "kill-me" signal within the cells [50–52]. Subsequently, the autophagy receptor proteins (p62, NDP52, OPTN, TAX1BP1) recognize ubiquitinated pathogens and deliver them to autophagosomes for subsequent degradation [53–56].

Ubiquitination of pathogens needs three enzyme cascade effects: ubiquitin activating (E1), ubiquitin conjugating (E2), and ubiquitin ligating (E3) [57]. The

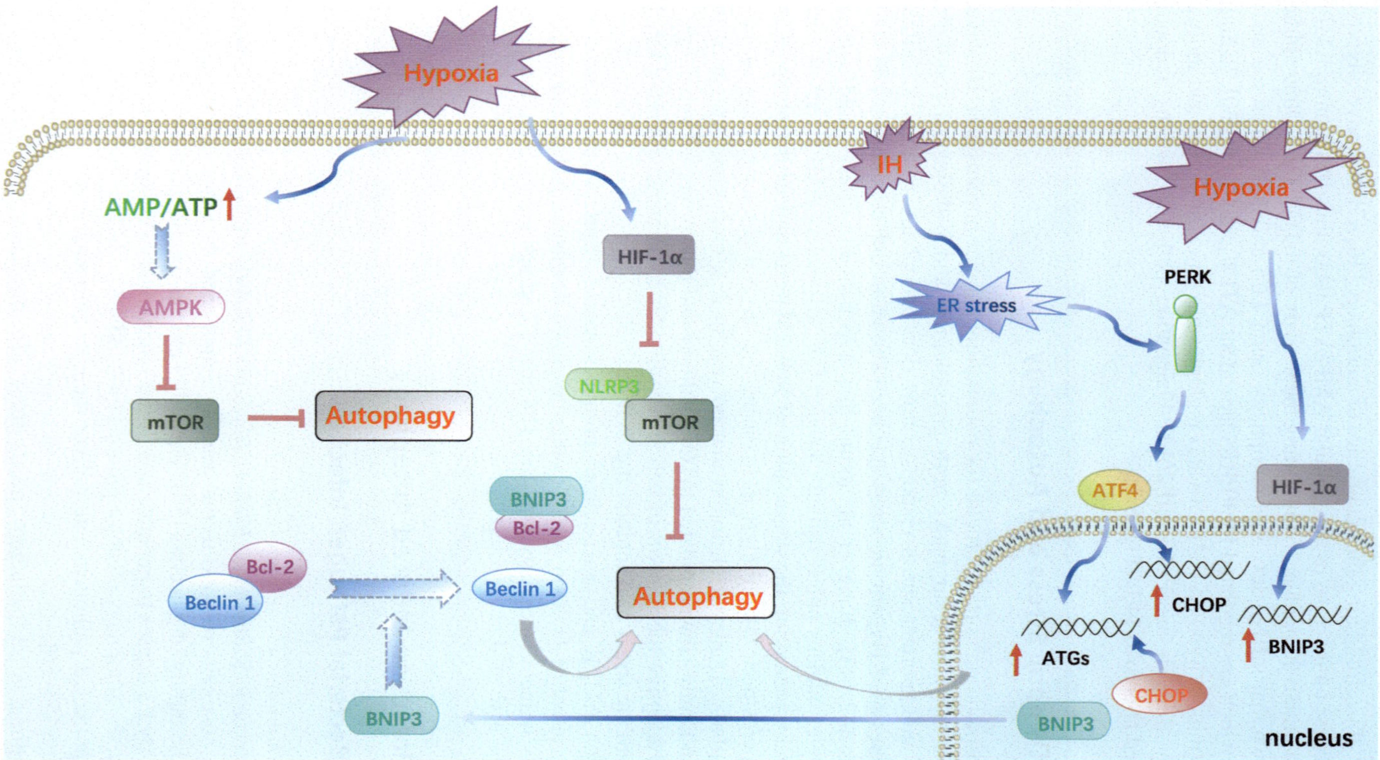

Fig. 6.3 Schematic presentation of hypoxia-induced autophagy. Hypoxia enhanced the activity of BNIP3 by promoting HIF-1α, which dissociates the interaction between Bcl-2 and beclin-1. On the other hand, hypoxia induces autophagy by inhibiting the activity of mTOR. An increase in AMP/ATP ratio is also observed in hypoxic cells. Meanwhile, mRNA levels of autophagy genes are increased through the regulation of ATF4 and CHOP. IH can stimulate ER stress and promote the PERK/eIF2α/ATF4 signaling pathway to induce autophagy

ubiquitin coat produced by E3 ligases can attract autophagy receptor proteins (p62, NDP52, OPTN, TAX1BP1), which increases the ability of autophagy to eliminate pathogens and limit bacterial proliferation [48, 58]. The ubiquitin coat consists of several different linkages such as linear K-63 and K-48, and all the linkages can perform as "eat me" signals in triggering autophagy to remove the pathogens [48, 59]. However, for proteins, the way of ubiquitin connection will affect subsequent degradation pathways of these proteins. The proteins wrapped by K-63 ubiquitin chains are mainly cleared by autophagy, while the proteins wrapped by K-48 ubiquitin chains are subjected to be degraded by proteasomes [60, 61]. Several E3 ligases have been identified to be involved in ubiquitinating bacterial pathogens and targeting them for selective autophagy [62–65]. Parkin is one of the most thoroughly studied E3 ligases, which is required for the ubiquitination of *Mycobacterium tuberculosis* and *Salmonella typhi* as well as subsequent autophagy-mediated degradation [49, 62]. The E3 ligase LRSAM1 is an important ligase in the process of clearing many intracellular bacterial pathogens such as *S. typhimurium*, *Escherichia coli*, and a *Shigella flexneri* [63]. Other E3 ligases such as ARIH, HOIPI, and LUBAC are also involved in the ubiquitination of bacterial pathogens [64, 65].

In selective autophagy, the autophagy adaptor proteins are a bridge between the ubiquitinated cargoes to be degraded and the newborn autophagosomes. The autophagy adaptor proteins, including ULK1/2, FYCO1, TBC1D5, PLEKHM1, and TIAM1, target the ubiquitinated substrate degradation by driving autophagy-lysosome pathway without degradation of themselves [66–71]. However, under the conditions caused by bacterial pathogen infection, the autophagy receptor proteins (such as p62, NDP52, OPTN, TAX1BP1) can act as the adaptor proteins to promote the maturation of autophagosomes and degrade along with the pathogens [53–56, 72, 73]. For example, during the *Salmonella* infection, (1) **p62** recognizes the targets of the E3 ligase ARIH, HOIP1, and LRSAM1 and then is recruited into *Salmonella* and used as a target for autophagy degradation [74]. (2) **NDP52** interacts with galectin 8 to help recruit TANK-binding kinase 1 (TNK1) to phosphorylate NDP52. NDP52-TBK1 is the first responder before E3 ligase LSRAM1 ubiquitinates bacteria, which can help NDP52 target bacterial autophagy [50, 75]. LC3 interacts with the (LIR) domain of NDP52 to deliver bacteria into autophagosomes and limit its proliferation [54]. (3) M1 polyubiquitination on the surface of bacteria is recruited by the E3 ligase complex LUBAC, and then **OPTN** recognizes LUBAC and targets the bacteria for autophagy degradation [64]. (4) **TAX1BP1** interacts with myosin VI to induce the fusion of autophagosome and lysosome in *Salmonella* [56].

The E3 ligase and autophagy receptor proteins are important in the selective autophagy of bacterial pathogens. Different autophagy receptor proteins display similar functions in targeting bacterial species for autophagy degradation. However, there are relatively few types of E3 ligase that have been clarified, and the ubiquitination (single ubiquitin or polyubiquitination) degree of ubiquitin chains (K-63, K-48, K-27) and the mechanisms in different infectious bacteria remain to be further studied.

6.6.2 Virus Infection

During viral infection, autophagy may be induced by adsorption, invasion, membrane fusion, protein synthesis, and other steps in the viral replication cycle [76–78]. At first, the adsorption of virions can induce autophagy [79]. For example, in measles virus (MeV) infection, human surface pathogen receptor CD46 induces autophagy by interacting with the Vps34-beclin-1 complex via the CD46-Cyt1/GOPC pathway [80–82]. MeV can also activate autophagy by targeting immunity-related GTPase family M (IRGM) [83, 84]. Moreover, autophagy is also activated during viral membrane fusion. The fusion of HIV-1 envelope glycoproteins gp120 and gp41 with the CD4 and CXCR4 receptors of uninfected $CD4^+$ T cells induces autophagy [78, 85]. Studies have shown that CXCR4 phosphorylates and activates p53, which increases the expression of Bax and activates the mitochondrial death pathway, thereby resulting in the death of uninfected $CD4^+$ T cells [86]. Another study suggested that autophagy activation may be also associated with increased bioactive lipids and ROS during membrane fusion [87].

After the membrane fusion, the viral genome activates different pattern recognition receptors (PRRs) such as retinoic acid-inducible gene I (RIG-I) and toll-like receptors (TLRs), and PRR induces autophagy by recognizing pathogen-associated molecular patterns (PAMPs) [88–90]. After TLRs recognize viral genome, the myeloid differentiation primary response protein MYD88 and adaptor molecule 1 (TRIF) will be recruited. TLR then stimulates MYD88 or TRIF to bind to beclin-1, which dissociates beclin-1 from the inhibitory complex Bcl-2 and induces autophagy [89, 91]. After being stimulated by viral dsRNA, RIG-I transmits signals to mitochondrial antiviral signaling proteins (MAVS) to activate interferon regulator 3 (IRF3), inducing the expression of IFNα/β and IFN target genes [88]. Double-stranded RNA-dependent protein kinase R (PKR) is activated by interferon and binds to beclin-1 in the Vps34-beclin-1 complex to initiate autophagosome formation. PKR can also induce autophagy by inhibiting protein translation through phosphorylating and inactivating eukaryotic initiation factor 2 (eIF2) [89, 92]. During DNA virus infection, the DNA sensor cyclic GMP-AMP synthase (cGAS) competes with the autophagy inhibitor Rubicon to bind to beclin-1, thereby triggering autophagy and stimulating the degradation of cytoplasmic viral DNA [93].

The viruses can induce autophagy not only directly through the interaction between its own proteins and autophagy-related proteins but also indirectly by inducing cellular stress [85, 94]. Take hepatitis C virus (HCV) infection as an example: the NS3/4A protease of HCV can bind to the autophagy-related protein IRGM to induce autophagy [95]. The NS4B protein of HCV has been shown to induce LC3 lipidation and autophagosome accumulation and may form complexes with Rab5, hVps34, and beclin-1 to induce autophagy [96, 97]. NS5ATP9 is a functional protein involved in DNA damage response, cellular signal transduction, and cell death. It has been reported that HCV NS5A upregulates the expression of beclin-1 via NS5ATP9 and may therefore induce autophagy [98–100]. A recent study shows that the host endoplasmic reticulum transmembrane protein SCOTIN interacts with NS5A of HCV and induces autophagic degradation, thereby inhibiting virus replication [101].

On the other hand, HCV infection may induce ER stress, which enables the unfolded protein response (UPR) to induce autophagy and enhance protein degradation, thus alleviating ER stress [102, 103]. At the same time, ER stress releases calcium ions from the endoplasmic reticulum, which increases the activity of mitochondria, leading to the production of excessive ROS and causing oxidative stress. The damaged mitochondria are then degraded by targeted mitophagy [104, 105]. Another study has shown that HCV infection induces the expression of the protein kinase PINK1 and its substrate parkin (an E3 ubiquitin ligase), which are meanwhile recruited to the mitochondrial membrane to induce mitophagy [94, 106].

In the autophagy induced by viral infection, a few viruses such as HSV-1 and SINV are targets of xenophagy [107, 108]. The mechanism is that the host recognizes the ubiquitin chain on the virus and targets the virus particles to the autophagosome for degradation [109]. This is similar to the mechanism by which a bacterial pathogen infection induces autophagy. But in most cases, autophagy is induced by viral components such as nucleic acids and proteins or indirectly stimulated by viral infection-mediated cellular stress.

6.7 Autophagy Under Heat and Cold Stress

Autophagy and the heat shock response are two cellular protein homeostasis systems, the former occurring only in eukaryotes while the latter existing in all prokaryotes and eukaryotes [110]. The heat shock protein (HSP) chaperone system can regulate the folding of newborn peptides and the transport of mature proteins [111, 112]. When cells are under heat stress, the expression of the molecular chaperone HSP increases. HSP promotes the degradation of irreversible protein aggregation during heat stress through ubiquitin-proteasome system (UPS) and chaperone-mediated autophagy pathways [111, 113]. For example, in chaperone-mediated autophagy, heat shock 70 kDa protein 8 (HSPA8/HSC70) recognizes the signal peptide (KFERQ) at the N-terminal of the cytosolic protein and then binds it to lysosomal-associated membrane protein 2A (LAMP2A) followed by subsequent degradation in lysosomes [114].

A new research suggests that in response to cellular stress, cells prioritize the heat shock response over autophagy in order to ensure protein homeostasis. When the two systems are activated, the heat shock response can inhibit autophagy [115]. Overexpression of HSP70 protein is the main executor of heat shock response [116]. In the cell culture model, overexpression of HSP70 inhibits starvation-mediated increase of LC3 and the degradation of p62, because LC3 and p62 are respectively involved in the elongation of autophagosome membranes and the degradation of autophagosomes, suggesting that overexpression of HSP70 significantly inhibits starvation-induced autophagy [115, 117, 118]. Moreover, this inhibition may be related to the activation of Akt/mTOR pathway. HSP70 phosphorylates mTOR by inducing the phosphorylation of Akt, leading to the inhibition of autophagy [19]. Knockdown of heat shock transcription factor 1 (HSF1) leads to accumulation of misfolded proteins, which increases the lipidation of LC3 associated with autophagosome formation and promotes hunger-induced autophagy [115].

According to the researches on cold stress, severe cold exposure can promote the oxidative stress of organisms and induce the increase of ROS and the damage of mitochondria. Meanwhile, cold stress also plays an important role in the regulation of autophagy [119–122]. Transient receptor potential vanilloid 1 (TRPV1) is a nonselective cation channel with high permeability to calcium ions, which can be activated by a variety of endogenous and exogenous physical or chemical stimuli and plays a key role in pain and temperature perception [123, 124]. In a study of cold stress and pressure overload [122], it was found that pressure overload produced by abdominal aortic constriction (AAC) treatment of mice led to upregulated TRPV1 expression, phosphorylated autophagy regulator AMPK, downregulated autophagy marker protein p62, and increased ratio of LC3 II/LC3 I. And subsequent continuous cold exposure not only did promote autophagy by significantly enhancing the regulation of TRPV1, AMPK, and p62 by AAC surgery but also increased the production of ROS and mitochondrial damage induced by pressure overload. Under pressure overload, cold stress-induced ROS production may help induce AMPK activation [125]. Moreover, TRPV1 promotes autophagy dependent of ROS and AMPK [126]. Therefore, it is speculated that in the state of pressure overload, the upregulation of TRPV1 expression induced by cold stress may promote excessive autophagy [122].

A study found that the mice exposed to intermittent cold stress (ICS) demonstrated induced mitochondrial dysfunction and autophagy, reflected by significantly increased transcription levels of autophagy-related genes *MAP 1LC3B* and *BECN1* that respectively encodes LC3 and beclin-1 [121]. Accordingly, another study showed increased LC3 II/LC3 I ratio and observable marks of autophagosome vesicles in cells exposed to mild cold shock, suggesting that the autophagy pathway can also be activated under mild cold stress [127].

There are few studies on autophagy-related cold stress, and the regulation of autophagy may not be directly affected by cold stress. Some have been accompanied by a combination of oxidative stress. For example, in studies on cold stress and pressure overload, cold exposure itself has little impact on the generation of autophagy regulators and ROS as well as can only enhance the effect of pressure overload on autophagy promotion. However, the mechanism of heat stress-induced autophagy is more complex. It has been reported that autophagy and the heat shock response are two systems in protein homeostasis complementing each other under certain conditions, but the underlying molecular mechanisms remain to be further explained.

6.8 Autophagy Inducers or Inhibitors

At present, some typical autophagy inhibitors including 3-methyladenine (3-MA), chloroquine (CQ), and hydroxychloroquine have been clearly clarified [128]. 3-MA inhibits autophagy by inactiviting PI3K [129], and CQ exerts inhibitory effect by disrupting the fusion of autophagy lysosomes and blocking autophagy flux [130]. Chemicals that target the autophagy-related gene themselves or key upstream autophagy regulatory proteins may alter the levels of autophagy. Sulfosalicylic acid has been shown to reduce autophagy of cachectic cancer mice in vivo [131]. On the contrary, there are also some chemicals or drugs showing a promotive effect on autophagy. For example, inflammatory bowel disease drug corticosteroids [132] could induce autophagy of bone cells, L6 myoblasts, and lymphocytes through mTORC1 inhibition [133–135]. By inhibiting mTOR, Akt, and ERK1/2 phosphorylation, harmaline can regulate B16 cell autophagy [136]. Similarly, sinomenine inhibits the PI3K/Akt/mTOR signaling pathway in melanoma cells, which reduces the proliferation of melanoma cells by stimulating autophagy and promoting the apoptosis of melanoma cells [137]. In conclusion, different chemicals or drugs have dual effects on regulation of autophagy, and the treatment of some diseases by targeting autophagy has already become the focus of researches.

6.9 Conclusions and Perspectives

As an important stress response, autophagy is closely related to the development of various diseases. In traditional view, autophagy serves as a pro-survival mechanism. However, there is also large quantity of evidence demonstrating that the excessive activation of autophagy is disadvantageous for cell survival. For example, hypoxia promotes autophagy by inhibiting the mTOR/NLRP3 pathway, which reduces the expression of inflammatory genes in Crohn's disease and improves the inflammatory response [138]. Another study also demonstrates that autophagy may play a protective role in ischemic hypoxia injury [139]. In a variety of tumor cells, autophagy is usually induced through HIF-1 [140, 141], ATG9A [142], AEG-1 [143], and other pathways under hypoxia environment, thereby promoting the proliferation and metastasis of tumor cells and reducing the chemical sensitivity of tumor cells. In summary, autophagy plays a role of double-edged sword in determining the fate of cells, which may change depending on different stress conditions, degrees of damage, and cell models. Nevertheless, the manipulation of stress-induced autophagy can become a treatment strategy for related diseases.

References

1. Mizushima N, Levine B, Cuervo AM, Klionsky DJ. Autophagy fights disease through cellular self-digestion. Nature. 2008;451:1069–75. https://doi.org/10.1038/nature06639.

2. Murrow L, Debnath J. Autophagy as a stress-response and quality-control mechanism: implications for cell injury and human disease. Annu Rev Pathol. 2013;8:105–37. https://doi.org/10.1146/annurev-pathol-020712-163918.
3. Maiuri MC, Zalckvar E, Kimchi A, Kroemer G. Self-eating and self-killing: crosstalk between autophagy and apoptosis. Nat Rev Mol Cell Biol. 2007;8:741–52. https://doi.org/10.1038/nrm2239.
4. Takeshige K, Baba M, Tsuboi S, Noda T, Ohsumi Y. Autophagy in yeast demonstrated with proteinase-deficient mutants and conditions for its induction. J Cell Biol. 1992;119:301–11. https://doi.org/10.1083/jcb.119.2.301.
5. Scherz-Shouval R, Elazar Z. ROS, mitochondria and the regulation of autophagy. Trends Cell Biol. 2007;17:422–7. https://doi.org/10.1016/j.tcb.2007.07.009.
6. Filomeni G, De Zio D, Cecconi F. Oxidative stress and autophagy: the clash between damage and metabolic needs. Cell Death Differ. 2015;22:377–88. https://doi.org/10.1038/cdd.2014.150.
7. Lee J, Giordano S, Zhang J. Autophagy, mitochondria and oxidative stress: cross-talk and redox signalling. Biochem J. 2012;441:523–40. https://doi.org/10.1042/bj20111451.
8. Li L, Tan J, Miao Y, Lei P, Zhang Q. ROS and autophagy: interactions and molecular regulatory mechanisms. Cell Mol Neurobiol. 2015;35:615–21. https://doi.org/10.1007/s10571-015-0166-x.
9. Filomeni G, Desideri E, Cardaci S, Rotilio G, Ciriolo MR. Under the ROS...thiol network is the principal suspect for autophagy commitment. Autophagy. 2010;6:999–1005. https://doi.org/10.4161/auto.6.7.12754.
10. Bensaad K, Cheung EC, Vousden KH. Modulation of intracellular ROS levels by TIGAR controls autophagy. EMBO J. 2009;28:3015–26. https://doi.org/10.1038/emboj.2009.242.
11. Scherz-Shouval R, et al. Reactive oxygen species are essential for autophagy and specifically regulate the activity of Atg4. EMBO J. 2007;26:1749–60. https://doi.org/10.1038/sj.emboj.7601623.
12. Scherz-Shouval R, Elazar Z. Regulation of autophagy by ROS: physiology and pathology. Trends Biochem Sci. 2011;36:30–8. https://doi.org/10.1016/j.tibs.2010.07.007.
13. Zoncu R, Efeyan A, Sabatini DM. mTOR: from growth signal integration to cancer, diabetes and ageing. Nat Rev Mol Cell Biol. 2011;12:21–35. https://doi.org/10.1038/nrm3025.
14. Filomeni G, Rotilio G, Ciriolo MR. Cell signalling and the glutathione redox system. Biochem Pharmacol. 2002;64:1057–64. https://doi.org/10.1016/s0006-2952(02)01176-0.
15. Bindoli A, Fukuto JM, Forman HJ. Thiol chemistry in peroxidase catalysis and redox signaling. Antioxid Redox Signal. 2008;10:1549–64. https://doi.org/10.1089/ars.2008.2063.
16. Cardaci S, Filomeni G, Ciriolo MR. Redox implications of AMPK-mediated signal transduction beyond energetic clues. J Cell Sci. 2012;125:2115–25. https://doi.org/10.1242/jcs.095216.
17. Zmijewski JW, et al. Exposure to hydrogen peroxide induces oxidation and activation of AMP-activated protein kinase. J Biol Chem. 2010;285:33154–64. https://doi.org/10.1074/jbc.M110.143685.
18. Wolff NC, et al. Cell-type-dependent regulation of mTORC1 by REDD1 and the tumor suppressors TSC1/TSC2 and LKB1 in response to hypoxia. Mol Cell Biol. 2017;37 https://doi.org/10.1128/mcb.00224-17.
19. Nazio F, Cecconi F. mTOR, AMBRA1, and autophagy: an intricate relationship. Cell Cycle. 2013;12:2524–5. https://doi.org/10.4161/cc.25835.
20. Fimia GM, Di Bartolomeo S, Piacentini M, Cecconi F. Unleashing the Ambra1-Beclin 1 complex from dynein chains: Ulk1 sets Ambra1 free to induce autophagy. Autophagy. 2011;7:115–7. https://doi.org/10.4161/auto.7.1.14071.
21. Vadlamudi RK, Joung I, Strominger JL, Shin J. p62, a phosphotyrosine-independent ligand of the SH2 domain of p56lck, belongs to a new class of ubiquitin-binding proteins. J Biol Chem. 1996;271:20235–7. https://doi.org/10.1074/jbc.271.34.20235.

22. Bjørkøy G, et al. p62/SQSTM1 forms protein aggregates degraded by autophagy and has a protective effect on huntingtin-induced cell death. J Cell Biol. 2005;171:603–14. https://doi.org/10.1083/jcb.200507002.
23. Giles NM, Gutowski NJ, Giles GI, Jacob C. Redox catalysts as sensitisers towards oxidative stress. FEBS Lett. 2003;535:179–82. https://doi.org/10.1016/s0014-5793(02)03890-5.
24. Scherz-Shouval R, et al. Reactive oxygen species are essential for autophagy and specifically regulate the activity of Atg4. Embo J. 2019;38 https://doi.org/10.15252/embj.2019101812.
25. Smith M, Wilkinson S. ER homeostasis and autophagy. Essays Biochem. 2017;61:625–35. https://doi.org/10.1042/ebc20170092.
26. Buchberger A, Bukau B, Sommer T. Protein quality control in the cytosol and the endoplasmic reticulum: brothers in arms. Mol Cell. 2010;40:238–52. https://doi.org/10.1016/j.molcel.2010.10.001.
27. Qi Z, Chen L. Endoplasmic reticulum stress and autophagy. Adv Exp Med Biol. 2019;1206:167–77. https://doi.org/10.1007/978-981-15-0602-4_8.
28. Senft D, Ronai Z, et al. UPR, autophagy, and mitochondria crosstalk underlies the ER stress response. Trends Biochem Sci. 2015;40:141–8. https://doi.org/10.1016/j.tibs.2015.01.002.
29. Cheng Y-C, Chang J-M, Chen C-A, Chen H-C. Autophagy modulates endoplasmic reticulum stress-induced cell death in podocytes: a protective role. Exp Biol Med (Maywood). 2015;240:467–76. https://doi.org/10.1177/1535370214553772.
30. Kouroku Y, et al. ER stress (PERK/eIF2alpha phosphorylation) mediates the polyglutamine-induced LC3 conversion, an essential step for autophagy formation. Cell Death Differ. 2007;14:230–9. https://doi.org/10.1038/sj.cdd.4401984.
31. Ogata M, et al. Autophagy is activated for cell survival after endoplasmic reticulum stress. Mol Cell Biol. 2006;26:9220–31. https://doi.org/10.1128/mcb.01453-06.
32. Hoyer-Hansen M, et al. Control of macroautophagy by calcium, calmodulin-dependent kinase kinase-beta, and Bcl-2. Mol Cell. 2007;25:193–205. https://doi.org/10.1016/j.molcel.2006.12.009.
33. Yang Y, et al. The ER-localized Ca(2+)-binding protein calreticulin couples ER stress to autophagy by associating with microtubule-associated protein 1A/1B light chain 3. J Biol Chem. 2019;294:772–82. https://doi.org/10.1074/jbc.RA118.005166.
34. Choi H, et al. Hypoxia promotes noncanonical autophagy in nucleus pulposus cells independent of MTOR and HIF1A signaling. Autophagy. 2016;12:1631–46. https://doi.org/10.1080/15548627.2016.1192753.
35. Wu H, et al. Hypoxia-induced autophagy contributes to the invasion of salivary adenoid cystic carcinoma through the HIF-1α/BNIP3 signaling pathway. Mol Med Rep. 2015;12:6467–74. https://doi.org/10.3892/mmr.2015.4255.
36. Shi R-Y, et al. BNIP3 interacting with LC3 triggers excessive mitophagy in delayed neuronal death in stroke. CNS Neurosci Ther. 2014;20:1045–55. https://doi.org/10.1111/cns.12325.
37. Bellot G, et al. Hypoxia-induced autophagy is mediated through hypoxia-inducible factor induction of BNIP3 and BNIP3L via their BH3 domains. Mol Cell Biol. 2009;29:2570–81. https://doi.org/10.1128/MCB.00166-09.
38. Gwinn DM, et al. AMPK phosphorylation of raptor mediates a metabolic checkpoint. Mol Cell. 2008;30:214–26. https://doi.org/10.1016/j.molcel.2008.03.003.
39. Rouschop KMA, et al. The unfolded protein response protects human tumor cells during hypoxia through regulation of the autophagy genes MAP 1LC3B and ATG5. J Clin Invest. 2010;120:127–41. https://doi.org/10.1172/JCI40027.
40. Song S, Tan J, Miao Y, Sun Z, Zhang Q. Intermittent-hypoxia-induced autophagy activation through the ER-stress-related PERK/eIF2α/ATF4 pathway is a protective response to pancreatic β-cell apoptosis. Cell Physiol Biochem. 2018;51:2955–71. https://doi.org/10.1159/000496047.
41. Peña-Oyarzun D, et al. Hyperosmotic stress stimulates autophagy via polycystin-2. Oncotarget. 2017;8:55984–97. https://doi.org/10.18632/oncotarget.18995.

42. Schutt KL, Moseley JB. Transient activation of fission yeast AMPK is required for cell proliferation during osmotic stress. Mol Biol Cell. 2017;28:1804–14. https://doi.org/10.1091/mbc.E17-04-0235.
43. Nunes P, et al. Hypertonic stress promotes autophagy and microtubule-dependent autophagosomal clusters. Autophagy. 2013;9:550–67. https://doi.org/10.4161/auto.23662.
44. Jiang LB, et al. Activation of autophagy via Ca(2+)-dependent AMPK/mTOR pathway in rat notochordal cells is a cellular adaptation under hyperosmotic stress. Cell Cycle. 2015;14:867–79. https://doi.org/10.1080/15384101.2015.1004946.
45. Soto-Burgos J, Bassham DC. SnRK1 activates autophagy via the TOR signaling pathway in Arabidopsis thaliana. PLoS One. 2017;12:e0182591. https://doi.org/10.1371/journal.pone.0182591.
46. Pu Y, Soto-Burgos J, Bassham DC. Regulation of autophagy through SnRK1 and TOR signaling pathways. Plant Signal Behav. 2017;12:e1395128. https://doi.org/10.1080/15592324.2017.1395128.
47. Gatica D, Lahiri V, Klionsky DJ. Cargo recognition and degradation by selective autophagy. Nat Cell Biol. 2018;20:233–42. https://doi.org/10.1038/s41556-018-0037-z.
48. McEwan DG. Host-pathogen interactions and subversion of autophagy. Essays Biochem. 2017;61:687–97. https://doi.org/10.1042/ebc20170058.
49. Sharma V, Verma S, Seranova E, Sarkar S, Kumar D. Selective autophagy and Xenophagy in infection and disease. Front Cell Dev Biol. 2018;6:147. https://doi.org/10.3389/fcell.2018.00147.
50. Thurston TL, Wandel MP, von Muhlinen N, Foeglein A, Randow F. Galectin 8 targets damaged vesicles for autophagy to defend cells against bacterial invasion. Nature. 2012;482:414–8. https://doi.org/10.1038/nature10744.
51. Boyle KB, Randow F. The role of 'eat-me' signals and autophagy cargo receptors in innate immunity. Curr Opin Microbiol. 2013;16:339–48. https://doi.org/10.1016/j.mib.2013.03.010.
52. Shaid S, Brandts CH, Serve H, Dikic I. Ubiquitination and selective autophagy. Cell Death Differ. 2013;20:21–30. https://doi.org/10.1038/cdd.2012.72.
53. Pankiv S, et al. p62/SQSTM1 binds directly to Atg8/LC3 to facilitate degradation of ubiquitinated protein aggregates by autophagy. J Biol Chem. 2007;282:24131–45. https://doi.org/10.1074/jbc.M702824200.
54. von Muhlinen N, et al. LC3C, bound selectively by a noncanonical LIR motif in NDP52, is required for antibacterial autophagy. Mol Cell. 2012;48:329–42. https://doi.org/10.1016/j.molcel.2012.08.024.
55. Wild P, et al. Phosphorylation of the autophagy receptor optineurin restricts Salmonella growth. Science. 2011;333:228–33. https://doi.org/10.1126/science.1205405.
56. Tumbarello DA, et al. The autophagy receptor TAX1BP1 and the molecular motor myosin VI are required for clearance of Salmonella Typhimurium by autophagy. PLoS Pathog. 2015;11:e1005174. https://doi.org/10.1371/journal.ppat.1005174.
57. Morreale FE, Walden H. Types of ubiquitin ligases. Cell. 2016;165:248–248.e241. https://doi.org/10.1016/j.cell.2016.03.003.
58. van Wijk SJL, et al. Linear ubiquitination of cytosolic Salmonella Typhimurium activates NF-kappaB and restricts bacterial proliferation. Nat Microbiol. 2017;2:17066. https://doi.org/10.1038/nmicrobiol.2017.66.
59. van Wijk SJ, et al. Fluorescence-based sensors to monitor localization and functions of linear and K63-linked ubiquitin chains in cells. Mol Cell. 2012;47:797–809. https://doi.org/10.1016/j.molcel.2012.06.017.
60. Grumati P, Dikic I. Ubiquitin signaling and autophagy. J Biol Chem. 2018;293:5404–13. https://doi.org/10.1074/jbc.TM117.000117.
61. Linares JF, et al. K63 polyubiquitination and activation of mTOR by the p62-TRAF6 complex in nutrient-activated cells. Mol Cell. 2013;51:283–96. https://doi.org/10.1016/j.molcel.2013.06.020.

62. Manzanillo PS, et al. The ubiquitin ligase parkin mediates resistance to intracellular pathogens. Nature. 2013;501:512–6. https://doi.org/10.1038/nature12566.
63. Huett A, et al. The LRR and RING domain protein LRSAM1 is an E3 ligase crucial for ubiquitin-dependent autophagy of intracellular Salmonella Typhimurium. Cell Host Microbe. 2012;12:778–90. https://doi.org/10.1016/j.chom.2012.10.019.
64. Noad J, et al. LUBAC-synthesized linear ubiquitin chains restrict cytosol-invading bacteria by activating autophagy and NF-kappaB. Nat Microbiol. 2017;2:17063. https://doi.org/10.1038/nmicrobiol.2017.63.
65. Polajnar M, Dietz MS, Heilemann M, Behrends C. Expanding the host cell ubiquitylation machinery targeting cytosolic Salmonella. EMBO Rep. 2017;18:1572–85. https://doi.org/10.15252/embr.201643851.
66. Stolz A, Ernst A, Dikic I. Cargo recognition and trafficking in selective autophagy. Nat Cell Biol. 2014;16:495–501. https://doi.org/10.1038/ncb2979.
67. Kraft C, et al. Binding of the Atg1/ULK1 kinase to the ubiquitin-like protein Atg8 regulates autophagy. EMBO J. 2012;31:3691–703. https://doi.org/10.1038/emboj.2012.225.
68. Pankiv S, et al. FYCO1 is a Rab7 effector that binds to LC3 and PI3P to mediate microtubule plus end-directed vesicle transport. J Cell Biol. 2010;188:253–69. https://doi.org/10.1083/jcb.200907015.
69. Popovic D, et al. Rab GTPase-activating proteins in autophagy: regulation of endocytic and autophagy pathways by direct binding to human ATG8 modifiers. Mol Cell Biol. 2012;32:1733–44. https://doi.org/10.1128/mcb.06717-11.
70. McEwan DG, et al. PLEKHM1 regulates autophagosome-lysosome fusion through HOPS complex and LC3/GABARAP proteins. Mol Cell. 2015;57:39–54. https://doi.org/10.1016/j.molcel.2014.11.006.
71. Genau HM, et al. CUL3-KBTBD6/KBTBD7 ubiquitin ligase cooperates with GABARAP proteins to spatially restrict TIAM1-RAC1 signaling. Mol Cell. 2015;57:995–1010. https://doi.org/10.1016/j.molcel.2014.12.040.
72. Verlhac P, et al. Autophagy receptor NDP52 regulates pathogen-containing autophagosome maturation. Cell Host Microbe. 2015;17:515–25. https://doi.org/10.1016/j.chom.2015.02.008.
73. Petkova DS, et al. Distinct contributions of autophagy receptors in measles virus replication. Viruses. 2017;9 https://doi.org/10.3390/v9050123.
74. Zheng YT, et al. The adaptor protein p62/SQSTM1 targets invading bacteria to the autophagy pathway. J Immunol. 2009;183:5909–16. https://doi.org/10.4049/jimmunol.0900441.
75. Thurston TL, et al. Recruitment of TBK1 to cytosol-invading Salmonella induces WIPI2-dependent antibacterial autophagy. Embo J. 2016;35:1779–92. https://doi.org/10.15252/embj.201694491.
76. Cheng CY, Chi PI, Liu HJ. Commentary on the regulation of viral proteins in autophagy process. Biomed Res Int. 2014;2014:962915. https://doi.org/10.1155/2014/962915.
77. Perot BP, Ingersoll MA, Albert ML. The impact of macroautophagy on CD8(+) T-cell-mediated antiviral immunity. Immunol Rev. 2013;255:40–56. https://doi.org/10.1111/imr.12096.
78. Mao J, et al. Autophagy and viral infection. Adv Exp Med Biol. 2019;1209:55–78. https://doi.org/10.1007/978-981-15-0606-2_5.
79. Chiramel AI, Brady NR, Bartenschlager R. Divergent roles of autophagy in virus infection. Cell. 2013;2:83–104. https://doi.org/10.3390/cells2010083.
80. Joubert PE, et al. Autophagy induction by the pathogen receptor CD46. Cell Host Microbe. 2009;6:354–66. https://doi.org/10.1016/j.chom.2009.09.006.
81. Meiffren G, et al. Pathogen recognition by the cell surface receptor CD46 induces autophagy. Autophagy. 2010;6:299–300. https://doi.org/10.4161/auto.6.2.11132.
82. Mouna L, et al. Analysis of the role of autophagy inhibition by two complementary human cytomegalovirus BECN1/Beclin 1-binding proteins. Autophagy. 2016;12:327–42. https://doi.org/10.1080/15548627.2015.1125071.

83. Grégoire IP, Rabourdin-Combe C, Faure M. Autophagy and RNA virus interactomes reveal IRGM as a common target. Autophagy. 2012;8:1136–7. https://doi.org/10.4161/auto.20339.
84. Petkova DS, Viret C, Faure M. IRGM in autophagy and viral infections. Front Immunol. 2012;3:426. https://doi.org/10.3389/fimmu.2012.00426.
85. Abdoli A, Alirezaei M, Mehrbod P, Forouzanfar F. Autophagy: the multi-purpose bridge in viral infections and host cells. Rev Med Virol. 2018;28:e1973. https://doi.org/10.1002/rmv.1973.
86. Castedo M, et al. Sequential involvement of Cdk1, mTOR and p53 in apoptosis induced by the HIV-1 envelope. EMBO J. 2002;21:4070–80. https://doi.org/10.1093/emboj/cdf391.
87. Denizot M, et al. HIV-1 gp41 fusogenic function triggers autophagy in uninfected cells. Autophagy. 2008;4:998–1008. https://doi.org/10.4161/auto.6880.
88. Chan ST, Ou JJ. Hepatitis C virus-induced autophagy and host innate immune response. Viruses. 2017;9 https://doi.org/10.3390/v9080224.
89. Choi Y, Bowman JW, Jung JU. Autophagy during viral infection - a double-edged sword. Nat Rev Microbiol. 2018;16:341–54. https://doi.org/10.1038/s41579-018-0003-6.
90. Saito T, Owen DM, Jiang F, Marcotrigiano J, Gale M Jr. Innate immunity induced by composition-dependent RIG-I recognition of hepatitis C virus RNA. Nature. 2008;454:523–7. https://doi.org/10.1038/nature07106.
91. Li K, et al. Activation of chemokine and inflammatory cytokine response in hepatitis C virus-infected hepatocytes depends on toll-like receptor 3 sensing of hepatitis C virus double-stranded RNA intermediates. Hepatology. 2012;55:666–75. https://doi.org/10.1002/hep.24763.
92. García MA, et al. Impact of protein kinase PKR in cell biology: from antiviral to antiproliferative action. Microbiol Mol Biol Rev. 2006;70:1032–60. https://doi.org/10.1128/mmbr.00027-06.
93. Liang Q, et al. Crosstalk between the cGAS DNA sensor and Beclin-1 autophagy protein shapes innate antimicrobial immune responses. Cell Host Microbe. 2014;15:228–38. https://doi.org/10.1016/j.chom.2014.01.009.
94. Wang L, Ou JH. Hepatitis C virus and autophagy. Biol Chem. 2015;396:1215–22. https://doi.org/10.1515/hsz-2015-0172.
95. Grégoire IP, et al. IRGM is a common target of RNA viruses that subvert the autophagy network. PLoS Pathog. 2011;7:e1002422. https://doi.org/10.1371/journal.ppat.1002422.
96. Su WC, et al. Rab5 and class III phosphoinositide 3-kinase Vps34 are involved in hepatitis C virus NS4B-induced autophagy. J Virol. 2011;85:10561–71. https://doi.org/10.1128/jvi.00173-11.
97. Wang L, Tian Y, Ou JH. HCV induces the expression of Rubicon and UVRAG to temporally regulate the maturation of autophagosomes and viral replication. PLoS Pathog. 2015;11:e1004764. https://doi.org/10.1371/journal.ppat.1004764.
98. Chatterji U, Bobardt M, Tai A, Wood M, Gallay PA. Cyclophilin and NS5A inhibitors, but not other anti-hepatitis C virus (HCV) agents, preclude HCV-mediated formation of double-membrane-vesicle viral factories. Antimicrob Agents Chemother. 2015;59:2496–507. https://doi.org/10.1128/aac.04958-14.
99. Quan M, et al. A functional role for NS5ATP9 in the induction of HCV NS5A-mediated autophagy. J Viral Hepat. 2014;21:405–15. https://doi.org/10.1111/jvh.12155.
100. Shrivastava S, Bhanja Chowdhury J, Steele R, Ray R, Ray RB. Hepatitis C virus upregulates Beclin1 for induction of autophagy and activates mTOR signaling. J Virol. 2012;86:8705–12. https://doi.org/10.1128/jvi.00616-12.
101. Kim N, et al. Interferon-inducible protein SCOTIN interferes with HCV replication through the autolysosomal degradation of NS5A. Nat Commun. 2016;7:10631. https://doi.org/10.1038/ncomms10631.
102. Joyce MA, et al. HCV induces oxidative and ER stress, and sensitizes infected cells to apoptosis in SCID/Alb-uPA mice. PLoS Pathog. 2009;5:e1000291. https://doi.org/10.1371/journal.ppat.1000291.

103. Sir D, et al. Induction of incomplete autophagic response by hepatitis C virus via the unfolded protein response. Hepatology. 2008;48:1054–61. https://doi.org/10.1002/hep.22464.
104. Vescovo T, et al. Autophagy in HCV infection: keeping fat and inflammation at bay. Biomed Res Int. 2014;2014:265353. https://doi.org/10.1155/2014/265353.
105. Ivanov AV, Bartosch B, Smirnova OA, Isaguliants MG, Kochetkov SN. HCV and oxidative stress in the liver. Viruses. 2013;5:439–69. https://doi.org/10.3390/v5020439.
106. Kim SJ, Syed GH, Siddiqui A. Hepatitis C virus induces the mitochondrial translocation of Parkin and subsequent mitophagy. PLoS Pathog. 2013;9:e1003285. https://doi.org/10.1371/journal.ppat.1003285.
107. Tallóczy Z, Virgin HWT, Levine B. PKR-dependent autophagic degradation of herpes simplex virus type 1. Autophagy. 2006;2:24–9. https://doi.org/10.4161/auto.2176.
108. Orvedahl A, et al. Autophagy protects against Sindbis virus infection of the central nervous system. Cell Host Microbe. 2010;7:115–27. https://doi.org/10.1016/j.chom.2010.01.007.
109. Levine B. Eating oneself and uninvited guests: autophagy-related pathways in cellular defense. Cell. 2005;120:159–62. https://doi.org/10.1016/j.cell.2005.01.005.
110. Dokladny K, Myers OB, Moseley PL. Heat shock response and autophagy--cooperation and control. Autophagy. 2015;11:200–13. https://doi.org/10.1080/15548627.2015.1009776.
111. Morimoto RI, Sarge KD, Abravaya K. Transcriptional regulation of heat shock genes. A paradigm for inducible genomic responses. J Biol Chem. 1992;267:21987–90.
112. Ellis RJ. The molecular chaperone concept. Semin Cell Biol. 1990;1:1–9.
113. Goloubinoff P. Mechanisms of protein homeostasis in health, aging and disease. Swiss Med Wkly. 2016;146:w14306. https://doi.org/10.4414/smw.2016.14306.
114. Chiang HL, Terlecky SR, Plant CP, Dice JF. A role for a 70-kilodalton heat shock protein in lysosomal degradation of intracellular proteins. Science. 1989;246:382–5. https://doi.org/10.1126/science.2799391.
115. Dokladny K, et al. Regulatory coordination between two major intracellular homeostatic systems: heat shock response and autophagy. J Biol Chem. 2013;288:14959–72. https://doi.org/10.1074/jbc.M113.462408.
116. Fernandez-Fernandez MR, Gragera M, Ochoa-Ibarrola L, Quintana-Gallardo L, Valpuesta JM. Hsp70 - a master regulator in protein degradation. FEBS Lett. 2017;591:2648–60. https://doi.org/10.1002/1873-3468.12751.
117. Mizushima N, Yoshimori T, Levine B. Methods in mammalian autophagy research. Cell. 2010;140:313–26. https://doi.org/10.1016/j.cell.2010.01.028.
118. Komatsu M, et al. Homeostatic levels of p62 control cytoplasmic inclusion body formation in autophagy-deficient mice. Cell. 2007;131:1149–63. https://doi.org/10.1016/j.cell.2007.10.035.
119. Cong P, et al. Cold exposure induced oxidative stress and apoptosis in the myocardium by inhibiting the Nrf2-Keap1 signaling pathway. BMC Cardiovasc Disord. 2018;18:36. https://doi.org/10.1186/s12872-018-0748-x.
120. Sharma A, et al. Cold environment exacerbates brain pathology and oxidative stress following traumatic brain injuries: potential therapeutic effects of Nanowired antioxidant compound H-290/51. Mol Neurobiol. 2018;55:276–85. https://doi.org/10.1007/s12035-017-0740-y.
121. Oezel L, et al. Fibromyalgia syndrome: metabolic and autophagic processes in intermittent cold stress mice. Pharmacol Res Perspect. 2016;4:e00248. https://doi.org/10.1002/prp2.248.
122. Lu S, Xu D. Cold stress accentuates pressure overload-induced cardiac hypertrophy and contractile dysfunction: role of TRPV1/AMPK-mediated autophagy. Biochem Biophys Res Commun. 2013;442:8–15. https://doi.org/10.1016/j.bbrc.2013.10.128.
123. Caterina MJ, et al. The capsaicin receptor: a heat-activated ion channel in the pain pathway. Nature. 1997;389:816–24. https://doi.org/10.1038/39807.
124. Caterina MJ, Rosen TA, Tominaga M, Brake AJ, Julius D. A capsaicin-receptor homologue with a high threshold for noxious heat. Nature. 1999;398:436–41. https://doi.org/10.1038/18906.

125. Rider MH, Hussain N, Horman S, Dilworth SM, Storey KB. Stress-induced activation of the AMP-activated protein kinase in the freeze-tolerant frog Rana sylvatica. Cryobiology. 2006;53:297–309. https://doi.org/10.1016/j.cryobiol.2006.08.001.
126. Farfariello V, Amantini C, Santoni G. Transient receptor potential vanilloid 1 activation induces autophagy in thymocytes through ROS-regulated AMPK and Atg4C pathways. J Leukoc Biol. 2012;92:421–31. https://doi.org/10.1189/jlb.0312123.
127. Neutelings T, Lambert CA, Nusgens BV, Colige AC. Effects of mild cold shock (25°C) followed by warming up at 37°C on the cellular stress response. PLoS One. 2013;8:e69687. https://doi.org/10.1371/journal.pone.0069687.
128. Chude CI, Amaravadi RK. Targeting autophagy in cancer: update on clinical trials and novel inhibitors. Int J Mol Sci. 2017;18 https://doi.org/10.3390/ijms18061279.
129. Wu YT, et al. Dual role of 3-methyladenine in modulation of autophagy via different temporal patterns of inhibition on class I and III phosphoinositide 3-kinase. J Biol Chem. 2010;285:10850–61. https://doi.org/10.1074/jbc.M109.080796.
130. Mauthe M, et al. Chloroquine inhibits autophagic flux by decreasing autophagosome-lysosome fusion. Autophagy. 2018;14:1435–55. https://doi.org/10.1080/15548627.2018.1474314.
131. Chacon-Cabrera A, et al. Pharmacological strategies in lung cancer-induced cachexia: effects on muscle proteolysis, autophagy, structure, and weakness. J Cell Physiol. 2014;229:1660–72. https://doi.org/10.1002/jcp.24611.
132. Hooper KM, Barlow PG, Stevens C, Henderson P. Inflammatory bowel disease drugs: a focus on autophagy. J Crohns Colitis. 2017;11:118–27. https://doi.org/10.1093/ecco-jcc/jjw127.
133. Kar R, Riquelme MA, Hua R, Jiang JX. Glucocorticoid-induced autophagy protects osteocytes against oxidative stress through activation of MAPK/ERK signaling. JBMR Plus. 2019;3:e10077. https://doi.org/10.1002/jbm4.10077.
134. Wang H, Kubica N, Ellisen LW, Jefferson LS, Kimball SR. Dexamethasone represses signaling through the mammalian target of rapamycin in muscle cells by enhancing expression of REDD1. J Biol Chem. 2006;281:39128–34. https://doi.org/10.1074/jbc.M610023200.
135. Fatkhullina AR, Abramov SN, Skibo Iu V, Abramova ZI. Dexamethasone affect on the expression of bcl-2 and mTOR genes in T-lymphocytes from healthy donors. Tsitologiia. 2014;56:459–61.
136. Zou N, et al. The inhibitory effects of compound Muniziqi granule against B16 cells and harmine induced autophagy and apoptosis by inhibiting Akt/mTOR pathway. BMC Complement Altern Med. 2017;17:517. https://doi.org/10.1186/s12906-017-2017-4.
137. Sun Z, Zheng L, Liu X, Xing W, Liu X. Sinomenine inhibits the growth of melanoma by enhancement of autophagy via PI3K/AKT/mTOR inhibition. Drug Des Devel Ther. 2018;12:2413–21. https://doi.org/10.2147/dddt.S155798.
138. Cosin-Roger J, et al. Hypoxia ameliorates intestinal inflammation through NLRP3/mTOR downregulation and autophagy activation. Nat Commun. 2017;8:98. https://doi.org/10.1038/s41467-017-00213-3.
139. Shi Z-Y, et al. Protective effect of autophagy in neural ischemia and hypoxia: negative regulation of the Wnt/β-catenin pathway. Int J Mol Med. 2017;40:1699–708. https://doi.org/10.3892/ijmm.2017.3158.
140. Zhang W, Yuan W, Song J, Wang S, Gu X. LncRNA CPS1-IT1 suppresses EMT and metastasis of colorectal cancer by inhibiting hypoxia-induced autophagy through inactivation of HIF-1α. Biochimie. 2018;144:21–7. https://doi.org/10.1016/j.biochi.2017.10.002.
141. Yang X, et al. Hypoxia-induced autophagy promotes gemcitabine resistance in human bladder cancer cells through hypoxia-inducible factor 1α activation. Int J Oncol. 2018;53:215–24. https://doi.org/10.3892/ijo.2018.4376.
142. Abdul Rahim SA, et al. Regulation of hypoxia-induced autophagy in glioblastoma involves ATG9A. Br J Cancer. 2017;117:813–25. https://doi.org/10.1038/bjc.2017.263.
143. Yan J, et al. AEG-1 is involved in hypoxia-induced autophagy and decreases chemosensitivity in T-cell lymphoma. Mol Med. 2018;24:35. https://doi.org/10.1186/s10020-018-0033-6.

Cysteine Metabolism in Cancer Progression and Therapy Resistance

7

Dawn Sijin Nin, Shabana Binte Idres, and Lih-Wen Deng

Abstract

Redox homeostasis and sustainable energy production are vital for cancer survival and growth. As cancer cells proliferate uncontrollably, metabolic substrate dwindles, while reactive chemical species rise. To overcome these challenges, cancer cells evolve to switch to alternative metabolic substrate available and boost its anti-oxidative defense. Emerging studies implicate nonessential amino acid cysteine in these two pro-cancer processes, which in turn promote resistance to therapy. Here, we will discuss the role of cysteine metabolism in cancer progression, focusing on how increased cysteine uptake enables metabolic flexibility and benefits cancer survival. The interplay between cysteine metabolism and other metabolic pathways and the regulation of cysteine homeostasis in cancer and the tumor microenvironment will also be addressed. Finally, we will also discuss current developments in targeting cysteine flux to impede tumor growth and therapy resistance.

Keywords

xCT · Glutathione · Cysteine · Metabolism · Ferroptosis

D. S. Nin · S. B. Idres
Department of Biochemistry, Yong Loo Lin School of Medicine, National University of Singapore, Singapore, Singapore

L.-W. Deng (✉)
Department of Biochemistry, Yong Loo Lin School of Medicine, National University of Singapore, Singapore, Singapore

National University Cancer Institute, National University Health System, Singapore, Singapore
e-mail: bchdlw@nus.edu.sg

C. Huang, Y. Zhang (eds.), *Oxidative Stress*,
https://doi.org/10.1007/978-981-16-0522-2_7

7.1 Introduction

Cancer cells need to generate sufficient energy to support cell proliferation; besides, they also need to meet the anabolic demands of macromolecular biosynthesis (nucleotides, lipids, and proteins) necessary for growth expansion [1, 2]. Further, it is also essential for cancer cells to maintain cellular redox homeostasis in response to the rapid production of toxic reactive oxygen species (ROS) [3]. Nutrient supplies are often in excess under in vitro culture conditions; however, nutrient availability within a tumor mass in in vivo conditions is generally limited due to poor tumor vasculature in the tumor microenvironment. Coupled with the increased innate nutritional needs of cancer cells, the lack of nutrient supply in a tumor mass often results in a state of metabolic stress in these cells [4]. To fulfill the energy and biomass requirements for survival and growth under conditions of high metabolic stress, cancer cells frequently undergo metabolic rewiring as a form of metabolic adaptation to increase their metabolic flexibility [1, 2, 4]. It is well known that cysteine plays a major role in the balance of intracellular redox and is also pivotal in cellular processes such as cell proliferation and survival [5, 6]. Over the last few years, an increasing body of evidence has emerged, demonstrating that cysteine metabolism can contribute to tumor progression and survival, via the maintenance of redox homeostasis either on its own or as a component of glutathione (GSH) synthesis, rewiring metabolic pathways through cross talk with glucose and glutamine metabolism and with the pentose phosphate pathway [6–12]. In the present review, emphasis will be given to the role of cysteine metabolism in cancer development, with a focus on how increased cysteine uptake and metabolic reliance on cysteine can benefit metabolic remodeling and cancer survival. The interplay between cysteine metabolism and other metabolic pathways, as well as the regulation of cysteine metabolism-related enzymes and transporters in cancer and the tumor microenvironment, will also be addressed. Finally, we will also discuss current developments in targeting cysteine flux and its associated metabolic vulnerability in tumor growth and therapy resistance.

7.2 Cysteine Homeostasis

The body meets its cysteine requirements through multiple sources, including dietary absorption, de novo synthesis through the transsulfuration pathway from methionine and serine, via the breakdown of extracellular GSH, and protein catabolism [6, 11, 12] (Fig. 7.1). The mammalian liver removes a substantial proportion of cysteine from the portal circulation, and the levels of intracellular cysteine pool and plasma cysteine are tightly controlled by the liver [11]. For instance, in rats, intracellular cysteine remains within a narrow window of 20–100 nmol/g even when dietary protein or sulfur amino acid intake ranges from sub-requirement to excess levels. The tight regulation of cysteine balance underscores the importance of maintaining homeostatic cysteine levels within the cell. While cysteine levels must be kept sufficiently high enough to meet multiple cellular needs such as for protein

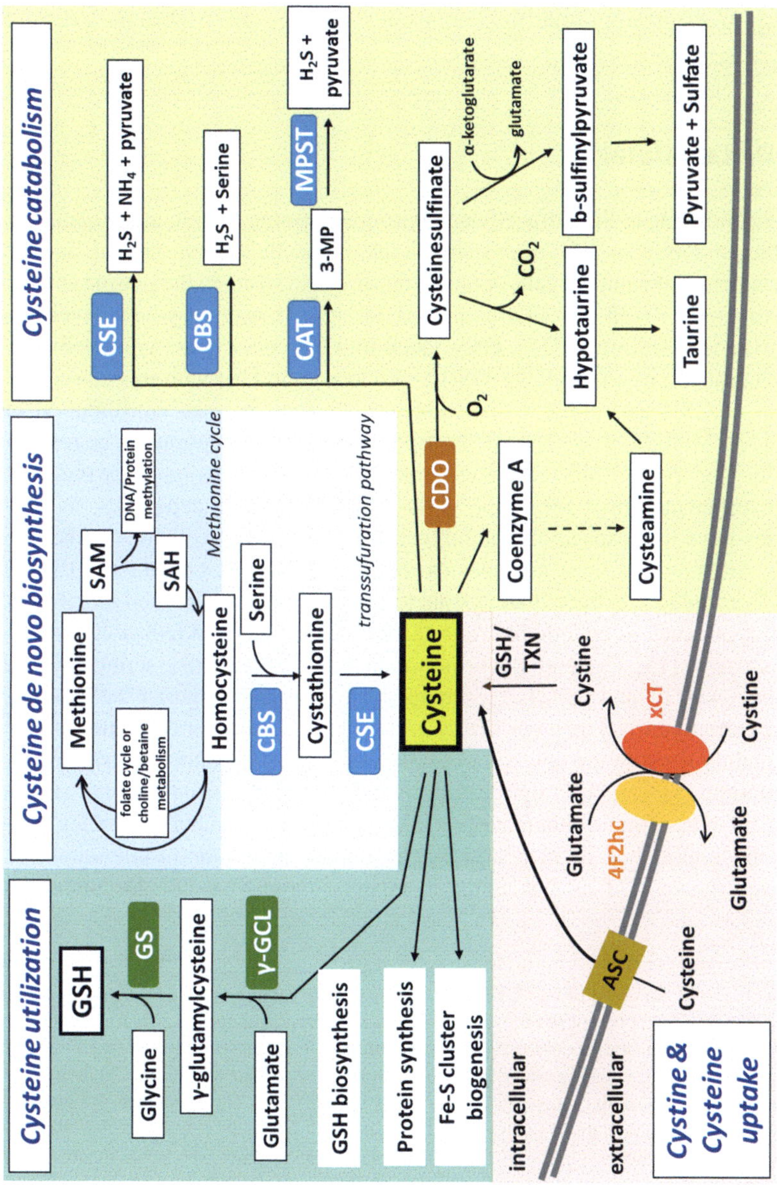

Fig. 7.1 Homeostasis of cysteine metabolism. Cellular cysteine is mainly from de novo synthesis via the transsulfuration pathway from methionine and serine, as well as from the uptake of cystine via system x_c^- antiporter or cysteine transporters like alanine-serine-cysteine (ASC) transporter. Cellular cysteine is utilized

synthesis and production of some essential molecules such as GSH, coenzyme A, and taurine, there is also a need to keep cysteine levels below the threshold of cytotoxicity. Pathologically, while low cysteine levels result in increased susceptibility to oxidative stress, chronically high cysteine levels have been associated with a higher risk of rheumatoid arthritis, Parkinson's disease, and Alzheimer's disease [13, 14].

7.2.1 Dietary Uptake

Oxidizing conditions in the extracellular environment promote cysteine oxidation to cystine, the predominant form of cysteine in the extracellular environment, and the major form being taken up by cells. Concentrations of cysteine in the plasma are low at approximately 10–25 μmol/L, compared to that of cystine where levels lie between 50 and 150 μmol/L [15]. Once taken into the cell, cystine is immediately reduced to cysteine by intracellular GSH and thioredoxin [16] with the subsequent restoration of GSH and thioredoxin redox balance requiring the consumption of NADPH [17]. Multiple tissue-specific transporters have been reported to be responsible for the import of cystine and/or cysteine [18, 19], and cells typically transport one more efficiently than the other. It is noted that some cell types such as hepatocytes have low capacity for the direct transport of extracellular cystine into their cells. Alternatively, these hepatocytes uptake cysteine after glutathione effluxed from the liver reduces extracellular cystine to cysteine on the outer cell membrane. Other cell types such as endothelial cells can take up cystine and reduce it intracellularly to cysteine [15]. Cysteine can be imported through the alanine-serine-cysteine transporters (ASCT1 and ASCT2), excitatory amino acid transporter 3 (EAAT3), and solute carrier family 3 member 1 (SLC3A1) [18–21]. In cancer cells, most obtain cysteine mainly through the uptake of cystine via a Na^+-independent plasma membrane antiporter, system x_c^-, that transports extracellular cystine into the cell in exchange for intracellular glutamate [22, 23]. A growing number of studies have revealed the importance of increased cystine uptake in tumor proliferation and survival and regard xCT as a potential therapeutic target [24, 25]. Details on its

Fig. 7.1 (continued) for glutathione (GSH) synthesis, protein synthesis, and iron-sulfur (Fe-S) cluster biogenesis. GSH is one of the most crucial cellular antioxidants responsible for the mitigation of ROS generation and xenobiotics detoxification. GSH synthesis requires two cytosolic enzymes, γ-glutamate-cysteine ligase (γ-GCL) and glutathione synthetase (GS). Catabolism of intracellular cysteine is mainly mediated through two pathways: (1) oxidation by cysteine dioxygenase (CDO) to generate cysteinesulfinate intermediate and subsequently metabolized to taurine and CO_2 via decarboxylation or pyruvate and inorganic sulfur via transamination and (2) generation of H_2S by transsulfuration pathway enzymes, cystathionine β-synthase (CBS), cystathionine γ-lyase (CSE), and mercaptopyruvate sulfurtransferase (MPST) along with L-cysteine aminotransferase (CAT). Cysteine can be used for the production of coenzyme A and its degradation end product, cysteamine, which can be oxidized to hypotaurine and eventually to taurine. *3-MP* 3-mercaptopyruvate, *SAM* S-adenosylmethionine, *SAH* S-adenosylhomocysteine

regulation and molecular mechanisms will be addressed in detail in subsequent sections.

7.2.2 De Novo Biosynthesis through Transsulfuration Pathway

In addition to cystine uptake, cysteine can be synthesized from methionine and serine through the de novo transsulfuration pathway, mainly in the liver [26]. It is estimated that approximately 20%–50% of cysteine/cystine within the body is produced in the liver through the transsulfuration pathway [6, 27]. Methionine is obtained from the extracellular environment (e.g., through the daily diet) and converted to S-adenosylmethionine (SAM) by methionine adenosyltransferase (MAT). SAM is a significant source of the methyl group and is converted to S-adenosylhomocysteine (SAH) after donating its methyl group for cellular methylation. SAH is then metabolized to homocysteine by SAH hydrolase. Homocysteine metabolism is the critical branch point between the methionine cycle and the transsulfuration pathway. Homocysteine can be remethylated and converted back to methionine by regaining the methyl group from the folate cycle or through choline/betaine metabolism (Fig. 7.1). Homocysteine can also irreversibly enter the transsulfuration pathway through the condensation of serine and homocysteine by cystathionine β-synthase (CBS) to produce cystathionine, which is further hydrolyzed by cystathionine γ-lyase (CSE, also named CTH and CGL) to produce cysteine, ammonia, and α-ketobutyrate [26, 28] (Fig. 7.1). The transsulfuration pathway is essentially the key pathway of methionine catabolism, resulting in the transfer of sulfur from methionine to serine to form cysteine. Thus, the sulfur from either methionine or cysteine ends up being oxidized via cysteine catabolic pathways to give the end products sulfate and taurine, which are excreted in the urine [26, 29].

Homocysteine is positioned at the branch point between the methionine cycle and the transsulfuration pathway. Homocysteine cellular levels are regulated by CBS, which controls the levels of homocysteine entering the methionine cycle or channeled into the transsulfuration pathway to produce cysteine. CBS is a cytosolic homotetrameric enzyme of approximately 64 kDa that binds to two cofactors, pyridoxal 5′-phosphate (PLP, the active form of vitamin B6) and haem [30]. It is allosterically activated by SAM and also by cleavage of its carboxyl-terminal at Arg413 to generate a 45 kDa monomer to give a highly active form of CBS, which is unresponsive to SAM regulation. It has been reported that TNFα stimulation can induce CBS cleavage, increasing transsulfuration pathway activity, promoting glutathione synthesis, and reducing ROS generation [31]. CSE is the second enzyme in the transsulfuration pathway and is the only enzyme that can directly generate de novo cysteine from cystathionine in mammals. CBS KO mice die postnatally at 5 weeks, while CSE KO mice are viable and show no notable phenotype in the presence of dietary cysteine [32, 33]. However, CSE KO mice present with similar symptoms, as those shown in CBS KO mice following cysteine starvation [32].

Cysteine is considered an essential amino acid in cells that do not express CSE. This includes the liver of newborns, especially in premature babies, and patients with cystathioninuria [34–36]. Even in healthy adults, naïve T lymphocytes depend on cysteine supply from antigen-presenting cells (APCs) for its activation and proliferation [37–39]. As such, the essence of "nonessential" cysteine is actually cell- and context-dependent. CSE can be induced by a wide variety of stimuli ranging from oxidative and endoplasmic reticulum stress to nutrient deprivation [40–42]. Oxidative stress elevates the expression of both CSE and CBS through stabilization of the master regulator of oxidative response, nuclear factor erythroid 2-related factor 2 (Nrf2). This leads to the activation of the transsulfuration pathway to synthesize two major antioxidants, cysteine and GSH, for oxidative cellular defense. Nutrient restriction and amino acid starvation can also upregulate CSE expression through activating transcription factor 4 (ATF4), which binds to the CSE promoter in addition to promoters of other amino acid biosynthetic and transport genes. Loss of CSE leads to oxidative stress, aberrant stress responses, vascular deficits, and hyperhomocysteinemia [41–43].

7.2.3 Cysteine Salvaged from GSH Degradation through γ-Glutamyl Cycle

Glutathione not only protects cells from oxidative stress and maintains the redox balance of the extracellular environment but also serves as a cysteine reservoir. The kidney and liver secrete large amounts of GSH in circulation. The breakdown of extracellular glutathione to produce cysteine is achieved by the γ-glutamyl cycle [44, 45], which is an alternative to the de novo biosynthetic pathway, and plays an important role in the maintenance of intracellular GSH and cysteine levels [46]. γ-Glutamyl transpeptidase (GGT) is a membrane-bound heterodimeric glycoprotein enzyme. Its extracellular-facing domain cleaves the γ-glutamyl bond of GSH and releases γ-glutamyl amino acid from the cysteinylglycine dipeptide, which is subsequently cleaved by a dipeptidase, aminopeptidase N, into cysteine and glycine (Fig. 7.2). Cysteine can then be readily taken up by cysteine transporters such as ASC transporters or oxidized and taken up by system x_c^- to replenish the intracellular cysteine pool and resynthesis of GSH. The importance of the γ-glutamyl cycle in the maintenance of the cysteine pool is evident in GGT KO mice studies [47]. GGT KO mice grow and mature slowly and begin to die at around 10–12 weeks of age. Its plasma and urine GSH levels are elevated sixfold and 2500-fold, respectively, and plasma cysteine levels are reduced to 20%, compared with wild-type mice. Tissue GSH and cysteine levels are markedly reduced [47, 48], and these phenotypes can be reversed by the addition of cysteine precursor N-acetylcysteine (NAC), suggesting cysteine as a rate-limiting amino acid in this context. This demonstrates that the catabolism of GSH by GGT and dipeptidase is a primary source of circulating cysteine. Without this GGT-initiated reutilization of cysteine, normal dietary intake and de novo biosynthesis of cysteine are insufficient to compensate for its loss as GSH.

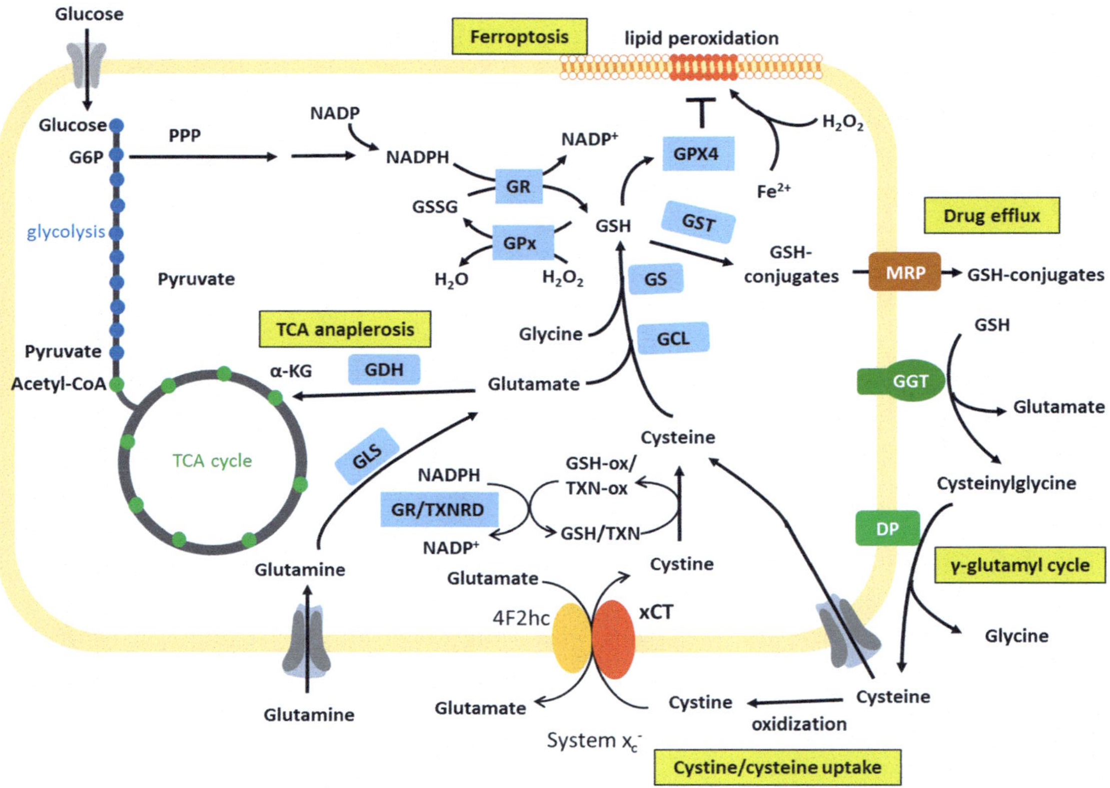

Fig. 7.2 Interplay between intracellular cysteine with glutathione homeostasis, glucose and glutamine metabolism, and pentose phosphate pathway. *GSH* glutathione, *GSH-ox* oxidized GSH, *TXN* thioredoxin, *TXN-ox* oxidized thioredoxin, *GR* glutathione reductase, *GPx* glutathione peroxidase, *TXNRD*

7.2.4 Scavenge from Protein Degradation

Exogenous protein in the extracellular space may also potentially be a reservoir of cysteine available to cells for proliferation and survival through micropinocytosis, a means by which eukaryotic cells nonselectively ingest extracellular liquid and dissolved molecules [49, 50]. Internalized macropinosomes containing proteins are shuttled to and fused with lysosomes. Proteins within the vesicles are then degraded into their constituent amino acids. Amino acids are then transported out of the lysosomes into the cytosol for use by the cell. However, cysteine is only found sparsely in proteins, and scavenging proteins from the external environment are a relatively poor route to acquire cysteine [51], Davidson and coworkers were able to demonstrate increased intracellular cysteine levels along with the other amino acids in pancreatic tumors in vivo from the catabolism of labeled albumin [52].

7.3 Utilization of Cysteine

The nonessential amino acid cysteine performs diverse biological roles in cells, including catalysis regulation, redox sensitivity, metabolic rewiring, and metal trafficking [5, 6, 11, 12, 26]. The majority of cellular cysteine is used for protein synthesis, and its incorporation into protein is highly conserved due to the importance of its thiol reactivity [51]. Interestingly, despite being one of the least abundant amino acids in protein synthesis, cysteine is the most acquired amino acid due to missense mutations in the onco-genome [53], highlighting the crucial roles of acquired cysteine in tumorigenesis. Another primary usage of cysteine is in the biosynthesis of the antioxidant GSH, which is crucial for the mitigation of ROS generation and in xenobiotics detoxification in the cells [15, 45, 54–56]. GSH consists of glutamate, cysteine, and glycine, and cysteine availability is known to be a rate-limiting step for GSH biosynthesis. Two cytosolic ATP-requiring enzymes synthesize glutathione: γ-glutamate-cysteine ligase (γ-GCL) (also named γ-glutamylcysteine synthetase, GCS) and glutathione synthetase (GS) (Fig. 7.1). The first reaction is rate-limiting, catalyzing the formation of the peptide bond between a cysteine and a glutamic acid by γ-GCL. The second step is catalyzed by GS where glycine is added to the C-terminus of γ-glutamylcysteine to produce GSH [57]. GSH acts as a scavenger of free radicals and elicits its anti-oxidative activity either through thiol oxidation producing oxidized glutathione (GSSG) or through its use by glutathione peroxidase (GPx) (Fig. 7.2). GSH is present in the cell mainly in the reduced form (approximately 1–10 mM), while the content of GSSG does not

Fig. 7.2 (continued) thioredoxin reductase, *α-KG* α-ketoglutarate, *G6P* glucose-6-phosphate, *PPP* pentose phosphate pathway, *GLS* glutaminase, *GST* glutathione S-transferase, *GDH* glutamate dehydrogenase, *NAD* nicotinamide adenine dinucleotide phosphate, *NADPH* reduced form of NADP, *GPX4* glutathione peroxidase 4, *GGT* γ-glutamyl transpeptidase, *DP* dipeptidase, *MRP* multidrug resistance protein, *GCL* γ-glutamate-cysteine ligase, *GS* glutathione synthetase

exceed 1% of its total intracellular level [45]. High levels of GSH or GSH/GSSG ratio are known to protect cells from oxidative stress for better cell survival or proliferation in response to stress [58]. Dysregulated glutathione homeostasis and metabolism have been linked to cancer progression and therapy resistance, and a growing body of evidence suggests GSH effects in cancer development can be attributed to the availability of the GSH precursor, cysteine [45, 56].

Besides its use for protein synthesis and GSH production, cysteine is known to be the sulfur source for iron-sulfur (Fe-S) cluster biogenesis. Fe-S clusters are key cofactors of proteins involved in diverse biological processes from energy metabolism, iron homeostasis, and amino acid and nucleotide metabolism to DNA synthesis and repair [59]. Interestingly, a recent study by Novera et al. [60] suggests that cysteine deprivation in respiring ovarian clear cell carcinoma (OCCC) can limit Fe-S cluster synthesis. This, in turn, results in an increase in iron accumulation and iron overload, cumulating to mitochondrial damage and eventual cell death by apoptosis. However, how cysteine availability affects Fe-S cluster biogenesis; how this subsequently impacts the functions of mitochondrial, cytosolic and nuclear Fe-S protein; and how it modulates apoptosis or even ferroptosis remain largely uncharacterized and require more in-depth studies.

Cysteine can also be utilized to synthesize coenzyme A [11], which is known for its role in the synthesis and oxidation of fatty acids and the oxidation of pyruvate in the trichloroacetic acid (TCA) cycle [61]. The first step of coenzyme A biosynthesis involves the phosphorylation of pantothenate (vitamin B5) by pantothenate kinase to form 4′-phosphopantothenate. Cysteine is then condensed with 4-′-phosphopantothenate by the enzyme phosphopantothenoylcysteine synthetase to form the intermediate for coenzyme A synthesis, 4′-phospho-N-pantothenoylcysteine. This step is coupled with ATP hydrolysis. Despite the pathway for coenzyme A synthesis being fully established, the extent of cysteine consumption for coenzyme A turnover and the impact of cysteine availability on coenzyme A synthesis remain largely uncharacterized. Recently, an interesting finding by Badgley and colleagues provided some clues as to how cysteine may impact coenzyme A synthesis [62]. They provided compelling evidence that pancreatic ductal adenocarcinoma (PDAC) cells used cysteine to synthesize glutathione and coenzyme A, which, together, downregulated ferroptosis, a new form of oxidative cell death which was recently discovered and associated with system x_c^- inhibition and suppression of glutathione peroxidase 4 (GPX4). Ferroptosis is an iron-dependent and non-apoptotic mode of cell death, usually accompanied by iron accumulation and lipid oxidation [63–66]. Impaired system x_c^- activity or depletion of cystine/cysteine triggered tumor-specific ferroptosis and reduced tumor growth [62]. Additional work is necessary to better understand the interplay between cysteine availability and metabolism and coenzyme A-related metabolic pathways such as lipid synthesis and the TCA cycle.

7.4 Cysteine Catabolism

Catabolism of intracellular cysteine is mainly mediated through two pathways: oxidation by cysteine dioxygenase (CDO) to generate taurine, pyruvate, and sulfate or generation of H_2S by transsulfuration pathway enzymes, CBS, and CSE [11, 12, 28, 29, 67]. Cysteine catabolism in mammals is dependent on CDO, which catalyzes the oxidation of the L-cysteine thiol group to form cysteinesulfinate, followed by further metabolism to either pyruvate and sulfate through the transamination pathway or taurine and CO_2 through the decarboxylation pathway (Fig. 7.1). CDO plays an essential and highly regulated role in the degradation of excess cysteine obtained either from the diet or synthesized from the transsulfuration pathway [11, 67]. CDO also regulates intracellular levels of cysteine in response to changes in cysteine availability. Previous studies in vivo demonstrated that the regulation of hepatic CDO is mainly at the protein turnover level rather than at the level of mRNA expression, with its concentration responding to changes in cysteine availability [68]. In cysteine-deprived conditions, there are increased CDO ubiquitination and degradation. The converse is true for high cysteine levels, which also increases hypotaurine/taurine production. Since CDO acts directly on cysteine, it can also modulate H_2S production by influencing substrate availability, and mice lacking CDO have been demonstrated to present with both elevated cysteine and H_2S production capacity [69, 70]. CDO can affect cysteine metabolism and determine its flux between GSH production and cysteine catabolism to taurine synthesis or transamination to form pyruvate and sulfate. Overexpression of CDO1 in breast cancer cells shifts glutathione synthesis toward cysteine catabolism. Reduction of GSH synthesis resulted in elevated ROS levels, in turn leading to reduction of viability and growth [71]. On the other hand, silencing CDO1 protects gastric cancer cells from erastin-induced ferroptosis by restoring cellular GSH levels to prevent ROS generation and reduce lipid peroxidation [72]. CDO1 is also reported to be preferentially silenced by promoter methylation in human non-small-cell lung cancers harboring mutations in KEAP1 [73].

The second catabolic pathway of intracellular cysteine produces hydrogen sulfide (H_2S) (Fig. 7.1). It is catalyzed by two pyridoxal 5′-phosphate (PLP)-dependent enzymatic reactions involving CBS and CSE and via a PLP-independent pathway utilizing mercaptopyruvate sulfurtransferase (MPST) along with L-cysteine aminotransferase [28, 74]. H_2S is now recognized as an important signaling molecule capable of modulating a plethora of physiological and pathological processes in a variety of human tissues [28, 29, 74, 75]. An emerging body of evidence now links the dysregulation of H_2S signaling to the pathogenesis of a wide range of diseases, including cancer. In cancer, the requirement of cysteine metabolism in tumor proliferation and survival is mainly attributed to its capacity as an antioxidant due to its role as a precursor of glutathione synthesis, as well as its ability to elicit metabolic reprogramming and generate H_2S.

7.5 Regulation of Cysteine Metabolism in Cancer Proliferation and Survival

7.5.1 De Novo Transsulfuration Pathway

Both CBS and CSE can catalyze cysteine to generate H_2S which has been shown to contribute to cancer proliferation and survival. This is particularly evident in cancers of the colon and ovaries, where malignant cells both overexpress CBS and produce increased amounts of H_2S [76]. In colorectal cancer, studies have shown that H_2S exerts pro-cancerous effects and helps in the survival of tumors by stimulating H_2S, which serves as a donor in the electron transport chain (ETC). It can also stimulate angiogenesis along with the promotion of cell proliferation, metastasis, and drug resistance [77–80]. Recent research suggests a role for CSE in prostate cancer progression and metastasis [81]. Clinical data revealed that elevated CSE expression correlated with poor survival. Mechanistically, CSE promoted NF-kB nuclear translocation through H_2S-mediated sulfhydration, resulting in increased IL-1β expression in vitro, and overexpression of CSE promoted tumor growth and metastasis in the prostate cancer xenograft mouse model [81]. CSE is also strongly expressed in melanoma cell lines, mostly in a Myc-dependent manner [82]. Myc is known to drive the evasion of reactive oxygen stress-induced melanocyte senescence, and its growth-suppressing interaction partner, Miz1, antagonizes this function. Leikam and colleagues identified CSE as a new Myc target gene, and blocking CSE enzymatic activity in human melanoma cells increased intracellular stress and senescence and thus reduced tumorigenic effects such as proliferation, H_2O_2 resistance, and soft agar growth [82]. Of note, unlike the effects of endogenous H_2S derived from the transsulfuration pathway, extensive studies have reported that exogenous H_2S donors exhibited antitumor effects in a variety of malignant cells [83, 84]. Bell-shaped pharmacology of H_2S, whereby lower (endogenous) H_2S production tends to promote growth and higher (exogenous) H_2S production leads to growth inhibition, has been proposed for the paradoxical actions of H_2S [76].

In the cancer model, due to the high cellular demand for cysteine in the maintenance of glutathione homeostasis, reliance on the transsulfuration pathway for cysteine is insufficient to meet cellular needs. To ramp up intracellular cysteine availability to meet their cellular requirements, cancer cells may perhaps need to adapt and increase cystine/cysteine uptake through the upregulation of expression of transporters such as system x_c^-. To this end, it was intriguing to observe that some tumors actually possess markedly reduced levels of transsulfuration enzymes, mostly due to transcriptional silencing [85, 86]. It remains unknown why some cancers are incapable of utilizing the transsulfuration pathway and rely heavily on system x_c^- for cysteine supply, while some cancers are able to use the transsulfuration pathway and are not reliant on system x_c^-.

7.5.2 Cystine Uptake by Cystine/Glutamate Antiporter System x_c^-

Cysteine can be imported into cells either directly or in its oxidized form, cystine, via the cystine/glutamate antiporter system x_c^- (Fig. 7.1). While several other amino acid transporters that can transport cysteine have been described, these transporters can also transport other amino acids and do not exhibit cysteine specificity, making it difficult to attribute its associated functions solely to the dysregulation of cysteine transportation. For example, ASC transporters can also transport alanine and serine, and EAAT3 has a higher affinity for glutamate and aspartate than for cysteine [19], while in breast cancers, the overexpression of SLC3A1 has been shown to increase the uptake of cysteine and activate AKT signaling through inhibition of PP2A phosphatase activity, enhancing tumor growth [21]. It has also been demonstrated that mutations in SLC3A1 lead to cystinuria [87], an inheritable disease which results in the accumulation of cystine in the kidneys and bladder, implying that SLC3A1 more likely transports cystine instead [6].

In contrast, uptake of the oxidized form of cystine through the Na^+-independent plasma membrane antiporter system x_c^- which takes up extracellular cystine in exchange for intracellular glutamate has been well acknowledged. System x_c^- consists of two core components: a catalytic light chain, xCT, encoded by solute carrier family 7 member 11 (*SLC7A11*), and an anchoring heavy chain 4F2 or CD98 encoded by *SLC3A2* [18, 22–25]. For clarity, we will refer to the functional transporter as "system x_c^-" and the light chain subunit as xCT. In line with a role in cystine import, *SLC7A11*-deficient mice exhibited increased cystine concentrations in their blood plasma and decreased intracellular GSH levels in comparison to wild-type littermates [88]. Cellular cysteine levels were comparable between KO and WT littermates, suggesting compensatory cysteine synthesis via the transsulfuration pathway [88].

7.5.3 Regulation of System x_c^-

The import of cystine by system x_c^- has been at the forefront of many cysteine metabolism studies in cancer cells as xCT has been reported to be upregulated in many cancers including lymphoma, melanoma, glioma, and breast, lung, ovarian, liver, colon, head and neck, prostate, and pancreatic cancers, suggesting the importance of increased cysteine uptake in tumor proliferation and survival [89–95]. Several mechanisms underlying the regulation of xCT expression and system x_c^- activity have been reported. Extensive studies have focused on the role of system x_c^- in cellular redox homeostasis as it imports cystine, the rate-limiting precursor for the synthesis of the antioxidant glutathione. In recent years, compelling evidence revealed that system x_c^- also plays a pivotal role in metabolic reprogramming and interplay with many metabolic pathways, including glucose and glutamine metabolism, and the pentose phosphate pathway.

7.5.3.1 Regulation of Expression by Transcription

xCT expression and system x_c^- activity can be induced by various stimuli, including oxidative stress, hypoxia, and nutrient starvation. Elevated xCT expression or system x_c^- activity leads to increased cystine uptake and enhanced intracellular GSH levels to promote cell proliferation and survival. In response to oxidative stress, the redox-sensitive transcription factor, Nrf2, dissociates from Kelch-like ECH-associated protein 1 (Keap1), enters the nucleus, and binds to the antioxidant response element (ARE) in the promoter of its target genes to attenuate ROS damage [96–98]. Expression of xCT and activity of system x_c^- can be regulated by Nrf2 in response to oxidative stress or in Keap1-mutated cells [99], thereby increasing levels of both xCT mRNA and protein promoting GSH synthesis. ARIDIA, a SWI/SNF chromatin-remodeling factor, was also recently shown to regulate xCT transcript level [100]. In the absence of ARID1A, impaired recruitment of the SWI/SNF chromatin-remodeling complex to the transcriptional start site of *SLC7A11* attenuates recruitment of Nrf2 and RNAPII, resulting in reduced transcription of *SLC7A11*. The resultant decrease in cystine uptake impairs GSH synthesis making ARID1A-deficient cancer cells vulnerable to the inhibition of the GSH metabolic pathway.

Similar to Nrf2, activating transcription factor 4 (ATF4) is part of the integrated stress response, which responds to various stimuli such as endoplasmic reticulum (ER) stress and amino acid starvation [101]. When cells are deprived of cysteine, xCT expression and system x_c^- activity are elevated [102], and ATF4 was shown to induce the expression of xCT in response to cystine starvation via its interaction with the amino acid response elements within the proximal region of the *SLC7A11* gene promoter [103]. Other stressors, such as proteasome inhibition, glucose starvation, and ER stress, have also been reported to activate Nrf2 and ATF4 either alone or in combination to activate xCT transcription. For example, Ye and colleague reported that both Nrf2 and ATF4 were upregulated by proteasome inhibition and cooperatively enhanced human *SLC7A11* gene expression [104]. Similarly, glucose starvation induces *SLC7A11* gene expression through the ATF4 and Nrf2 transcription factors, and correspondingly, ATF4 or Nrf2 deficiencies render cancer cells more resistant to glucose starvation [7]. Further, stress-induced inhibition of nonsense-mediated mRNA decay (NMD) protects cells against oxidative stress via stabilization and upregulation of xCT directly or through ATF4 activation [105].

Apart from their roles in the activation of xCT expression, several transcription factors have also been implicated in the repression of xCT expression and activity. In an elegant study, Jian and colleagues described a negative regulatory role of the tumor suppressor p53 on xCT expression [106]. p53 is the most commonly mutated tumor suppressor, which may, in part, explain why xCT is highly expressed in many human tumors. Its overexpression inhibited ROS-induced ferroptosis, promoting tumor proliferation and survival. Interestingly, cytosolic accumulation of p53 mutants typically found in cancer mutants was also shown to bind and sequester Nrf2, preventing Nrf2 from nuclear localization and induction of target genes such as xCT [107]. Other than p53, Linher-Melville et al. provide evidence that the signal transducer and activator of transcription 3 and 5 (STAT3/5) proteins also negatively regulate basal xCT expression. Inhibition of the STAT3/5-mediated signaling pathway induced an adaptive compensatory mechanism protecting breast cancer cells from stress and reactive oxygen species, through the upregulation of xCT expression and the function of system x_c^- [108].

7.5.3.2 Posttranslational Modifications Affecting System x_c^- Activity

System x_c^- transporter activity can be regulated not only through transcriptional regulation of *SLC7A11* gene expression but also through posttranslational modifications. Gu et al. identified the mammalian target of rapamycin complex 2 (mTORC2) as an important regulator of glutamate and glutathione metabolism through its ability to phosphorylate and regulate xCT activity [109]. It was demonstrated that the phosphorylation of serine 26 on the cytosolic N-terminus of xCT by mTORC2 inhibited xCT activity and suppressed glutamate secretion in glioblastomas. This discovery linked growth factor receptor signaling to amino acid uptake and utilization. When nutrient levels in the microenvironment are sufficient to support tumor cell proliferation, mTORC2-mediated inhibition of xCT ensures that glutamine-derived glutamate is used primarily for tumor growth, and secretion of glutamate is not favored. However, when nutrients become limited, it would be advantageous for tumor cells to increase xCT-dependent cystine uptake at the expense of glutamate efflux, enabling tumor cells to buffer cellular redox through the synthesis of glutathione from xCT-derived cysteine [109]. The same Ser26 residue has also been identified to be phosphorylated by Protein Kinase B, also known as AKT, in breast cancers upon activation by oncogenic phosphatidylinositol 3-kinase (PIK3CA) mutations, inhibiting xCT activity [110]. The reduction in cysteine uptake due to the suppression of xCT activity subsequently induced methionine dependence in these cells, shifting the balance toward cysteine production via the transsulfuration pathway to compensate for decreased cysteine uptake as a result of reduced xCT activity.

7.5.3.3 Other Modes of System x_c^- Regulation

Some interacting partners of xCT have been reported to regulate its expression or activity. In gastric cancer and breast cancer, variant isoforms of CD44 (CD44v), but not the standard isoform (CD44s), have been reported to interact with and enhance cell surface expression and stability of xCT, thereby increasing GSH synthesis for better protection against ROS [111–113]. Another study by Hasegawa and colleagues suggests that mucin 1 transmembrane C-terminal (MUC1-C) forms a complex with xCT and the CD44 variant (CD44v), thereby controlling GSH levels [114]. MUC1-C can protect cells against erastin treatment-induced ferroptosis in triple-negative breast cancer (TNBC) cells. In glioma, instead of CD44v, epidermal growth factor receptor (EGFR) was reported to interact with xCT, promoting its cell surface expression and function. Inhibition of xCT suppressed EGFR-dependent anti-oxidative capacity and reduced glioma growth and invasiveness [115]. Song and coworkers recently reported a non-autophagic role of beclin 1 in promoting ferroptosis by binding to xCT [116]. Notably, AMP-activated protein kinase (AMPK)-mediated phosphorylation of beclin 1 at Ser90/93/96 is required for beclin

1-xCT complex formation, which in turn blocked xCT activity promoting lipid peroxidation and ferroptosis. More work needs to be done to understand which additional signals act on beclin 1 (beyond AMPK activation) to better determine if other proteins will preferentially engage the beclin 1-xCT or beclin 1-PIK3C3 complex to stimulate ferroptosis and autophagy, respectively [116].

Expression and function of xCT can also be responsive to extracellular amino acid conditions and growth factors. In TNBC, high levels of extracellular glutamate inhibited xCT through a paracrine mechanism. Suppression of xCT activity results in intracellular cysteine depletion, which directly inhibits hypoxia-inducible factor (HIF) prolyl hydroxylases activity, thereby inducing HIF to promote tumor growth [117]. In estrogen receptor-positive breast cancer cells, insulin-like growth factor I (IGF-I) stimulates cell growth in part by activating expression and function of xCT in an IGF receptor substrate-1-dependent manner, thereby increasing cystine uptake and GSH synthesis to protect cells from ROS [118]. Such mechanisms enable tumor cells to adapt to changing nutrient levels, linking proliferative signals to environmental conditions.

7.5.4 Cross Talk Between Cysteine Metabolism and Other Metabolic Pathways

Many studies have shown that the overexpression of xCT benefits tumor survival and proliferation by increasing glutathione production and anti-oxidative capacity. Interestingly, Herris et al. demonstrate that although GSH is critical for tumor initiation, it is dispensable for subsequent tumor growth due to the activation of the thioredoxin antioxidant pathway through elevated xCT activity. This is supported by the notion that treatment of cancer cells with γ-GCL inhibitor, DL-buthionine-SR-sulfoximine (BSO), sulfasalazine (system x_c^- inhibitor), or auranofin (known to reduce thioredoxin regeneration) alone had minimal or no effects on the cancer cell survival, while the combination of BSO with either sulfasalazine or auranofin induced remarkable cell death [119], suggesting the multifaceted roles of xCT activity in tumorigenesis. Recently, the importance of xCT-mediated glutamate export and metabolic reprogramming in tumor progression began to emerge. xCT activity requires the obligatory export of glutamate to import extracellular cystine. Therefore, extracellular and intracellular levels of glutamate would have an impact on xCT activity and vice versa. When extracellular glutamate concentrations are high, for example, in patients with advanced cancers or in cell culture media containing high levels of glutamate, cystine uptake is competitively inhibited by glutamate, leading to reduction of GSH synthesis. Glutamate is produced intracellularly from glutamine by glutaminase. Many cancer cells consume large amounts of glutamine, which catabolism replenishes intermediates of the mitochondrial TCA cycle (a process termed anaplerosis) (Fig. 7.2) and provides nitrogen for the synthesis of nonessential amino acids and nucleotides [1, 120]. Reports have shown that higher xCT activity can promote glutamine dependency and increase sensitivity to glutaminase inhibition [9, 121]. A drop in intracellular glutamate due to elevated extracellular cystine uptake may increase glutamine uptake and glutaminase activity. This will, in turn, enhance the glutamate-dependent conversion of α-KG production and contribute to glutamine anaplerosis and glutaminase dependency (Fig. 7.2). Similarly, activation of Nrf2 or mutation of KEAP1 can rewire cellular metabolism to glutamine dependence due to a similar glutamine anaplerosis mechanism mediated by high xCT activity [121].

Nutrient supplies are often in excess when tumor cells are cultured in vitro in cell culture conditions but are often limited in vivo because of poor tumor vasculature restricting nutrient uptake by cancer cells from the tumor microenvironment. The restriction of nutrient availability eventually induces metabolic stress in cancer cells of solid tumors [4]. To overcome such metabolic stresses, cancer cells often develop metabolic adaptation strategies for better survival and growth. For instance, when glucose supply is limited, cancer cells can enhance glutamine metabolism via its deamidation to glutamate and subsequent conversion to α-ketoglutarate (α-KG), followed by entry into the TCA cycle to support bioenergetics and biosynthesis [122] (Fig. 7.2). Recent studies in glioblastoma and breast cancer cells have shown that elevated xCT-mediated efflux of intracellular glutamate as a result of xCT overexpression reduces metabolic flexibility in cancer cells increasing their dependence on glucose for survival [7, 10]. Under conditions of glucose withdrawal, xCT depletion maintains intracellular glutamate levels and glutamine/glutamate metabolism which is vital for the survival of system x_c^--dependent cells in conditions of low glucose [7, 10]. Similarly, a recent study showed that under glucose deprivation, lysosomal degradation of xCT is promoted when glioblastoma cells are cultured at high cell densities. Reduced xCT protein levels led to improved cell viability in these cells [123]. These studies show that despite the upregulation of xCT conferring protection to oxidative stress, it also antagonizes glutamine metabolism and restricts nutrient flexibility, pushing cells to rely more on glucose metabolism. Therefore, targeting of glucose metabolism may be an attractive therapeutic strategy for glioblastomas with high expressions of xCT.

Another metabolic vulnerability associated with high xCT expression was suggested by a recent study where the researchers identified cross talk between xCT-associated cystine metabolism and the pentose phosphate pathway [8]. In cancer cells with high xCT expression, the increased cystine uptake is beneficial for cells in its defense against oxidative stress but at a significant cost. The influx of a high amount of oxidized cystine is toxic due to its poor solubility and prompt reduction to cysteine. As this reduction step consumes cellular NADPH, this induces a strong pentose phosphate pathway dependency in xCT-overexpressing cells (Fig. 7.2). As long as glucose supply is sufficient and pentose phosphate pathway is efficient, xCT overexpression can be tolerated with the sustenance of NADPH levels and redox balance. However, when glucose is deprived, the limited NADPH produced can no longer cope with the massive accumulation of cystine resulting in the collapse of the redox system, eventually leading to cell death. A similar notion has been reported in glioblastoma, where cystine uptake through xCT under glucose starvation rapidly induces NADPH depletion, ROS accumulation, and cell death

[124]. Taken together, these studies underlie the vulnerability of xCT-overexpressing cancer cells to glucose limitation.

Catabolism of cysteine to cysteinesulfinate by CDO can undergo further decarboxylation to taurine and transamination to give pyruvate and sulfate. A recent study by Kang and coworkers demonstrated cysteine dioxygenase as a metabolic liability for non-small-cell lung cancer (NSCLC). Higher CDO1 activity induced by xCT-mediated elevated intracellular cysteine promotes cysteine catabolism to wasteful and toxic cysteinesulfinate and sulfite and also depletes cellular NADPH [73]. This antagonizes the growth and survival of NSCLC cells harboring mutations in KEAP1. Silencing of CDO1, on the other hand, promotes the accumulation of intracellular cysteine and NADPH, preventing wastage of cysteine due to its conversion to cysteinesulfinate and sulfite and improving tumor growth and survival. Another report suggests that during knockdown of a key glycolic enzyme, pyruvate kinase, pancreatic ductal adenocarcinoma rewires their metabolic pathway by increasing cysteine metabolism and serine biosynthesis to support pyruvate production and maintains proliferation [125]. It might, therefore, be intriguing to see if CDO activity is increased in these pyruvate kinase knockdown cells. Moreover, the finding that, when glycolysis is impaired, cysteine can act as a source of pyruvate to fulfil energy and biomass requirement, suggests the possible connection between cysteine metabolism and glucose metabolism (Fig. 7.2).

System x_c^- activity has also been linked to the interplay between cysteine and methionine metabolism through homocysteine, a principal junction metabolite that lies at the branch point of the two pathways involving methionine and cysteine metabolism (Fig. 7.1). In the methionine cycle, homocysteine can be remethylated to methionine. Alternatively, homocysteine can be catabolized to cysteine through the transsulfuration pathway. Depending on cellular demand, homocysteine can be directed toward the methionine cycle to increase methylation potential or the transsulfuration pathway to contribute to antioxidant metabolism. In breast cancer, the oncogenic PIK3C mutant's methionine-dependent phenotype is associated with the reduction of xCT activity and cystine uptake [110]. In these cells, homocysteine is used for cysteine synthesis to restore cysteine levels, rendering less homocysteine available for the production of methionine.

7.5.5 Cysteine Salvaged from Glutathione Degradation

Cysteine availability can limit tumor growth, and expression of GGT can provide additional cysteine through the breakdown of glutathione via γ-glutamyl cycle (Fig. 7.2). The role of GGT in enhancing cysteine availability and promoting tumor growth has been shown in in vivo mouse models. Nude mice injection with GGT-overexpressing skin cell lines or the human prostate cancer PC3 cell line grew much more rapidly than their mock-transfected controls [126, 127]. It is worth mentioning that cell culture studies on the role of GGT under oxidative stress may be misleading if the cysteine levels are not adjusted to physiologically relevant levels [128]. The concentration of cysteine and cystine in standard tissue culture media

(200 μM cystine, equivalent to 400 μM cysteine in DMEM or RPMI-1640) is far higher than the concentration present in the human serum (around 12 μM) [129]. As mentioned above, the GGT-overexpressing PC3 cells grew faster in mice; however, no notable differences in growth were observed when cultured in RPMI-1640 [126]. Therefore, potential complications may arise when one attempts to study the effects of GGT or extracellular GSH in the cell culture system.

Nonetheless, the upregulation of GGT has been reported in ovarian, liver, and nasopharyngeal cancers [126, 130, 131]. Moreover, high GTT expression or elevated GGT activity has been shown to be associated with resistance to alkylating agents such as L-phenylalanine mustard [132] and cisplatin [133, 134]. Taken together, GGT inhibition may represent an alternative means to reduce intracellular GSH levels for sensitizing tumor cells to chemotherapeutics.

7.6 Cysteine Metabolism in Tumor Microenvironment

Extracellular free amino acids can contribute to the interaction between tumor and the microenvironment through effects on cellular metabolism and malignant behavior [135, 136]. Remarkably, the anti-oxidative property of cysteine and glutathione extends beyond the cancer cell population. The importance of cystine/cysteine metabolism on the tumor microenvironment was reported in the chronic lymphocytic leukemia (CLL) model [137]. The study found that bone marrow stromal cells promoted CLL survival and resistance to fludarabine and oxaliplatin in vitro and in vivo. Bone marrow stromal cells have high expression of system x_c^- that allows them to take up extracellular cystine and reduce it to cysteine. CLL within the vicinity of bone marrow stromal cells, in turn, take up the released cysteine, presumably via system ASC, for glutathione synthesis. Accordingly, treatment with sulfasalazine, a system x_c^- inhibitor, reduced tumor burden and rendered CLL more sensitive to the conventional cytotoxic drugs. The role of cysteine and glutathione in the tumor microenvironment has also been reported in ovarian cancers [138]. Similar to bone marrow stromal cells in CLL, ovarian cancer-associated fibroblasts (CAFs) were shown to take up cystine and supply reduced cysteine and glutathione to ovarian cancer cells. Glutathione is then used for platinum conjugation, which in turn facilitates its export out of the cancer cells. As such, CAFs confer platinum chemoresistance to ovarian cancer. The study further found that effector CD8+ T cells abrogated the symbiosis between ovarian cancer and CAFs by secreting interferon-gamma (IFN-γ) that binds to the IFN-γ receptor on CAFs. This results in suppressed system x_c^- expression and induced γ-glutamyl transpeptidase 5 (GGT5), which functions to break down glutathione. Consequently, the amount of cysteine and glutathione supplied by CAFs is diminished.

Perhaps, the importance of cysteine in the tumor microenvironment is better appreciated in the immunology field. Here, myeloid-derived suppressor cells (MDSCs) are shown to be potent inhibitors of T-cell-mediated immune responses against cancer [39]. The study found MDSCs, having expressed system x_c^-, compete with antigen-presenting cells (APCs), such as macrophages and dendritic cells

(DCs) for cystine. However, unlike APCs, MDSCs do not express system ASC that allows APCs to export and supply cysteine to T lymphocytes. Naïve T lymphocytes are known to lack CSE and system x_c^- expression, making them dependent on APCs for cysteine supply. As such, MDSCs suppress the immune response by sequestering cystine and limiting cysteine availability for T cells, thereby blocking its activation.

Similarly, in a separate study using tumor mesenchymal stem cells (MSCs) that were conditioned to murine B16 melanoma tumor supernatant, it was found that tumor MSCs secreted immunosuppressive interleukin-10 (IL-10) that binds to the IL-10 receptor in DCs [37]. This led to the activation of the STAT3 signaling cascade, which repressed CSE expression in DCs. With CSE expression repressed in DCs, cysteine synthesis in DCs becomes defective, and DCs can no longer provide cysteine to naïve T cells. Together, these studies demonstrate the significant contributions of rather simple metabolites such as cysteine and glutathione in cancer progression and response not only to chemotherapy but also to the host immune system.

7.7 Therapeutic Potential of Targeting Cysteine Metabolism in Cancer

7.7.1 Targeting System x_c^- and GSH Homeostasis

xCT has been regarded as a therapeutic target for cancer based on the observation that xCT is upregulated in many patient samples and tumor cell lines, and inhibition of system x_c^- suppresses tumor growth in vivo and in vitro [89, 90, 92, 95, 139–141]. Therefore, the exploitation of specific and potent inhibitors of system x_c^- is potentially considered to be of great benefit for cancer chemotherapy. Among the many potent inhibitors for system x_c^- which includes α-aminoadipate, α-aminopimelate, homocysteate, (S)-4-carboxyphenylglycine, L-serine O-sulfate, ibotenate, (RS)-4-bromohomoibotenate, and quisqualate, sulfasalazine is probably one of the most commonly used inhibitor for system x_c^- activity [24, 142]. Sulfasalazine is a Food and Drug Administration (FDA)-approved drug. It was originally and is still being used in the treatment of rheumatoid arthritis and chronic ulcerative colitis. Notably, sulfasalazine is the first system x_c^- inhibitor to be used as an anticancer agent. In the early 2000s, sulfasalazine was found to target system x_c^- in lymphoma and suppressed its growth [140]. Since then, sulfasalazine has been repurposed in several studies as a potential therapeutic strategy for various cancers including non-Hodgkin's lymphoma, glioblastomas, and prostate, pancreatic, and small cell lung cancer [25, 89, 139–141]. Many reports have demonstrated that inhibition or downregulation of system x_c^- function attenuates proliferation, invasion, and metastasis of cancer cells both in vitro and in vivo [24, 25].

Since the uptake of cystine is inevitably coupled to the release of glutamate, a major neurotransmitter in the central nervous system, system x_c^-, has been linked to a variety of normal functions and neurological diseases, such as Parkinson's disease, Alzheimer's disease, and amyotrophic lateral sclerosis [143]. The effect of system

x_c^- inhibition using sulfasalazine in male rats was investigated by Lutgen and colleagues. They reported no noticeable impact on emotional and cognitive function upon sulfasalazine treatment [144]. While sulfasalazine inhibited xCT-mediated uptake of cystine [140], it was also reported to inhibit other proteins including NF-kB, cyclooxygenase 2, sepiapterin reductase, and reduced folate carrier [145–148]. Therefore, when interpreting the inhibitory action of sulfasalazine on system x_c^-, one should also consider the lack of specificity for system x_c^-.

Pharmacological modulation of cysteine has been gaining medicinal interest owing to the discovery of ferroptosis in various cancers, including in therapy-resistant cancers [149, 150]. An inducer of ferroptosis, erastin (named after the eradicator of RAS- and ST-expressing cells), is a selective and irreversible inhibitor of system x_c^- and is the most potent system x_c^- inhibitor known so far with IC_{50} of about 1.4 μM [151, 152]. The persistence of its inhibitory effects makes a short pre-treatment of erastin suffice to synergize with cisplatin to efficiently induce cancer cell death, providing a promise of novel cancer treatment paradigms [152]. Despite its high potency, erastin does not seem to meet the pharmacokinetic standards for clinical application as yet due to poor water solubility and metabolic instability. More metabolic stable variants of erastin such as piperazine erastin and imidazole-ketone erastin have been demonstrated to possess antitumor activity in fibrosarcoma and diffuse histiocytic lymphoma, respectively [66, 153]. Another approach is to package erastin in folate-encapsulated exosomes to target erastin to folate receptor overexpressing TNBC specifically [154]. Cancer cells carrying oncogenic mutations appear to be more sensitive to erastin-induced ferroptosis. Human mammary epithelial cells expressing EGFR mutation were more susceptible to cystine deprivation-induced ferroptosis [155]. Sensitivity was related to the activation of MAPK signaling and the generation and release of hydrogen peroxide. Similar effects were observed in non-small-cell lung cancer cell lines [155]. Whether mitochondria activity is required for the execution of ferroptosis appears to differ depending on the strategy by which the ferroptosis is induced [63]. When triggered by cystine starvation or by erastin, the activity of the mitochondrial TCA cycle was necessary for ferroptosis induction in contrast to those induced by GPX4 inhibitor or deletion [156, 157]. Cancer cells deficient in a metabolic TCA cycle enzyme, fumarate hydratase, were incapable of inducing ferroptosis upon cystine deprivation [157]. Cysteine desulfurase (NFS1) is an essential enzyme in eukaryotes that harvest sulfur from cysteine for the biosynthesis of iron-sulfur clusters. Inhibition of NFS1 induces an iron starvation response, which in combination with inhibition of glutathione biosynthesis or inhibition of cysteine transport can trigger ferroptosis in lung adenocarcinoma [158].

The importance of system x_c^- in immunotherapy has been suggested in recent studies [159, 160]. Interferon-gamma (IFN-γ) released from immunotherapy-activated CD8+ T cell downregulates the expression of two subunits of system x_c^- (*SLC3A2* and *SLC7A11*) and impairs the uptake of cystine by tumor cells, promoting tumor cell lipid peroxidation and ferroptosis. Importantly, transcriptome analysis before and during nivolumab therapy revealed that clinical benefits correlated with reduced expression of *SLC3A2* and increased IFN-γ and CD8 [160]. Interestingly, an extension of the work by the same group suggests that system x_c^- can be downregulated through both immunotherapy-activated IFN-γ and radiotherapy-induced ataxia-telangiectasia mutated gene (ATM), leading to an increase in tumor lipid peroxidation and subsequently ferroptosis [159]. This highlights the interplay between radiotherapy and immunotherapy in cancer and suggests ferroptosis agonists such as system x_c^- inhibitors as a novel class of radiotherapy and immunotherapy sensitizer.

7.7.2 Targeting Extracellular Cyst(e)ine

Starving cancer cells of nutrients are an effective strategy for cancer treatment. For example, asparaginase treatment aimed at the depletion of L-asparagine in the circulation in combination chemotherapy has been used in the clinics for the treatment of acute lymphocytic leukemia, a cancer that has shown critical reliance on asparagine for growth for decades [161]. Intriguingly, Jones and colleagues recently reported that leukemia stem cells (LSCs) in patients with de novo acute myeloid leukemia heavily relied on cysteine to drive oxidative phosphorylation [162]. When cysteine was deprived, impaired GSH synthesis reduced glutathionylation of succinate dehydrogenase A. This suggests cysteine depletion may also be a potential therapeutic strategy for the eradication of LSCs. Cysteine depletion by utilizing CSE and cysteine desulfhydrase, which are known to use cysteine as their substrate for the inhibition of growth in cysteine-dependent leukemia, has been demonstrated both in vitro and in vivo [163]. Recently, Cramer and colleagues demonstrated the systemic depletion of cysteine using an engineered and pharmacologically optimized human cyst(e)inase enzyme in mice and nonhuman primates [164]. The engineered mutant form of CSE, with mutations at E59T and E339V, exhibits 25- and 50-fold higher catalytic efficiency than wild-type CSE for L-cysteine and cystine. This was attributed to the increased hydrophobicity of the active sites improving the binding of cyst(e)ine. The engineered cyst(e)inase resulted in the depletion of intracellular GSH, elevated ROS, and induction of ferroptosis. The engineered enzyme has been shown to suppress the growth of prostate, breast, and non-small-cell lung cancer, as well as in pancreatic cancer xenografts and selectively eradicated human leukemia stem cells [62, 155, 162, 164]. An engineered cyst(e)inase represents a safe and effective therapeutic modality as suppression of tumor growth in multiple mice and primate tumor models was shown to be mediated by the depletion of serum cysteine by cyst(e)inase treatment. Moreover, during treatment, the depletion of the cystine pool is well tolerated for prolonged periods with no observable toxicity, weight loss, or reduction in appetite. A recent study demonstrated that the combination of cyst(e)inase with immune checkpoint blockade (e.g., anti-PD-L1 antibody) synergistically enhanced T-cell-mediated antitumor immunity and induced ferroptosis in tumor cells [160], highlighting its potential application as an adjuvant in cancer immunotherapy.

7.7.3 Targeting Enzymes of the Transsulfuration Pathway

Cysteine modulates not only GSH-mediated antioxidant capacity but also H_2S-mediated chemoresistance. For instance, in the context of ovarian cancer, CBS is often overexpressed and has been shown to promote cellular bioenergetics, tumor growth, and drug resistance [165]. CSE was also reported to be overexpressed in ovarian clear cell carcinoma, which is known to be the most platinum-resistant subtype of ovarian cancer in the clinics [166]. A recent study reported that OCCC exhibited higher dependence on cysteine compared to methionine. This was evidenced by the observation that the most severe cell death occurs upon cysteine deprivation compared to the depletion of the other 19 amino acids, including methionine [60]. Consistent with this notion, metabolic analysis showed significant differences in cysteine and methionine metabolism between platinum-sensitive and platinum-resistant ovarian cancer cells [167]. These data suggest that the poor prognosis of OCCC could be due to a higher dependence of such cancers on cysteine metabolism. The most commonly used and best studied pharmacological inhibitors for CBS and CSE are aminooxyacetic acid (AOAA) and DL-propargyl glycine (PAG), respectively, despite the poor selectivity and low potency of such drugs [168]. There are extensive studies reporting the neuroprotective and cardioprotective functions of cysteine-derived H_2S. These studies implicate CBS and CSE as the endogenous H_2S source on account of their high expressions in the neuronal and cardiovascular systems, respectively [29, 74, 75]. However, there are relatively lesser reports implicating these enzymes and the application of their inhibitors in cancer. AOAA has been shown to suppress the proliferation of colon cancer cells in vitro and reduce tumor growth in vivo [76, 79]. It has also been shown to sensitize colon cancer cells to oxaliplatin via the potentiation of apoptosis induced by ROS [169]. Similarly, a more selective CBS inhibitor CH004 was also shown to retard tumor growth in vivo in xenograft models of liver cancer by triggering ferroptosis [170].

On the other hand, inhibition of CSE enzymatic activity using PAG was reported to reduce cell invasion mainly through H_2S-mediated signaling pathways [81]. A novel inhibitor of CSE, I157172, has recently been reported to inhibit growth and migration of breast cancer through sirtuin 1-mediated deacetylation of STAT and the subsequent inactivation of the STAT3 pathway [171]. AOAA and PAG are more often used to demonstrate their activity on H_2S production, and less emphasis has been placed on their actions on cysteine metabolism. PAG inhibits CTH irreversibly by conferring steric hindrance at the substrate-binding site [172] which covalently binds to tyrosine 114 residue of CTH, thereby inhibiting substrate release [173]; hence, it can inhibit the conversion of cystathionine to cysteine, reducing intracellular cysteine levels. In principle, depletion of extracellular cysteine via engineered cyst(e)inase or inhibition of cystine import, as well as depletion of intracellular cysteine synthesis via PAG, can serve to reduce the amount of cysteine that cancer cells can utilize for GSH synthesis, thereby increasing tumor cell susceptibility to oxidative stress and sensitivity toward chemotherapy. A recent study demonstrating the most profound reduction of tumor growth in OCCC was revealed by the combination of sulfasalazine and PAG, but not either one alone may support this notion [60].

7.7.4 Cysteine Metabolism and Therapy Resistance

Cancer cells can become resistant to anticancer drugs via several mechanisms, including alterations in drug metabolism and transport and activation of detoxifying pathways such as those mediated by GSH, glutathione S-transferases (GSTs), and multidrug resistance proteins (MRPs) [174] (Fig. 7.2). The increase in GSH levels has been recognized as a major contributing factor in the drug-resistant phenotype through its conjugation with drugs, interactions with ROS, prevention of damage to protein or DNA, and the modulation of the DNA repair pathway. For instance, a conformational change in MRP1 induced by GSH impairs its interaction with doxorubicin and consequently reduces its drug-efflux function [175]. The anti-oxidative property of glutathione has been implicated in cancer resistance to chemo- and radiotherapy for more than two decades. The radioprotective property of glutathione in cancer was first reported in human lymphoid cells, where glutathione depletion sensitized cells to irradiation [176]. Reports on the chemo-sensitizing effect of glutathione depletion then followed. Depletion of glutathione by BSO or by incubating cells in cystine-free medium was demonstrated to sensitize ovarian cancer cells to melphalan [177]. Enhanced melphalan cytotoxicity upon BSO treatment was also observed in the ovarian cancer mice xenograft model [178]. Interestingly, stimulation of glutathione synthesis by oxothiazolidine-4-carboxylate in a normal human lung fibroblast cell line, CCL-210, conferred protection against melphalan, cisplatin, and bleomycin [179]. Through enhanced biosynthesis of intracellular GSH, xCT also protects cancer cells from drug-induced oxidative stress. Increase of xCT expression and function of system x_c^- were associated with chemoresistance of tumor cells [24]. Studies have shown that xCT expression negatively correlated with cytotoxic drug potency for compounds susceptible to GSH-mediated chemoresistance, across sixty cell lines available at the National Cancer Institute [91, 180]. Of note, the number of significant *SLC7A11*-drug correlations was much greater than other known chemoresistance genes such as *ABCB1* (MDR1) and *GCL* and *GSTP1* [180], suggesting xCT plays a vital role in drug resistance. Further, a positive correlation between intracellular glutathione levels and response to chemotherapy has been reported in a panel of platinum-resistant ovarian cancers found to have high levels of glutathione synthesis [133]. Overexpression of system x_c^- has been shown to confer cisplatin resistance in human ovarian cancer cell lines through enhanced glutathione biosynthesis [181]. In support of this notion, many studies demonstrate that inhibition of system x_c^- function can sensitize resistant cells to anticancer drugs through the depletion of GSH in multiple malignant cells including ovarian cancer, melanoma, bladder cancer, glioma, and breast cancer [181–188]. Besides utilizing pharmacological system x_c^- inhibitor to block cysteine uptake and impair GSH synthesis, a recently developed cysteine-degrading cyst(e)inase can also be a promising therapeutic agent for the sensitization of tumor cells to chemotherapy.

A recent study from Bekeschus et al. [189] suggests that the expression of xCT affects intrinsic resistance to cold physical plasma treatment in tumor cells. Physical plasma, a prooxidant therapy, is an emerging treatment strategy employing partially ionized gas expulsion of reactive oxygen and nitrogen species to selectively kill tumor cells via the disturbance of cellular redox balance. Comparing the differences between resistant and sensitive cell lines, there is surprisingly no correlation between basal GSH levels and therapy sensitivity. Instead, xCT was found to be significantly more abundant in the resistant cells, and pharmacological inhibition of system x_c^- activity restored sensitivity to cold plasma in resistant cells. This novel feature of xCT makes it an ideal marker for predicting cold physical plasma treatment outcomes. Furthermore, xCT overexpression has been associated with poor prognosis [90, 92, 95]. Interestingly, it was found that there are higher circulating cysteine levels in the blood serum of ovarian cancer patients than healthy people [184]. Being highly dependent on cysteine, the presence of cysteine in ovarian cancer patients may lead to poor prognosis [190]. Collectively, the targeting of system x_c^- not only presents as a promising therapeutic strategy and potential therapy sensitizer; xCT expression or plasma cysteine level may also hold a promise as a novel diagnostic tool for prediction of therapeutic outcome.

7.8 Conclusions and Perspectives

Cysteine, though categorized as a “nonessential” amino acid, clearly has a non-negligible role in both normal and cancer cell biology. Cancer cells particularly seem to possess a certain level of cysteine addiction, making them vulnerable to cysteine availability. The multifaceted roles of cysteine metabolism in cellular energetics, cellular redox maintenance, and response to therapeutic challenges in cancer cells further underscore the potential of manipulating cysteine availability in cancer treatment. The biology of cysteine metabolism has been well characterized, and pathways of cysteine acquisition by the cell have also been elucidated. Recent studies have also implicated cysteine in other pathways such as in Fe-S cluster biogenesis, and ferroptosis and more can be done to better understand the role of cysteine in these pathways. Emerging studies are beginning to show the promise of either targeting components of the cell’s cysteine acquisition pathways by pharmacological compounds or by limiting the availability of cysteine in the extracellular environment as strategies for cancer therapy or as adjuvants to sensitize cancer cells to currently available therapeutics. Several pharmacological inhibitors of the cysteine acquisition pathway have been identified. However, apart from sulfasalazine, other compounds have yet to be proven to show efficacy in the clinics. Further, strategies to limit extracellular cysteine availability are also emerging. The use of recombinant enzymes to deplete cysteine in the tumor microenvironment holds great promise, and the development of techniques to achieve targeted delivery of these enzymes to tumor sites will be an area of great clinical interest.

References

1. Cantor JR, Sabatini DM. Cancer cell metabolism: one hallmark, many faces. Cancer Discov. 2012;2(10):881–98. https://doi.org/10.1158/2159-8290.Cd-12-0345.
2. Vander Heiden MG, Cantley LC, Thompson CB. Understanding the Warburg effect: the metabolic requirements of cell proliferation. Science. 2009;324(5930):1029–33. https://doi.org/10.1126/science.1160809.
3. Cairns RA, Harris IS, Mak TW. Regulation of cancer cell metabolism. Nat Rev Cancer. 2011;11(2):85–95. https://doi.org/10.1038/nrc2981.
4. Jones RG, Thompson CB. Tumor suppressors and cell metabolism: a recipe for cancer growth. Genes Dev. 2009;23(5):537–48. https://doi.org/10.1101/gad.1756509.
5. Choi B-H, Coloff JL. The diverse functions of non-essential amino acids in cancer. Cancers. 2019;11(5) https://doi.org/10.3390/cancers11050675.
6. Combs JA, DeNicola GM. The non-essential amino acid cysteine becomes essential for tumor proliferation and survival. Cancers. 2019;11(5) https://doi.org/10.3390/cancers11050678.
7. Koppula P, Zhang Y, Shi J, Li W, Gan B. The glutamate/cystine antiporter SLC7A11/xCT enhances cancer cell dependency on glucose by exporting glutamate. J Biol Chem. 2017;292(34):14240–9. https://doi.org/10.1074/jbc.M117.798405.
8. Liu X, Olszewski K, Zhang Y, Lim EW, Shi J, Zhang X, Zhang J, Lee H, Koppula P, Lei G, Zhuang L, You MJ, Fang B, Li W, Metallo CM, Poyurovsky MV, Gan B. Cystine transporter regulation of pentose phosphate pathway dependency and disulfide stress exposes a targetable metabolic vulnerability in cancer. Nat Cell Biol. 2020;22(4):476–86. https://doi.org/10.1038/s41556-020-0496-x.
9. Muir A, Danai LV, Gui DY, Waingarten CY, Lewis CA, Vander Heiden MG. Environmental cystine drives glutamine anaplerosis and sensitizes cancer cells to glutaminase inhibition. elife. 2017;6 https://doi.org/10.7554/eLife.27713.
10. Shin C-S, Mishra P, Watrous JD, Carelli V, D'Aurelio M, Jain M, Chan DC. The glutamate/cystine xCT antiporter antagonizes glutamine metabolism and reduces nutrient flexibility. Nat Commun. 2017;8(1) https://doi.org/10.1038/ncomms15074.
11. Stipanuk MH, Dominy JE, Lee J-I, Coloso RM. Mammalian cysteine metabolism: new insights into regulation of cysteine metabolism. J Nutr. 2006;136(6):1652S–9S. https://doi.org/10.1093/jn/136.6.1652S.
12. Yin J, Ren W, Yang G, Duan J, Huang X, Fang R, Li C, Li T, Yin Y, Hou Y, Kim SW, Wu G. l-cysteine metabolism and its nutritional implications. Mol Nutr Food Res. 2016;60(1):134–46. https://doi.org/10.1002/mnfr.201500031.
13. Bradley H, Gough A, Sokhi RS, Hassell A, Waring R, Emery P. Sulfate metabolism is abnormal in patients with rheumatoid arthritis. Confirmation by in vivo biochemical findings. J Rheumatol. 1994;21(7):1192–6.
14. Heafield MT, Fearn S, Steventon GB, Waring RH, Williams AC, Sturman SG. Plasma cysteine and sulphate levels in patients with motor neurone, Parkinson's and Alzheimer's disease. Neurosci Lett. 1990;110(1-2):216–20. https://doi.org/10.1016/0304-3940(90)90814-p.
15. Wu G, Fang Y-Z, Yang S, Lupton JR, Turner ND. Glutathione metabolism and its implications for health. J Nutr. 2004;134(3):489–92. https://doi.org/10.1093/jn/134.3.489.
16. Mandal PK, Seiler A, Perisic T, Kölle P, Banjac Canak A, Förster H, Weiss N, Kremmer E, Lieberman MW, Bannai S, Kuhlencordt P, Sato H, Bornkamm GW, Conrad M. System xc−and thioredoxin reductase 1 cooperatively rescue glutathione deficiency. J Biol Chem. 2010;285(29):22244–53. https://doi.org/10.1074/jbc.M110.121327.
17. Vučetić M, Cormerais Y, Parks SK, Pouysségur J. The central role of amino acids in cancer redox homeostasis: vulnerability points of the cancer redox code. Front Oncol. 2017;7 https://doi.org/10.3389/fonc.2017.00319.

18. Conrad M, Sato H. The oxidative stress-inducible cystine/glutamate antiporter, system x (c) (−) : cystine supplier and beyond. Amino Acids. 2012;42(1):231–46. https://doi.org/10.1007/s00726-011-0867-5.
19. Hyde R, Taylor PM, Hundal HS. Amino acid transporters: roles in amino acid sensing and signalling in animal cells. Biochem J. 2003;373(1):1–18. https://doi.org/10.1042/bj20030405.
20. Chen Y, Swanson RA. Astrocytes and brain injury. J Cereb Blood Flow Metab. 2003;23(2):137–49. https://doi.org/10.1097/01.WCB.0000044631.80210.3C.
21. Jiang Y, Cao Y, Wang Y, Li W, Liu X, Lv Y, Li X, Mi J. Cysteine transporter SLC3A1 promotes breast cancer tumorigenesis. Theranostics. 2017;7(4):1036–46. https://doi.org/10.7150/thno.18005.
22. Bannai S, Kitamura E. Transport interaction of L-cystine and L-glutamate in human diploid fibroblasts in culture. J Biol Chem. 1980;255(6):2372–6.
23. Sato H, Tamba M, Ishii T, Bannai S. Cloning and expression of a plasma membrane cystine/glutamate exchange transporter composed of two distinct proteins. J Biol Chem. 1999;274(17):11455–8. https://doi.org/10.1074/jbc.274.17.11455.
24. Lewerenz J, Hewett SJ, Huang Y, Lambros M, Gout PW, Kalivas PW, Massie A, Smolders I, Methner A, Pergande M, Smith SB, Ganapathy V, Maher P. The cystine/glutamate antiporter system xc− in health and disease: from molecular mechanisms to novel therapeutic opportunities. Antioxidants Redox Signal. 2013;18(5):522–55. https://doi.org/10.1089/ars.2011.4391.
25. Lo M, Wang Y-Z, Gout PW. The xc− cystine/glutamate antiporter: a potential target for therapy of cancer and other diseases. J Cell Physiol. 2008;215(3):593–602. https://doi.org/10.1002/jcp.21366.
26. Stipanuk MH. Sulfur amino acid metabolism: pathways for production and removal of homocysteine and cysteine. Annu Rev Nutr. 2004;24(1):539–77. https://doi.org/10.1146/annurev.nutr.24.012003.132418.
27. Mosharov E, Cranford MR, Banerjee R. The quantitatively important relationship between homocysteine metabolism and glutathione synthesis by the transsulfuration pathway and its regulation by redox changes†. Biochemistry. 2000;39(42):13005–11. https://doi.org/10.1021/bi001088w.
28. Kimura H. Production and physiological effects of hydrogen sulfide. Antioxid Redox Signal. 2014;20(5):783–93. https://doi.org/10.1089/ars.2013.5309.
29. Szabo C, Papapetropoulos A, Ohlstein EH. International union of basic and clinical pharmacology. CII: pharmacological modulation of H2S levels: H2S donors and H2S biosynthesis inhibitors. Pharmacol Rev. 2017;69(4):497–564. https://doi.org/10.1124/pr.117.014050.
30. Meier M. Structure of human cystathionine beta-synthase: a unique pyridoxal 5′-phosphate-dependent heme protein. EMBO J. 2001;20(15):3910–6. https://doi.org/10.1093/emboj/20.15.3910.
31. Zou C-G, Banerjee R. Tumor necrosis factor-α-induced targeted proteolysis of cystathionine β-synthase modulates redox homeostasis. J Biol Chem. 2003;278(19):16802–8. https://doi.org/10.1074/jbc.M212376200.
32. Ishii I, Akahoshi N, Yamada H, Nakano S, Izumi T, Suematsu M. Cystathionine gamma-Lyase-deficient mice require dietary cysteine to protect against acute lethal myopathy and oxidative injury. J Biol Chem. 2010;285(34):26358–68. https://doi.org/10.1074/jbc.M110.147439.
33. Watanabe M, Osada J, Aratani Y, Kluckman K, Reddick R, Malinow MR, Maeda N. Mice deficient in cystathionine beta-synthase: animal models for mild and severe homocyst(e)-inemia. Proc Natl Acad Sci U S A. 1995;92(5):1585–9. https://doi.org/10.1073/pnas.92.5.1585.
34. Pascal TA, Gaull GE, Beratis NG, Gillam BM, Tallan HH. Cystathionase deficiency: evidence for genetic heterogeneity in primary cystathioninuria. Pediatr Res. 1978;12(2):125–33. https://doi.org/10.1203/00006450-197802000-00012.

35. Viña J, Vento M, García-Sala F, Puertes IR, Gascó E, Sastre J, Asensi M, Pallardó FV. L-cysteine and glutathione metabolism are impaired in premature infants due to cystathionase deficiency. Am J Clin Nutr. 1995;61(5):1067–9. https://doi.org/10.1093/ajcn/61.4.1067.
36. Zlotkin SH, Anderson GH. The development of cystathionase activity during the first year of life. Pediatr Res. 1982;16(1):65–8. https://doi.org/10.1203/00006450-198201001-00013.
37. Ghosh MK, Nguyen V, Muller HK, Walker AM. Maternal milk T cells drive development of transgenerational Th1 immunity in offspring thymus. J Immunol. 2016;197(6):2290. https://doi.org/10.4049/jimmunol.1502483.
38. Levring TB, Hansen AK, Nielsen BL, Kongsbak M, von Essen MR, Woetmann A, Ødum N, Bonefeld CM, Geisler C. Activated human CD4+ T cells express transporters for both cysteine and cystine. Sci Rep. 2012;2(1):266. https://doi.org/10.1038/srep00266.
39. Srivastava MK, Sinha P, Clements VK, Rodriguez P, Ostrand-Rosenberg S. Myeloid-derived suppressor cells inhibit T-cell activation by depleting cystine and cysteine. Cancer Res. 2010;70(1):68–77. https://doi.org/10.1158/0008-5472.Can-09-2587.
40. Dickhout JG, Carlisle RE, Jerome DE, Mohammed-Ali Z, Jiang H, Yang G, Mani S, Garg SK, Banerjee R, Kaufman RJ, Maclean KN, Wang R, Austin RC. Integrated stress response modulates cellular redox state via induction of cystathionine γ-lyase. J Biol Chem. 2012;287(10):7603–14. https://doi.org/10.1074/jbc.M111.304576.
41. Sbodio JI, Snyder SH, Paul BD. Transcriptional control of amino acid homeostasis is disrupted in Huntington's disease. Proc Natl Acad Sci U S A. 2016;113(31):8843–8. https://doi.org/10.1073/pnas.1608264113.
42. Sbodio JI, Snyder SH, Paul BD. Golgi stress response reprograms cysteine metabolism to confer cytoprotection in Huntington's disease. Proc Natl Acad Sci. 2018;115(4):780–5. https://doi.org/10.1073/pnas.1717877115.
43. Yang G, Wu L, Jiang B, Yang W, Qi J, Cao K, Meng Q, Mustafa AK, Mu W, Zhang S, Snyder SH, Wang R. H2S as a physiologic vasorelaxant: hypertension in mice with deletion of cystathionine -lyase. Science. 2008;322(5901):587–90. https://doi.org/10.1126/science.1162667.
44. Orlowski M, Meister A. The gamma-glutamyl cycle: a possible transport system for amino acids. Proc Natl Acad Sci U S A. 1970;67(3):1248–55. https://doi.org/10.1073/pnas.67.3.1248.
45. Traverso N, Ricciarelli R, Nitti M, Marengo B, Furfaro AL, Pronzato MA, Marinari UM, Domenicotti C. Role of glutathione in cancer progression and chemoresistance. Oxidative Med Cell Longev. 2013;2013:972913. https://doi.org/10.1155/2013/972913.
46. Zhang H, Forman HJ. Redox regulation of gamma-glutamyl transpeptidase. Am J Respir Cell Mol Biol. 2009;41(5):509–15. https://doi.org/10.1165/rcmb.2009-0169TR.
47. Lieberman MW, Wiseman AL, Shi ZZ, Carter BZ, Barrios R, Ou CN, Chevez-Barrios P, Wang Y, Habib GM, Goodman JC, Huang SL, Lebovitz RM, Matzuk MM. Growth retardation and cysteine deficiency in gamma-glutamyl transpeptidase-deficient mice. Proc Natl Acad Sci. 1996;93(15):7923–6. https://doi.org/10.1073/pnas.93.15.7923.
48. Zhang H, Forman HJ, Choi J. Gamma-glutamyl transpeptidase in glutathione biosynthesis. Methods Enzymol. 2005;401:468–83. https://doi.org/10.1016/s0076-6879(05)01028-1.
49. Bloomfield G, Kay RR. Uses and abuses of macropinocytosis. J Cell Sci. 2016;129(14):2697–705. https://doi.org/10.1242/jcs.176149.
50. Commisso C, Davidson SM, Soydaner-Azeloglu RG, Parker SJ, Kamphorst JJ, Hackett S, Grabocka E, Nofal M, Drebin JA, Thompson CB, Rabinowitz JD, Metallo CM, Vander Heiden MG, Bar-Sagi D. Macropinocytosis of protein is an amino acid supply route in Ras-transformed cells. Nature. 2013;497(7451):633–7. https://doi.org/10.1038/nature12138.
51. Bak DW, Bechtel TJ, Falco JA, Weerapana E. Cysteine reactivity across the subcellular universe. Curr Opin Chem Biol. 2019;48:96–105. https://doi.org/10.1016/j.cbpa.2018.11.002.
52. Davidson SM, Jonas O, Keibler MA, Hou HW, Luengo A, Mayers JR, Wyckoff J, Del Rosario AM, Whitman M, Chin CR, Condon KJ, Lammers A, Kellersberger KA, Stall BK,

Stephanopoulos G, Bar-Sagi D, Han J, Rabinowitz JD, Cima MJ, Langer R, Vander Heiden MG. Direct evidence for cancer-cell-autonomous extracellular protein catabolism in pancreatic tumors. Nat Med. 2017;23(2):235–41. https://doi.org/10.1038/nm.4256.
53. Visscher M, Arkin MR, Dansen TB. Covalent targeting of acquired cysteines in cancer. Curr Opin Chem Biol. 2016;30:61–7. https://doi.org/10.1016/j.cbpa.2015.11.004.
54. Ballatori N, Krance SM, Notenboom S, Shi S, Tieu K, Hammond CL. Glutathione dysregulation and the etiology and progression of human diseases. Biol Chem. 2009;390 (3):191–214. https://doi.org/10.1515/BC.2009.033.
55. Kalinina EV, Chernov NN, Novichkova MD. Role of glutathione, glutathione transferase, and glutaredoxin in regulation of redox-dependent processes. Biochemistry (Mosc). 2014;79 (13):1562–83. https://doi.org/10.1134/s0006297914130082.
56. Simon MC, Bansal A. Glutathione metabolism in cancer progression and treatment resistance. J Cell Biol. 2018;217(7):2291–8. https://doi.org/10.1083/jcb.201804161.
57. Lu SC. Glutathione synthesis. Biochim Biophys Acta (BBA) Gen Subjects. 2013;1830 (5):3143–53. https://doi.org/10.1016/j.bbagen.2012.09.008.
58. Lee Z-W, Low Y-L, Huang S, Wang T, Deng L-W. The cystathionine γ-lyase/hydrogen sulfide system maintains cellular glutathione status. Biochem J. 2014;460(3):425–35. https://doi.org/10.1042/BJ20131434.
59. Stehling O, Lill R. The role of mitochondria in cellular iron-sulfur protein biogenesis: mechanisms, connected processes, and diseases. Cold Spring Harbor Perspect Biol. 2013;5 (8):a011312. https://doi.org/10.1101/cshperspect.a011312.
60. Novera W, Lee Z-W, Nin DS, Dai MZ-Y, Binte Idres S, Wu H, Damen MA, Tan TZ, Sim AYL, Long YC, Wu W, Huang RY-J, Deng L-W. Cysteine deprivation targets ovarian clear cell carcinoma via oxidative stress and iron−sulfur cluster biogenesis deficit. Antioxid Redox Signal. 2020; https://doi.org/10.1089/ars.2019.7850.
61. Martinez David L, Tsuchiya Y, Gout I. Coenzyme a biosynthetic machinery in mammalian cells. Biochem Soc Trans. 2014;42(4):1112–7. https://doi.org/10.1042/bst20140124.
62. Badgley MA, Kremer DM, Maurer HC, DelGiorno KE, Lee H-J, Purohit V, Sagalovskiy IR, Ma A, Kapilian J, Firl CEM, Decker AR, Sastra SA, Palermo CF, Andrade LR, Sajjakulnukit P, Zhang L, Tolstyka ZP, Hirschhorn T, Lamb C, Liu T, Gu W, Seeley ES, Stone E, Georgiou G, Manor U, Iuga A, Wahl GM, Stockwell BR, Lyssiotis CA, Olive KP. Cysteine depletion induces pancreatic tumor ferroptosis in mice. Science. 2020;368 (6486):85–9. https://doi.org/10.1126/science.aaw9872.
63. Bebber CM, Müller F, Prieto Clemente L, Weber J, von Karstedt S. Ferroptosis in cancer cell biology. Cancers. 2020;12(1) https://doi.org/10.3390/cancers12010164.
64. Dixon Scott J, Lemberg Kathryn M, Lamprecht Michael R, Skouta R, Zaitsev Eleina M, Gleason Caroline E, Patel Darpan N, Bauer Andras J, Cantley Alexandra M, Yang Wan S, Morrison B, Stockwell Brent R. Ferroptosis: an iron-dependent form of nonapoptotic cell death. Cell. 2012;149(5):1060–72. https://doi.org/10.1016/j.cell.2012.03.042.
65. Li J, Cao F, H-l Y, Z-j H, Lin Z-t, Mao N, Sun B, Wang G. Ferroptosis: past, present and future. Cell Death Dis. 2020;11(2):88. https://doi.org/10.1038/s41419-020-2298-2.
66. Yang Wan S, SriRamaratnam R, Welsch Matthew E, Shimada K, Skouta R, Viswanathan Vasanthi S, Cheah Jaime H, Clemons Paul A, Shamji Alykhan F, Clish Clary B, Brown Lewis M, Girotti Albert W, Cornish Virginia W, Schreiber Stuart L, Stockwell Brent R. Regulation of ferroptotic cancer cell death by GPX4. Cell. 2014;156(1-2):317–31. https://doi.org/10.1016/j.cell.2013.12.010.
67. Stipanuk MH, Ueki I, Dominy JE, Simmons CR, Hirschberger LL. Cysteine dioxygenase: a robust system for regulation of cellular cysteine levels. Amino Acids. 2008;37(1) https://doi.org/10.1007/s00726-008-0202-y.
68. Dominy JE, Hwang J, Guo S, Hirschberger LL, Zhang S, Stipanuk MH. Synthesis of amino acid cofactor in cysteine dioxygenase is regulated by substrate and represents a novel post-translational regulation of activity. J Biol Chem. 2008;283(18):12188–201. https://doi.org/10.1074/jbc.M800044200.

69. Jurkowska H, Roman HB, Hirschberger LL, Sasakura K, Nagano T, Hanaoka K, Krijt J, Stipanuk MH. Primary hepatocytes from mice lacking cysteine dioxygenase show increased cysteine concentrations and higher rates of metabolism of cysteine to hydrogen sulfide and thiosulfate. Amino Acids. 2014;46(5):1353–65. https://doi.org/10.1007/s00726-014-1700-8.
70. Rose P, Moore PK, Zhu YZ. H(2)S biosynthesis and catabolism: new insights from molecular studies. Cell Mol Life Sci. 2017;74(8):1391–412. https://doi.org/10.1007/s00018-016-2406-8.
71. Jeschke J, Hagan HM, Zhang W, Vatapalli R, Calmon MF, Danilova L, Nelkenbrecher C, Van Neste L, Bijsmans ITGW, Van Engeland M, Gabrielson E, Schuebel KE, Winterpacht A, Baylin SB, Herman JG, Ahuja N. Frequent inactivation of em cysteine dioxygenase type 1/em contributes to survival of breast cancer cells and resistance to anthracyclines. Clin Cancer Res. 2013;19(12):3201. https://doi.org/10.1158/1078-0432.CCR-12-3751.
72. Hao S, Yu J, He W, Huang Q, Zhao Y, Liang B, Zhang S, Wen Z, Dong S, Rao J, Liao W, Shi M. Cysteine dioxygenase 1 mediates erastin-induced ferroptosis in human gastric cancer cells. Neoplasia. 2017;19(12):1022–32. https://doi.org/10.1016/j.neo.2017.10.005.
73. Kang YP, Torrente L, Falzone A, Elkins CM, Liu M, Asara JM, Dibble CC, DeNicola GM. Cysteine dioxygenase 1 is a metabolic liability for non-small cell lung cancer. elife. 2019;8 https://doi.org/10.7554/eLife.45572.
74. Huang CW, Moore PK. H2S synthesizing enzymes: biochemistry and molecular aspects. Handb Exp Pharmacol. 2015;230:3–25. https://doi.org/10.1007/978-3-319-18144-8_1.
75. Wang R. Physiological implications of hydrogen sulfide: a whiff exploration that blossomed. Physiol Rev. 2012;92(2):791–896. https://doi.org/10.1152/physrev.00017.2011.
76. Hellmich MR, Szabo C. Hydrogen sulfide and cancer. Handb Exp Pharmacol. 2015;230:233–41. https://doi.org/10.1007/978-3-319-18144-8_12.
77. Jurkowska H, Placha W, Nagahara N, Wróbel M. The expression and activity of cystathionine-γ-lyase and 3-mercaptopyruvate sulfurtransferase in human neoplastic cell lines. Amino Acids. 2011;41(1):151–8. https://doi.org/10.1007/s00726-010-0606-3.
78. Sonke E, Verrydt M, Postenka CO, Pardhan S, Willie CJ, Mazzola CR, Hammers MD, Pluth MD, Lobb I, Power NE, Chambers AF, Leong HS, Sener A. Inhibition of endogenous hydrogen sulfide production in clear-cell renal cell carcinoma cell lines and xenografts restricts their growth, survival and angiogenic potential. Nitric Oxide. 2015;49:26–39. https://doi.org/10.1016/j.niox.2015.06.001.
79. Szabo C, Coletta C, Chao C, Modis K, Szczesny B, Papapetropoulos A, Hellmich MR. Tumor-derived hydrogen sulfide, produced by cystathionine- -synthase, stimulates bioenergetics, cell proliferation, and angiogenesis in colon cancer. Proc Natl Acad Sci. 2013;110(30):12474–9. https://doi.org/10.1073/pnas.1306241110.
80. Untereiner AA, Pavlidou A, Druzhyna N, Papapetropoulos A, Hellmich MR, Szabo C. Drug resistance induces the upregulation of H2S-producing enzymes in HCT116 colon cancer cells. Biochem Pharmacol. 2018;149:174–85. https://doi.org/10.1016/j.bcp.2017.10.007.
81. Wang YH, Huang JT, Chen WL, Wang RH, Kao MC, Pan YR, Chan SH, Tsai KW, Kung HJ, Lin KT, Wang LH. Dysregulation of cystathionine γ-lyase promotes prostate cancer progression and metastasis. EMBO Rep. 2019c;20(10) https://doi.org/10.15252/embr.201845986.
82. Leikam C, Hufnagel A, Walz S, Kneitz S, Fekete A, Muller MJ, Eilers M, Scharti M, Meierjohann S. Cystathionase mediates senescence evasion in melanocytes and melanoma cells. Oncogene. 2014;33:771.
83. Lee ZW, Deng LW. Role of H2S donors in cancer biology. Handb Exp Pharmacol. 2015;230:243–65. https://doi.org/10.1007/978-3-319-18144-8_13.
84. Nin DS, Idres SB, Song ZJ, Moore PK, Deng LW. Biological effects of morpholin-4-ium 4 methoxyphenyl (morpholino) phosphinodithioate and other phosphorothioate-based hydrogen sulfide donors. Antioxid Redox Signal. 2020;32(2):145–58. https://doi.org/10.1089/ars.2019.7896.
85. Kim J, Hong SJ, Park JH, Park SY, Kim SW, Cho EY, Do IG, Joh JW, Kim DS. Expression of cystathionine beta-synthase is downregulated in hepatocellular carcinoma and associated with poor prognosis. Oncol Rep. 2009;21(6):1449–54. https://doi.org/10.3892/or_00000373.

86. Zhao H, Li Q, Wang J, Su X, Ng KM, Qiu T, Shan L, Ling Y, Wang L, Cai J, Ying J. Frequent epigenetic silencing of the folate-metabolising gene cystathionine-beta-synthase in gastrointestinal cancer. PLoS One. 2012;7(11):e49683. https://doi.org/10.1371/journal.pone.0049683.
87. Pras E, Raben N, Golomb E, Arber N, Aksentijevich I, Schapiro JM, Harel D, Katz G, Liberman U, Pras M, et al. Mutations in the SLC3A1 transporter gene in cystinuria. Am J Hum Genet. 1995;56(6):1297–303.
88. Sato H, Shiiya A, Kimata M, Maebara K, Tamba M, Sakakura Y, Makino N, Sugiyama F, Yagami K, Moriguchi T, Takahashi S, Bannai S. Redox imbalance in cystine/glutamate transporter-deficient mice. J Biol Chem. 2005;280(45):37423–9. https://doi.org/10.1074/jbc.M506439200.
89. Chung WJ, Lyons SA, Nelson GM, Hamza H, Gladson CL, Gillespie GY, Sontheimer H. Inhibition of cystine uptake disrupts the growth of primary brain tumors. J Neurosci. 2005;25(31):7101–10. https://doi.org/10.1523/JNEUROSCI.5258-04.2005.
90. Guo W, Zhao Y, Zhang Z, Tan N, Zhao F, Ge C, Liang L, Jia D, Chen T, Yao M, Li J, He X. Disruption of xCT inhibits cell growth via the ROS/autophagy pathway in hepatocellular carcinoma. Cancer Lett. 2011;312(1):55–61. https://doi.org/10.1016/j.canlet.2011.07.024.
91. Huang Y, Dai Z, Barbacioru C, Sadée W. Cystine-glutamate transporter SLC7A11 in cancer chemosensitivity and chemoresistance. Cancer Res. 2005;65(16):7446–54. https://doi.org/10.1158/0008-5472.Can-04-4267.
92. Ji X, Qian J, Rahman SMJ, Siska PJ, Zou Y, Harris BK, Hoeksema MD, Trenary IA, Heidi C, Eisenberg R, Rathmell JC, Young JD, Massion PP. xCT (SLC7A11)-mediated metabolic reprogramming promotes non-small cell lung cancer progression. Oncogene. 2018;37(36):5007–19. https://doi.org/10.1038/s41388-018-0307-z.
93. Liu XX, Li XJ, Zhang B, Liang YJ, Zhou CX, Cao DX, He M, Chen GQ, He JR, Zhao Q. MicroRNA-26b is underexpressed in human breast cancer and induces cell apoptosis by targeting SLC7A11. FEBS Lett. 2011;585(9):1363–7. https://doi.org/10.1016/j.febslet.2011.04.018.
94. Takeuchi S, Wada K, Toyooka T, Shinomiya N, Shimazaki H, Nakanishi K, Nagatani K, Otani N, Osada H, Uozumi Y, Matsuo H, Nawashiro H. Increased xCT expression correlates with tumor invasion and outcome in patients with glioblastomas. Neurosurgery. 2013;72(1):33–41; discussion 41. https://doi.org/10.1227/NEU.0b013e318276b2de.
95. Timmerman Luika A, Holton T, Yuneva M, Louie Raymond J, Padró M, Daemen A, Hu M, Chan Denise A, Ethier Stephen P, van 't Veer Laura J, Polyak K, McCormick F, Gray Joe W. Glutamine sensitivity analysis identifies the xCT antiporter as a common triple-negative breast tumor therapeutic target. Cancer Cell. 2013;24(4):450–65. https://doi.org/10.1016/j.ccr.2013.08.020.
96. Menegon S, Columbano A, Giordano S. The dual roles of NRF2 in cancer. Trends Mol Med. 2016;22(7):578–93. https://doi.org/10.1016/j.molmed.2016.05.002.
97. Rojo de la Vega M, Chapman E, Zhang DD. NRF2 and the hallmarks of cancer. Cancer Cell. 2018;34(1):21–43. https://doi.org/10.1016/j.ccell.2018.03.022.
98. Tonelli C, Chio IIC, Tuveson DA. Transcriptional regulation by Nrf2. Antioxid Redox Signal. 2018;29(17):1727–45. https://doi.org/10.1089/ars.2017.7342.
99. Habib E, Linher-Melville K, Lin H-X, Singh G. Expression of xCT and activity of system xc− are regulated by NRF2 in human breast cancer cells in response to oxidative stress. Redox Biol. 2015;5:33–42. https://doi.org/10.1016/j.redox.2015.03.003.
100. Ogiwara H, Takahashi K, Sasaki M, Kuroda T, Yoshida H, Watanabe R, Maruyama A, Makinoshima H, Chiwaki F, Sasaki H, Kato T, Okamoto A, Kohno T. Targeting the vulnerability of glutathione metabolism in ARID1A-deficient cancers. Cancer Cell. 2019;35(2):177–190.e178. https://doi.org/10.1016/j.ccell.2018.12.009.
101. Pakos-Zebrucka K, Koryga I, Mnich K, Ljujic M, Samali A, Gorman AM. The integrated stress response. EMBO Rep. 2016;17(10):1374–95. https://doi.org/10.15252/embr.201642195.

102. Bannai S, Kitamura E. Adaptive enhancement of cystine and glutamate uptake in human diploid fibroblasts in culture. Biochim Biophys Acta (BBA) Mol Cell Res. 1982;721(1):1–10. https://doi.org/10.1016/0167-4889(82)90017-9.
103. Sato H, Nomura S, Maebara K, Sato K, Tamba M, Bannai S. Transcriptional control of cystine/glutamate transporter gene by amino acid deprivation. Biochem Biophys Res Commun. 2004;325(1):109–16. https://doi.org/10.1016/j.bbrc.2004.10.009.
104. Ye P, Mimura J, Okada T, Sato H, Liu T, Maruyama A, Ohyama C, Itoh K. Nrf2- and ATF4-dependent upregulation of xCT modulates the sensitivity of T24 bladder carcinoma cells to proteasome inhibition. Mol Cell Biol. 2014;34(18):3421–34. https://doi.org/10.1128/mcb.00221-14.
105. Martin L, Gardner LB. Stress-induced inhibition of nonsense-mediated RNA decay regulates intracellular cystine transport and intracellular glutathione through regulation of the cystine/glutamate exchanger SLC7A11. Oncogene. 2015;34(32):4211–8. https://doi.org/10.1038/onc.2014.352.
106. Jiang L, Kon N, Li T, Wang S-J, Su T, Hibshoosh H, Baer R, Gu W. Ferroptosis as a p53-mediated activity during tumour suppression. Nature. 2015;520(7545):57–62. https://doi.org/10.1038/nature14344.
107. Liu DS, Duong CP, Haupt S, Montgomery KG, House CM, Azar WJ, Pearson HB, Fisher OM, Read M, Guerra GR, Haupt Y, Cullinane C, Wiman KG, Abrahmsen L, Phillips WA, Clemons NJ. Inhibiting the system xC−/glutathione axis selectively targets cancers with mutant-p53 accumulation. Nat Commun. 2017;8(1) https://doi.org/10.1038/ncomms14844.
108. Linher-Melville K, Haftchenary S, Gunning P, Singh G. Signal transducer and activator of transcription 3 and 5 regulate system xc- and redox balance in human breast cancer cells. Mol Cell Biochem. 2015;405(1-2):205–21. https://doi.org/10.1007/s11010-015-2412-4.
109. Gu Y, Albuquerque CP, Braas D, Zhang W, Villa GR, Bi J, Ikegami S, Masui K, Gini B, Yang H, Gahman TC, Shiau AK, Cloughesy TF, Christofk HR, Zhou H, Guan K-L, Mischel PS. mTORC2 regulates amino acid metabolism in cancer by phosphorylation of the cystine-glutamate antiporter xCT. Mol Cell. 2017;67(1):128–138.e127. https://doi.org/10.1016/j.molcel.2017.05.030.
110. Lien EC, Ghisolfi L, Geck RC, Asara JM, Toker A. Oncogenic PI3K promotes methionine dependency in breast cancer cells through the cystine-glutamate antiporter xCT. Sci Signal. 2017;10(510) https://doi.org/10.1126/scisignal.aao6604.
111. Ishimoto T, Nagano O, Yae T, Tamada M, Motohara T, Oshima H, Oshima M, Ikeda T, Asaba R, Yagi H, Masuko T, Shimizu T, Ishikawa T, Kai K, Takahashi E, Imamura Y, Baba Y, Ohmura M, Suematsu M, Baba H, Saya H. CD44 variant regulates redox status in cancer cells by stabilizing the xCT subunit of system xc− and thereby promotes tumor growth. Cancer Cell. 2011;19(3):387–400. https://doi.org/10.1016/j.ccr.2011.01.038.
112. Nagano O, Okazaki S, Saya H. Redox regulation in stem-like cancer cells by CD44 variant isoforms. Oncogene. 2013;32(44):5191–8. https://doi.org/10.1038/onc.2012.638.
113. Yae T, Tsuchihashi K, Ishimoto T, Motohara T, Yoshikawa M, Yoshida GJ, Wada T, Masuko T, Mogushi K, Tanaka H, Osawa T, Kanki Y, Minami T, Aburatani H, Ohmura M, Kubo A, Suematsu M, Takahashi K, Saya H, Nagano O. Alternative splicing of CD44 mRNA by ESRP1 enhances lung colonization of metastatic cancer cell. Nat Commun. 2012;3(1) https://doi.org/10.1038/ncomms1892.
114. Hasegawa M, Takahashi H, Rajabi H, Alam M, Suzuki Y, Yin L, Tagde A, Maeda T, Hiraki M, Sukhatme VP, Kufe D. Functional interactions of the cystine/glutamate antiporter, CD44v and MUC1-C oncoprotein in triple-negative breast cancer cells. Oncotarget. 2016;7(11):11756–69. https://doi.org/10.18632/oncotarget.7598.
115. Tsuchihashi K, Okazaki S, Ohmura M, Ishikawa M, Sampetrean O, Onishi N, Wakimoto H, Yoshikawa M, Seishima R, Iwasaki Y, Morikawa T, Abe S, Takao A, Shimizu M, Masuko T, Nagane M, Furnari FB, Akiyama T, Suematsu M, Baba E, Akashi K, Saya H, Nagano O. The EGF receptor promotes the malignant potential of glioma by regulating amino acid transport

system xc(−-). Cancer Res. 2016;76(10):2954–63. https://doi.org/10.1158/0008-5472.Can-15-2121.
116. Song X, Zhu S, Chen P, Hou W, Wen Q, Liu J, Xie Y, Liu J, Klionsky DJ, Kroemer G, Lotze MT, Zeh HJ, Kang R, Tang D. AMPK-mediated BECN1 phosphorylation promotes ferroptosis by directly blocking system xc– activity. Curr Biol. 2018;28(15):2388–2399.e2385. https://doi.org/10.1016/j.cub.2018.05.094.
117. Briggs Kimberly J, Koivunen P, Cao S, Backus Keriann M, Olenchock Benjamin A, Patel H, Zhang Q, Signoretti S, Gerfen Gary J, Richardson Andrea L, Witkiewicz Agnieszka K, Cravatt Benjamin F, Clardy J, Kaelin William G. Paracrine induction of HIF by glutamate in breast cancer: EglN1 senses cysteine. Cell. 2016;166(1):126–39. https://doi.org/10.1016/j.cell.2016.05.042.
118. Yang Y, Yee D. IGF-I regulates redox status in breast cancer cells by activating the amino acid transport molecule xC−. Cancer Res. 2014;74(8):2295–305. https://doi.org/10.1158/0008-5472.Can-13-1803.
119. Harris Isaac S, Treloar Aislinn E, Inoue S, Sasaki M, Gorrini C, Lee Kim C, Yung Ka Y, Brenner D, Knobbe-Thomsen Christiane B, Cox Maureen A, Elia A, Berger T, Cescon David W, Adeoye A, Brüstle A, Molyneux Sam D, Mason Jacqueline M, Li Wanda Y, Yamamoto K, Wakeham A, Berman Hal K, Khokha R, Done Susan J, Kavanagh Terrance J, Lam C-W, Mak Tak W. Glutathione and thioredoxin antioxidant pathways synergize to drive cancer initiation and progression. Cancer Cell. 2015;27(2):211–22. https://doi.org/10.1016/j.ccell.2014.11.019.
120. Hensley CT, Wasti AT, DeBerardinis RJ. Glutamine and cancer: cell biology, physiology, and clinical opportunities. J Clin Investig. 2013;123(9):3678–84. https://doi.org/10.1172/jci69600.
121. Sayin VI, LeBoeuf SE, Singh SX, Davidson SM, Biancur D, Guzelhan BS, Alvarez SW, Wu WL, Karakousi TR, Zavitsanou AM, Ubriaco J, Muir A, Karagiannis D, Morris PJ, Thomas CJ, Possemato R, Vander Heiden MG, Papagiannakopoulos T. Activation of the NRF2 antioxidant program generates an imbalance in central carbon metabolism in cancer. elife. 2017;6 https://doi.org/10.7554/eLife.28083.
122. DeBerardinis RJ, Lum JJ, Hatzivassiliou G, Thompson CB. The biology of cancer: metabolic reprogramming fuels cell growth and proliferation. Cell Metab. 2008;7(1):11–20. https://doi.org/10.1016/j.cmet.2007.10.002.
123. Yamaguchi I, Yoshimura SH, Katoh H. High cell density increases glioblastoma cell viability under glucose deprivation via degradation of the cystine/glutamate transporter xCT (SLC7A11). J Biol Chem. 2020;295(20):6936–45. https://doi.org/10.1074/jbc.RA119.012213.
124. Goji T, Takahara K, Negishi M, Katoh H. Cystine uptake through the cystine/glutamate antiporter xCT triggers glioblastoma cell death under glucose deprivation. J Biol Chem. 2017;292(48):19721–32. https://doi.org/10.1074/jbc.M117.814392.
125. Yu L, Teoh ST, Ensink E, Ogrodzinski MP, Yang C, Vazquez AI, Lunt SY. Cysteine catabolism and the serine biosynthesis pathway support pyruvate production during pyruvate kinase knockdown in pancreatic cancer cells. Cancer Metabol. 2019a;7(1) https://doi.org/10.1186/s40170-019-0205-z.
126. Hanigan MH, Gallagher BC, Townsend DM, Gabarra V. Gamma-glutamyl transpeptidase accelerates tumor growth and increases the resistance of tumors to cisplatin in vivo. Carcinogenesis. 1999;20(4):553–9. https://doi.org/10.1093/carcin/20.4.553.
127. Slaga TJ, Budunova IV, Gimenez-Conti IB, Aldaz CM. The mouse skin carcinogenesis model. J Investig Dermatol Symp Proc. 1996;1(2):151–6.
128. Hanigan MH. Gamma-glutamyl transpeptidase: redox regulation and drug resistance. Adv Cancer Res. 2014;122:103–41. https://doi.org/10.1016/b978-0-12-420117-0.00003-7.
129. Blanco RA, Ziegler TR, Carlson BA, Cheng PY, Park Y, Cotsonis GA, Accardi CJ, Jones DP. Diurnal variation in glutathione and cysteine redox states in human plasma. Am J Clin Nutr. 2007;86(4):1016–23. https://doi.org/10.1093/ajcn/86.4.1016.

130. Gerber MA, Thung SN. Enzyme patterns in human hepatocellular carcinoma. Am J Pathol. 1980;98(2):395–400.
131. Luo M, Sun W, Wu C, Zhang L, Liu D, Li W, Mei Q, Hu G. High pretreatment serum gamma-glutamyl transpeptidase predicts an inferior outcome in nasopharyngeal carcinoma. Oncotarget. 2017;8(40):67651–62. https://doi.org/10.18632/oncotarget.18798.
132. Ahmad S, Okine L, Wood R, Aljian J, Vistica DT. Gamma-Glutamyl transpeptidase (gamma-GT) and maintenance of thiol pools in tumor cells resistant to alkylating agents. J Cell Physiol. 1987;131(2):240–6. https://doi.org/10.1002/jcp.1041310214.
133. Godwin AK, Meister A, O'Dwyer PJ, Huang CS, Hamilton TC, Anderson ME. High resistance to cisplatin in human ovarian cancer cell lines is associated with marked increase of glutathione synthesis. Proc Natl Acad Sci. 1992;89(7):3070–4. https://doi.org/10.1073/pnas.89.7.3070.
134. Oguchi H, Kikkawa F, Kojima M, Maeda O, Mizuno K, Suganuma N, Kawai M, Tomoda Y. Glutathione related enzymes in cis-diamminedichloroplatinum (II)-sensitive and-resistant human ovarian carcinoma cells. Anticancer Res. 1994;14(1a):193–200.
135. Lettieri-Barbato D, Aquilano K. Pushing the limits of cancer therapy: the nutrient game. Front Oncol. 2018;8 https://doi.org/10.3389/fonc.2018.00148.
136. Stepka P, Vsiansky V, Raudenska M, Gumulec J, Adam V, Masarik M. Metabolic and amino acid alterations of the tumor microenvironment. Curr Med Chem. 2020; https://doi.org/10.2174/0929867327666200207114658.
137. Zhang W, Trachootham D, Liu J, Chen G, Pelicano H, Garcia-Prieto C, Lu W, Burger JA, Croce CM, Plunkett W, Keating MJ, Huang P. Stromal control of cystine metabolism promotes cancer cell survival in chronic lymphocytic leukaemia. Nat Cell Biol. 2012;14(3):276–86. https://doi.org/10.1038/ncb2432.
138. Wang W, Kryczek I, Dostál L, Lin H, Tan L, Zhao L, Lu F, Wei S, Maj T, Peng D, He G, Vatan L, Szeliga W, Kuick R, Kotarski J, Tarkowski R, Dou Y, Rattan R, Munkarah A, Liu JR, Zou W. Effector T cells abrogate stroma-mediated chemoresistance in ovarian cancer. Cell. 2016;165(5):1092–105. https://doi.org/10.1016/j.cell.2016.04.009.
139. Doxsee DW, Gout PW, Kurita T, Lo M, Buckley AR, Wang Y, Xue H, Karp CM, Cutz J-C, Cunha GR, Wang Y-Z. Sulfasalazine-induced cystine starvation: potential use for prostate cancer therapy. Prostate. 2007;67(2):162–71. https://doi.org/10.1002/pros.20508.
140. Gout PW, Buckley AR, Simms CR, Bruchovsky N. Sulfasalazine, a potent suppressor of lymphoma growth by inhibition of the xc − cystine transporter: a new action for an old drug. Leukemia. 2001;15(10):1633–40. https://doi.org/10.1038/sj.leu.2402238.
141. Guan J, Lo M, Dockery P, Mahon S, Karp CM, Buckley AR, Lam S, Gout PW, Wang YZ. The xc- cystine/glutamate antiporter as a potential therapeutic target for small-cell lung cancer: use of sulfasalazine. Cancer Chemother Pharmacol. 2009;64(3):463–72. https://doi.org/10.1007/s00280-008-0894-4.
142. Patel SA, Warren BA, Rhoderick JF, Bridges RJ. Differentiation of substrate and non-substrate inhibitors of transport system xc(−): an obligate exchanger of L-glutamate and L-cystine. Neuropharmacology. 2004;46(2):273–84. https://doi.org/10.1016/j.neuropharm.2003.08.006.
143. Massie A, Boillée S, Hewett S, Knackstedt L, Lewerenz J. Main path and byways: non-vesicular glutamate release by system xc−as an important modifier of glutamatergic neurotransmission. J Neurochem. 2015;135(6):1062–79. https://doi.org/10.1111/jnc.13348.
144. Lutgen V, Resch J, Qualmann K, Raddatz NJ, Panhans C, Olander EM, Kong L, Choi S, Mantsch JR, Baker DA. Behavioral assessment of acute inhibition of system xc (−) in rats. Psychopharmacology. 2014;231(24):4637–47. https://doi.org/10.1007/s00213-014-3612-4.
145. Chidley C, Haruki H, Pedersen MG, Muller E, Johnsson K. A yeast-based screen reveals that sulfasalazine inhibits tetrahydrobiopterin biosynthesis. Nat Chem Biol. 2011;7(6):375–83. https://doi.org/10.1038/nchembio.557.
146. Jansen G, van der Heijden J, Oerlemans R, Lems WF, Ifergan I, Scheper RJ, Assaraf YG, Dijkmans BAC. Sulfasalazine is a potent inhibitor of the reduced folate carrier: implications

for combination therapies with methotrexate in rheumatoid arthritis. Arthritis Rheum. 2004;50 (7):2130–9. https://doi.org/10.1002/art.20375.

147. Pruzanski W, Stefanski E, Vadas P, Ramamurthy NS. Inhibition of extracellular release of proinflammatory secretory phospholipase A2 (sPLA2) by sulfasalazine: a novel mechanism of anti-inflammatory activity. Biochem Pharmacol. 1997;53(12):1901–7. https://doi.org/10.1016/s0006-2952(97)00137-8.
148. Wahl C, Liptay S, Adler G, Schmid RM. Sulfasalazine: a potent and specific inhibitor of nuclear factor kappa B. J Clin Investig. 1998;101(5):1163–74. https://doi.org/10.1172/jci992.
149. Dixon SJ, Stockwell BR. The role of iron and reactive oxygen species in cell death. Nat Chem Biol. 2014;10(1):9–17. https://doi.org/10.1038/nchembio.1416.
150. Viswanathan VS, Ryan MJ, Dhruv HD, Gill S, Eichhoff OM, Seashore-Ludlow B, Kaffenberger SD, Eaton JK, Shimada K, Aguirre AJ, Viswanathan SR, Chattopadhyay S, Tamayo P, Yang WS, Rees MG, Chen S, Boskovic ZV, Javaid S, Huang C, Wu X, Tseng YY, Roider EM, Gao D, Cleary JM, Wolpin BM, Mesirov JP, Haber DA, Engelman JA, Boehm JS, Kotz JD, Hon CS, Chen Y, Hahn WC, Levesque MP, Doench JG, Berens ME, Shamji AF, Clemons PA, Stockwell BR, Schreiber SL. Dependency of a therapy-resistant state of cancer cells on a lipid peroxidase pathway. Nature. 2017;547(7664):453–7. https://doi.org/10.1038/nature23007.
151. Dixon SJ, Patel DN, Welsch M, Skouta R, Lee ED, Hayano M, Thomas AG, Gleason CE, Tatonetti NP, Slusher BS, Stockwell BR. Pharmacological inhibition of cystine-glutamate exchange induces endoplasmic reticulum stress and ferroptosis. elife. 2014;3:e02523. https://doi.org/10.7554/eLife.02523.
152. Sato M, Kusumi R, Hamashima S, Kobayashi S, Sasaki S, Komiyama Y, Izumikawa T, Conrad M, Bannai S, Sato H. The ferroptosis inducer erastin irreversibly inhibits system xc− and synergizes with cisplatin to increase cisplatin's cytotoxicity in cancer cells. Sci Rep. 2018;8(1) https://doi.org/10.1038/s41598-018-19213-4.
153. Zhang Y, Tan H, Daniels JD, Zandkarimi F, Liu H, Brown LM, Uchida K, O'Connor OA, Stockwell BR. Imidazole ketone erastin induces ferroptosis and slows tumor growth in a mouse lymphoma model. Cell Chem Biol. 2019;26(5):623–633.e629. https://doi.org/10.1016/j.chembiol.2019.01.008.
154. Yu M, Gai C, Li Z, Ding D, Zheng J, Zhang W, Lv S, Li W. Targeted exosome-encapsulated erastin induced ferroptosis in triple negative breast cancer cells. Cancer Sci. 2019b;110 (10):3173–82. https://doi.org/10.1111/cas.14181.
155. Poursaitidis I, Wang X, Crighton T, Labuschagne C, Mason D, Cramer SL, Triplett K, Roy R, Pardo OE, Seckl MJ, Rowlinson SW, Stone E, Lamb RF. Oncogene-selective sensitivity to synchronous cell death following modulation of the amino acid nutrient cystine. Cell Rep. 2017;18(11):2547–56. https://doi.org/10.1016/j.celrep.2017.02.054.
156. Gao M, Monian P, Quadri N, Ramasamy R, Jiang X. Glutaminolysis and transferrin regulate ferroptosis. Mol Cell. 2015;59(2):298–308. https://doi.org/10.1016/j.molcel.2015.06.011.
157. Gao M, Yi J, Zhu J, Minikes AM, Monian P, Thompson CB, Jiang X. Role of mitochondria in ferroptosis. Mol Cell. 2019;73(2):354–363.e353. https://doi.org/10.1016/j.molcel.2018.10.042.
158. Alvarez SW, Sviderskiy VO, Terzi EM, Papagiannakopoulos T, Moreira AL, Adams S, Sabatini DM, Birsoy K, Possemato R. NFS1 undergoes positive selection in lung tumours and protects cells from ferroptosis. Nature. 2017;551(7682):639–43. https://doi.org/10.1038/nature24637.
159. Lang X, Green MD, Wang W, Yu J, Choi JE, Jiang L, Liao P, Zhou J, Zhang Q, Dow A, Saripalli AL, Kryczek I, Wei S, Szeliga W, Vatan L, Stone EM, Georgiou G, Cieslik M, Wahl DR, Morgan MA, Chinnaiyan AM, Lawrence TS, Zou W. Radiotherapy and immunotherapy

promote tumoral lipid oxidation and ferroptosis via synergistic repression of SLC7A11. Cancer Discov. 2019;9(12):1673–85. https://doi.org/10.1158/2159-8290.Cd-19-0338.

160. Wang W, Green M, Choi JE, Gijón M, Kennedy PD, Johnson JK, Liao P, Lang X, Kryczek I, Sell A, Xia H, Zhou J, Li G, Li J, Li W, Wei S, Vatan L, Zhang H, Szeliga W, Gu W, Liu R, Lawrence TS, Lamb C, Tanno Y, Cieslik M, Stone E, Georgiou G, Chan TA, Chinnaiyan A, Zou W. CD8+ T cells regulate tumour ferroptosis during cancer immunotherapy. Nature. 2019b;569(7755):270–4. https://doi.org/10.1038/s41586-019-1170-y.
161. Keating MJ, Holmes R, Lerner S, Ho DH. L-asparaginase and PEG asparaginase--past, present, and future. Leuk Lymphoma. 1993;10(Suppl):153–7. https://doi.org/10.3109/10428199309149129.
162. Jones CL, Stevens BM, D'Alessandro A, Culp-Hill R, Reisz JA, Pei S, Gustafson A, Khan N, DeGregori J, Pollyea DA, Jordan CT. Cysteine depletion targets leukemia stem cells through inhibition of electron transport complex II. Blood. 2019;134(4):389–94. https://doi.org/10.1182/blood.2019898114.
163. Uren JR, Lazarus H. L-cyst(e)ine requirements of malignant cells and progress toward depletion therapy. Cancer Treat Rep. 1979;63(6):1073–9.
164. Cramer SL, Saha A, Liu J, Tadi S, Tiziani S, Yan W, Triplett K, Lamb C, Alters SE, Rowlinson S, Zhang YJ, Keating MJ, Huang P, DiGiovanni J, Georgiou G, Stone E. Systemic depletion of L-cyst(e)ine with cyst(e)inase increases reactive oxygen species and suppresses tumor growth. Nat Med. 2016;23(1):120–7. https://doi.org/10.1038/nm.4232.
165. Bhattacharyya S, Saha S, Giri K, Lanza IR, Nair KS, Jennings NB, Rodriguez-Aguayo C, Lopez-Berestein G, Basal E, Weaver AL, Visscher DW, Cliby W, Sood AK, Bhattacharya R, Mukherjee P. Cystathionine Beta-synthase (CBS) contributes to advanced ovarian Cancer progression and drug resistance. PLoS One. 2013;8:e79167.
166. Cochrane DR, Tessier-Cloutier B, Lawrence KM, Nazeran T, Karnezis AN, Salamanca C, Cheng AS, McAlpine JN, Hoang LN, Gilks CB, Huntsman DG. Clear cell and endometrioid carcinomas: are their differences attributable to distinct cells of origin? J Pathol. 2017;243(1):26–36. https://doi.org/10.1002/path.4934.
167. Poisson LM, Munkarah A, Madi H, Datta I, Hensley-Alford S, Tebbe C, Buekers T, Giri S, Rattan R. A metabolomic approach to identifying platinum resistance in ovarian cancer. J Ovarian Res. 2015;8(1):13. https://doi.org/10.1186/s13048-015-0140-8.
168. Asimakopoulou A, Panopoulos P, Chasapis CT, Coletta C, Zhou Z, Cirino G, Giannis A, Szabo C, Spyroulias GA, Papapetropoulos A. Selectivity of commonly used pharmacological inhibitors for cystathionine β synthase (CBS) and cystathionine γ lyase (CSE). Br J Pharmacol. 2013;169(4):922–32. https://doi.org/10.1111/bph.12171.
169. Yue T, Zuo S, Bu D, Zhu J, Chen S, Ma Y, Ma J, Guo S, Wen L, Zhang X, Hu J, Wang Y, Yao Z, Chen G, Wang X, Pan Y, Wang P, Liu Y. Aminooxyacetic acid (AOAA) sensitizes colon cancer cells to oxaliplatin via exaggerating apoptosis induced by ROS. J Cancer. 2020;11(7):1828–38. https://doi.org/10.7150/jca.35375.
170. Wang L, Cai H, Hu Y, Liu F, Huang S, Zhou Y, Yu J, Xu J, Wu F. A pharmacological probe identifies cystathionine β-synthase as a new negative regulator for ferroptosis. Cell Death Dis. 2018;9(10):1005. https://doi.org/10.1038/s41419-018-1063-2.
171. Wang L, Shi H, Zhang X, Zhang X, Liu Y, Kang W, Shi X, Wang T. I157172, a novel inhibitor of cystathionine γ-lyase, inhibits growth and migration of breast cancer cells via SIRT1-mediated deacetylation of STAT3. Oncol Rep. 2019a;41(1):427–36. https://doi.org/10.3892/or.2018.6798.
172. Sun Q, Collins R, Huang S, Holmberg-Schiavone L, Anand GS, Tan C-H, van den Berg S, Deng L-W, Moore PK, Karlberg T, Sivaraman J. Structural basis for the inhibition mechanism of human cystathionine γ-lyase, an enzyme responsible for the production of H2S. J Biol Chem. 2009;284(5):3076–85. https://doi.org/10.1074/jbc.M805459200.

173. Clausen T, Huber R, Prade L, Wahl MC, Messerschmidt A. Crystal structure of Escherichia coli cystathionine γ-synthase at 1.5 Å resolution. EMBO J. 1998;17(23):6827–38. https://doi.org/10.1093/emboj/17.23.6827.
174. Gottesman MM, Fojo T, Bates SE. Multidrug resistance in cancer: role of ATP–dependent transporters. Nat Rev Cancer. 2002;2(1):48–58. https://doi.org/10.1038/nrc706.
175. Manciu L, Chang X-B, Buyse F, Hou Y-X, Gustot A, Riordan JR, Ruysschaert JM. Intermediate structural states involved in MRP1-mediated drug transport. J Biol Chem. 2003;278(5):3347–56. https://doi.org/10.1074/jbc.M207963200.
176. Jensen GL, Meister A. Radioprotection of human lymphoid cells by exogenously supplied glutathione is mediated by gamma-glutamyl transpeptidase. Proc Natl Acad Sci. 1983;80(15):4714–7. https://doi.org/10.1073/pnas.80.15.4714.
177. Green JA, Vistica DT, Young RC, Hamilton TC, Rogan AM, Ozols RF. Potentiation of melphalan cytotoxicity in human ovarian cancer cell lines by glutathione depletion. Cancer Res. 1984;44(11):5427–31.
178. Ozols RF, Louie KG, Plowman J, Behrens BC, Fine RL, Dykes D, Hamilton TC. Enhanced melphalan cytotoxicity in human ovarian cancer in vitro and in tumor-bearing nude mice by buthionine sulfoximine depletion of glutathione. Biochem Pharmacol. 1987;36(1):147–53. https://doi.org/10.1016/0006-2952(87)90392-3.
179. Russo A, DeGraff W, Friedman N, Mitchell JB. Selective modulation of glutathione levels in human normal versus tumor cells and subsequent differential response to chemotherapy drugs. Cancer Res. 1986;46(6):2845–8.
180. Dai Z, Huang Y, Sadee W, Blower P. Chemoinformatics analysis identifies cytotoxic compounds susceptible to chemoresistance mediated by glutathione and cystine/glutamate transport system xc. J Med Chem. 2007;50(8):1896–906. https://doi.org/10.1021/jm060960h.
181. Okuno S, Sato H, Kuriyama-Matsumura K, Tamba M, Wang H, Sohda S, Hamada H, Yoshikawa H, Kondo T, Bannai S. Role of cystine transport in intracellular glutathione level and cisplatin resistance in human ovarian cancer cell lines. Br J Cancer. 2003;88(6):951–6. https://doi.org/10.1038/sj.bjc.6600786.
182. Hansson J, Edgren M, Ehrsson H, Ringborg U, Nilsson B. Effect of D,L-buthionine-S,R-sulfoximine on cytotoxicity and DNA cross-linking induced by bifunctional DNA-reactive cytostatic drugs in human melanoma cells. Cancer Res. 1988;48(1):19–26.
183. Lien EC, Lyssiotis CA, Juvekar A, Hu H, Asara JM, Cantley LC, Toker A. Glutathione biosynthesis is a metabolic vulnerability in PI(3)K/Akt-driven breast cancer. Nat Cell Biol. 2016;18(5):572–8. https://doi.org/10.1038/ncb3341.
184. Nunes SC, Ramos C, Lopes-Coelho F, Sequeira CO, Silva F, Gouveia-Fernandes S, Rodrigues A, Guimarães A, Silveira M, Abreu S, Santo VE, Brito C, Félix A, Pereira SA, Serpa J. Cysteine allows ovarian cancer cells to adapt to hypoxia and to escape from carboplatin cytotoxicity. Sci Rep. 2018;8(1):9513. https://doi.org/10.1038/s41598-018-27753-y.
185. Pendyala L, Velagapudi S, Toth K, Zdanowicz J, Glaves D, Slocum H, Perez R, Huben R, Creaven PJ, Raghavan D. Translational studies of glutathione in bladder cancer cell lines and human specimens. Clin Cancer Res. 1997;3(5):793–8.
186. Révész L, Edgren MR, Wainson AA. Selective toxicity of buthionine sulfoximine (BSO) to melanoma cells in vitro and in vivo. Int J Radiat Oncol Biol Phys. 1994;29(2):403–6. https://doi.org/10.1016/0360-3016(94)90298-4.
187. Rocha CRR, Garcia CCM, Vieira DB, Quinet A, de Andrade-Lima LC, Munford V, Belizário JE, Menck CFM. Glutathione depletion sensitizes cisplatin- and temozolomide-resistant glioma cells in vitro and in vivo. Cell Death Dis. 2014;5(10):–e1505. https://doi.org/10.1038/cddis.2014.465.

188. Zheng Z-G, Xu H, Suo S-S, Xu X-L, Ni M-W, Gu L-H, Chen W, Wang L-Y, Zhao Y, Tian B, Hua Y-J. The essential role of H19 contributing to cisplatin resistance by regulating glutathione metabolism in high-grade serous ovarian cancer. Sci Rep. 2016;6(1):26093. https://doi.org/10.1038/srep26093.
189. Bekeschus S, Eisenmann S, Sagwal SK, Bodnar Y, Moritz J, Poschkamp B, Stoffels I, Emmert S, Madesh M, Weltmann K-D, von Woedtke T, Gandhirajan RK. xCT (SLC7A11) expression confers intrinsic resistance to physical plasma treatment in tumor cells. Redox Biol. 2020;30:101423. https://doi.org/10.1016/j.redox.2019.101423.
190. Nunes S, Serpa J. Glutathione in ovarian cancer: a double-edged sword. Int J Mol Sci. 2018;19 (7) https://doi.org/10.3390/ijms19071882.

Stress and Circadian Rhythms

8

Meng-Juan Sun, Hai-Yan Yin, Xin Cao, Liang Kang, Yu-Shi Hu, and Yong Tang

Abstract

Our lives are dominated by circadian rhythms (which last about a whole day), most notably through the sleep-wake cycle. Almost all important physiological and metabolic processes are controlled by circadian rhythms. The ability to predict day-night changes in the environment gives most species on earth an evolutionary advantage. As a result, from plants to higher mammals, organisms form endogenous biological clocks in order to adapt to circadian rhythm changes. In the absence of external time indicators, the internal clock can automatically run on a cycle of about 24 h. Stress responses begin with a local physical (such as skeletal muscle contusion) or mental (such as loss of a loved one) stressor but always end up with a broad, systematic process of response that affects many organs and systems. It is normal, then, that disorders of circadian rhythm put the body in a state of stress that leads to various mental, neurological, and metabolic disorders.

Keywords

Circadian rhythm · Stress · Molecular clock · HPA axis

Plants to higher mammals on earth have adapted to the 24 h cycle by the evolution of internal rhythms, named circadian clocks, which adjusts physiology and behavior to

M.-J. Sun · H.-Y. Yin · X. Cao · Y. Tang (✉)
Chengdu University of Traditional Chinese Medicine, Chengdu, China

Acupuncture and Chronobiology Key Laboratory of Sichuan Province, Chengdu, China
e-mail: tangyong@cdutcm.edu.cn

L. Kang · Y.-S. Hu
Sports Medicine Key Laboratory of Sichuan Province, Chengdu Sport University, Chengdu, China

C. Huang, Y. Zhang (eds.), *Oxidative Stress*,
https://doi.org/10.1007/978-981-16-0522-2_8

the repeating changes in around environmental conditions. In mammals, the main pacemaker in the suprachiasmatic nucleus (SCN) of the hypothalamus receives ambient light changes in order to synchronize peripheral tissue and central non-SCN clocks with circadian rhythm changes. Regulatory systems, such as the autonomic nervous system (ANS) and the thalamic-pituitary-adrenal (HPA) axis, are most important in modifying stress responses that receive strong circadian input.

8.1 Stress

Stress, which was proposed by Hans Selye about 80 years ago, is a physiological response of the body to changes in the environment and is now a basic scientific, social, and clinical concept [1, 2]. Stress response is the typical response of attention, such as arousal and emotion, emphasizing the stress, awakening, and emotional change caused by environmental stimulus stressor. Stressors include psychogenic (e.g., social stimuli) or neurogenic (e.g., painful stimuli) or a combination of both; this issue comprises results to which connect with psychogenic stressors. Stressors can affect mammals in the womb. Postnatally, environmental stressors can affect external stress that can be transmitted from a lactating mother to her offspring.

Some studies focus on the intrinsic factors that form individual stress response. It suggests that any stressor is heterogeneous, depending on internal factors such as gender, genes, and age. This was confirmed in a retrospective study of rodents. Some researches show to the fore the essential role of the circadian clock in intermediate between the individual's stress response and the environmental stressor. The main pacemaker located in the suprachiasmatic nucleus of the hypothalamus affects not only the HPA axis but also the autonomic nervous system, making both stress response systems receive strong circadian input. Therefore, the interaction between circadian rhythm and environmental stress is universal and has pathological and physiological effects.

8.2 Circadian Rhythms

Circadian rhythm is regulated by the circadian clock inside the body, which enables the body to maintain a circadian rhythm 24 h a day, and the complexity of its structure varies according to the corresponding organism [3]. The mammalian physiological system consists mainly of a central pacemaker located in the suprachiasmatic nucleus (SCN) of the hypothalamus, also known as the central clock, which activates ganglion cells by the eye to provide melanin to regulate the light/dark changes of the day. These light/darkness cycles are the principal external stimulus to the SCN; however, there are other periodic synchronizers, such as the schedule of ingestion/fasting and the exercise (activity/rest). SCN is the main external stimulation light/dark cycle but also other periodic synchronizers, this exercise, for example (the activity/rest), and the schedule of ingestion/fasting. These synchronize with the central pacemaker by affecting other clocks in the body's tissues and organs, such as the heart, fatty tissues, pancreas, and lungs, through autonomic system activity and periodic hormone secretion [4]. Some molecular researches have demonstrated oscillations in the transcription of clock-specific genes that play a key part in the generation of the circadian rhythms. Many of these clock genes work as transcriptional agents that stimulated or disable their own expression in a series of feedback reactions [4].

8.3 Interaction Between Stress and Circadian Rhythms

Disruption of circadian rhythm is closely related to stressors. Repeated or long-term exposure to stressors may contribute to the development of mental illness or metabolism in rodents and humans and can result to long-lasting adaptations, such as energy metabolism [5, 6]. Stressors are vital to the characteristics of the stress response. The main stressors were social interaction, environmental factors, constant light exposure, hot and cold stimulation, and increased shift work. It is important to realize that these changes not only affect the activity of the hypothalamus but may also lead to the development of a range of stress-related diseases Light is of the most important direct causes of circadian dysrhythmias. In rodent animals after 60 min short pulse of light exposure can be found to increase plasma corticosterone. This rise has both been characterized as ACTH-independent in rodents. However, there was ACTH-dependent contribution by the HPA axis in mice [7]. However, light is not the only cause of high corticosterone. The day cycle is shortened by 1 h (9.5 h light vs. 13.5 h dark), and the GCs in the cycle are not aligned with the adrenal ACTH but with behavior and the SCN clock [8]. Similarly, the effect of food on circadian rhythm cannot be ignored. Restricting mice's food intake during the day promoted GC levels to reach a second peak. Chronic circadian disruption owing to frequent and inappropriate food or light intake can effectively alter diurnal levels of secreted GCs and pressure-induced GC reactions. When suffering for a long time in stressors and disruption of circadian rhythm can result in similar pathological outcomes, such as impaired immune response, metabolic disorders [9], accelerated growth and increased death rate [10], and accelerated aging [11].

The effectors of the stress system also impact the adjustment of circadian rhythm. How does the stress system affect circadian rhythm? GCs and epinephrine are major stress effectors that act as synchronizers of the body clock and act through specific receptors. In the case of GCs, it acts as both a major stress signal and a major circadian signal throughout the body [12, 13]. Two kinds of mammalian intracelluar receptors for CORT have been characterized and isolated – the glucocorticoid receptor (GR) and the mineralocorticoid receptor (MR), which also binds aldosterone with high affinity [14, 15] – mainly GR, because even at the lowest point of GC's circadian rhythm, MR is constrained by GCs, which makes MR signals inefficient in transmitting time information. GR is expressed in almost all organs and tissues [16–18] except SCN, where no GR expression has been detected [19]. Therefore, SCN is not controlled by direct feedback synchronization of GCs. SCN is periodic information that perceives light as darkness. But even in the absence of light information, it can independently produce an internal circadian rhythm [20]. The increase in CORT is both a signal of stress in tissues and an entrapment signal of circadian rhythm throughout the body. Some studies believe that stress is an exaggerated arousal state, and the temporal correlation of the determinants of stress-induced CORT activity is not the temporal correlation between stress-induced CORT and the basic circadian phase of CORT. Appropriate stress response depends on the circadian rhythm changes, and this association is the initial factor in the quantitative and qualitative reaction of the cell system to stress-induced CORT secretion [21]. Many rodent studies have investigated the effects of stress on the circadian rhythm. Using the behavior failure paradigm, Tahara et al. [22] reported that stress at the start of photology leads an early phase shift in mRNA expression rhythm of several core clock genes in peripheral organs of rats. When the rats were stressed at other times of the day, the effect was phase delay and even loss of synchronicity, indicating that the effect of stress on the peripheral clock was time dependent. After a few weeks of long-term exposure to stressors, these effects are more subtle or less noticeable, suggesting habituation effects. In a longer-term approach, the effects of repeated social failures over 19 days were analyzed, both in the early dark and the early light. The amplitude of Per2 oscillations in SCN increased, while the expressions of Cry1 and Per2 were downregulated in adrenal gland under predark pressure. Conversely, stress in early light does not affect the SCN clock but causes the phase of the adrenal oscillator to advance. Phase changes in the adrenal clock were similar to those in Tahara et al. 's study.

8.3.1 Molecular Clock in Stress and Circadian Rhythms

The Earth rotate its axis 24 h a day; as a result, organisms on Earth have the ability to use molecular clocks to predict changes in the environment, and this has passed on the dominant genes for most species' evolution. As a result, all living things, whether mammals or plants, have evolved an internal clock, a circadian rhythm, that allows these clocks to run almost 24 h a day without cues from external time. The internal body clock allows the body to accurately estimate the time of day, in order to align internal biological functions with environmental changes. The biological clock combines with external zeitgebers, such as light and temperature, of which light is the most favorable. The circadian clock is temperature-compensated, a feature that is particularly important for poikilotherms [23]. In 2017, Michael W. Young, Jeffrey C. Hall, and Michael Rosbash were awarded the Nobel Prize in Physiology or Medicine due to their discovery of what was the molecular machinery underpinning the biological clock 1–4 (Box 1). The molecular clocks are important to elucidate the mechanism of circadian rhythms in disease and physiology. In mammals, circadian clocks control circadian rhythmicity. These clocks, which consist of about 20,000 neurons in the suprachiasmatic nucleus of the hypothalamus, can be divided into central or peripheral clocks [24], and the peripheral clock can exist in almost any organization [25]. Light signal is received and transmitted to the master clock through the retina and is input to the peripheral clock of the whole body by the master clock through a variety of neurohumoral signals [26].

In mammals, the biological clock is associated with a molecular oscillator in every cell in the body. The molecular oscillator is built on self-sustaining oscillations generated by transcription-translation feedback loops (TTFLs). The clock's core TTFL is consist of the muscle ant-like 1 (Bamal 1) and genes brian, cryptochrome (Cry) 1/2, period (Per)1–3, and circadian locomotor output cycles kaput (Clock) [27, 28]. In mammals of the TTFLs involves transcriptional activation by Clock/Bmal1 heterodimers, which drive 24 h expression of Cry and Per genes by E-box regulatory sequences. The suprachiasmatic nucleus (SCN) of the hypothalamus is the body's main biological clock and the fulcrum of the gene-to-cell circuit. The SCN determines the body's physiological cycle and daily rhythm of life [29].

There are innumerable cellular clocks across the organisms [30, 31], but the principal circadian pacemaker in mammals is the SCN of the hypothalamus [32]. Its highly coordinated multicellular oscillations can continue indefinitely to lead to internal synchronization of the body's cellular clocks [33]. Cry genes including Cry1 and Cry2 play a negative part of limb in the clock feedback loop. However, Per proteins have subtle effect on their cellular location Per and cry proteins accumulate in the cytoplasm and form complexes that translocate back into the nucleus where they are able to inhibit BMAL 1/CLOCK.-mediated transcription [34]. The time-keeping of cellular molecular clock in other organizations and the SCN pivots around self-sustaining TTFLs [35] in which, starting at the brain and muscle ARNT-like 1 (BMAL1; also known as ARNTL), circadian dawn (i.e., circadian time (CT) 0), heterodimers of circadian locomotor output cycles protein kaput (CLOCK), the positive regulators of the loop, drives the expression of the CRY and PER proteins, the negative regulators, via enhancer box (E- box) regulatory sequences [29]. By the end of the circadian day, the BMAL1/CLOCK heterodimer has been reactivated to start a new cycle. The cycle is stabilized by accessory loops in which BMAL1 and CLOCK drive E-box-mediated circadian expression of the nuclear receptors RORα, REV-ERBα, and REV-ERBβ, which in turn act via REV response element (RRE) sequences to suppress and activate BMAL1 transcription, respectively. Thus, the core TTFL of circadian expression of REV-ERB proteins is an output and also an input.

Besides, a novel mechanism that TFE3 and TFEB directly bind to the promoter of Nr1d1, a negative regulator of metabolism (e.g., Srebf1/Srebp1, Fasn), autophagy (e.g., Atg5, Ulk1) and several clock (e.g., Arntl/Bmal1, Npas2) genes binding to autophagy and nutrient availability with the cell-autonomous circadian clock through the regulation of gene expression [36, 37]. An additional loop regulates BMAL1–CLOCK activity by the transcriptional regulation of Bmal1. The core of hormone receptors REV-ERBβ (also known as Nr1d2) and retinoic acid-related orphan response (ROR) α, β, and γ and REV-ERBα (also known asNr1d1) include E-box elements in their promoter sequences and are under transcriptional control by CLOCK–BMAL1 [38, 39]. In fact, RORs are positive regulators of Bmal1 transcription and compare with REV-ERBs for retinoid orphan receptor response element

(RORE) connecting sites within the Bmal1 promoter; in contrast, REV-ERB proteins exert a negative feedback, banning Bmal1 transcription [40].

An important discovery in circadian rhythms is that SCN is not the only biological clock in the body. In fact, most cells and tissues, including other peripheral organs except SCN and brain regions, have circadian oscillations. Moreover, these oscillators can operate independently of the SCN [41, 42]. For example, in mammalian livers, these rhythms are regulated by the core circadian clock of TTFLs and fasting. The circadian rhythm occurring in the liver is regulated in two ways; one by the said autonomic diurnal oscillation combination of swing, on the other hand, can be affected by feeding or eating both at different times during the feeding and fasting cycle. mRNA can be produced and the transcription results have circadian rhythm and the expressed circadian rhythm is different in the periodic phase that shifts forward or backward [43].

8.3.2 HPA Axis in Stress and Circadian Rhythms

In addition to the core body temperature cycle, the circadian clock is synchronized with the environment on a 24 h cycle, and the adrenal gland plays a major role in coordinating the molecular oscillations of all biological clocks [44]. The stress system was regulated by the circadian system to make animals regulate cyclic challenges they encounter during day and night. The primary roles of the stress system are the hypothalamus–pituitary–adrenal (HPA) axis and the sympathetic nervous system. HPA axis mediates the adrenals to release glucocorticoids (GCs) when they are exposed to stress [44]. Stress is an internal or external challenge that needs the body to respond adequately to discomfort in less severe situations or to avoid pain or to survive. When the body is stressed, a complex response system is activated. It involves delayed response via HPA axis-mediated release of GCs and immediate response via activation of the autonomic nervous system (ANS) [45]. The adrenal gland among all peripheral oscillators is important to play a special part since the adrenal circadian clock can regularize release of hormones with clock-modulating properties to influence rhythms in other tissues. GCs from the intermediate zona fasciculata of adrenal gland. It is all known that corticosterone and cortisol are the main GCs in rodents and humans, respectively [46]. When organisms exposure to short- or long-term stress, it can rapidly activate the HPA axis to secret GCs. Glucocorticoid circadian rhythms are also closely related to the levels of arousal; the time of awakening of the rapid increase in glucocorticoid levels is thought to promote cognitive activity together with the activation of peripheral metabolic pathways in the muscle, liver, and adipose tissue, while at the end of the day, glucocorticoid levels fall, which helps you fall asleep [13]. Studies have shown that glucocorticoids can affect PER gene in peripheral tissues but are insensitive to the direct effect of glucocorticoids on rhythmic expression in SCN [47].

Hypothalamic paraventricular nucleus (PVH) stimulates the anterior pituitary to release corticotrophin under stress, thus promoting the synthesis and release of adrenal hormone [45, 48]. Glucocorticoids, such as corticosterone (CORT), not only affect the brain through cellular signaling systems but also act on cellular stress by controlling gene expression [49, 50]. Furthermore, GCs improve the vigilance of the brain and prevent inflammatory processes in the body. Besides, GCs act by altering the activity of various kinases, such as mitogen-activated protein kinase, phosphoinositide 3-kinases (PI3K), and RAC-alpha serine/threonine kinase (AKT) [51, 52]; however, it is still not sure if between rhythmic HPA axis activity and circadian rhythm of GC secretion is relative. On one hand, adrenal rhythms persist after hypophysectomy, when no ACTH is present [53]; on the other hand, ACTH is able to phase-dependently reset GC rhythms [54]. Deliberately disturbing the adrenal clock can eliminate the circadian pattern of GC secretion, suggesting that the peripheral tissue clock ultimately dominates the GC secretion pattern [55]. Therefore, GC rhythmic and regular secretion plays an important role in time regulation. Given these different mechanisms of action, glucocorticoids can rapidly affect neural circuits during stress exposure and have long-term effects in the form of structural or functional plasticity. These various changes in neurological function can significantly affect the body's response to stress on a behavioral level [56, 57].

Information about stress can be gathered by all the sensory systems in the body (such as changes in blood composition, decreased blood volume, or encounters with predators) and pass it on to the brain stem [45]. Subsequent activation of the ANS and HPA axes was regulated. In terms of the HPA axis, stress-mediated activation triggers the production and release of adrenal GCs. Stress signals in amygdala, hippocampus, and prefrontal cortex are transmitted to the para-ventral nucleus (PVN), which stimulates CRH secretion and activates the HPA axis. The GCs need to be resynthesized after each trigger, resulting in a delay in the final effect response. Thus, this dynamic process is slower than (within minutes) the activation of ANS, which occurs within a few seconds of the beginning of the stress. Sympathetic preganglionic neurons in the spinal cord can be activated by stressors, leading to an increase in catecholamine secretion [58, 59]. Signals are transmitted to post-ganglia neurons and projected into peripheral effector organs, translating into the classic "fight or flight," or the preganglionic nerve, such as the visceral peripheral effector. Catecholamines are released through a shortcut between the visceral nerve and the adrenal medulla. When the organism receives the stressor, the HPA axis inhibits or stimulates transcription through GCs binding to the cell kernel receptor (GR) as shown in Fig. 8.1.

GCs bind to intracellular nuclear receptors (GR) to activate or repress transcription. GCs diffuse through the cell membrane and bind to cytoplasmic receptors (GR). This binding mediates dissociation of GR from heat shock proteins (hsp) and dimerization of GR molecules bound to GC. GR/GC complexes translocate into the nucleus where they bind to (negative) glucocorticoid response promoter elements ((n)GRE) and activate transcription, or they bind to nGRE, leading to transrepression. Alternatively, they affiliate with other transcription factors (TF) and regulate their action on corresponding responsive DNA elements (RE).

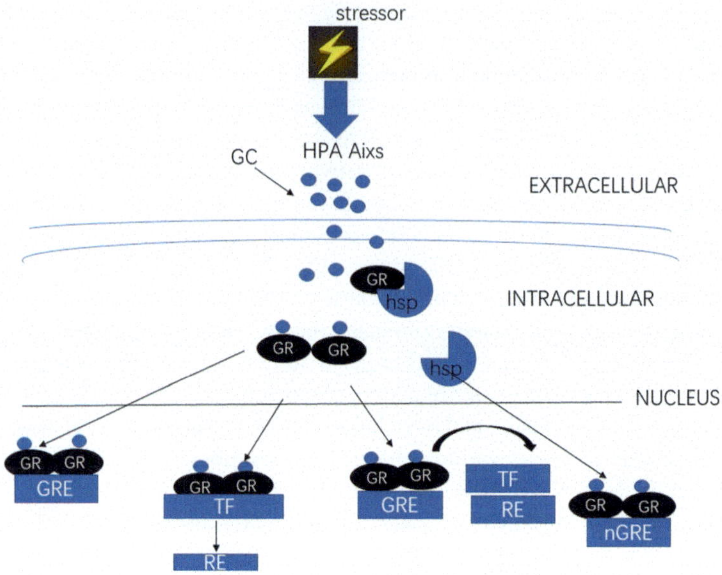

Fig. 8.1 Stress-mediated activation of the HPA axis triggers the production and release of adrenal GCs

8.4 Diseases Related to Circadian Rhythm under Stress

The term stress was first coined by Selye to explain the body's nonspecific response to any physical or psychological event that results from the transfer of physiological states within the body. He believes that stress should be divided into benign stress and malignant stress. Benign stress can positively regulate the body's function, while malignant stress will lead to the body's dysfunction and lead to disease [60]. Thus, an appropriate stress response to a variety of serious external threats is undoubtedly critical to the survival of the host [61]. The concepts of allostatic and allostatic load also enable a more detailed understanding of the mechanisms that determine an individual's susceptibility and adaptability to different physiological stresses [62, 63]. Researches have revealed that stress recovery involves a coordinated system of interactions between peripheral (body) and central (brain) signaling mechanisms [64]. Studies in rodent models and humans have clarified that SNS and the HPA axis play an important role in stress adaptation and that their dynamics also affect host flexibility to new challenges [65, 66]. Maintaining a proper time relationship between the organisms' various physiological signaling systems leads to optimal allocation of energy resources and anticipates predictable (daily) changes in the environment, thereby promoting homeostasis [67]. Thus, circadian rhythm desynchrony is related to the pathological regulation of various physiological signaling systems and is representative of the accumulation of allostatic load. This 24 h periodic rhythm is controlled by a layered circadian rhythm system, in which a master pacemaker located in the suprachiasmatic nucleus (SCN) of the hypothalamus synchronizes physiological and behavioral rhythms by regulating the rhythmic activity of various neuronal and humoral allostatic mediators [41, 68]. Metabolic hormone – such as ghrelin, secreted by the stomach, and leptin secreted by adipose tissue, as an afferent signal – participates in the hypothalamus and other brain centers, thus regulating the physiological process of regulating energy homeostasis [69–71].

According to past studies, there is a clear bidirectional link between the HPA axis of stress response and the biological clock mechanism [47]. Circadian rhythms directly influence the systemic regulation of cardiovascular functions [72] and metabolic, hormonal [41], immune pathways [73]. The main effect of circadian rhythm is that chronic disruption of homeostasis increases susceptibility to systemic inflammation and metabolic disorders [73, 74]. There is a growing awareness of the negative health effects of modern lifestyles, including sleep deprivation, the mismatch between the light/dark cycle and the body clock during air travel and shift work, and also increased exposure to light at night [75]. Animal experiments have shown that chronic circadian rhythm disorders are associated with pathological disorders of many important physiological signaling systems, including neural, metabolic, and immune pathways, and behavioral changes [76–78]. Chronic circadian dysregulation in these animals results in obesity, weight gain, and disruption of metabolic hormones. In addition, the dysregulation of circadian rhythm greatly changes the response of cytokines to immune stress [79–81]. The HPA axis is necessary for maintaining circadian rhythm homeostasis not only in response to stress but also for appropriate physiological regulation of target tissues sensitive to systemic glucocorticoids [13, 44, 82].

Next, this chapter will focus on circadian rhythm-related diseases under stress, as shown in the Fig. 8.2.

The circadian rhythm cycle is the main external synchronizer of the central circadian pacemaker, but other external stimuli can affect the phase and amplitude of peripheral oscillators. The molecular mechanism generating a self-sustained circadian oscillation in SCN neurons is a complex transcriptional-translational feedback loop comprising core transcriptional activators BMAL1/CLOCK and two sets of repressors PER and CRY. The core transcriptional activators BMAL1/CLOCK regulate numerous genes. The master clock in the SCN serves to synchronize central and peripheral oscillators to optimize the function of the organism relative to the 24 h periodicities in the environment. Signals from peripheral tissues can affect the phase and amplitude of the central pacemaker.

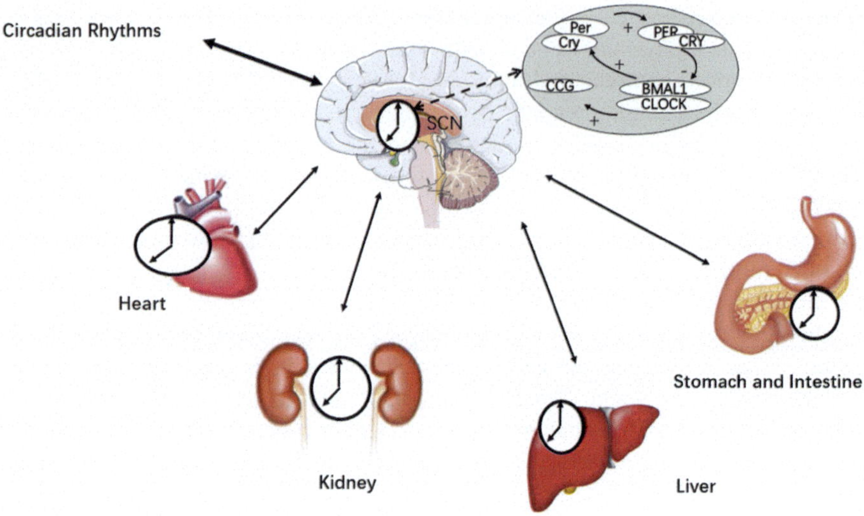

Fig. 8.2 Relationship between circadian rhythm and disease

8.4.1 Cardiovascular Disease in Circadian Rhythms and the Molecular Clock

Circadian rhythm is an important regulator of cardiovascular physiology and diseases. Each type of cardiovascular cell has a peripheral clock [83–86] that regulates various physiological functions, such as heart rate, blood pressure and endothelial function [87, 88], as well as acute myocardial infarction and the onset of arrhythmia. Almost all cell types have a functional circadian clock [25]. Core elements of the clock, such as CLOCK and BMAL1 – CLOCK and BMAL1 proteins form a complex heterodimer – function not only to connect to the E-boxes but also to sustain the clock turning of clock-controlled genes to realize the rhythmic activation of the various parts of the genome, consequently leading to a negative feedback loop with a period of approximately 24 h in bodily functions [82]. The expression levels of core clock genes in human embryonic stem cells, including BMAL1 and PER2, gradually increased during cardiac differentiation [89, 90]. The molecular clock has specific roles within each type of cardiovascular tissue, such as cardiac progenitor-like cells, endothelial cells, vascular smooth muscle cells, and fibroblasts [91]. In cardiovascular system, the circadian clock was discovered in veins and arteries, mouse aortas, vascular smooth muscle cells cultured from cultured cardio-fibroblasts, and transgenic rats [92–94]. In addition to the blood system, circadian rhythms have been exploited in myocardial stromal fibroblasts, cardiac progenitor-like cells, and cardiomyocytes [86, 95]. In the vascular system, the circadian clock is involved in thrombosis, signaling resident cells, and vascular function [96]. These observed daily fluctuations are associated not only with fluctuations in intrinsic vascular properties but also with daily sleep and wake cycles [97, 98]. Many studies have highlighted the important physiological role of the biological clock in the cardiovascular system [99–101].

When a rodent's circadian rhythm is out of synchrony with its environment, the animal can develop myocardial fibrosis, cardiomyopathy, and systolic dysfunction, leading to cardiovascular death [102, 103]. Many studies have shown a clear relationship between various cardiovascular risk factors and circadian rhythm disorders and vice versa [104–107]. Myocardial infarction is a good example of disruption of circadian rhythm and circadian clock. Studies have shown that myocardial infarction occurs early in the morning and causes more damage and dysfunction than in the afternoon [108–111].

8.4.2 The Relationship Between Circadian Rhythm and Brain Gut Axis Diseases

Studies have shown that intestinal microbes not only affect the metabolism, digestion, and immune functions of the host but also regulate the mental state and sleep of the host through the microbial-gut-brain axis, including *Firmicutes* and *Bacteroidetes*, *Proteus*, *Actinomycete*, *Clostridium*, *Verrucomicrobia*, and *Cyanobacteria*. There are about 1000 species of microbiota in the adult gut [112]. The microbiome-visceral-brain (MGB) axis has become a hot topic of research. In this axis, the microbiomes in the intestinal tract affect brain function in three ways to generate bidirectional information flow [113–117], including immunoregulatory pathway, neuroendocrine pathway, and vagus nerve pathway. The gut nervous system connects the gut to the brain through the vagus nerve [118] and forms information transmission pathway, which can be called the gut microbe-intestinal nervous system (ENS)-vagus-brain pathway. In addition, neurotoxic metabolites such as ammonia produced and d-lactic acid by intestinal flora may enter the central nervous system through the vagus nerve, thereby affecting stress response, brain function, and circadian rhythm [119, 120]. The intestinal microbiota showed circadian rhythm in population functional and structure activity. Studies have shown that the daily variation of 60% flora, including *Clostridiales*, *Lactobacillus*, and *Bacteroidales*, leads to specific taxonomic configuration [121]. Studies have shown that sleep deprivation, circadian clock dysregulation, and shift experience alter microbial community structure and circadian clock gene expression [122–125]. Changes in sleep patterns in mice also affected the structure and diversity of intestinal microbiota [126]. Therefore, there must be a specific link between the intestinal microbiota and the host circadian clock. The body's biological clock works in concert with the microbial clock, and when the host's circadian rhythm is disrupted, it changes the balance of the intestinal microbiome, similar to changes found in human shift work [127, 128]. Studies have shown that some host clock genes such as Per1, Per2, and Bmal1 are closely related to the change of intestinal microbial rhythm [121, 129].

8.4.3 The Relationship Between Circadian Rhythm and Liver Diseases

The liver is an important metabolic center, and many of its functions must adapt to changes in circadian rhythms. As a result, the liver also has an internal clock that maintains physiological processes throughout the day. In fact, all metabolic activities in the gastrointestinal tract and liver have a biological clock. A daily rhythm is regulated by the gastrointestinal tract, liver, and food. And the process of food and clock control is interactive in many cases not easily separated. Under the action of the cell circadian oscillator, a large number of genes produce regular daily changes in physiology and behavior through rhythmic expression [130]. PER and CRY's negative feedback loop is at the heart of the mammalian molecular clock. Over time, PER and CRY accumulate in cells, eventually inhibiting transcription at their own loci. The stability of core oscillator is realized by the nuclear receptor of ROR, REV-ERBA, and REV-ERBB through regulating the expression of Bmal1. With posttranscriptional regulation of these gene products, the clock can be fine-tuned [131]. Peripheral oscillators such as the liver clock are coupled to cells and physiological mechanisms of the body through energy- and nutrient-sensing systems such as AMPK [132], PPARGC1A [133], metabolic feedback loops involving metabolites such as polyamines, and nuclear hormone receptors and nuclear hormone receptor signaling pathways [134]. Therefore, peripheral clocks are obviously affected by cell metabolism, the physiological and metabolic state of the surrounding tissue, and serum-borne signals [135]. Genome-wide expression studies have shown the importance of biological clock in the physiological activities of the liver [86, 136–138]. At the end of the light and dark phases, there are two peaks of rhythmically regulated transcripts that are respectively observed in mouse liver. Metabolic enzyme cycle and transcription are relatively constant throughout the day, indicating that the body also participates in the regulation of liver circadian rhythm after transcription. The liver metabolome indicates that the transcription of some circulating metabolic enzymes is relatively stable throughout the day, suggesting that posttranscriptional mechanisms are also involved in the diurnal regulation of liver function [139, 140].

Transcriptome analyses illustrated that the liver are rhythmically controlled in most core functions, such as metabolic pathways [136–138]. The liver may be able to maintain energy stability by regulating liver gene expression in liver cells through circadian mechanisms [141, 142]. The liver plays an important role in regulating glucose homeostasis. The biological clock seems to provide a rhythmical baseline for regular and repeated changes in glucose supply. Damage to the essential clock gene Bmal1 in mouse liver leads to excessive fluctuations in blood glucose levels, mainly in the post-absorption phase controlled by the current dominant clock in the SCN [143].

8.5 Conclusion

With the change of lifestyle, the relationship between stress and circadian rhythm has been paid more and more attention. There is a growing understanding of stress. Stress is not only a local stimulus to the body but also affects the whole body function. When the internal molecular clock of the body is out of sync with the circadian cycle of the environment, it will cause pathological changes in the body, leading to a variety of diseases such as metabolic diseases, anxiety and depression, tumors, and so on. However, the daily life of modern people can be seen everywhere because of overtime work duty, and high-intensity social, work, and life pressure, which leads to the disruption of circadian rhythm. At the same time, due to the development of the Internet era, more and more people are addicted to online games all day, many office workers are used to shopping online at night, and so on, so the internal clock is inconsistent with the external environment. Now more and more people are in a state of chronic stress, which seriously affects their work and life.

By analyzing the literature of the past decades, it is not difficult to find that the body is in a chronic state of stress due to the long-term inconsistency of circadian rhythm, from the nerve, endocrine, immune, and other aspects of body function. The most obvious one is that it affects the function of HPA axis and causes the secretion of GCs and ACTH to be out of balance, so the body is in a state of stress.

References

1. Fink G. In retrospect: eighty years of stress. Nature. 2016;539:175e176.
2. Selye H. A syndrome produced by diverse nocuous agents. Nature. 1936;138:32.
3. Xu T, Lu B. The effects of phytochemicals on circadian rhythm and related diseases. Crit Rev Food Sci Nutr. 2018;12:1–30.
4. Corbalan-Tutau D, Madrid JA, Nicolas F, Garaulet M. Daily profile in two circadian markers 'melatonin and cortisol' and associations with metabolic syndrome components. Physiol Behav. 2014;123:231–5.
5. Foster RG, Leon K. The rhythms of life: what your body clock means to you! Exp Physiol. 2014;99:599–606.
6. Jones SG, Benca RM. Circadian disruption in psychiatric disorders. Sleep Med Clin. 2015;10:481–93.
7. Mohawk JA, Pargament JM, Lee TM. Circadian dependence of corticosterone release to light exposure in the rat. Physiol Behav. 2007;92:800–6.
8. Sollars PJ, Weiser MJ, Kudwa AE, Bramley JR, Ogilvie MD, Spencer RL, Handa RJ, Pickard GE. Altered entrainment to the day/night cycle attenuates the daily rise in circulating corticosterone in the mouse. PLoS One. 2014;9:e111944.
9. Bartlang MS, Neumann ID, Slattery DA, Nicole U-S, Dominik K, Charlotte H-F, Reber SO. Time matters: pathological effects of repeated psychosocial stress during the active, but not inactive, phase of male mice. J Endocrinol. 2012;215:425–37.
10. Van Dycke KCG, Wendy R, van Oostrom CTM, van Kerkhof LWM, Pennings Jeroen LA, Till R, van Harry S, van der Horst Gijsbertus TJ. Chronically alternating light cycles increase breast Cancer risk in mice. Curr Biol. 2015;25:1932–7.
11. Kondratov RV, Kondratova AA, Gorbacheva VY, Vykhovanets OV, Antoch MP. Early aging and age-related pathologies in mice deficient in BMAL1, the core component of the circadian clock. Genes Dev. 2006;20:1868–73.

12. Dickmeis T, Weger BD, Weger M. The circadian clock and glucocorticoids–interactions across many time scales. Mol Cell Endocrinol. 2013;380:2–15.
13. Oster H, Challet E, Ott V, Arvat E, de Kloet ER, Dijk DJ, Lightman S, Vgontzas A, Van Cauter E. The functional and clinical significance of the 24-hour rhythm of circulating glucocorticoids. Endocr Rev. 2017;38:3–45.
14. Arriza JL, Weinberger C, Cerelli G, Glaser TM, Handelin BL, Housman DE, Evans RM. Cloning of human mineralocorticoid receptor complementary DNA: structural and functional kinship with the glucocorticoid receptor. Science. 1987;237:268–75.
15. Hollenberg SM, Weinberger C, Ong ES, Cerelli G, Oro A, Lebo R, Thompson EB, Rosenfeld MG, Evans RM. Primary structure and expression of a functional human glucocorticoid receptor cDNA. Nature. 1985;318:635–41.
16. Chrousos GP, Tomoshige K. Intracellular glucocorticoid signaling: a formerly simple system turns stochastic. Sci STKE. 2005;2005:e48.
17. Kino T, Chrousos GP. Glucocorticoid and mineralocorticoid receptors and associated diseases. Essays Biochem. 2004;40:137–55.
18. Nader N, Chrousos GP, Kino T. Interactions of the circadian CLOCK system and the HPA axis. Trends Endocrinol Metab. 2010;21:277–86.
19. Balsalobre A, Brown SA, Marcacci L, Tronche F, Kellendonk C, Reichardt HM, Schütz G, Schibler U. Resetting of circadian time in peripheral tissues by glucocorticoid signaling. Science. 2000;289:2344–7.
20. Welsh DK, Takahashi JS, Kay SA. Suprachiasmatic nucleus: cell autonomy and network properties. Annu Rev Physiol. 2010;72:551–77.
21. Spencer RL, Chun LE, Hartsock MJ, Woodruff ER. Glucocorticoid hormones are both a major circadian signal and major stress signal: how this shared signal contributes to a dynamic relationship between the circadian and stress systems. Front Neuroendocrinol. 2018;49:52–71.
22. Tahara Y, Takuya S, Yosuke K, Atsushi H, Daisuke K, Hiroyuki S, Hiroaki M, Tomoko S, Shigenobu S. Entrainment of the mouse circadian clock by sub-acute physical and psychological stress. Sci Rep. 2015;5:11417.
23. Buhr ED, Takahashi JS (2013) Molecular components of the Mammalian circadian clock. Handb Exp Pharmacol, undefined, 3-27
24. Brown TM, Piggins HD. Electrophysiology of the suprachiasmatic circadian clock. Prog Neurobiol. 2007;82:229–55.
25. Zhang R, Lahens NF, Ballance HI, Hughes ME, Hogenesch JB. A circadian gene expression atlas in mammals: implications for biology and medicine. Proc Natl Acad Sci U S A. 2014;111:16219–24.
26. Cajochen C, Kräuchi K, Wirz-Justice A. Role of melatonin in the regulation of human circadian rhythms and sleep. J Neuroendocrinol. 2003;15:432–7.
27. Fustin JM, O'Neill JS, Hastings MH, Hazlerigg DG, Dardente H. Cry1 circadian phase in vitro: wrapped up with an E-box. J Biol Rhythm. 2009;24:16–24.
28. Gekakis N, Staknis D, Nguyen HB, Davis FC, Wilsbacher LD, King DP, Takahashi JS, Weitz CJ. Role of the CLOCK protein in the mammalian circadian mechanism. Science. 1998;280:1564–9.
29. Hastings MH, Maywood ES, Marco B. Generation of circadian rhythms in the suprachiasmatic nucleus. Nat Rev Neurosci. 2018;19:453–69.
30. Reppert SM, Weaver DR. Coordination of circadian timing in mammals. Nature. 2002;418:935–41.
31. Dallmann R, Brown SA, Gachon F. Chronopharmacology: new insights and therapeutic implications. Annu Rev Pharmacol Toxicol. 2014;54:339–61.
32. Weaver DR. The suprachiasmatic nucleus: a 25-year retrospective. J Biol Rhythm. 1998;13:100–12.
33. Brancaccio M, Edwards MD, Patton AP, Smyllie NJ, Chesham JE, Maywood ES, Hastings MH. Cell-autonomous clock of astrocytes drives circadian behavior in mammals. Science. 2019;363:187–92.

34. Kume K, Zylka MJ, Sriram S, Shearman LP, Reppert SM. Mcry1 and mcry2 are essential components of the negative limb of the circadian clock feedback loop. Cell. 1999;98(2):193–205.
35. Takahashi JS. Transcriptional architecture of the mammalian circadian clock. Nat Rev Genet. 2017;18:164–79.
36. Pastore N, Vainshtein A, Herz NJ, Huynh T, Brunetti L, Klisch TJ, Mutarelli M, Annunziata P, Kinouchi K, Brunetti-Pierri N, Sassone-Corsi P. Nutrient-sensitive transcription factors TFEB and TFE 3 couple autophagy and metabolism to the peripheral clock. EMBO J. 2019;38(12): e101347.
37. Pastore N, Ballabio A. Keeping the autophagy tempo. Autophagy. 2019;15:1854–6.
38. Preitner N, Damiola F, Lopez-Molina L, Zakany J, Duboule D, Albrecht U, et al. The orphan nuclear receptor rev-erbα controls circadian transcription within the positive limb of the mammalian circadian oscillator. Cell. 2002;110(2):251–60.
39. Streuli CH, Qing-Jun M. Influence of the extracellular matrix on cell-intrinsic circadian clocks. J Cell Sci. 2019;132, undefined
40. Akashi M, Takumi T. The orphan nuclear receptor ROR alpha regulates circadian transcription of the mammalian core-clock Bmal1. Nat Struct Mol Biol. 2005;12:441–8.
41. Buijs RM, van Eden CG, Goncharuk VD, Kalsbeek A. The biological clock tunes the organs of the body: timing by hormones and the autonomic nervous system. J Endocrinol. 2003;177:17–26.
42. Yoo SH, Yamazaki S, Lowrey PL, Shimomura K, Ko CH, Buhr ED, Siepka SM, Hong HK, Oh WJ, Yoo OJ, Menaker M. PERIOD2:: LUCIFERASE real-time reporting of circadian dynamics reveals persistent circadian oscillations in mouse peripheral tissues. Proc Natl Acad Sci. 2004;101(15):5339–46.
43. Berge KE, Tian H, Graf GA, Yu L, Grishin NV, Schultz J, Kwiterovich P, Shan B, Barnes R, Hobbs HH. Accumulation of dietary cholesterol in sitosterolemia caused by mutations in adjacent ABC transporters. Science. 2000;290:1771–5.
44. Koch CE, Leinweber B, Drengberg BC, Blaum C, Oster H. Interaction between circadian rhythms and stress. Neurobiol Stress. 2017;6:57–67.
45. Ulrich-Lai YM, Herman JP. Neural regulation of endocrine and autonomic stress responses. Nat Rev Neurosci. 2009;10:397–409.
46. Windle RJ, Wood SA, Shanks N, Lightman SL, Ingram CD. Ultradian rhythm of basal corticosterone release in the female rat: dynamic interaction with the response to acute stress. Endocrinology. 1998;139:443–50.
47. Nicolaides NC, Evangelia C, Tomoshige K, Chrousos George P. Stress-related and circadian secretion and target tissue actions of glucocorticoids: impact on health. Front Endocrinol. 2017;8
48. Herman JP, McKlveen JM, Sriparna G, Brittany K, Aynara W, Ryan M, Jessie S, Brent M. Regulation of the hypothalamic-pituitary-adrenocortical stress response. Compr Physiol. 2016;6:603–21.
49. Datson NA, Morsink MC, Meijer OC, De Ronald KE. Central corticosteroid actions: search for gene targets. Eur J Pharmacol. 2008;583:272–89.
50. Joëls M, Sarabdjitsingh RA, Karst H. Unraveling the time domains of corticosteroid hormone influences on brain activity: rapid, slow, and chronic modes. Pharmacol Rev. 2012;64:901–38.
51. Groeneweg FL, Henk K, De Kloet ER, Marian J. Mineralocorticoid and glucocorticoid receptors at the neuronal membrane, regulators of nongenomic corticosteroid signalling. Mol Cell Endocrinol. 2012;350:299–309.
52. Oakley RH, Cidlowski JA. The biology of the glucocorticoid receptor: new signaling mechanisms in health and disease. J Allergy Clin Immunol. 2013;132:1033–44.
53. Fahrenkrug J, Hannibal J, Georg B. Diurnal rhythmicity of the canonical clock genes Per1, Per2 and Bmal1 in the rat adrenal gland is unaltered after hypophysectomy. J Neuroendocrinol. 2008;20:323–9.
54. Yoder JM, Brandeland M, Engeland WC. Phase-dependent resetting of the adrenal clock by ACTH in vitro. Am J Physiol Regul Integr Comp Physiol. 2014;306:R387–93.

55. Oster H, Damerow S, Hut RA, Eichele G. Transcriptional profiling in the adrenal gland reveals circadian regulation of hormone biosynthesis genes and nucleosome assembly genes. J Biol Rhythms. 2006;21:350–61.
56. McEwen BS, Bowles NP, Gray JD, Hill MN, Hunter RG, Karatsoreos IN, Carla N. Mechanisms of stress in the brain. Nat Neurosci. 2015a;18:1353–63.
57. McEwen BS. Neurobiological and systemic effects of chronic stress. Chronic Stress. 2017;1, undefined
58. Nygren LG, Olson L. A new major projection from locus coeruleus: the main source of noradrenergic nerve terminals in the ventral and dorsal columns of the spinal cord. Brain Res. 1977;132(1):85–93.
59. Westlund KN, Bowker RM, Ziegler MG, Coulter JD. Noradrenergic projections to the spinal cord of the rat. Brain Res. 1983;263(1):15–31.
60. Selye H. The stress of life. Rev ed: McGraw Hill; 1978.
61. McEwen BS. Protective and damaging effects of stress mediators: central role of the brain. Dialogues Clin Neurosci. 2006;8:367–81.
62. Karatsoreos IN, McEwen BS. Psychobiological allostasis: resistance, resilience and vulnerability. Trends Cogn Sci (Regul Ed). 2011;15:576–84.
63. McEwen BS, Jason G, Carla N. Recognizing resilience: learning from the effects of stress on the brain. Neurobiol Stress. 2015b;1:1–11.
64. Pfau ML, Russo SJ. Peripheral and central mechanisms of stress resilience. Neurobiol Stress. 2015;1:66–79.
65. Charney DS. Psychobiological mechanisms of resilience and vulnerability: implications for successful adaptation to extreme stress. Am J Psychiatry. 2004;161:195–216.
66. Russo SJ, Murrough JW, Ming-Hu H, Charney DS, Nestler EJ. Neurobiology of resilience. Nat Neurosci. 2012;15:1475–84.
67. Riede SJ, van der Vincent V, Hut Roelof A. The flexible clock: predictive and reactive homeostasis, energy balance and the circadian regulation of sleep-wake timing. J Exp Biol. 2017;220:738–49.
68. Urbanski HF. Role of circadian neuroendocrine rhythms in the control of behavior and physiology. Neuroendocrinology. 2011;93:211–22.
69. Diéguez C, Vazquez MJ, Romero A, López M, Nogueiras R. Hypothalamic control of lipid metabolism: focus on leptin, ghrelin and melanocortins. Neuroendocrinology. 2011;94:1–11.
70. Howick K, Griffin BT, Cryan JF, Schellekens H. From belly to brain: targeting the ghrelin receptor in appetite and food intake regulation. Int J Mol Sci. 2017;18:273.
71. Park HK, Ahima RS. Physiology of leptin: energy homeostasis, neuroendocrine function and metabolism. Metab Clin Exp. 2015;64:24–34.
72. Lemmer B. Importance of circadian rhythms for regulation of the cardiovascular system–studies in animal and man. Conf Proc IEEE Eng Med Biol Soc. 2006;1:168–70.
73. Cermakian N, Westfall S, Kiessling S. Circadian clocks and inflammation: reciprocal regulation and shared mediators. Arch Immunol Ther Exp. 2014;62:303–18.
74. Takahashi JS, Hee-Kyung H, Ko CH, McDearmon EL. The genetics of mammalian circadian order and disorder: implications for physiology and disease. Nat Rev Genet. 2008;9:764–75.
75. Fonken LK, Nelson RJ. The effects of light at night on circadian clocks and metabolism. Endocr Rev. 2014;35:648–70.
76. Karatsoreos IN, Sarah B, Bloss Erik B, Morrison John H, McEwen BS. Disruption of circadian clocks has ramifications for metabolism, brain, and behavior. Proc Natl Acad Sci USA. 2011;108:1657–62.
77. Karatsoreos IN. Links between circadian rhythms and psychiatric disease. Front Behav Neurosci. 2014;8:162.
78. West AC, Bechtold DA. The cost of circadian desynchrony: evidence, insights and open questions. BioEssays. 2015;37:777–88.
79. Stevens RG, Yong Z. Electric light, particularly at night, disrupts human circadian rhythmicity: is that a problem? Philos Trans R Soc Lond B Biol Sci. 2015;370, undefined
80. Wang X-S, Armstrong MEG, Cairns BJ, Key TJ, Travis RC. Shift work and chronic disease: the epidemiological evidence. Occup Med (Lond). 2011;61:78–89.

81. Spiegel K. Impact of sleep debt on metabolic and endocrine function. Lancet. 1999;354
82. Shearman LP, Sriram S, Weaver DR, Maywood ES, Chaves I, Zheng B, Kume K, Lee CC, van der Horst GT, Hastings MH, Reppert SM. Interacting molecular loops in the mammalian circadian clock. Science. 2000;288:1013–9.
83. Lin C, Xiao T, Zhu Z, Liao X, Ran Z, Weiguo F, Bin C, Junhao J, Ruizhe Q, Guo D. The rhythmic expression of clock genes attenuated in human plaque-derived vascular smooth muscle cells. Lipids Health Dis. 2014;13:14.
84. Beesley S, Noguchi T, Welsh DK. Cardiomyocyte circadian oscillations are cell-autonomous, amplified by β-adrenergic signaling, and synchronized in cardiac ventricle tissue. PLoS One. 2016;11:e0159618.
85. Du Pré BC, Demkes EJ, Feyen DA, Dierickx P, Crnko S, Kok BJ, Sluijter JP, Doevendans PA, Vos MA, Van Veen TA, Van Laake LW. SCA1 cells from the heart possess a molecular circadian clock and display circadian oscillations in cellular functions. Stem Cell Rep. 2017;9:762–9.
86. Takeda N, Maemura K, Horie S, Oishi K, Imai Y, Harada T, Saito T, Shiga T, Amiya E, Manabe I, Ishida N. Thrombomodulin is a clockcontrolled gene in vascular endothelial cells. J Biol Chem. 2007;282:32561–7.
87. Degaute JP, van de Borne P, Linkowski P, Van Cauter E. Quantitative analysis of the 24-hour blood pressure and heart rate patterns in young men. Hypertension. 1991;18:199–210.
88. Kollias GE, Stamatelopoulos KS, Papaioannou TG, Zakopoulos NA, Alevizaki M, Alexopoulos GP, Kontoyannis DA, Karga H, Koroboki E, Lekakis JP, Papamichael CM. Diurnal variation of endothelial function and arterial stiffness in hypertension. J Hum Hypertens. 2009;23:597–604.
89. Dierickx P, Vermunt MW, Muraro MJ, Creyghton MP, Doevendans PA, van Oudenaarden A, Geijsen N, Van Laake LW. Circadian networks in human embryonic stem cell-derived cardiomyocytes. EMBO Rep. 2017;18:1199–212.
90. Kowalska E, Moriggi E, Bauer C, Dibner C, Brown SA. The circadian clock starts ticking at a developmentally early stage. J Biol Rhythms. 2010;25:442–9.
91. Davidson AJ, London B, Block GD, Menaker M. Cardiovascular tissues contain independent circadian clocks. Clin Exp Hypertens. 2005;27:307–11.
92. Chalmers JA, Martino TA, Nazneen T, Ralph MR, Sole MJ, Belsham DD. Vascular circadian rhythms in a mouse vascular smooth muscle cell line (Movas-1). Am J Physiol Regul Integr Comp Physiol. 2008;295:R1529–38.
93. McNamara P, Seo SB, Rudic RD, Sehgal A, Chakravarti D, FitzGerald GA. Regulation of CLOCK and MOP4 by nuclear hormone receptors in the vasculature: a humoral mechanism to reset a peripheral clock. Cell. 2001;105:877–89.
94. Nonaka H, Emoto N, Ikeda K, Fukuya H, Rohman MS, Raharjo SB, Yagita K, Okamura H, Yokoyama M. Angiotensin II induces circadian gene expression of clock genes in cultured vascular smooth muscle cells. Circulation. 2001;104:1746–8.
95. Panza JA, Epstein SE, Quyyumi AA. Circadian variation in vascular tone and its relation to alpha-sympathetic vasoconstrictor activity. N Engl J Med. 1991;325:986–90.
96. Bastianini S, Silvani A, Berteotti C, Martire VL, Zoccoli G. Mice show circadian rhythms of blood pressure during each wake-sleep state. Chronobiol Int. 2012;29:82–6.
97. Sei H, Oishi K, Chikahisa S, Kitaoka K, Takeda E, Ishida N. Diurnal amplitudes of arterial pressure and heart rate are dampened in Clock mutant mice and adrenalectomized mice. Endocrinology. 2008;149:3576–80.
98. Viswambharan H, Carvas JM, Antic V, Marecic A, Jud C, Zaugg CE, Ming XF, Montani JP, Albrecht U, Yang Z. Mutation of the circadian clock gene Per2 alters vascular endothelial function. Circulation. 2007;115:2188–95.
99. Curtis AM, Yan C, Shiv K, Dermot R, Price Tom S, Fitzgerald Garret A. Circadian variation of blood pressure and the vascular response to asynchronous stress. Proc Natl Acad Sci USA. 2007;104:3450–5.
100. Martino TA, Young ME. Influence of the cardiomyocyte circadian clock on cardiac physiology and pathophysiology. J Biol Rhythm. 2015;30:183–205.

101. Young ME. The circadian clock within the heart: potential influence on myocardial gene expression, metabolism, and function. Am J Physiol Heart Circ Physiol. 2006;290:H1–16.
102. Paschos GK, FitzGerald GA. Circadian clocks and vascular function. Circ Res. 2010;106:833–41.
103. Penev PD, Kolker DE, Zee PC, Turek FW. Chronic circadian desynchronization decreases the survival of animals with cardiomyopathic heart disease. Am J Phys. 1998;275:H2334–7.
104. Thosar SS, Butler MP, Shea SA. Role of the circadian system in cardiovascular disease. J Clin Invest. 2018;128:2157–67.
105. Portaluppi F, Tiseo R, Smolensky MH, Hermida RC, Ayala DE, Fabbian F. Circadian rhythms and cardiovascular health. Sleep Med Rev. 2012;16:151–66.
106. Reutrakul S, Knutson KL. Consequences of circadian disruption on cardiometabolic health. Sleep Med Clin. 2015;10:455–68.
107. Rüger M, Scheer Frank AJL. Effects of circadian disruption on the cardiometabolic system. Rev Endocr Metab Disord. 2009;10:245–60.
108. Ammirati E, Maseri A, Cannistraci CV. Still need for compelling evidence to support the circadian dependence of infarct size after STelevation myocardial infarction. Circ Res. 2013;113:e43–4.
109. Eckle T, Hartmann K, Bonney S, Reithel S, Mittelbronn M, Walker LA, Lowes BD, Han J, Borchers CH, Buttrick PM, Kominsky DJ, Colgan SP, Eltzschig HK. Adora2b-elicited Per2 stabilization promotes a HIF-dependent metabolic switch crucial for myocardial adaptation to ischemia. Nat Med. 2012;18:774–82.
110. Fournier S, Eeckhout E, Mangiacapra F, Trana C, Lauriers N, Beggah AT, Monney P, Cook S, Bardy D, Vogt P, Muller O. Circadian variations of ischemic burden among patients with myocardial infarction undergoing primary percutaneous coronary intervention. Am Heart J. 2012;163:208–13.
111. Reiter R, Swingen C, Moore L, Henry TD, Traverse JH. Circadian dependence of infarct size and left ventricular function after ST elevation myocardial infarction. Circ Res. 2012;110:105–10.
112. Cénit MC, Matzaraki V, Tigchelaar EF, Zhernakova A. Rapidly expanding knowledge on the role of the gut microbiome in health and disease. Biochim Biophys Acta. 2014;1842:1981–92.
113. Cryan JF, Dinan TG. Mind-altering microorganisms: the impact of the gut microbiota on brain and behaviour. Nat Rev Neurosci. 2012;13:701–12.
114. Heijtz RD, Wang S, Anuar F, Qian Y, Björkholm B, Samuelsson A, et al. Normal gut microbiota modulates brain development and behavior. Proc Natl Acad Sci U S A. 2011;108(7):3047–52.
115. Macpherson AJ, Harris NL. Interactions between commensal intestinal bacteria and the immune system. Nat Rev Immunol. 2004;4:478–85.
116. Bäckhed F, Ley RE, Sonnenburg JL, Peterson DA, Gordon JI. Host-bacterial mutualism in the human intestine. Science. 2005;307:1915–20.
117. Breit S, Kupferberg A, Rogler G, Hasler G. Vagus nerve as modulator of the brain-gut axis in psychiatric and inflammatory disorders. Front Psychiatry. 2018;9:44.
118. Powley TL, Wang X-Y, Fox EA, Phillips RJ, Liu LWC, Huizinga JD. Ultrastructural evidence for communication between intramuscular vagal mechanoreceptors and interstitial cells of Cajal in the rat fundus. Neurogastroenterol Motil. 2008;20:69–79.
119. Bonaz B, Picq C, Sinniger V, Mayol JF, Clarençon D. Vagus nerve stimulation: from epilepsy to the cholinergic anti-inflammatory pathway. Neurogastroenterol Motil. 2013;25:208–21.
120. Wang Y, Kasper LH. The role of microbiome in central nervous system disorders. Brain Behav Immun. 2014;38:1–12.
121. Thaiss CA, David Z, Maayan L, Gili Z-S, Jotham S, Tengeler AC, Lior A, Katz MN, Tal K, Niv Z, Yael K, Inbal B, Shlomit G, Alon H, Hagit S, Zamir H, Eran S, Eran E. Transkingdom control of microbiota diurnal oscillations promotes metabolic homeostasis. Cell. 2014;159:514–29.
122. Davies SK, Ern AJ, Revell Victoria L, Ben H, Anuska M, Robertson Francesca P, Nanyi C, Benita M, Katrin A, Manfred K, Thumser Alfred E, Raynaud Florence I, Skene Debra J. Effect of sleep deprivation on the human metabolome. Proc Natl Acad Sci U S A. 2014;111:10761–6.

123. Johnston JD, Ordovás JM, Scheer FA, Turek FW. Circadian rhythms, metabolism, and chrononutrition in rodents and humans. Adv Nutr. 2016;7:399–406.
124. Kunze KN, Hanlon EC, Prachand VN, Brady MJ. Peripheral circadian misalignment: contributor to systemic insulin resistance and potential intervention to improve bariatric surgical outcomes. Am J Physiol Regul Integr Comp Physiol. 2016;311:R558–63.
125. Wu T, Yang L, Jiang J, Ni Y, Zhu J, Zheng X, Wang Q, Lu X, Fu Z. Chronic glucocorticoid treatment induced circadian clock disorder leads to lipid metabolism and gut microbiota alterations in rats. Life Sci. 2018;192:173–82.
126. Thaiss Christoph A, Maayan L, Tal K, Lenka D, Hagit S, Jaitin Diego A, Eyal D, Winter Deborah R, Meital G-BA, Evgeny T, Timur T, Sara F, Niv Z, David Z, Mally D-B, Meirav P-F, Elena K, Alexander B, Alon H, Oren S, Zamir H, Kenya H, Ido A, Eran S, Eran E. Microbiota diurnal rhythmicity programs host transcriptome oscillations. Cell. 2016;167:1495–1510.e12.
127. Reynolds AC, Josiane B, Paterson JL, Wright KP, Ferguson SA. Sleepy, circadian disrupted and sick: could intestinal microbiota play an important role in shift worker health? Mol Metab. 2017a;6:12–3.
128. Reynolds AC, Paterson JL, Ferguson SA, Dragana S, Wright KP, Drew D. The shift work and health research agenda: considering changes in gut microbiota as a pathway linking shift work, sleep loss and circadian misalignment, and metabolic disease. Sleep Med Rev. 2017b;34:3–9.
129. Liang X, Bushman FD, FitzGerald GA. Rhythmicity of the intestinal microbiota is regulated by gender and the host circadian clock. Proc Natl Acad Sci U S A. 2015;112:10479–84.
130. Dibner C, Schibler U. Circadian timing of metabolism in animal models and humans. J Intern Med. 2015;277:513–27.
131. Brown SA, Elzbieta K, Robert D. (Re)inventing the circadian feedback loop. Dev Cell. 2012;22:477–87.
132. Lamia KA, Sachdeva UM, Luciano DT, Williams EC, Alvarez JG, Egan DF, Vasquez DS, Henry J, Satchidananda P, Shaw RJ, Thompson CB, Evans RM. AMPK regulates the circadian clock by cryptochrome phosphorylation and degradation. Science. 2009;326:437–40.
133. Liu C, Li S, Liu T, Borjigin J, Lin JD. Transcriptional coactivator PGC-1alpha integrates the mammalian clock and energy metabolism. Nature. 2007;447:477–81.
134. Teboul M, Guillaumond F, Gréchez-Cassiau A, Delaunay F. The nuclear hormone receptor family round the clock. Mol Endocrinol. 2008;22:2573–82.
135. Gerber A, Esnault C, Aubert G, Treisman R, Pralong F, Schibler U. Blood-borne circadian signal stimulates daily oscillations in actin dynamics and SRF activity. Cell. 2013;152:492–503.
136. Akhtar RA, Reddy AB, Maywood ES, Clayton JD, King VM, Smith AG, et al. Circadian cycling of the mouse liver transcriptome, as revealed by cdna microarray, is driven by the suprachiasmatic nucleus. Curr Biol. 2002;12(7):540–50.
137. Panda S, Antoch MP, Miller BH, Su AI, Schook AB, Straume M, Schultz PG, Kay SA, Takahashi JS, Hogenesch JB. Coordinated transcription of key pathways in the mouse by the circadian clock. Cell. 2002;109:307–20.
138. Storch KF, Lipan O, Leykin I, Viswanathan N, Davis FC, Wong WH, Weitz CJ. Extensive and divergent circadian gene expression in liver and heart. Nature. 2002;417:78–83.
139. Mauvoisin D, Wang J, Jouffe C, Martin E, Atger F, Waridel P, Quadroni M, Gachon F, Naef F. Circadian clock-dependent and -independent rhythmic proteomes implement distinct diurnal functions in mouse liver. Proc Natl Acad Sci U S A. 2014;111:167–72.
140. Robles MS, Cox J, Mann M. In-vivo quantitative proteomics reveals a key contribution of posttranscriptional mechanisms to the circadian regulation of liver metabolism. PLoS Genet. 2014;10:e1004047.
141. Kornmann B, Schaad O, Bujard H, Takahashi JS, Schibler U. System-driven and oscillator-dependent circadian transcription in mice with a conditionally active liver clock. PLoS Biol. 2007;5:e34.

142. Vollmers C, Gill S, DiTacchio L, Pulivarthy SR, Le Hiep D, Panda S. Time of feeding and the intrinsic circadian clock drive rhythms in hepatic gene expression. Proc Natl Acad Sci U S A. 2009;106:21453–8.
143. Lamia KA, Kai-Florian S, Weitz Charles J. Physiological significance of a peripheral tissue circadian clock. Proc Natl Acad Sci USA. 2008;105:15172–7.

Reactive Oxygen Species, Glucose Metabolism, and Lipid Metabolism

9

Ying Yang, Yue Wu, Xiao-Dong Sun, and Yuanyuan Zhang

Abstract

Reactive oxygen species (ROS) are free radicals produced by the reduction of molecular oxygen. The mitochondrial respiratory chain complex is regarded as a major ROS producer, which is composed of four protein complexes, the complex I, complex II, complex III, and complex IV. In recent years, a large amount of evidence has demonstrated that ROS plays a pivotal role in regulating cell glucose and lipid metabolism and participates in the occurrence and development of disorders related to glucose and lipid metabolism. Generally, glucose metabolism increases the production of ROS and enhances oxidative stress by glucose auto-oxidation and glycosylation of proteins and activated the polyalcohol pathway, etc. This review aims to clarify the role of ROS in regulating glucose metabolism (glucose uptake, glycolysis, gluconeogenesis, glycogen synthesis, glycogenolysis, pentose phosphate pathway).

As we all know, ROS plays a very important role in the pathogenesis of metabolic diseases induced by dysregulation of lipid metabolism, such as obesity, nonalcoholic fatty liver disease, and atherosclerosis. In this chapter, we focus on the interaction between ROS and lipid metabolism, especially the effects of ROS on the synthesis and oxidation of fatty acids, as well as phospholipids, cholesterol, and browning of beige fat.

Ying Yang and Yue Wu contributed equally with all other contributors.

Y. Yang · Y. Wu · X.-D. Sun (✉) · Y. Zhang (✉)
West China School of Basic Medical Sciences and Forensic Medicine, Sichuan University, Chengdu, People's Republic of China
e-mail: sunxd@scu.edu.cn; zhangyy@scu.edu.cn

C. Huang, Y. Zhang (eds.), *Oxidative Stress*,
https://doi.org/10.1007/978-981-16-0522-2_9

Keywords

Reactive oxygen species (ROS) · Oxidative stress · Glucose metabolism · Glycolysis · Gluconeogenesis · Glycogen synthesis · Glycogenolysis · Pentose phosphate pathway · Lipid metabolism

9.1 ROS Generation and ROS Signaling

Reactive oxygen species (ROS) are free radicals produced by the reduction of molecular oxygen. ROS are a large class of molecules, including superoxide anion ($O2^{\cdot -}$), hydrogen peroxide (H_2O_2), hydroxyl radical (OH), and singlet oxygen ($^1O2$) [1]. ROS are mainly produced by mitochondrial oxidative phosphorylation, in which electrons are transferred along the respiratory chain enzymes and leaked to molecular oxygen [2]. Reactions generating the different ROS members are shown in Fig. 9.1a.

The mitochondrial respiratory chain complex is regarded as a major ROS producer, especially at several sites of the complex I enzyme, complex III, coenzyme Q

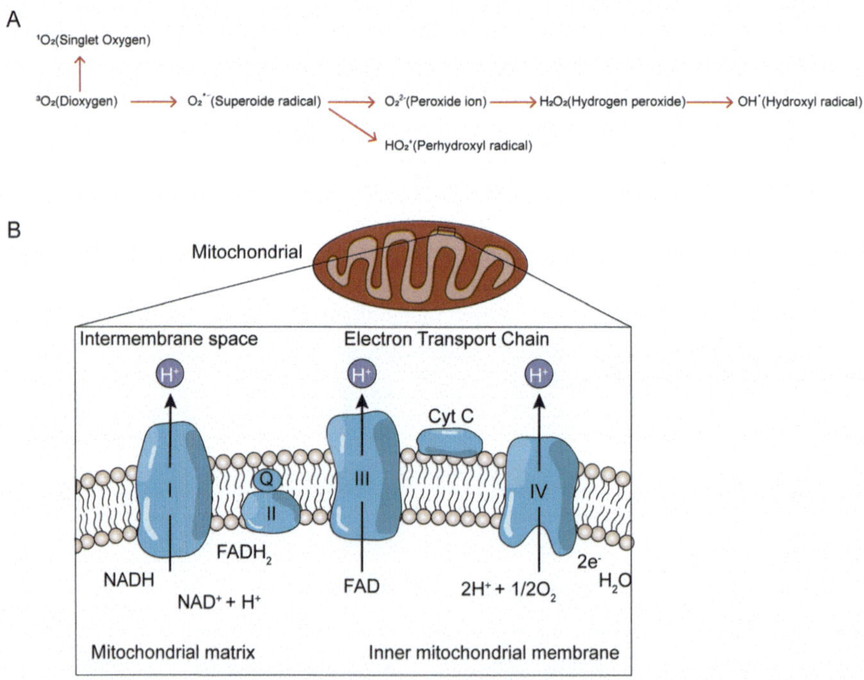

Fig. 9.1 (**a**) Generation of ROS by energy transfer. (**b**) The electron transport chain is responsible for mitochondrial respiration and the main source of ROS within the cell

interacting subunits, and complex II under low substrate conditions [3]. ROS is also generated by nicotinamide adenine dinucleotide phosphate (NADPH) oxidase, xanthine oxidase, peroxidase, cytochrome P450, various cytosolic enzymes such as cyclooxygenase and lipoxygenase, and hemoglobin in red blood cells [4–7]. Among them, the electron transport chain (ETC) on the inner lining of the mitochondria, also known as the respiratory chain, is responsible for about 95% of the intracellular ROS production [8]. Mitochondrial ROS production is the result of oxidative phosphorylation (OXPHOS), a process that uses nicotinamide adenine dinucleotide (NADH) or flavin amine adenine dinucleotide ($FADH_2$) to control the oxidation to produce proton potential energy across the mitochondrial inner membrane. Electrons from NADH or $FADH_2$ can directly react with oxygen or other electron acceptors to generate ROS [5, 9–11].

Each complex of the electron transport chain receives electrons from one of the coenzymes or one of the other complexes of the chain. Complex I is responsible for releasing the hydrogen and electrons of NADH [12]. The energy obtained by absorbing the electrons causes the complex I to pump hydrogen atoms through the mitochondrial inner membrane and to concentrate hydrogen in the intermembrane space. The electrons are transferred to the coenzyme Q (CoQ). Then the CoQ absorbs hydrogen and electrons and be reduced to coenzyme QH2. Then coenzymes transfer electrons to the complex III [13]. At the same time, the complex II accepts electrons and protons from $FADH_2$ and the citric acid cycle. The complex II releases the electrons of $FADH_2$ and passes these electrons to CoQ. Coenzymes transfer them to the complex III, which receives electrons and their energy, and allow the complex III to pump large amounts of hydrogen through the membrane. Cytochrome c (Cyt c) allows electrons to be transferred to the complex IV [14]. The complex IV transfers electrons to oxygen molecules to form H_2O [15]. Another proton passes through the membrane using these energy. Figure 9.1b shows the actions of each complex.

In recent years, extensive evidence has shown that ROS plays an important role in multiple cell signaling pathways. Most of the risk factors related to chronic diseases, such as stress, radiation, virus, and diet, cause damage to cells through generating ROS [16]. ROS activates various transcription factors, such as B cell nuclear factor-kappa B (NF-κB), which leads to the transcription of many proteins regulating inflammation, transformation, survival, proliferation and invasion, angiogenesis, and metastasis [17, 18]. ROS also regulates the expression of various tumor suppressor genes such as *PTEN* [19, 20]. Radiation and chemotherapeutic agents can be used to treat cancer through the generation of ROS [21]. The generation of ROS is involved in the oxidative deterioration of food and more than 100 human diseases, such as hypertension, atherosclerosis, etc. [22–25]. Other desirous diseases are also correlated with ROS, such as diabetes [26], chronic inflammation [27], neurodegenerative diseases [28], cancer [29, 30], chronic immune [31], and aging [32].

9.2 ROS and Glucose Metabolism

9.2.1 Dysregulated Glucose Metabolism Leads to Increased Production of ROS

Generally, glucose metabolism increases the production of ROS and enhances oxidative stress by glucose auto-oxidation [33] and glycosylation of proteins and activated the polyalcohol pathway [34], etc. Glucose and its metabolites react with hydrogen peroxide to produce enediols and dihydroxy compounds, known as glucose auto-oxidation, which produce a lot of ROS [35, 36]. ROS can also be produced by nonenzymatic glycosylation of proteins. The aldehyde or carbonyl group of the sugar molecules undergoes a nonenzymatic glycosylation with the protein, lipid, or nucleic acid amino groups and resulted in an irreversible covalent chemical substance which is called advanced glycosylation end products (AGEs) [36]. This process is accompanied by a series of oxidation reactions and produces a large amount of ROS [37]. The polyalcohol of glucose is activated due to enhanced activity of aldose reductase under hyperglycemia and converts glucose to sorbitol and fructose. The accumulation of sorbitol results in increased cellular osmotic pressure and consumes a large amount of NADPH and glutathione (GSH) [38, 39]. Then intracellular oxidative stress is enhanced due to impaired ability to scavenge free radicals. ROS can also be produced by activating protein kinase C (PKC) pathway [40–42]. Hyperglycemia increases the production of diesteryl glycerol, activates PKC, and then activates cellular NADPH oxidase, which enhances the production of ROS, subsequent lipid peroxidation, and nonenzymatic glycosylation of proteins [43]. And ROS also activates PKC, which further increases the production of ROS [44]. Mitochondrial oxidative phosphorylation also contributes to the production of ROS [45] (Figs. 9.2, 9.3, and 9.4).

During normal oxidative phosphorylation, 0.4–4% of oxygen molecules are converted into superoxide anions [46]. The production of superoxide anions in the respiratory chain depends on the interaction between respiratory chain complexes and coenzyme Q10 [13], also known as superoxide. Hyperglycemia changes the formation of superoxide anions. Then the respiratory chain complex becomes the main provider of electrons and promotes the formation of peroxide. Hyperglycemia results in decreased ROS clearance and increased the production of ROS [47]. Hyperglycemia leads to the glycosylation of antioxidant enzymes, which inhibits activities of many antioxidant enzymes such as superoxide dismutase [48], catalase [49], and glutathione peroxidase [50]. Impaired glucose metabolism reduces levels of vitamin C, vitamin E, GSH, etc. So, the antioxidant ability is significantly weakened.

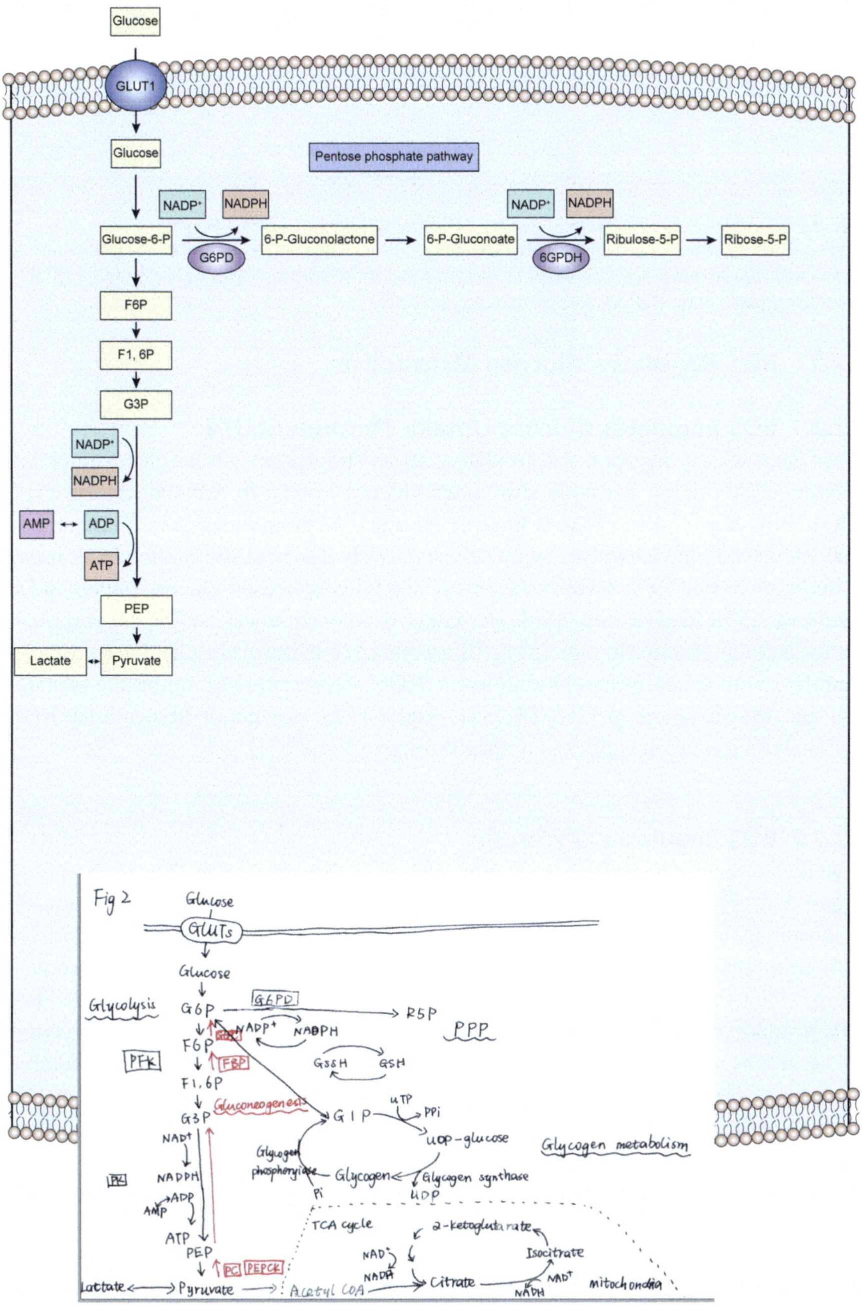

Fig. 9.2 Reactive oxygen species (ROS) regulates the involved process of glucose metabolism

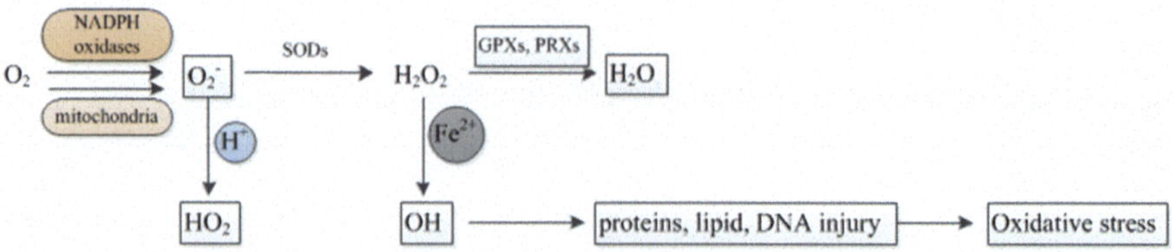

Fig. 9.3 Cellular reactive oxygen species (ROS) production. There are mainly two ways to produce ROS in cells. One is to activate NADPH oxidase. The other is to utilize the mitochondria to generate superoxide (O2•−). And ROS is scavenged by activating superoxide dismutase (SOD), glutathione peroxidase (GPX), and peroxiredoxin (PRX)

9.2.2 ROS Regulates Glucose Metabolism

9.2.2.1 ROS Regulates Glucose Uptake Through GLUT4

It has been widely reported that oxidative stress stimulates cellular glucose uptake. Exogenous hydrogen peroxide stimulates glucose uptake in skeletal muscle [51], C2C12 myoblasts [52], clone 9 liver cells, and 3T3 fibroblasts [53]. Endogenous ROS enhanced glucose uptake in muscle cells with electrical stimulation [54]. Interestingly, increased ROS levels were associated with decreased glucose uptake in L6 myoblasts [55], hepG-2 cells [56], etc. Glucose transporters (GLUTs) are responsible for cellular glucose uptake [57]. All together, 14 isoforms of GLUTs have been identified with different regulation modes. ROS was reported to inhibit the production and translocation of GLUT4 [58]. And NOX_2 is a major myocellular ROS source, which regulates glucose uptake during exercise [57].

9.2.2.2 ROS Regulates Glycolysis

Glycolysis converts glucose into pyruvate and a hydrogen ion. The free energy released in glycolysis is utilized to form the high-energy molecules adenosine triphosphate (ATP) and reduced NADH [34]. Glycolysis creates ATP with or without oxygen. In cells which primarily depend on aerobic respiration to produce energy, the pyruvate formed from glycolysis is used in the citric acid cycle [59]. For cells primarily depending on oxidative phosphorylation to produce energy, glycolysis serves as an emergent energy backup or the preparation step for oxidative phosphorylation. In highly oxidative organs such as the heart, pyruvate is important for the synthesis of acetyl-CoA and L-malate [60], which serves as a precursor to many molecules such as oxaloacetate and lactate [61]. Normal hematopoietic stem cells rely more on glycolysis than on OXPHOS to generate energy. The conversion of oxidative stress-induced glycolysis to OXPHOS is mediated by the synthesis of cytochrome c oxidase 2 (SCO2) [62]. The upregulation of glycolysis prevents myeloid-derived suppressor cells (MDSC) from generating excessive ROS, thereby protecting MDSC from apoptosis. Most importantly, researchers determined that phosphoenolpyruvate (PEP), a glycolytic metabolite, is an important antioxidant that prevents excessive ROS production and contributes to the survival of MDSC [63]. The metabolic phenotype changes from mitochondrial oxidative

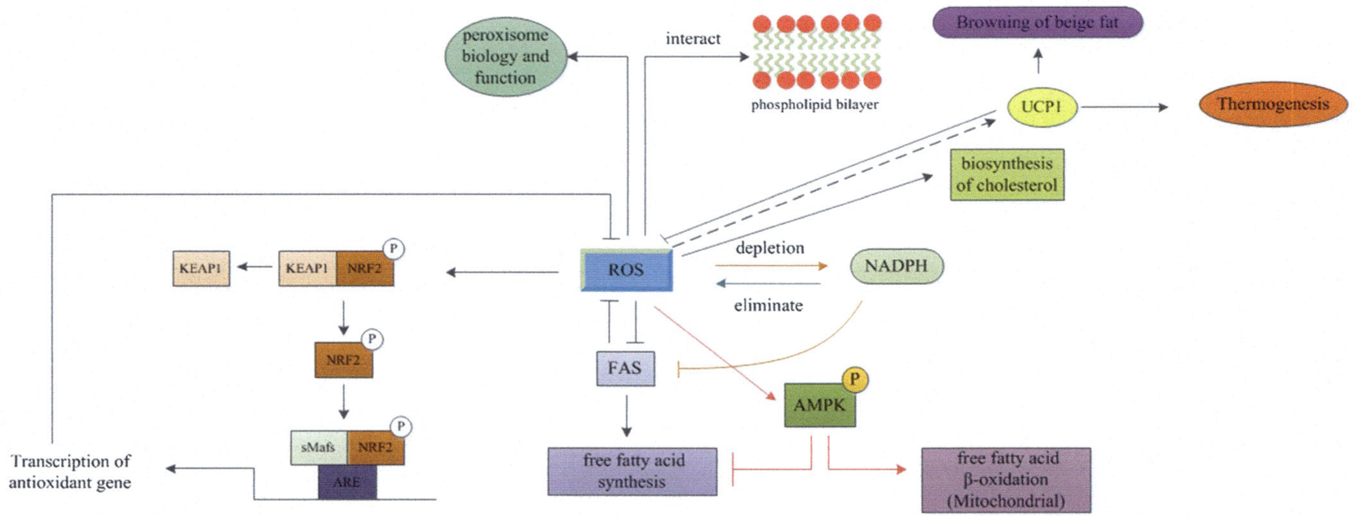

Fig. 9.4 ROS regulates lipid metabolism. There is a mutual regulation correlation among ROS, FAS, and NADPH. ROS activates AMPK, enhances the β-oxidation of free fatty acids, and influences the biology of peroxisome. At the same time, ROS interacts with phospholipid bilayer to result in cell death. As for browning of beige fat, UCP1 regulates the redox state of the respiratory chain and reduces the production of ROS. The accumulated ROS will enhance the expression of Nrf2 and improve the clearance of ROS

phosphorylation to aerobic glycolysis and changes the cell phenotype to malignancy, which may result in chemical resistance and poor prognosis [64–66].

9.2.2.3 ROS Regulates Gluconeogenesis

Gluconeogenesis is the process of forming glucose or glycogen from non-glucose precursors such as sulfuric acid, lactic acid, glycerol, and glycogenic amino acids, which mainly occur in the liver [67]. Gluconeogenesis plays a key role in maintaining glucose homeostasis [68]. Glucose in the liver mainly comes from liver glycogen breakdown and gluconeogenesis. When liver glycogen breakdown is insufficient to meet the need, gluconeogenesis is activated to provide glucose [69]. Four key enzymes in gluconeogenesis, pyruvate carboxylase (PC), phosphoenolpyruvate carboxykinase (PEPCK), fructose-1,6-bisphosphatase (FBPase), and glucose-6-phosphatase (G-6-Pase), are mainly induced by allosteric modification of hormones and related transcription factors. Transcriptional activators and co-activators involved in gluconeogenesis include cAMP-response element binding protein (CREB), CREB regulated transcription coactivator 2 (CRTC2) [70], PGC-1α [71], Forkhead Box Class O proteins (FoxOs) [72], etc. These factors have been widely reported to be regulated by oxidative stress.

Two key genes, GLUT4 and G6Pase, play important roles in regulating gluconeogenesis. GLUT4 has a stress-transporting ability similar to skeletal muscle and myocardium, which can rapidly transport glucose under stress [73]. Studies have shown that GLUT4 is the only member of the glucose transporter family (Gluts) responsible for transporting glucose under stress [74]. Total GLUT4 protein level is inversely correlated with liver glycogen concentration [75]. GLUT4 is overexpressed when liver glycogen concentration is low, while GLUT4 is underexpressed when liver glycogen concentration is high [76]. G6Pase plays essential roles in gluconeogenesis and glycogen decomposition. When the liver is resistant to insulin, glucose cannot be used to provide energy for cells [77]. Then glycogen production is decreased, and gluconeogenesis is significantly increased. Triglyceride output is also enhanced [78]. Compensatory hyperglycemia and hyperlipidemia increase the oxidative burden of mitochondria and lead to increased production of free radicals, enhanced oxidative stress, and aggravated insulin resistance [79]. In hepatocytes from fasting mice, reduced βOHB/Acoc led to enhanced gluconeogenesis and increased glycogen synthesis [80].

9.2.2.4 ROS Regulates Glycogen Synthesis

Glycogen synthase kinase-3β (Gsk-3β), a serine/threonine protein kinase, plays key roles in glycogen synthesis. Gsk-3β regulates glycogen synthesis by regulating mitochondrial biogenesis, permeability, motility, and apoptosis. The multifunctional function of GSK-3β may be related to its wide range of substrates [81]. Many growth factors, such as epidermal growth factor, insulin growth factor, and hepatocyte growth factor, inhibit Gsk-3β via phosphorylation on ser9 of Gsk-3β through PI3K/Akt/mTOR pathway [82]. Following phosphorylation on the serine residue, N-terminal tail of Gsk-3 fits into the Gsk-3 substrate-binding pocket and prevents substrate binding [83]. Phosphorylated serine resulted in loss of inhibitory effects of

Gsk-3β. Gsk-3β is a direct substrate of Akt [84]. Interestingly, the expression of Akt1 and Gsk-3β ser9 was downregulated due to oxidative stress in transient global cerebral ischemic rats [85]. And the expression of Gsk-3β tyr216 was gradually increased in oxygen glucose deprivation/reperfusion (OGD/R) and MCAO/R rats [86]. These evidences suggest that Gsk-3β is activated under hypoxia by enhancing the phosphorylation on tyr216. Gsk-3β also participates in many signaling pathways, such as regulating Nrf2 [87, 88]. Nrf2 controls the expression of phase II detoxifying enzyme and antioxidant proteins. So NRF2 protects cells from oxidative stress [89, 90]. Gsk-3β plays a key role in the translocation of NRF2 from cytoplasm to nucleus [91]. Gsk-3β was activated in cerebral ischemia and reperfusion, which in turn inhibited the expression of Nrf2 and Nrf2/ARE downstream proteins such as HQ-1 and NOQ-1 [92].

Oxidative stress regulates Gsk-3β mainly through PI3K/AKt/Gsk-3β signaling pathway. By regulating the activity of Gsk-3β, the tilt of the antioxidant/pro-oxidant equilibrium system is changed [93]. Besides PI3K/AKT/Gsk-3β axis, oxidative stress activates multiple signaling pathways such as mitogen-activated protein kinase (MAPK), nuclear factor-κB (NF-κB), p53, and β-catenin/Wnt to regulate cell damage and death [94]. H_2O_2/antimycin A-induced oxidative stress enhances the activity of Gsk-3β, alleviates oxidative stress, and inhibits cell death caused by oxidative stress [95]. It has been reported that blocking oxidative stress in nerves increases the blood flow to the neurotrophic vessels, promotes nerve conduction, and alleviates the pathological changes of peripheral nerves in patients with diabetes. Activation of the PI3K/AKT/Gsk-3β pathway obviously inhibits oxidative stress, promotes cell proliferation, and inhibits inflammation [96]. The phosphorylation of protein kinase B (Akt) further inhibits the phosphorylation of Gsk-3β and Tau [97].

9.2.2.5 ROS Regulates Glycogenolysis

Glycogen phosphorylase is the key enzyme for glycogen decomposition. It acts on the α-1,4-glycosidic bond at the nonreducing end of the sugar chain to catalyze the decomposition of a glucose group to produce glucose-1-phosphate [98]. There are mainly three isoforms of glycogen phosphorylase, muscle type (muscle-type glycogen phosphorylase, PYGM), brain type (brain-type glycogen phosphorylase, PYGB), and liver type (liver-type glycogen phosphorylase, PYGL). PYGL is responsible for catalyzing the decomposition of hepatic glycogen [99]. The functions of PYGB remain unclear, which may catalyze glycogen decomposition to provide a short-term energy supply to the brain under hypoxia and ischemia [100]. The role of PYGM is responsible for mobilizing muscle glycogen to provide energy for muscles [101, 102]. PYGL depletion and the consequent glycogen accumulation increased the generation of ROS and induced senescence dependent on p53 [103]. Deletion of glycogen phosphorylase, gph1Δ, leads to a shortened life span, stress intolerance, and higher levels of ROS [104].

9.2.2.6 ROS Regulates Pentose Phosphate Pathway

Increased glucose uptake leads to the accumulation of glycolytic intermediates and the enhanced metabolism of the pentose phosphate pathway (PPP) [105]. The PPP, also known as the hexose phosphate branch, is a branch of sugar metabolism originated after the first step of glycolysis to produce glucose 6-phosphate (G6P) [106]. More and more evidence show that PPP provides not only ribose as a raw material for DNA synthesis but also generates NADPH which is an important hydrogen donor for the synthesis of biological macromolecules [107, 108]. The latter function may be more important than the former, since NADPH plays key roles in maintaining intracellular redox balance [109, 110]. NADPH is the donor of reductive potential to thioredoxins and glutathione, which in turn neutralize ROS [111]. NADPH acts as the ultimate donor of reductive power for a large number of ROS-detoxifying enzymes. Several metabolic pathways produce NADPH, such as reactions catalyzed by the malic enzymes (ME), folate dehydrogenases (FDH), and isocitrate dehydrogenases (IDH) [112]. However, cellular NADPH are mainly produced by two enzymes of the oxidative branch of PPP, the PPP rate-limiting enzyme G6PD and 6PGDH [113]. PPP includes two branches, the oxidized branch and the non-oxidized branch. The oxidative branch produces NADPH and ribulose 5-phosphate, including three irreversible reactions [113]. In the first reaction, G6P is catalyzed by glucose-6-phosphate dehydrogenase (G6PDH) to produce one molecule of NADPH and one molecule of 6-phosphate gluconolactone, which is the rate-limiting reaction of the oxidative branch [114]. In the non-oxidative branch, PPP consists of several steps of reversible reactions utilizing glycolytic intermediates as raw materials. The main role of non-oxidizing branch is to produce ribose without NADPH [106]. The production of ROS is obviously increased in rapidly proliferating or metabolically active cells, which in turn need abundant NADPH. NADPH is key in maintaining the reduced state of glutathione, which protects biological macromolecules from damage by oxidants, especially peroxides [115]. The ROS level in damaged cells is higher than that of normal cells due to defects in the antioxidant system. It's a life and death issue for cells to keep the ROS level below the threshold [116]. Therefore, the regulation of the antioxidant system is especially crucial. In conclusion, G6PDH is the rate-limiting enzyme of the oxidative branching reaction [117].

9.3 ROS and Lipid Metabolism

9.3.1 Introduction to Lipid Metabolism and ROS

There are many forms of lipids, such as fatty acids, cholesterol, and phospholipids. The role of fatty acids is complex and varies significantly depending on the amount and nature of the molecule [118]. Lipid metabolism homeostasis plays important roles in humans. Unbalanced lipid metabolism resulted in many metabolic diseases, such as obesity, NAFLD, atherosclerosis, etc. The excess intake of fatty acids and inordinate β-oxidation resulted in obesity and diabetes and caused great damage to the function of the heart [119]. Similarly, insufficient production of cholesterol leads to Smith-Lemli-Opitz syndrome, while cholesterol accumulation in the cornea results in Schnyder corneal dystrophy [120]. NAFLD has become a leading cause of liver-related morbidity and mortality [121].

Oxidative stress is caused by the imbalance between the generation of ROS and the availability of antioxidants or free radical scavengers [122]. Oxidative stress is associated with many metabolic diseases and age-related diseases, such as cardiovascular disease, chronic obstructive pulmonary disease, chronic kidney disease, neurodegenerative disease, diabetes, atherosclerosis, etc. [32]. Oxidative stress plays critical roles in lipid metabolism. Frataxin deficiency regulates oxidative stress, which leads to abnormal protein function and metabolic changes. The deficiency of frataxin and changes in lipid metabolism have been observed in several diseases [123]. Lipid droplet accumulation and aggregated lipid oxidation were induced by high glucose in ARPE-19 cells. Adding free fatty acids (FFAs) to ARPE-19 cells resulted in the production of hydrogen peroxide and senescence [124].

Lipids are fats absorbed from food or synthesized by the liver [125]. Lipids include triglycerides (TGs), cholesterol, cholesterol esters, and phospholipids [126]. Although all lipids are physiologically important, TGs and cholesterol make the most contribution to many metabolic disease [125]. Dietary triglycerides are digested by gastric lipase and pancreatic lipase into glycerol monolipids (MGs) and FFAs in the stomach and duodenum [125]. FFAs cause insulin resistance and inflammation in multiple organs, which is closely correlated with various diseases, such as obesity, T2DM, hypertension, and hyperlipidemia [127].

9.3.2 Dysregulated Lipid Metabolism Leads to Increased Production of ROS

There are numerous similarities between the metabolic phenotypes of diabetes and obese hearts. For example, fatty acid intake and fatty acid oxidation (FAO) are increased. Glucose uptake and glucose oxidation are enhanced. And levels of triacylglycerol and circulating free fatty acids (FFAs) in the myocardium and circulation are increased [128]. Metabolic changes have also been observed in obese or diabetic hearts, such as increased FAO content, glucose autoxidation, mitochondrial dysfunction, impaired polyol metabolism, or increased hexosamine metabolism, which may lead to increased ROS release [128]. During insulin resistance, obesity, and/or diabetes, excessive fatty acids were transported into cardiomyocytes, which may significantly impair cardiac function through increasing fatty acid oxidation, decreasing cardiac efficiency, and promoting lipid accumulation [129].

Increased carbonylation of proteins, including one-third of mitochondrial proteins, was observed in diet-induced adipose tissue of obese mice, subcutaneous adipose tissue of overweight human, and cultured adipose cells treated with TNF-α, which results in decreased activity and increased ROS formation [130].

9.3.3 ROS Regulates Lipid Metabolism

9.3.3.1 Fatty Acid Synthesis

Excessive FFA severely enhances the production of ROS, leading to lipid toxicity associated with endoplasmic reticulum (ER) stress, calcium imbalance, mitochondrial dysfunction, and cell death [131].

The fatty acid synthase (FAS) is responsible for synthesizing saturated long-chain fatty acids from acetyl CoA, malonyl CoA, and NADPH [132]. There is a mutual regulation between ROS and FAS. Capsaicin inhibited the expression and activity of FAS through increasing the production of intracellular ROS [133]. And FAS is a key component in maintaining oxidative stress balance in breast cancer cells [134]. Inhibition of FAS results in the production of ROS, thereby influencing the growth of HER-2 overexpressed cells [134]. Cytoplasmic NADPH plays important roles in the biosynthesis of biomass components and the balance of oxidative molecules [135]. Increased production of ROS leads to exhaustion of cytoplasmic NADPH, which in turn pauses the production of FAS [135]. If oxidative stress increases sharply, the production of cytosolic NADPH is immediately enhanced to scavenge ROS and stop lipid synthesis [135].

In pancreatic beta cells, the KEAP1/NRF2 complex releases KEAP1 through redox signaling, causing NRF2 to translocate to the nucleus and interact with the antioxidant response element (ARE) in the promoter of target genes encoding the antioxidant enzymes. Eventually, the genes of antioxidant enzymes are expressed. Cells can be protected from lipotoxicity [136].

Through the combined actions of glycolysis, TCA cycle, and fatty acid biosynthesis, hypertrophic oxidation reduces the ratio of NAD^+/NADH in adipocytes, thereby weakening the activity of all NAD^+-dependent enzymes including sirtuins. Sirt1 is the most widely investigated member among all of the seven sirtuins in mammals. The NAD^+/NADH ratio of adipocytes can be used as an important indicator of the energy state of cells and the whole body, which is one of the key redox modulators of white adipose tissue (WAT) function [136].

Adipogenesis results in a metabolically healthy obese WAT phenotype. ROS play important roles in the production of NOX4-generated H_2O_2 to trigger and regulate adipogenesis [136]. The elimination of ROS by n-acetyl-l-cysteine results in the abolition of fat production in vitro [137]. Abnormal functioning obese WAT causes significantly increased protein carbonylation, suggesting that excessive ROS and oxidative stress play a key role in obesity-induced insulin resistance [138].

9.3.3.2 Fatty Acid Oxidation

Besides FAS, the fatty acid oxidation (FAO) also plays a pivotal role in the survival and proliferation of cancer cells [135]. The oxidative degradation of lipids is known as lipid peroxidation. During lipid peroxidation, oxidants such as free radicals attack lipids containing carbon-carbon double bonds especially polyunsaturated fatty acids (PUFA) [139].

As one of three substrates that organisms can use to maintain metabolic homeostasis, fatty acid is required for the generation of energy [140]. The major way of fatty acid degradation is fatty acid β-oxidation in mitochondria [141]. Oxidation of FFA in plasma or lipoprotein-associated TGs occurs mainly in mitochondria. So lipids have to get across the cell membrane and mitochondria membrane with the help of transport proteins [140]. Before oxidative degradation, FFA needs to be activated to acyl-coenzyme A (-CoA) ester [142]. Medium-chain fatty acids are free to spread to the mitochondria, while long-chain fatty acids are firstly esterified in the cytoplasm and then transported to the mitochondria [143].

Beta-oxidation occurs in both mitochondria and peroxisome in higher eukaryotes [144]. In fact, it has been clearly established that most dietary fatty acids (FA) are oxidized by oxygen in the mitochondria. However, peroxisomes catalyze a range of FA and fatty acid derivatives that mitochondria cannot handle, including ultra-long chain FA, fatty acids, and bile acid intermediates dihydroxy and trihydroxy cholesteric acids [145].

Hepatitis C virus (HCV) induced growth arrest and activated AMPK through ROS in Huh7.5 cells. Metabolic labeling experiments provided direct evidence that lipid synthesis was attenuated, while β-oxidation enhanced in these cells [146]. In kidney tubule in short-term diabetes, mitochondrial fatty acid oxidation is responsible for the increased production of net ROS. And the electron leakage site is located at the proximal end of coenzyme Q on the electron transfer flavin protein, which transfers electrons from acyl-coa dehydrogenase to coenzyme Q [147].

In addition, relative ROS production from different organelles was also compared. The study on rat livers suggests that endoplasmic reticulum (ER) and peroxisomes may have even stronger ROS production capacity than mitochondria [148]. ROS and nitrogen species are produced by peroxisomal β-oxidation as byproducts [149, 150]. It is reported that long-term administration of peroxisomal proliferative agent induces oxidative stress in liver cells in rodents [151].

At the same time, cellular ROS may influence the biology of peroxisome [150]. There is considerable evidence that the function and morphology of peroxisomes change dramatically when cells are exposed to a variety of oxidative stress stimuli [150]. For example, exposure to ultraviolet light induced peroxisomes to proliferate and form tubular peroxisomes [152]. The increase of ambient oxygen concentration leads to a moderate increase in the volume density of peroxisomes and peroxisomal enzymes involved in ROS scavenging [152]. With the accumulation of peroxisomal ROS, the cellular proteins, lipids, and DNA are destroyed by oxidation, organelles are impaired, and metabolism is slowed down [153].

9.3.3.3 Phospholipid

Phospholipids are an important part of all mammalian cells and have a variety of biological functions. Phospholipids form a lipid bilayer providing the structural integrity of organelles and cells. Phospholipids are used as a precursor to various lipid second messengers [154]. Some therapies, such as plasma therapy, produce extracellular ROS which interacts with the cell membrane and results in the oxidized head group and the lipid tail of phospholipids. The decreased lipid sequence may allow them to penetrate into the cell and cause oxidative damages, leading to cell death [155].

Hypertension is often accompanied by hyperlipidemia, which leads to ROS-mediated lipid peroxidation. ROS generated during hypertension alters enzyme-dependent phospholipid metabolism. Oxidative cleavage of electrophilic aldehydes such as malondialdehyde (MDA) and oxidative cycloprostaglandin derivatives induces vasoconstriction which aggravates hypertension [156].

Nearly 22 phospholipase A2 (PLA2) enzymes have been reported in mammalian tissues. They are roughly classified into three families according to their calcium dependence and calcium catalytic activity, namely, cytoplasmic (cPLA2), secretory (sPLA2) types, and calcium-independent forms (iPLA2) [154, 157]. Mitochondrial sPLA2 hydrolyzed cardiolipids to impair the mitochondrial respiratory chain which results in more ROS [154].

9.3.3.4 Cholesterol

Cholesterol has evolved to meet the complex biophysical, cellular signaling and endocrine requirements of eukaryotes. Cholesterol in the cell membrane increases both the hardness of the double layer and the impermeability to water and ions. Cholesterol is also the precursor of all steroid hormones. Excessive or insufficient cholesterol leads to multiple diseases in embryos or adults [158].

ROS biophysically modify membrane lipids and proteins to exert effects, which in turn regulate the biosynthesis of cholesterol or other lipid metabolic pathways [159]. Low-density lipoprotein cholesterol (LDL cholesterol) promotes intestinal inflammation and colorectal cancer (CRC) progression by activating redox signaling pathways, such as the MAPK pathway [160].

9.3.3.5 Browning of Beige Fat

Adipose tissue is divided into energy-storing white adipose tissue (WAT) and heat-producing brown adipose tissue (BAT). BAT burns fatty acids to generate heat, which is important to maintain body temperature and normal physiological function. Therefore, adipose tissue plays important roles in controlling energy homeostasis through regulating the balance between energy storage and consumption [161]. Brown fats have high mitochondrial content and high mitochondrial UCP1 [162]. Over the years, it has been known that certain cells in WAT also express high levels of UCP1 and exhibit a multiocular appearance after prolonged exposure to coldness or increased intracellular cyclic AMP [163]. This "brown-like" cells within WAT have been called beige cells or brite (brown in white) cells, iBAT (induced BAT) [162, 164].

Brown fat increases energy expenditure through generating heat. So the BAT and beige fat are regarded as therapeutic targets in fighting against obesity and metabolic diseases [165]. Thermogenesis of BAT is activated by the β3-epinephrine pathway and occurs in mitochondria though UCP1 [161]. The rapid activation of thermogenesis in BAT is influenced by the significantly increased mitochondrial ROS [166]. Thermogenic ROS alters the redox state of cysteine mercaptan in BAT to enhance respiration [166]. At the same time, UCP1 regulates the redox state of the respiratory chain and reduces the production of ROS through influencing proton dynamics [167]. Besides UCP1, peroxisome-proliferator activator receptor γ (PPARγ) agonists lead to increased expression of adiponectin and alleviated insulin resistance in type 2 diabetes mellitus (T2DM) [168, 169]. PPARγ directly regulates the expression of some antioxidant and pro-oxidant genes [170]. Rosiglitazone enhanced the antioxidant defense ability of hepatocytes and improved the clearance of ROS through enhancing the expression of Nrf2 and heme oxygenase 1 (HO-1) [171]. When ROS increases, NRF2 is released from Keap1 and heterodimerizes with small Mafs, and then the transcription of many antioxidant and cytoprotective genes simultaneously begins [172].

9.4 Conclusion

ROS are free radicals produced by the reduction of molecular oxygen. The mitochondrial respiratory chain complex is regarded as a major ROS producer. In recent years, a large amount of evidence has demonstrated that ROS plays important roles in regulating glucose and lipid metabolism and participates in the occurrence and development of disorders related to glucose and lipid metabolism. Generally, glucose metabolism increases the production of ROS and enhances oxidative stress by glucose auto-oxidation and glycosylation of proteins and activated the polyalcohol pathway, etc. Hyperglycemia changes the formation of superoxide anions. Then the respiratory chain complex becomes the main provider of electrons and promotes the formation of peroxide. Hyperglycemia results in decreased ROS clearance and increased the production of ROS. Both exogenous and endogenous ROS stimulation lead to glucose uptake in muscle cells and liver cells. The upregulation of glycolysis prevents myeloid-derived suppressor cells (MDSC) from generating excessive ROS, thereby protecting MDSC from apoptosis. Researchers determined that phosphoenolpyruvate (PEP), a glycolytic metabolite, is an important antioxidant that prevents excessive ROS production and contributes to the survival of MDSC. Gluconeogenesis is very important in maintaining glucose homeostasis. Glucose in the liver mainly comes from liver glycogen breakdown and gluconeogenesis. When liver glycogen breakdown is insufficient to meet the need, gluconeogenesis is activated to provide glucose. Two key genes, GLUT4 and G6Pase, play important roles in regulating gluconeogenesis. GLUT4 has a stress-transporting ability similar to skeletal muscle and myocardium, which can rapidly transport glucose under stress. G6Pase plays essential roles in gluconeogenesis and glycogen decomposition. When the liver is resistant to insulin, glucose cannot be used to provide energy for cells. Increased glucose uptake leads to the accumulation of glycolytic intermediates and the enhanced metabolism of the pentose phosphate pathway. More and more evidence show that PPP provides not only ribose as a raw material for DNA synthesis but also generates NADPH which is an important hydrogen donor for the synthesis of biological macromolecules. The production of ROS is obviously increased in rapidly proliferating or metabolically active cells, which in turn need abundant NADPH. NADPH is key in maintaining the reduced state of glutathione, which protects biological macromolecules from damage by oxidants, especially peroxides. The means of ROS antagonist must be investigated in depth since ROS has been linked to increased incidence of glucose and lipid metabolism disorders, such as metabolic syndrome, hyperlipidemia, NAFLD, etc. We have to better understand the role of elevated ROS production in glucose and lipid metabolism disorders, identify specific ROS targets, and explore new strategies to treat those disorders more effectively.

Acknowledgments This work was financially supported by the National Natural Science Foundation of China (Grant No. 81770580).

References

1. Moloney JN, Cotter TG. ROS signalling in the biology of cancer. Semin Cell Dev Biol. 2018;80:50–64.
2. Das K, Roychoudhury A. Reactive oxygen species (ROS) and response of antioxidants as ROS-scavengers during environmental stress in. Plan Theory. 2014;2:53.
3. Holzerová E, Prokisch H. Mitochondria: much ado about nothing? How dangerous is reactive oxygen species production? Int J Biochem Cell Biol. 2015;63:16–20.
4. Shadel GS, Horvath TL. Mitochondrial ROS signaling in organismal homeostasis. Cell. 2015;163(3):560–9.
5. Forrester SJ, et al. Reactive oxygen species in metabolic and inflammatory signaling. Circ Res. 2018;122(6):877–902.
6. Rodic S, Vincent MD. Reactive oxygen species (ROS) are a key determinant of cancer's metabolic phenotype. Int J Cancer. 2018;142(3):440–8.
7. Sies H, Jones DP. Reactive oxygen species (ROS) as pleiotropic physiological signalling agents. Nat Rev Mol Cell Biol. 2020;21(7):363–83.
8. Cogliati S, et al. Regulation of mitochondrial electron transport chain assembly. J Mol Biol. 2018;430(24):4849–73.
9. Bigarella CL, Liang R, Ghaffari S. Stem cells and the impact of ROS signaling. Development. 2014;141(22):4206–18.
10. Speijer D. Can all major ROS forming sites of the respiratory chain be activated by high FADH2 /NADH ratios?: ancient evolutionary constraints determine mitochondrial ROS formation. BioEssays. 2019;41(1):e1800180.
11. Kang SW, Lee S, Lee EK. ROS and energy metabolism in cancer cells: alliance for fast growth. Arch Pharm Res. 2015;38(3):338–45.
12. Hardeland R. Melatonin and the electron transport chain. Cell Mol Life Sci. 2017;74 (21):3883–96.
13. Guo R, et al. Structure and mechanism of mitochondrial electron transport chain. Biom J. 2018;41(1):9–20.
14. Letts JA, Sazanov LA. Clarifying the supercomplex: the higher-order organization of the mitochondrial electron transport chain. Nat Struct Mol Biol. 2017;24(10):800–8.
15. Sirey TM, Ponting CP. Insights into the post-transcriptional regulation of the mitochondrial electron transport chain. Biochem Soc Trans. 2016;44(5):1491–8.
16. Ray PD, Huang BW, Tsuji Y. Reactive oxygen species (ROS) homeostasis and redox regulation in cellular signaling. Cell Signal. 2012;24(5):981–90.
17. Cordeiro RM. Reactive oxygen species at phospholipid bilayers: distribution, mobility and permeation. Biochim Biophys Acta. 2014;1838(1 Pt B):438–44.
18. Long YC, et al. The biochemistry and cell biology of aging: metabolic regulation through mitochondrial signaling. Am J Physiol Endocrinol Metab. 2014;306(6):E581–91.

19. Zhang J, et al. ROS and ROS-mediated cellular signaling. Oxidative Med Cell Longev. 2016;2016:4350965.
20. Gupta SC, et al. Upsides and downsides of reactive oxygen species for cancer: the roles of reactive oxygen species in tumorigenesis, prevention, and therapy. Antioxid Redox Signal. 2012;16(11):1295–322.
21. Robello E, Bonetto JG, Puntarulo S. Cellular oxidative/antioxidant balance in γ-irradiated brain: an update. Mini Rev Med Chem. 2016;16(12):937–46.
22. Xu DP, et al. Natural antioxidants in foods and medicinal plants: extraction, assessment and resources. Int J Mol Sci. 2017;18(1):96.
23. Santos CX, et al. Endoplasmic reticulum stress and Nox-mediated reactive oxygen species signaling in the peripheral vasculature: potential role in hypertension. Antioxid Redox Signal. 2014;20(1):121–34.
24. Förstermann U, Xia N, Li H. Roles of vascular oxidative stress and nitric oxide in the pathogenesis of atherosclerosis. Circ Res. 2017;120(4):713–35.
25. Chan SHH, Chan JYH. Mitochondria and reactive oxygen species contribute to neurogenic hypertension. Physiology (Bethesda). 2017;32(4):308–21.
26. Panigrahy SK, Bhatt R, Kumar A. Reactive oxygen species: sources, consequences and targeted therapy in type 2 diabetes. J Drug Target. 2017;25(2):93–101.
27. Tafani M, et al. The interplay of reactive oxygen species, hypoxia, inflammation, and sirtuins in cancer initiation and progression. Oxidative Med Cell Longev. 2016;2016:3907147.
28. Bhat AH, et al. Oxidative stress, mitochondrial dysfunction and neurodegenerative diseases; a mechanistic insight. Biomed Pharmacother. 2015;74:101–10.
29. Prasad S, Gupta SC, Tyagi AK. Reactive oxygen species (ROS) and cancer: role of antioxidative nutraceuticals. Cancer Lett. 2017;387:95–105.
30. Kim J, Kim J, Bae JS. ROS homeostasis and metabolism: a critical liaison for cancer therapy. Exp Mol Med. 2016;48(11):e269.
31. Yang Y, et al. Reactive oxygen species in the immune system. Int Rev Immunol. 2013;32 (3):249–70.
32. Liguori I, et al. Oxidative stress, aging, and diseases. Clin Interv Aging. 2018;13:757–72.
33. Aghadavod E, et al. Role of oxidative stress and inflammatory factors in diabetic kidney disease. Iran J Kidney Dis. 2016;10(6):337–43.
34. Mulukutla BC, et al. Glucose metabolism in mammalian cell culture: new insights for tweaking vintage pathways. Trends Biotechnol. 2010;28(9):476–84.
35. Ghanbari Movahed Z, et al. Cancer cells change their glucose metabolism to overcome increased ROS: one step from cancer cell to cancer stem cell? Biomed Pharmacother. 2019;112:108690.
36. Liemburg-Apers DC, et al. Interactions between mitochondrial reactive oxygen species and cellular glucose metabolism. Arch Toxicol. 2015;89(8):1209–26.
37. Mulukutla BC, et al. Regulation of glucose metabolism - a perspective from cell bioprocessing. Trends Biotechnol. 2016;34(8):638–51.
38. Yin B, Barrionuevo G, Weber SG. Mitochondrial GSH systems in CA1 pyramidal cells and astrocytes react differently during oxygen-glucose deprivation and reperfusion. ACS Chem Neurosci. 2018;9(4):738–48.
39. Achari AE, Jain SK. l-Cysteine supplementation increases insulin sensitivity mediated by upregulation of GSH and adiponectin in high glucose treated 3T3-L1 adipocytes. Arch Biochem Biophys. 2017;630:54–65.
40. Quan Y, Du J, Wang X. High glucose stimulates GRO secretion from rat microglia via ROS, PKC, and NF-kappaB pathways. J Neurosci Res. 2007;85(14):3150–9.
41. Zhang Y, et al. High glucose-induced RhoA activation requires caveolae and PKCβ1-mediated ROS generation. Am J Physiol Renal Physiol. 2012;302(1):F159–72.
42. Volpe CMO, et al. Cellular death, reactive oxygen species (ROS) and diabetic complications. Cell Death Dis. 2018;9(2):119.

43. Preetha Rani MR, et al. Chlorogenic acid attenuates glucotoxicity in H9c2 cells via inhibition of glycation and PKC α upregulation and safeguarding innate antioxidant status. Biomed Pharmacother. 2018;100:467–77.
44. Graham S, et al. Abundance of TRPC6 protein in glomerular mesangial cells is decreased by ROS and PKC in diabetes. Am J Physiol Cell Physiol. 2011;301(2):C304–15.
45. Kizub IV, Klymenko KI, Soloviev AI. Protein kinase C in enhanced vascular tone in diabetes mellitus. Int J Cardiol. 2014;174(2):230–42.
46. Fazakerley DJ, et al. Mitochondrial oxidative stress causes insulin resistance without disrupting oxidative phosphorylation. J Biol Chem. 2018;293(19):7315–28.
47. Román-Pintos LM, et al. Diabetic polyneuropathy in type 2 diabetes mellitus: inflammation, oxidative stress, and mitochondrial function. J Diabetes Res. 2016;2016:3425617.
48. Ahangarpour A, et al. Antioxidant effect of myricitrin on hyperglycemia-induced oxidative stress in C2C12 cell. Cell Stress Chaperones. 2018;23(4):773–81.
49. Li W, et al. Hyperglycemia promotes the epithelial-mesenchymal transition of pancreatic cancer via hydrogen peroxide. Oxidative Med Cell Longev. 2016;2016:5190314.
50. Wang G, et al. Baicalin administration attenuates hyperglycemia-induced malformation of cardiovascular system. Cell Death Dis. 2018;9(2):234.
51. Jahn MP, et al. The effect of long-term DHEA treatment on glucose metabolism, hydrogen peroxide and thioredoxin levels in the skeletal muscle of diabetic rats. J Steroid Biochem Mol Biol. 2010;120(1):38–44.
52. Ding H, et al. Chronic reactive oxygen species exposure inhibits glucose uptake and causes insulin resistance in C2C12 myotubes. Biochem Biophys Res Commun. 2016;478 (2):798–803.
53. Prasad RK, Ismail-Beigi F. Mechanism of stimulation of glucose transport by H2O2: role of phospholipase C. Arch Biochem Biophys. 1999;362(1):113–22.
54. Peiró C, et al. High glucose induces cell death of cultured human aortic smooth muscle cells through the formation of hydrogen peroxide. Br J Pharmacol. 2001;133(7):967–74.
55. Dhanya R, et al. Preconditioning L6 muscle cells with naringin ameliorates oxidative stress and increases glucose uptake. PLoS One. 2015;10(7):e0132429.
56. Ren D, et al. Chemical characterization of Pleurotus eryngii polysaccharide and its tumor-inhibitory effects against human hepatoblastoma HepG-2 cells. Carbohydr Polym. 2016;138:123–33.
57. Henríquez-Olguin C, et al. Cytosolic ROS production by NADPH oxidase 2 regulates muscle glucose uptake during exercise. Nat Commun. 2019;10(1):4623.
58. Maeda A, et al. Piperine promotes glucose uptake through ROS-dependent activation of the CAMKK/AMPK signaling pathway in skeletal muscle. Mol Nutr Food Res. 2018;62(11): e1800086.
59. Mullarky E, Cantley LC. Diverting glycolysis to combat oxidative stress. Tokyo: Springer; 2015.
60. Karpusas M, Holland D, Remington SJ. 1.9-A structures of ternary complexes of citrate synthase with D- and L-malate: mechanistic implications. Biochemistry. 1991;30 (24):6024–31.
61. Kerbey AL, Radcliffe PM, Randle PJ. Diabetes and the control of pyruvate dehydrogenase in rat heart mitochondria by concentration ratios of adenosine triphosphate/adenosine diphosphate, of reduced/oxidized nicotinamide-adenine dinucleotide and of acetyl-coenzyme A/coenzyme A. Biochem J. 1977;164(3):509–19.
62. Du W, et al. SCO2 mediates oxidative stress-induced glycolysis to oxidative phosphorylation switch in hematopoietic stem cells. Stem Cells. 2016;34(4):960–71.
63. Jian SL, et al. Glycolysis regulates the expansion of myeloid-derived suppressor cells in tumor-bearing hosts through prevention of ROS-mediated apoptosis. Cell Death Dis. 2017;8 (5):e2779.
64. Yu L, et al. Modeling the genetic regulation of cancer metabolism: interplay between glycolysis and oxidative phosphorylation. Cancer Res. 2017;77(7):1564–74.

65. Hou GX, et al. Elimination of stem-like cancer cell side-population by auranofin through modulation of ROS and glycolysis. Cell Death Dis. 2018;9(2):89.
66. Qin W, et al. Inhibition of autophagy promotes metastasis and glycolysis by inducing ROS in gastric cancer cells. Oncotarget. 2015;6(37):39839–54.
67. Han HS, et al. Regulation of glucose metabolism from a liver-centric perspective. Exp Mol Med. 2016;48(3):e218.
68. Petersen MC, Vatner DF, Shulman GI. Regulation of hepatic glucose metabolism in health and disease. Nat Rev Endocrinol. 2017;13(10):572–87.
69. Yan H, et al. Estrogen improves insulin sensitivity and suppresses gluconeogenesis via the transcription factor Foxo1. Diabetes. 2019;68(2):291–304.
70. Li Y, et al. Thyroid stimulating hormone increases hepatic gluconeogenesis via CRTC2. Mol Cell Endocrinol. 2017;446:70–80.
71. Zhu X, et al. CREB-upregulated lncRNA MEG3 promotes hepatic gluconeogenesis by regulating miR-302a-3p-CRTC2 axis. J Cell Biochem. 2019;120(3):4192–202.
72. Dong XC. SCP4: a small nuclear phosphatase having a big effect on FoxOs in gluconeogenesis. Diabetes. 2018;67(1):23–5.
73. Na RS, et al. Itraconazole attenuates hepatic gluconeogenesis and promotes glucose uptake by regulating AMPK pathway. Exp Ther Med. 2018;15(2):2165–71.
74. Satoh T. Molecular mechanisms for the regulation of insulin-stimulated glucose uptake by small guanosine triphosphatases in skeletal muscle and adipocytes. Int J Mol Sci. 2014;15 (10):18677–92.
75. Gheibi S, et al. Effects of long-term nitrate supplementation on carbohydrate metabolism, lipid profiles, oxidative stress, and inflammation in male obese type 2 diabetic rats. Nitric Oxide. 2018;75:27–41.
76. Ahn YH. A journey to understand glucose homeostasis: starting from rat glucose transporter type 2 promoter cloning to hyperglycemia. Diabetes Metab J. 2018;42(6):465–71.
77. Titchenell PM, Lazar MA, Birnbaum MJ. Unraveling the regulation of hepatic metabolism by insulin. Trends Endocrinol Metab. 2017;28(7):497–505.
78. Patel BM, Goyal RK. Liver and insulin resistance: new wine in old bottle!!! Eur J Pharmacol. 2019;862:172657.
79. Ji L, et al. MICU1 alleviates diabetic cardiomyopathy through mitochondrial ca(2+)-dependent antioxidant response. Diabetes. 2017;66(6):1586–600.
80. Nocito L, et al. The extracellular redox state modulates mitochondrial function, gluconeogenesis, and glycogen synthesis in murine hepatocytes. PLoS One. 2015;10(3):e0122818.
81. Yang K, et al. The key roles of GSK-3β in regulating mitochondrial activity. Cell Physiol Biochem. 2017;44(4):1445–59.
82. Li BG, Hasselgren PO, Fang CH. Insulin-like growth factor-I inhibits dexamethasone-induced proteolysis in cultured L6 myotubes through PI3K/Akt/GSK-3beta and PI3K/Akt/mTOR-dependent mechanisms. Int J Biochem Cell Biol. 2005;37(10):2207–16.
83. Aoki M, et al. Structural insight into nucleotide recognition in tau-protein kinase I/glycogen synthase kinase 3 beta. Acta Crystallogr D Biol Crystallogr. 2004;60(Pt 3):439–46.
84. Mannoury la Cour C, et al. Signaling pathways leading to phosphorylation of Akt and GSK-3β by activation of cloned human and rat cerebral D_2 and D_3 receptors. Mol Pharmacol. 2011;79 (1):91–105.
85. Endo H, et al. Activation of the Akt/GSK3beta signaling pathway mediates survival of vulnerable hippocampal neurons after transient global cerebral ischemia in rats. J Cereb Blood Flow Metab. 2006;26(12):1479–89.
86. Chen X, et al. GSK-3β downregulates Nrf2 in cultured cortical neurons and in a rat model of cerebral ischemia-reperfusion. Sci Rep. 2016;6(1):20196.
87. Ranea-Robles P, et al. Aberrant regulation of the GSK-3β/NRF2 axis unveils a novel therapy for adrenoleukodystrophy. EMBO Mol Med. 2018;10(8):e8604.
88. Chen X, et al. GSK-3β downregulates Nrf2 in cultured cortical neurons and in a rat model of cerebral ischemia-reperfusion. Sci Rep. 2016;6:20196.

89. Shin JH, et al. Nrf2-Heme Oxygenase-1 attenuates high-glucose-induced epithelial-to-mesenchymal transition of renal tubule cells by inhibiting ROS-mediated PI3K/Akt/GSK-3β signaling. J Diabetes Res. 2019;2019:2510105.
90. Liu B, et al. GSPE reduces lead-induced oxidative stress by activating the Nrf2 pathway and suppressing miR153 and GSK-3β in rat kidney. Oncotarget. 2017;8(26):42226–37.
91. Li Q, et al. Gastrodin and isorhynchophylline synergistically inhibit MPP(+)-induced oxidative stress in SH-SY5Y cells by targeting ERK1/2 and GSK-3β pathways: involvement of Nrf2 nuclear translocation. ACS Chem Neurosci. 2018;9(3):482–93.
92. Armagan G, et al. Regulation of the Nrf2 pathway by glycogen synthase kinase-3β in MPP^+-induced cell damage. Molecules. 2019;24(7):1377.
93. Wang J, et al. Panax Quinquefolius Saponin of stem and leaf attenuates intermittent high glucose-induced oxidative stress injury in cultured human umbilical vein endothelial cells via PI3K/Akt/GSK-3 β pathway. Evid Based Complement Alternat Med. 2013;2013:196283.
94. Filomeni G, De Zio D, Cecconi F. Oxidative stress and autophagy: the clash between damage and metabolic needs. Cell Death Differ. 2015;22(3):377–88.
95. Zhang X, et al. Oxidative stress promotes hepatocyte apoptosis mediated by glycogen synthase kinase 3β. Xi Bao Yu Fen Zi Mian Yi Xue Za Zhi. 2015;31(1):27–31.
96. Wang Y, et al. Inactivation of GSK-3beta by metallothionein prevents diabetes-related changes in cardiac energy metabolism, inflammation, nitrosative damage, and remodeling. Diabetes. 2009;58(6):1391–402.
97. Yao Y, et al. Osthole decreases tau protein phosphorylation via PI3K/AKT/GSK-3β signaling pathway in Alzheimer's disease. Life Sci. 2019;217:16–24.
98. Adeva-Andany MM, et al. Glycogen metabolism in humans. BBA Clin. 2016;5:85–100.
99. Nogales-Gadea G, et al. Expression of glycogen phosphorylase isoforms in cultured muscle from patients with McArdle's disease carrying the p.R771PfsX33 PYGM mutation. PLoS One. 2010;5(10):e13164.
100. Obel LF, et al. Brain glycogen-new perspectives on its metabolic function and regulation at the subcellular level. Front Neuroenerg. 2012;4:3.
101. Greenberg CC, et al. Glycogen branches out: new perspectives on the role of glycogen metabolism in the integration of metabolic pathways. Am J Physiol Endocrinol Metab. 2006;291(1):E1–8.
102. Hearris MA, et al. Regulation of muscle glycogen metabolism during exercise: implications for endurance performance and training adaptations. Nutrients. 2018;10(3):298.
103. Favaro E, et al. Glucose utilization via glycogen phosphorylase sustains proliferation and prevents premature senescence in cancer cells. Cell Metab. 2012;16(6):751–64.
104. Favre C, Aguilar PS, Carrillo MC. Oxidative stress and chronological aging in glycogen-phosphorylase-deleted yeast. Free Radic Biol Med. 2008;45(10):1446–56.
105. Jiang P, Du W, Wu M. Regulation of the pentose phosphate pathway in cancer. Protein Cell. 2014;5(8):592–602.
106. Ge T, et al. The role of the pentose phosphate pathway in diabetes and cancer. Front Endocrinol (Lausanne). 2020;11:365.
107. Stanton RC. Glucose-6-phosphate dehydrogenase, NADPH, and cell survival. IUBMB Life. 2012;64(5):362–9.
108. Patra KC, Hay N. The pentose phosphate pathway and cancer. Trends Biochem Sci. 2014;39(8):347–54.
109. Sun S, et al. Cannabidiol attenuates OGD/R-induced damage by enhancing mitochondrial bioenergetics and modulating glucose metabolism via pentose-phosphate pathway in hippocampal neurons. Redox Biol. 2017;11:577–85.
110. Kwolek-Mirek M, Maslanka R, Molon M. Disorders in NADPH generation via pentose phosphate pathway influence the reproductive potential of the Saccharomyces cerevisiae yeast due to changes in redox status. J Cell Biochem. 2019;120(5):8521–33.
111. Fernandez-Marcos PJ, Nóbrega-Pereira S. NADPH: new oxygen for the ROS theory of aging. Oncotarget. 2016;7(32):50814–5.

112. Chen L, et al. NADPH production by the oxidative pentose-phosphate pathway supports folate metabolism. Nat Metab. 2019;1:404–15.
113. Stincone A, et al. The return of metabolism: biochemistry and physiology of the pentose phosphate pathway. Biol Rev Camb Philos Soc. 2015;90(3):927–63.
114. Benito A, et al. Glucose-6-phosphate dehydrogenase and transketolase modulate breast cancer cell metabolic reprogramming and correlate with poor patient outcome. Oncotarget. 2017;8 (63):106693.
115. Nita M, Grzybowski A. The role of the reactive oxygen species and oxidative stress in the pathomechanism of the age-related ocular diseases and other pathologies of the anterior and posterior eye segments in adults. Oxidative Med Cell Longev. 2016;2016:3164734.
116. Schieber M, Chandel NS. ROS function in redox signaling and oxidative stress. Curr Biol. 2014;24(10):R453–62.
117. Lemire J, et al. Metabolic defence against oxidative stress: the road less travelled so far. J Appl Microbiol. 2017;123(4):798–809.
118. Lecerf JM. Fatty acids and cardiovascular disease. Nutr Rev. 2009;67(5):273–83.
119. Lopaschuk GD, et al. Myocardial fatty acid metabolism in health and disease. Physiol Rev. 2010;90(1):207–58.
120. Chen L, et al. Regulation of glucose and lipid metabolism in health and disease. Sci China Life Sci. 2019;62(11):1420–58.
121. Whalley S, et al. Hepatology outpatient service provision in secondary care: a study of liver disease incidence and resource costs. Clin Med (Lond). 2007;7(2):119–24.
122. Chatterjee S. Chapter two - oxidative stress, inflammation, and disease. In: Dziubla T, Butterfield DA, editors. Oxidative stress and biomaterials. Cambridge, MA: Academic Press; 2016. p. 35–58.
123. Tamarit J, Obis È, Ros J. Oxidative stress and altered lipid metabolism in Friedreich ataxia. Free Radic Biol Med. 2016;100:138–46.
124. Chen Q, et al. Oxidative stress mediated by lipid metabolism contributes to high glucose-induced senescence in retinal pigment epithelium. Free Radic Biol Med. 2019;130:48–58.
125. Davidson MH. Overview of lipid metabolism. 2019. Available from https://www.merckmanuals.com/professional/endocrine-and-metabolic-disorders/lipid-disorders/overview-of-lipid-metabolism.
126. Thompson TE. Lipid. 2020. Available from https://www.britannica.com/science/lipid. Accessed 05 May 2020.
127. Boden G. Obesity and free fatty acids. Endocrinol Metab Clin N Am. 2008;37(3):635–46.
128. Niemann B, et al. Oxidative stress and cardiovascular risk: obesity, diabetes, smoking, and pollution: part 3 of a 3-part series. J Am Coll Cardiol. 2017;70(2):230–51.
129. Kim TT, Dyck JRB. The role of CD36 in the regulation of myocardial lipid metabolism. BBA Mol Cell Biol Lipids. 2016;1861(10):1450–60.
130. Quijano C, et al. Interplay between oxidant species and energy metabolism. Redox Biol. 2016;8:28–42.
131. Ly LD, et al. Oxidative stress and calcium dysregulation by palmitate in type 2 diabetes. Exp Mol Med. 2017;49(2):e291.
132. Berg JM, Tymoczko JL, Stryer L. Fatty acids are synthesized and degraded by different pathways. In: Biochemistry. 5th ed. New York: WH Freeman; 2002.
133. Impheng H, et al. The selective target of capsaicin on FASN expression and de novo fatty acid synthesis mediated through ROS generation triggers apoptosis in HepG2 cells. PLoS One. 2014;9(9):e107842.
134. Li B, et al. Inhibition of fatty acid synthase induces reactive oxygen species (ROS) to inhibit HER2 overexpressing breast cancer cell growth. Cancer Res. 2007;67:4462.
135. Mikalayeva V, et al. Fatty acid synthesis and degradation interplay to regulate the oxidative stress in cancer cells. Int J Mol Sci. 2019;20(6):1348.

136. Egea J, et al. European contribution to the study of ROS: a summary of the findings and prospects for the future from the COST action BM1203 (EU-ROS). Redox Biol. 2017;13:94–162.
137. Kanda Y, et al. Reactive oxygen species mediate adipocyte differentiation in mesenchymal stem cells. Life Sci. 2011;89(7–8):250–8.
138. Ruskovska T, Bernlohr DA. Oxidative stress and protein carbonylation in adipose tissue - implications for insulin resistance and diabetes mellitus. J Proteome. 2013;92:323–34.
139. Ayala A. Lipid peroxidation: production, metabolism, and signaling mechanisms of malondialdehyde and 4-hydroxy-2-nonenal. Oxidative Med Cell Longev. 2014;2014:1–31.
140. Houten SM, Wanders RJA. A general introduction to the biochemistry of mitochondrial fatty acid β-oxidation. J Inherit Metab Dis. 2010;33(5):469–77.
141. Bartlett K, Eaton S. Mitochondrial beta-oxidation. Eur J Biochem. 2004;271(3):462–9.
142. Grevengoed TJ, Klett EL, Coleman RA. Acyl-CoA metabolism and partitioning. Annu Rev Nutr. 2014;34:1–30.
143. Knottnerus SJG, et al. Disorders of mitochondrial long-chain fatty acid oxidation and the carnitine shuttle. Rev Endocr Metab Disord. 2018;19(1):93–106.
144. Wanders RJA, et al. The enzymology of mitochondrial fatty acid beta-oxidation and its application to follow-up analysis of positive neonatal screening results. J Inherit Metab Dis. 2010;33(5):479–94.
145. Wanders RJ, Waterham HR. Biochemistry of mammalian peroxisomes revisited. Annu Rev Biochem. 2006;75:295–332.
146. Douglas DN, et al. Oxidative stress attenuates lipid synthesis and increases mitochondrial fatty acid oxidation in hepatoma cells infected with hepatitis C virus. J Biol Chem. 2016;291:1974–90.
147. Rosca MG, et al. Oxidation of fatty acids is the source of increased mitochondrial reactive oxygen species production in kidney cortical tubules in early diabetes. Diabetes. 2012;61 (8):2074–83.
148. Boveris A, Oshino N, Chance B. The cellular production of hydrogen peroxide. Biochem J. 1972;128(3):617–30.
149. Reddy JK, Hashimoto T. Peroxisomal beta-oxidation and peroxisome proliferator-activated receptor alpha: an adaptive metabolic system. Annu Rev Nutr. 2001;21:193–230.
150. Fransen M, et al. Role of peroxisomes in ROS/RNS-metabolism: implications for human disease. Biochim Biophys Acta. 2012;1822(9):1363–73.
151. Kasai H, et al. Formation of 8-hydroxydeoxyguanosine in liver DNA of rats following long-term exposure to a peroxisome proliferator. Cancer Res. 1989;49(10):2603–5.
152. Schrader M, Fahimi HD. Peroxisomes and oxidative stress. Biochim Biophys Acta. 2006;1763 (12):1755–66.
153. Terlecky SR, Terlecky LJ, Giordano CR. Peroxisomes, oxidative stress, and inflammation. World J Biol Chem. 2012;3(5):93–7.
154. Adibhatla RM, Hatcher JF. Phospholipase A(2), reactive oxygen species, and lipid peroxidation in CNS pathologies. BMB Rep. 2008;41(8):560–7.
155. Yusupov M, et al. Effect of head group and lipid tail oxidation in the cell membrane revealed through integrated simulations and experiments. Sci Rep. 2017;7(1):5761.
156. Biernacki M, et al. Redox system and phospholipid metabolism in the kidney of hypertensive rats after FAAH inhibitor URB597 administration. Redox Biol. 2018;15:41–50.
157. Rao MA, Hatcher JF. Phospholipase A2, reactive oxygen species, and lipid peroxidation in cerebral ischemia. Free Radic Biol Med. 2006;40(3):376–87.
158. Cortes VA, et al. Physiological and pathological implications of cholesterol. Front Biosci (Landmark Ed). 2014;19:416–28.
159. Murphy RC, Johnson KM. Cholesterol, reactive oxygen species, and the formation of biologically active mediators. J Biol Chem. 2008;283(23):15521–5.
160. Wang C, et al. Cholesterol enhances colorectal cancer progression via ROS elevation and MAPK signaling pathway activation. Cell Physiol Biochem. 2017;42(2):729–42.

161. Alcalá M, et al. Increased inflammation, oxidative stress and mitochondrial respiration in brown adipose tissue from obese mice. Sci Rep. 2017;7(1):16082.
162. Wu J, et al. Beige adipocytes are a distinct type of thermogenic fat cell in mouse and human. Cell. 2012;150(2):366–76.
163. Cousin B, et al. Occurrence of brown adipocytes in rat white adipose tissue: molecular and morphological characterization. J Cell Sci. 1992;103(Pt 4):931–42.
164. Harms M, Seale P. Brown and beige fat: development, function and therapeutic potential. Nat Med. 2013;19(10):1252–63.
165. Kim SH, Plutzky J. Brown fat and browning for the treatment of obesity and related metabolic disorders. Diabetes Metab J. 2016;40(1):12–21.
166. Chouchani ET, et al. Mitochondrial ROS regulate thermogenic energy expenditure and sulfenylation of UCP1. Nature. 2016;532(7597):112–6.
167. Jastroch M. Uncoupling protein 1 controls reactive oxygen species in brown adipose tissue. Proc Natl Acad Sci. 2017;114(30):7744.
168. Jeremic N, Chaturvedi P, Tyagi SC. Browning of white fat: novel insight into factors, mechanisms, and therapeutics. J Cell Physiol. 2017;232(1):61–8.
169. Koppen A, Kalkhoven E. Brown vs white adipocytes: the PPARgamma coregulator story. FEBS Lett. 2010;584(15):3250–9.
170. Polvani S, Tarocchi M, Galli A. PPARγ and oxidative stress: con(β) catenating NRF2 and FOXO. PPAR Res. 2012;2012:641087.
171. Corona JC, Duchen MR. PPARγ as a therapeutic target to rescue mitochondrial function in neurological disease. Free Radic Biol Med. 2016;100:153–63.
172. Gonzalez-Donquiles C, et al. The NRF2 transcription factor plays a dual role in colorectal cancer: a systematic review. PLoS One. 2017;12(5):e0177549.

Drug Repurposing: An Avenue Toward Stress Medicine in Cancer Therapy

10

Na Xie and Canhua Huang

Abstract

Cancer cells harbor a higher level of radical species owing to the metabolic reprogramming and signaling aberrations required for the continuously uncontrolled proliferation. The high basal level of reactive oxygen species (ROS) in cancer cells renders them more susceptible to reagents that further augment ROS generation or that weaken antioxidant defenses, proposing that the augment of oxidative stress exclusively in cancer cells provides a rational therapeutic strategy for anticancer drug development. Taking advantage of an attractive strategy termed drug repurposing, which describes identifying new medical indications for existing drugs with less risk, lower cost, and shorter development timeline, a portfolio of non-oncology drugs with the potential of boosting oxidative stress is repurposed to extend the arsenal of cancer therapies. In this chapter, we will present the growing data that repurposing of old drugs, especially agents modulating oxidative stress, for anticancer activity, provides an opportunity to rapidly advance therapeutic strategies into clinical trials at a relatively low cost.

N. Xie
State Key Laboratory of Biotherapy and Cancer Center, West China Hospital, and West China School of Basic Medical Sciences & Forensic Medicine, Sichuan University, and Collaborative Innovation Center for Biotherapy, Chengdu, China

C. Huang (✉)
State Key Laboratory of Biotherapy and Cancer Center, West China Hospital, and West China School of Basic Medical Sciences & Forensic Medicine, Sichuan University, and Collaborative Innovation Center for Biotherapy, Chengdu, China

School of Basic Medical Sciences, Chengdu University of Traditional Chinese Medicine, Chengdu, China
e-mail: hcanhua@scu.edu.cn

C. Huang, Y. Zhang (eds.), *Oxidative Stress*,
https://doi.org/10.1007/978-981-16-0522-2_10

Keywords

Oxidative stress · Pro-oxidant · Antioxidant · Drug repurposing · Stress medicine · Cancer

10.1 Introduction

Cancer cells undergo metabolic reprogramming switching that provides the substrates and energy required for biomass generation to survive and unrestrained proliferate [1]. Inevitably, hyper-metabolism induces abundant ROS production from NADPH oxidases, mitochondria, endoplasmic reticulum (ER), and peroxisome [2, 3]. It is noted that oxidative stress causes biomolecular damage (DNA, lipids, and proteins), genomic instability, and chronic inflammation in the tumor niche, leading to malignant transformation [4–6]. Moreover, ROS functions as the second messengers that stimulate growth factor-dependent pathways and oncogene activity to support uncontrolled cell survival and proliferation, contributing to tumor initiation, development, and angiogenesis [7–9]. To maintain the redox balance with a higher basal ROS level, cancer cells build a powerful antioxidant response system. Thus, cancer cells are vulnerable to extra ROS insults aroused by agents that boost ROS generation or blunt antioxidant capacity above a survival threshold level [10, 11]. The fine-tune of redox homeostasis in cancer cells would provide a biological hint for the selective killing of malignant cells [10]. Therefore, manipulating the oxidative damage by delivering ROS-generating agents to tumor cells directly, suppressing the antioxidant systems, or combining use of pro-oxidants and antioxidants could disturb the redox homeostasis in cancer cells, providing a promising strategy to efficiently and selectively annihilate cancer cells without causing toxicity to normal cells.

To date, oxidative stress as key player in the molecular mechanism by which old drugs exert their anticancer effect has been recently investigated using drug repurposing (also called drug reprofiling, re-tasking, or repositioning). Drug repurposing, a strategy to identify new therapeutic purposes for approved or investigational drugs that are outside the scope of the original medical indication, has become an attractive area of cancer research for a variety of reasons than novel drug discovery, including more affordable and achievable, shorter developmental time, lower cost, and less safety risk [12, 13]. Various computational approaches and biological experimental approaches have been developed and increasingly being used along with or synergistically in drug repositioning [14]. Data-driven approaches mostly involve three general types of data, including omics data (transcriptomic, proteomic, or metabolomic data), chemical structures, medical and pharmacological databases, leading to the formulation of repurposing hypotheses [15]. Along with the accumulation of various databases, more systemic biological technologies, such as a molecular barcoding approach PRISM, have been established to map the landscape of the drug effects on cell viability. Using the PRISM method, a public resource containing the growth inhibitory activity of 4518 drugs across 578 human cancer cell lines has been created. Of note, an unexpectedly large number of non-oncology drugs selectively inhibit subsets of cancer cell lines, providing more effective treatment options for patients with cancer [16].

A broad spectrum of old drugs are applied for cancer treatment owing to their ability to increase oxidative stress and consequently contribute to cell death via disturbing the redox balance [10]. With the ability to exacerbate the oxidative stress-induced cell death, these reagents may open up new avenues to develop novel therapeutic strategies for cancer (Fig. 10.1).

10.2 Redox Homeostasis

The cellular redox homeostasis is maintained by the balance between the production of oxidants and their detoxification by antioxidant systems. Oxidants, especially ROS, could be produced from either metabolic reaction as a byproduct or a range of enzymes as a primary product, leading to oxidative stress. To counteract the oxidative damage, organisms have developed an antioxidant defense system consisting of both enzymatic and nonenzymatic scavengers [17, 18].

10.2.1 Production of Oxidants

In a physiological context, a majority of the cellular ROS are produced by mitochondria [19, 20]. Electrons from nicotinamide adenine dinucleotide (NADH) and flavin adenine dinucleotide ($FADH_2$) are transported through the mitochondrial electron transport chain (ETC) to generate a gradient of protons across the inner mitochondrial membrane, enabling the ATP synthesis. The complex I and complex III of ETC are able to produce the superoxide anion radical (O_2^-) and release it to both the matrix and the intermembrane space [19–21]. Additionally, NADH or FADH2 serves as the electron carriers for the mitochondrial membrane proteins, such as GPDM (the mitochondrial membrane forms of glycerol 3-phosphate dehydrogenase) and FQR (electron transfer flavoprotein-ubiquinone oxidoreductase), and the metabolic enzymes in the matrix, such as OGDH (2-oxoglutarate dehydrogenase) and PDH (pyruvate dehydrogenase), all of which contribute to ROS production [20]. Another essential source of ROS is the NADPH oxidases (NOXs) which generate ROS as a primary product rather than a byproduct, especially in response to physiological stimuli [19]. The enzymatically active complex of NADPH oxidases is formed by diverse cytosolic factors, including $p40^{phox}$ (NCF4), $p67^{phox}$ (NCF2), $p47^{phox}$ (NCF1), and small Rac GTPases, and the membrane-bound factors NOX2 and $p22^{phox}$. The catalytic subunit of NADPH oxidases has five isoforms, including NOX1–5, and two related enzymes, dual oxidase 1/2 (DUOX1/2) [22]. The presence of the FAD-binding region and transmembrane haems enables NOXs to act as an electron-transportation chain that transfers electrons from cytosolic NADPH to extracellular O_2, resulting in the generation of O_2^- [21, 23].

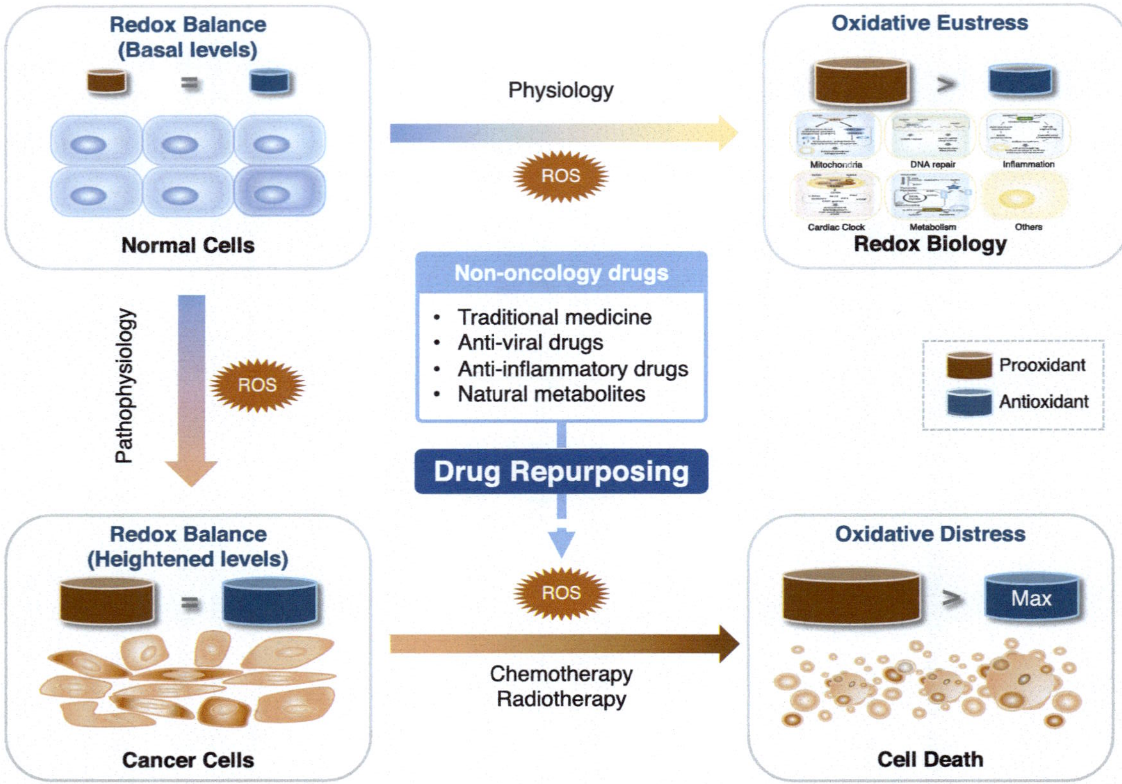

Fig. 10.1 Repurposing the anticancer potential of non-oncology drugs targeting redox homeostasis. ROS are conventionally classified as deleterious with oxidative damage. However, the physiologically controlled ROS, termed "oxidative eustress," is needed to maintain normal cellular processes and sustain the growth and proliferation of cancer cells. Cancer cells adapt to their inherently increased oxidative stress by developing a powerful antioxidant capacity to counteract with the toxic effects of ROS. Given the advances in drug repurposing, a variety of non-oncology drugs are repurposed for cancer therapy applications owing to their potential of boosting the excessive load of ROS, termed "oxidative distress"

Peroxisomes are multifunctional organelles with a pivotal role in the generation and decomposition of ROS. These organelles enclose large amounts of enzymes that produce oxidative species as part of their normal catalytic activity, including xanthine oxidase that generates O_2^- and NOS2 that releases nitric oxide (NO). Meanwhile, peroxisomes contain a broad spectrum of antioxidant enzymes that can degrade $ONOO^-$ (e.g., Prx5), O_2^- (e.g., SOD1), and lipid peroxides (e.g., Prx5 and GSTK1). The peroxisomal catabolism of very long-chain and branched-chain fatty acids, D-amino acids, and polyamines also results in the production of ROS and reactive nitrogen species (RNS) [24–26]. Persistent endoplasmic reticulum stress enhances ROS production through the oxidoreductase Ero1 and NOXs. The oxidation of Ero1 by O_2 initiates disulfide relays resulting in the formation of disulfide bonds during oxidative protein folding with H_2O_2 generated as a byproduct. More recently, increasing evidence suggests that NOXs, especially NOX4, are important ROS sources during the unfolded protein response (UPR). The luminal H_2O_2 arising from Ero1 and NOXs are scavenged by ER peroxidases, such as Prx4, GPx7, and GPx8, which transfer disulfide bond to PDI, promoting oxidative protein folding [27–30].

10.2.2 Antioxidant System

Organisms have developed a complicated antioxidant defense system to fine-tune the redox homeostasis [17]. Nonenzymatic components of antioxidant defense comprise ascorbic acid, alpha-tocopherol (vitamin E), beta carotenoids, retinol, coenzyme Q, thiol groups containing glutathione (GSH) and lipoic acid, as well as transition-metal ions, such as Mn^{2+}, Cu^{2+}, Fe^{2+}, and Zn^{2+} [31, 32]. Among these nonenzymatic antioxidants, glutathione, ascorbic acid, vitamin E, uric acid, and ubiquinol could quench the reactivity of free radicals by donating electrons to them while becoming much lesser reactive free radicals themselves. GSH is a cysteine-containing peptide with the thiol group that can be reversibly oxidized and reduced, providing the antioxidant properties. Given its high concentration and the crucial role in maintaining cellular redox homeostasis, GSH is one of the most crucial nonenzymatic antioxidants [33–36].

In addition to nonenzymatic scavengers, cells contain a portfolio of antioxidant enzymes with the capability of inhibiting oxidation reactions and attenuating ROS generation, including catalases (CATs), superoxide dismutases (SODs), glutathione peroxidases (GPxs), glutathione reductases (GRs), peroxiredoxins (Prxs), glutaredoxins (Grxs), and thioredoxins (Trxs) [18, 31, 32]. Cytosolic copper/zinc SOD (Cu/ZnSOD) and mitochondrially located manganese SOD (MnSOD) dismutate superoxide radicals produced from mitochondria and NOXs into H_2O_2, which are catalytically broken down to harmless H_2O by catalase, GPxs and Prxs [37, 38]. The Trx system, composing of thioredoxin reductase (TrxR), Trx, and NADPH, is a crucial disulfide reductase system that can provide reducing equivalents to a broad spectrum of antioxidant enzymes [39]. GPxs are selenocysteine-containing antioxidant enzymes that effectively reduce lipid peroxides and H_2O_2 and to lipid alcohols and water, respectively [40]. Meanwhile, metal ion binding proteins, including transferrin, lactoferrin, and caeruloplasmin, prevent the free radical formation via scavenging the redox-active metals, such as iron and copper [41–43]. Furthermore, both enzymatic thioredoxin and glutaredoxin systems and nonenzymatic GSH system coordinate with each other to manipulate redox homeostasis and cell fate using NADPH as the ultimate donor of reductive power (Fig. 10.2) [21, 44].

10.2.3 Antioxidant Pathway

Nuclear factor erythroid 2-related factor 2 (Nrf2) is a redox-sensitive transcription factor that governs the redox signaling events by promoting antioxidant enzymes and phase II detoxifying enzymes [45, 46]. In response to the oxidative stress triggered by diverse stimuli, Nrf2 is phosphorylated and dissociated from the cytoplasmic inhibitor, Keap1. Stabilized Nrf2 then translocates to the nucleus and transactivates the antioxidant response element (ARE)-bearing genes that encode antioxidant-detoxifying proteins, including SOD, GPx, γ-glutamylcysteine ligase (GCL), heme oxygenase-1 (HO-1), and NAD(P)H:quinone oxidoreductase 1 (NQO1) [47]. Through modulating the transcription of various categories of target genes, the Nrf2/Keap1 signaling pathway modulates the antioxidant defense system against oxidative stress [46, 48, 49].

10.3 Drug Repurposing Based on Modulation of Redox Homeostasis

10.3.1 Redirecting Oxidizing Agents for Cancer Treatment

10.3.1.1 Rediscovery of Traditional Chinese Medicines (TCM) for Cancer Therapy

Arsenic trioxide (As_2O_3) is one of the arsenic compounds found in nature that has been used medicinally for over 2400 years [50]. The arsenic-containing traditional Chinese medicine has been successfully used to treat psoriasis, syphilis, and rheumatosis [51]. Meanwhile, arsenic is recorded as both a therapeutic agent and poison in ancient Greek and Roman times. In the 1970s, As_2O_3 was introduced into the treatment of acute promyelocytic leukemia (APL) and showed striking effectiveness and was relatively safe [52]. The present studies indicate that As_2O_3 may also be a good candidate for treating various cancers, including breast cancer, HPV-associated tumors, esophageal carcinoma, and colorectal cancer [51, 53–55]. As_2O_3 triggers cytotoxicity to cells and tissues via inducing oxidative stress and intracellular calcium mobilization, which can be ameliorated by resveratrol, a well-known polyphenolic compound [56, 57]. However, the toxicity of As_2O_3 on cells highlights the potential for the repurposing of this compound in cancer therapy, as evidenced by the observation that As_2O_3 significantly triggers oxidative stress, DNA damage, and apoptosis in HL-60 cells in a dose-dependent manner. Moreover,

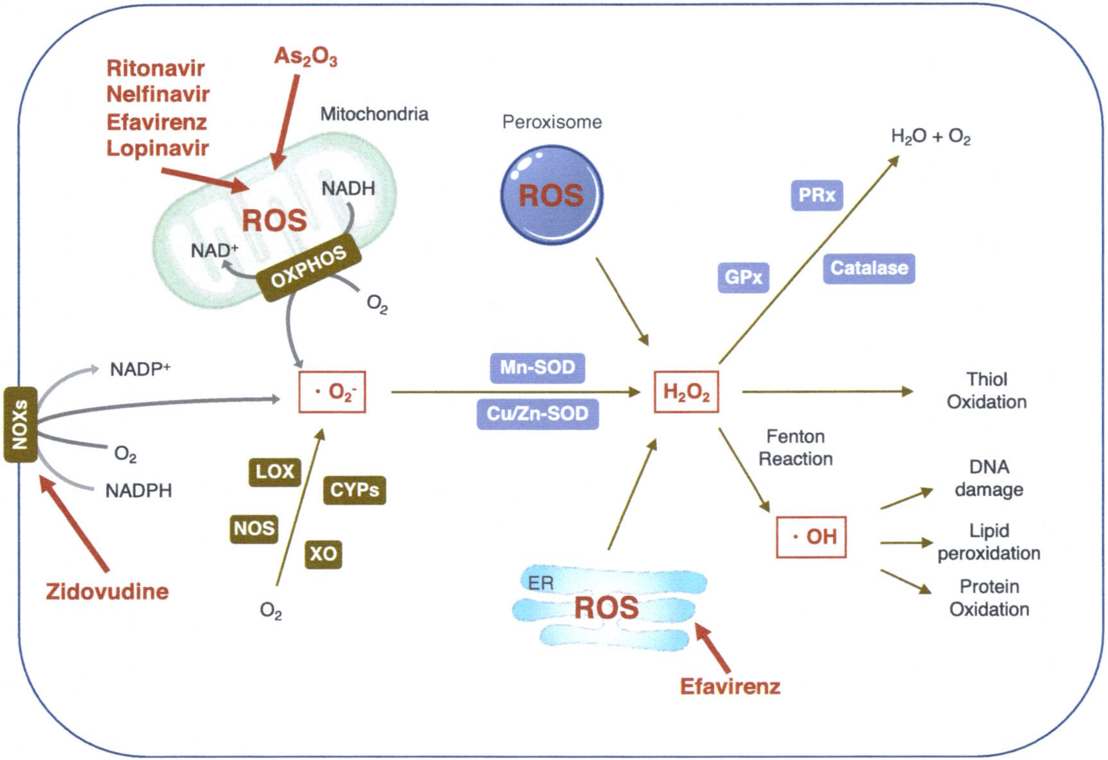

Fig. 10.2 Repurposing of drugs boosting ROS production for cancer therapies. The endogenous ROS can be generated from various cellular organelles such as mitochondria, peroxisomes, and endoplasmic reticulum (ER) with a high level of oxygen consumption. Besides, enzyme membrane-bound NADPH oxidases (NOXs) contribute to the production of cellular ROS. A variety of antiviral drugs, e.g., zidovudine, ritonavir, nelfinavir, efavirenz, and traditional drugs, e.g., As_2O_3, could exert their anticancer via stimulating the production of ROS. Abbreviation: *CYPs* Cytochromes P450, *LOX* lipoxygenases, *NOXs* NADPH oxidases, *NOS* nitric oxide synthases, *OXPHOS* oxidative phosphorylation, *XO* xanthine oxidase

low doses of As_2O_3 can arrest solid tumor growth via oxidative and nitrosative bursts-mediated depletion of Tregs, suggesting that As_2O_3 may be used to strengthen the antitumor effect of the adoptive immunotherapy strategies in human cancer [58]. Consistently, As_2O_3 is able to increase the thermosensitivity of tumor cells via inducing the expression of adhesion molecules in the vasculature by oxidative stress [59]. Mechanically, As_2O_3-enabled apoptosis in leukemic cells is dependent on the activity of enzymes, including GPxs, that regulate cellular redox homeostasis [60, 61].

10.3.1.2 Repositioning of Antiviral Drugs to Cancer Treatment

The antiviral drugs used for the treatment of viral infections, including hepatitis C virus (HCV) and human immunodeficiency virus (HIV), have a variety of side effects, such as disturbed lipid and protein metabolism, oxidative stress, and mitochondrial damage, all of which contribute to the cytotoxicity to the host cells [62]. However, all these drug-induced toxicities contribute to the repurposing of these drugs in the antitumor effect. Antiretroviral protease inhibitors including ritonavir, lopinavir, nelfinavir for HIV treatment, non-nucleoside reverse transcriptase inhibitor efavirenz, and purine nucleoside ribavirin for HCV all exhibit antitumor activity coupled with increased production of ROS limited to cancer cells (Fig. 10.3) [63–68]. Through increasing mitochondrial ROS by decreasing MnSOD protein levels or inducing the export of GSH, ritonavir, nelfinavir, and lopinavir induce oxidative stress, DNA damage, and mitochondrial dysfunction and arrested the cell cycle at G1 phase and apoptosis, resulting in inhibited growth of cancer cell lines [63, 65, 66, 69–71].

Another example is that zidovudine (AZT), an analog of the thymidine deoxynucleoside used to treat HIV-infected individuals, elevates oxidative stress via increasing the expression of NOX2 and $p47^{phox}$ and reducing the antioxidant power of plasma and cardiac homogenates. The increased oxidative stress, therefore, contributes to the side effects of zidovudine, muscle contractile dysfunction and hepatic steatosis [72–76]. However, exposure of cancer cells to zidovudine induces significant sensitization to cisplatin (CIS)-aroused toxicity via disrupting thiol metabolism and oxidative stress [77]. Besides, the widely prescribed antiretroviral drug, efavirenz (EFV), directly induces oxidative stress via decreasing the mitochondrial membrane potential, increasing the mitochondrial superoxide production, as well as reducing the cellular glutathione content [78, 79]. Oxidative stress and mitochondrial membrane depolarization appear in the pancreatic cancer cells immediately after efavirenz treatment, but not in the fibroblasts, indicating the specifically anticancer activity of efavirenz. Meanwhile, efavirenz-induced oxidative stress enhances the radiosensitization via causing DNA damage [80]. Efavirenz could also cause cytotoxicity owing to the cellular insult involving ER stress and mitochondria/ER interplay [79, 81, 82]. Ribavirin (1-β-d-ribofuranosyl-1,2,4-triazole-3-carboxamide) is a widely used broad-spectrum antiviral drug, but its clinical application is limited because of the ROS-induced toxicity. Given its genotoxic and cytotoxic effects, ribavirin can be explored as a therapy against HPV-18-induced cervical cancer [83–85].

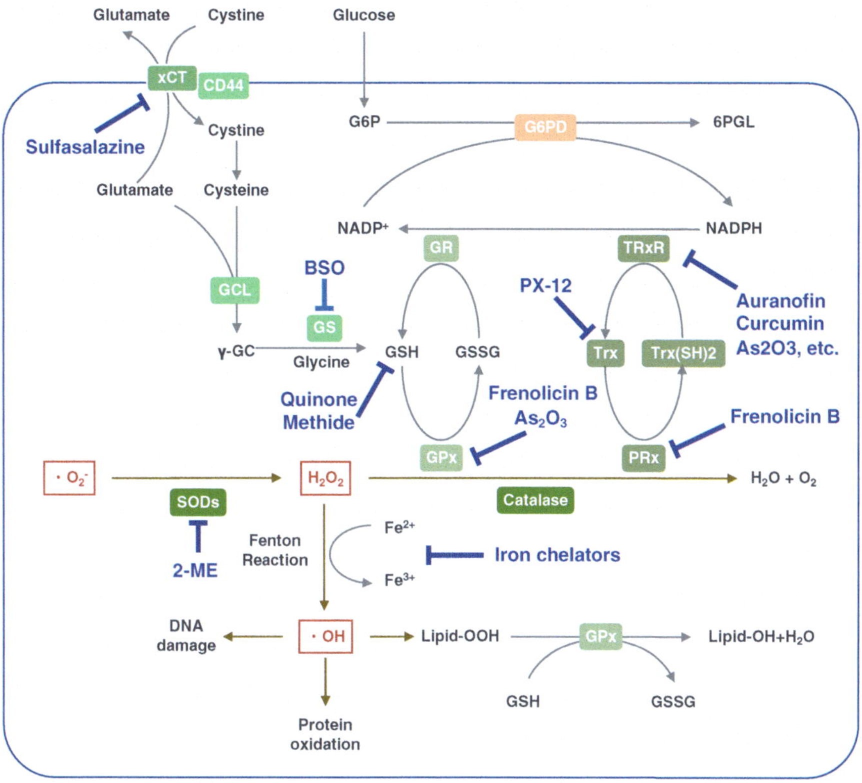

Fig. 10.3 Repurposing of reagents targeting antioxidant system for cancer treatment. To fine-tune the redox homeostasis, organisms have evolved an integrated antioxidant system, including enzymatic and nonenzymatic antioxidants. The antioxidant enzymes, including SODs, glutathione peroxidase (GPx), peroxiredoxin (Prx), thioredoxin peroxidases, and catalase, convert dangerous oxidative products to hydrogen peroxide (H_2O_2) and then to harmless water. As one of the most critical cellular nonenzymatic antioxidants, glutathione (GSH) is synthesized in cells from the precursor amino acid cysteine, glutamate, and glycine catalyzed by the sequential action of γ-glutamylcysteine ligase (GCL) and glutathione synthetase (GS). Moreover, GSH synthesis can be limited by the activity of xCT antiporter (system xc^-) that exports glutamate for cystine. Abbreviation: *2-ME* 2-methoxyestradiol, *BSO* buthionine sulfoximine, *G6PD* glucose-6-phosphate dehydrogenase (G6PD), *GCL* γ-glutamylcysteine ligase, *GPxs* glutathione peroxidases, *GRs* glutathione reductases, *Prxs* peroxiredoxins, *Grxs* glutaredoxins, *GS* glutathione synthetase, *GSH* glutathione, *QM* quinone methide, *SODs* superoxide dismutases, *Trxs* thioredoxins, *TrxR* thioredoxin reductase

10.3.2 Repurposing Molecules Targeting Antioxidant System

10.3.2.1 Re-Tasking of SOD Inhibitors for Cancer Treatment

The intrinsic powerful antioxidant capacity in cancer cells is associated with the upregulated expression of SOD and catalase proteins, rendering them to tolerate the increased ROS levels [86]. SODs are crucial antioxidant enzymes that eradicate superoxide radicals (O_2^-) to prevent cells from free radical-induced damage. The prolonged active O_2^- production and low SODs activity may render the malignant cells highly dependent on SODs for survival and sensitive to inhibition of SODs, indicating that targeting SODs may be a promising approach to the selective killing of cancer cells [86, 87].

2-Methoxyestradiol (2-ME(2)), an endogenous metabolite of 17beta-estradiol (E (2)), is reported as an inhibitor of the SODs [88]. Inhibition of SOD activity by 2-ME is an effective treatment modality capable of potentiating the antitumor, radiosensitizing, and thermosensitizing effects [89, 90]. 2-ME-mediated ROS production and mitochondrial membrane potential ($\Delta\Psi m$) loss induce the mitochondrial apoptosis pathway via decreasing the Bcl-2/Bax ratio, increasing the expression of apoptosis-inducing factor (AIF) and the activation of caspase-9 and caspase-3 [91, 92]. Meanwhile, 2-ME targeting of SODs impairs the DNA double-strand break (DSB) repair induced by oxidative stress in response to ionizing radiation (IR) [90, 93, 94]. Inhibition of Akt by specific inhibitor abolishes 2-ME-induced radiosensitization [90]. Furthermore, 2-ME induces autophagic cell death in the transformed cell line HEK293 and the cancer cell lines U87 and HeLa. The oxidative stress-induced autophagic cell death independent of apoptosis in transformed cells but not in nontransformed mouse astrocytes further expands the use of 2-ME in cancer treatment [95]. Similarly, malignant cells, including primary leukemia and ovarian cancer cells, are more sensitive to 2-ME than their normal counterparts, as indicated by the remarkable accumulation of ROS and subsequent apoptosis. The administration of ROS scavengers abolishes the ROS accumulation and apoptosis induction by 2-ME [86]. Additionally, SOD activity may be important in protecting cells exposed to heat. The additional inhibition of SOD by 2-ME could improve the local tumor control rate of ROS-generating hyperthermia [96, 97].

10.3.2.2 Repurposing the Trx/TrxR Inhibitors as Anticancer Drugs

The thioredoxin (Trx) system, consisting of Trx reductase (TrxR), Trx, and NADPH, is essential for maintaining a reduced intracellular environment, providing protection against excessive levels of ROS [39, 98, 99]. The increased expression of Trx1 and TrxR1 in cancer cells promotes cell proliferation, invasion, and migration [100, 101]. Overexpression of Trx1 or addition of exogenous Trx1 enhances cancer cell invasion, while expressing a redox inactive mutant Trx1, depleting Trx1 with the antisense mRNA, or inhibiting the redox function of Trx1 by a specific monoclonal antibody decreases cell invasion [101]. The cisplatin-resistant cells with a reduced cellular Trx1 due to increased secretion exhibit higher ROS production and increased dependency on oxidative metabolism (OXMET) [102]. High expression of Trx/TrxR also contributes to chemotherapy resistance, demonstrating Trx/TrxR as therapeutic targets for improving therapeutic potency [103]. An increasing number of natural molecules have been identified as TrxR inhibitors, including curcumin, shikonin, adenanthin, As_2O_3, parthenolide, myricetin, securinine, gambogic acid, mitomycin C, and alantolactone [103]. Moreover, there are a variety of synthetic molecules, including ebselen, ethaselen, PX-12, dimesna, irofulven, and auranofin, which remarkably inhibit TrxR/Trx with anticancer activity.

Drug-mediated, irreversible inhibition of cytosolic TrxR can yield selective anticancer efficacy without overt systemic toxicity [104]. Auranofin (Ridaura) is one of the potent inhibitors of TrxR1 and has been approved for the clinical application in rheumatoid arthritis treatment [105]. Intriguingly, it has been investigated for potential therapeutic applications for a variety of other diseases, including neurodegenerative disorders, cancer, parasitic infections, bacterial infections, and HIV/AIDS [106]. Auranofin is currently being repurposed for potential anticancer activity including small lymphocytic lymphoma (SLL), chronic lymphocytic leukemia (CLL), chronic myelogenous leukemia (CML), prolymphocytic lymphoma (PLL), gastrointestinal stromal tumor, and ovarian cancer [106]. By inhibiting TrxR activity, auranofin induces an intracellular ROS accumulation and a lethal ER stress response in CLL cells and gastric cancer (GC) cells. The combining use of selenocysteine (SeC), a naturally available selenoamino acid, with auranofin can be a highly efficient approach to achieve anticancer activity by synergetic inhibition of TrxR1 [107]. In addition, auranofin displayed synergistic lethality with HO-1 and γ-glutamylcysteine ligase (GCL) inhibitors against CLL cells [108, 109]. Moreover, inhibition of TrxR1 by auranofin overcomes the conventional chemotherapeutic drug resistance, including cisplatin and adriamycin (ADM), in human cancer cells [110]. The chemosensitive effect of auranofin is achieved by eliminating the stem-like cancer side population (SP) responsible for drug resistance, via simultaneous increasing ROS and depleting ATP by inhibition of glycolysis [111].

10.3.2.3 Repositioning the GSH-Depleting Pro-Oxidant to Cancer Treatment

GSH is the most abundant nonenzymatic oxidant scavenger that is produced from its constituent amino acids by the sequential step of γ-glutamylcysteine ligase (GCL), a heterodimer of catalytic (GCLC) and modifier (GCLM) subunits, and glutathione synthetase (GS) [112, 113]. Through scavenging free radicals, GSH supports the cellular defenses against oxidative stress and toxic compounds [114]. Sustaining GSH levels via GSH biosynthesis is crucial for the growth and survival of cancer cells via maintaining intracellular redox balance, providing a promising target in cancer therapy. Thus, a variety of cancer therapies based on GSH depletion have been approached [115–117].

xCT

The xCT antiporter (system xc^-) exhibits an antioxidant ability by exchanging glutamate for cystine, a central precursor for GSH synthesis. Inhibition of the xc^- cystine/glutamate antiporter can reduce cystine uptake, leading to subsequent depletion of intracellular GSH levels and the growth arrest of cancer cells [118]. It has been demonstrated that the expression of xCT is higher in both colorectal cancer (CRC) cell lines and human colorectal carcinoma tissues than in their normal counterparts [119]. Strikingly, xCT is reported to be crucial for oncogenic KRAS-mediated transformation in vitro and in vivo by alleviating oxidative stress, as knockdown of xCT significantly suppresses the growth of tumor xenografts developed from KRAS-transformed cells [120].

Targeting of xCT can be achieved pharmacologically using drugs such as sulfasalazine. Sulfasalazine that is first synthesized in the 1940s as a conjugate of 5-aminosalicylic acid and sulfapyridine linked by an azo bridge is commonly used to treat rheumatoid arthritis and inflammatory bowel disease. The transcription factor NF-κB, a central mediator of inflammation, is also a target of sulfasalazine-mediated immunosuppression [121]. The identification of sulfasalazine as a potent inhibitor of the xc^- cystine transporter raises the potential for its application in the cancer treatment, including lymphoma, prostate cancer, pancreatic cancer, head and neck squamous cell carcinoma, and mammary tumor [115, 122–126]. Suppression of xc^--mediated cystine uptake by sulfasalazine causes an impaired GSH synthesis and enhanced mitochondrial metabolism, resulting in ROS generation and culminating in oxidative damage [122, 126, 127]. As a result, sulfasalazine effectively improves the anticancer effects of pharmacological drugs, including vitamin C (VC), which triggers the accumulation of ROS [119, 127]. Together, these findings strongly indicate that sulfasalazine, a relatively nontoxic anti-inflammatory drug targeting cystine transporters, in combination with pharmacological drugs, may provide a superior effective therapy in cancers.

γ-GCS

γ-Glutamylcysteine synthetase (γ-GCS) is a rate-limiting enzyme metabolizing the first step of GSH biosynthesis: the adenosine triphosphate (ATP)-dependent ligation of cysteine and glutamate [128]. Buthionine sulphoximine (BSO) is an inhibitor of γ-GCS that consists of a catalytic (γ-GCS_H) and a modifier (γ-GCS_L) subunit [129]. Selective suppression of GSH synthesis by BSO lowers the GSH levels, resulting an enhanced cytotoxicity of chemo- and radiotherapy [130–134]. BSO treatment of MCF-7/ADRR cells results in 80–90% depletion in total GSH concentrations and at least twofold more ·OH formation induced by doxorubicin [130]. The reduction of GSH by BSO restores the sensitivity of resistant tumors to melphalan (L-PAM) in vitro and in vivo [132, 135]. BSO is found to be toxic to some AML blast populations, presented as the high levels of ROS, the loss of mitochondrial membrane potential, and increased intracellular calcium in GSH-depleted cells [133]. Combination treatment with ATO and BSO enhances apoptosis through increasing ROS production in hepatocellular carcinoma cells compared to ATO treated alone [136]. Moreover, cell lines with genomic amplification of MYCN are more susceptible to BSO than MYCN-nonamplified cell lines [137].

Alkylation of GSH

Quinone methide (QM) can elevate oxidative stress via rapidly alkylating GSH. A wide range of drugs executes their anticancer activity through generating QM intermediates. Nitric oxide-donating aspirin (NO-ASA) is one of the nitric-oxide-donating nonsteroidal anti-inflammatory drugs (NO-NSAIDs), comprising of a traditional aspirin and a NO-donating moiety covalently bound by a spacer molecule. NO-ASA inhibits the cell growth of cancers, including colon, pancreas, and prostate, which is several hundredfold more potent than its corresponding parent aspirin [138–142]. The potent anticancer activity of NO-ASA is attributed to QM, which selectively reacts with cellular GSH and, in turn, induces cell death, but not nitric oxide [143]. In vivo data reveals that NO-ASA reduces the number of aberrant crypt foci that are the earliest premalignant lesion in the colon. There are a number of tumors in Min mice that are highly vulnerable to spontaneous intestinal adenoma formation due to a nonsense mutation of the murine Apc gene, raising the possibility that NO-ASA has both chemopreventive and chemotherapeutic effects [144, 145]. Benzoyloxy dibenzyl carbonate (B2C) is another GSH-depleting pro-oxidant, which is designed to release QMs. The rapid depletion of GSH by B2C through esterase-mediated hydrolysis causes an overwhelming level of oxidative stress, leading to increased apoptotic cell death and significantly inhibited tumor growth [146].

10.3.2.4 Reprofiling Inhibitors of Prxs and Grxs for Cancer Therapy

Peroxiredoxins (Prxs) and glutaredoxins (Grxs) are antioxidant enzymes, known to catalyze the reduction of peroxide and the removal of protein-bound GSH, respectively [147]. Prxs can convert H_2O_2, ONOO-, and other peroxide substrates to H_2O via the oxidation of its conserved thiolate to a -SOH intermediate [148]. It has been demonstrated that the expression of Prxs is elevated in various cancers and is associated with the resistance of cancer cells to chemo- and radiotherapies [149–152]. Grx system is formed by NADPH, glutathione reductase (GR), GSH, and glutaredoxins (Grxs) [153]. Grxs mainly catalyze the removal of GSH from protein to maintain the highly reduced state of thiols in protein without exerting direct antioxidant properties [147].

Frenolicin B, a metabolite of *Streptomyces roseofulvus* that belongs to the family of naturally occurring naphthopyranquinones, is an effective anticoccidial and antimalarial agent [154]. Recently, frenolicin B is identified as a selective inhibitor of Grx3 and Prx1 via covalent modification of active-site cysteines. Frenolicin B-induced repression of Grx3 and Prx1 causes a reduced cellular GSH level, an increased ROS concentration, and concomitant arrested growth of cancer. ROS-mediated cytotoxicity is positively correlated with the repression of tumor growth by frenolicin B, establishing frenolicin B as a potent pro-oxidant via inhibiting Prx1/Grx3 [155]. Additionally, adenanthin, a diterpenoid isolated from the leaves of *R. adenantha*, induces the differentiation of APL cells via directly restricting the antioxidant activities of Prx I and Prx II [156]. Adenanthin can also kill the malignant liver cells in vitro and xenografts through augmenting ROS contents, proposing that the therapeutic significance of adenanthin targeting Prx I/II warrants to be further explored in solid tumor patients [157].

10.3.3 Repurposing Drugs Targeting Redox Signaling for Cancer Therapy

10.3.3.1 Extending Nrf2-Targeting Drugs to Cancer-Related Conditions

The transcription factor Nrf2 is a principal regulator of the cellular antioxidant defense [158] (Fig. 10.4). It has been demonstrated that Nrf2 exerts both pro-tumorigenic and anti-tumorigenic actions [159]. Notably, prolonged Nrf2 activation exhibits oncogenic properties in response to the activated PI3K pathway, a bona fide driver of tumorigenesis, but not under homeostatic conditions [160]. Under physiological conditions, the cytoprotective functions of Nrf2 suppress tumor initiation and metastasis via eradicating ROS and detoxifying DNA-damaging agents. Accumulating evidence has established that constitutive hyperactivity of Nrf2 signaling during tumorigenesis creates an environment that protects malignant cells against oxidative stress, chemotherapeutic agents, and radiotherapy, favoring their survival [161]. Constitutive activation of Nrf2 drives metabolic reprogramming, resulting in accelerated recurrence of dormant breast cancer [162]. Therefore, Nrf2 provides a compelling putative therapeutic target for cancer treatment, either using Nrf2 activators for cancer prevention in the early stages of tumorigenesis or Nrf2 inhibitors for anticancer therapy in the later stages [163].

The drug edaravone (EDA) is approved worldwide to treat ALS or acute cerebral infarction (stroke). This synthetic pyrazolone derivative regulates transcription factors, such as repressing NF-κB and activating Nrf2, to modulate oxidative stress. The anti-inflammatory and antioxidant activity of edaravone can be best implemented to prevent non-tumor cells from damages caused by radiation and chemotherapeutic drugs [164–166]. Edaravone increases the expression of Nrf2 and HO-1, which significantly improves the neurologic symptoms, including brain injury, oral mucositis, and neurobehavioral and neurochemical deficits, in cancer cells after chemotherapy and radiotherapy [165, 167]. Moreover, edaravone is found to mitigate autoimmune thyroiditis, a common side effect observed in patients treated with anti-PD-1 mAbs [168–170]. Altogether, these results suggest that edaravone's clinical application, a prescription drug for brain-related human diseases, can be extended to cancer-related conditions [164].

All-trans-retinoic acid (ATRA, a natural derivative of vitamin A) reduces the binding of Nrf2 to the ARE enhancer as a consequence of forming a complex with RARα, which increases the oxidative stress [161, 171]. ATRA is able to induce in vivo differentiation of APL blasts, exhibiting complete remission in almost all patients. Maintenance treatment with low-dose chemotherapy and ATRA can further reduce the incidence of relapse [172]. As a prospective remedial agent, ATRA reverses arsenic-induced free radical generation, activates the antioxidant defense, and subsequently inhibits p53-dependent apoptosis via suppressing the MAPK signaling [173]. Moreover, ATRA can counteract DOX-induced cardiomyopathy not only due to the antioxidative and anti-inflammatory properties but also via suppressing the activation of the mitochondrial apoptotic pathway [174]. The chemoprotective effect of ATRA on oxidative stress and tumorigenesis is also

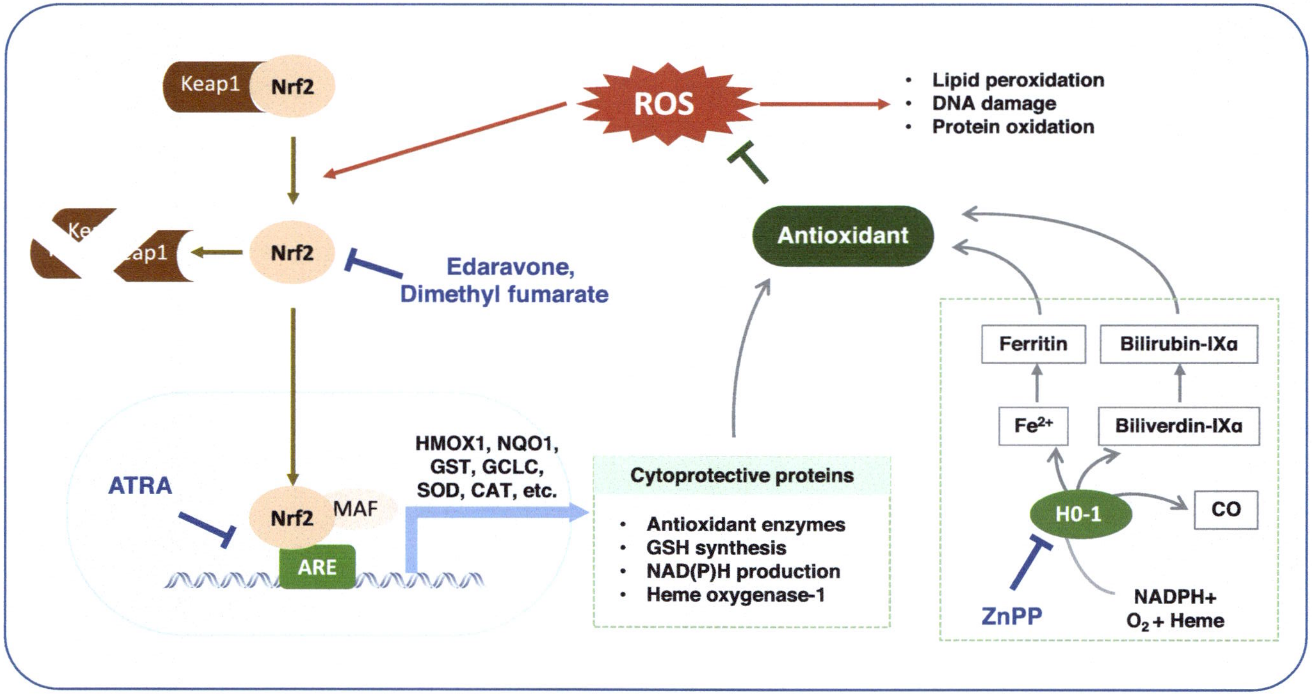

Fig. 10.4 Repurposing of reagents targeting redox signaling for the antitumor effect. Nrf2-antioxidant response element signaling pathway is a major mechanism responsible for regulating of cellular redox balance through controlling the expression of detoxification enzymes and stress proteins. One of the genes regulated through Nrf2 is HO-1, a rate-limiting enzyme that degrades heme to carbon monoxide (CO), free iron (which leads to the induction of ferritin), and biliverdin (which is rapidly converted to bilirubin); the last two can function as antioxidants. Abbreviation: *CAT* catalase, *GCLC* γ-glutamylcysteine ligase, *HMOX1/HO-1* heme oxygenase-1, *KEAP1* Kelch-like ECH-associated protein 1, *NQO1* NAD(P)H-quinone oxidoreductase 1, *GST* glutathione-S-transferase, *Nrf2* nuclear factor erythroid 2-related factor 2, *SOD* superoxide dismutase

proved by the study showing that ATRA attenuates the oxidative stress induced by benzo(a)pyrene in a mouse model of lung cancer [175].

Dimethyl fumarate (DMF) is a first-line treatment for relapsing-remitting multiple sclerosis (RRMS) [176]. Recent studies have demonstrated the potential anticancer properties of DMF in melanoma and glioblastoma cells [161, 177]. In contrast to the activation of both Nrf2 and NF-κB pathways in noncancer models, such as in multiple sclerosis, a high dose of DMF represses both Nrf2 and NF-κB signaling in cancer cells, resulting in exacerbated oxidative stress-induced cell cycle arrest and apoptosis [178, 179]. Moreover, in contrast to malignant cells, the same concentration of DMF has no significant cytotoxicity on non-tumorigenic cells, further opening up further important prospects for the therapeutic applications of DMF [180].

10.3.3.2 Reusing HO-Targeting Reagents as Anticancer Drugs

Heme oxygenase-1 (HO-1) is a rate-limiting enzyme catalyzing the degradation of heme to iron, carbon monoxide (CO), and biliverdin in mammalian cells [181] (Fig. 10.4). This stress-inducible enzyme is extensively expressed in a broad spectrum of cancers, providing the selective ability for cancer cells to overcome the amplified oxidative stress during tumorigenesis and treatment. Consequently, several anticancer therapies based on targeting HO-1, rendering tumor cells more sensitive to chemotherapy and radiation therapy [182]. It has been shown that the combination of a fasting-mimicking diet (FMD) and chemotherapy leads to a significant delay in the progression of melanoma and breast cancer, partially induced by the downregulation of HO-1 [183]. The water-soluble form of the HO-1 inhibitor, zinc protoporphyrin (ZnPP), is a metabolite that is formed in trace amounts during heme biosynthesis [184]. It has seemed promising in different in vivo models, in which it has induced growth arrest in tumor cells with few, if any, side effects [185]. Water-soluble polymer conjugates of ZnPP (PEG-ZnPP and SMA-ZnPP) exerts antitumor effect both by HO-1 inhibition and singlet oxygen-generated photodynamic activity [186]. Pegylated hemin-induced HO-1 significantly increases tumor blood flow and albumin accumulation in tumors, suggesting that HO-1 and/or CO is the essential regulator of the enhanced permeability and retention (EPR) effect. Though enhancing the EPR effect, pegylated hemin can improve anticancer chemotherapy's efficacy via increasing macromolecular drug accumulation in solid tumors [187].

10.3.4 Synergetic Therapy through Weakening Antioxidant Defense Combined with Elevating ROS Production

The synergistic use of ROS-generating agents with scavengers of antioxidant systems has also been reported to magnify oxidative stress and subsequent ROS-mediated cell death preferentially in cancer cells [10]. The GCLM-dependent synthesis of the antioxidant GSH is essential for cancer initiation, which could be inhibited by BSO, whereas GSH becomes dispensable potentially for advanced-stage cancer due to compensation from alternative antioxidant pathways. Thus, the upregulation of alternative antioxidant systems, such as the TXN pathway, in malignant cells renders treatment with BSO alone ineffective. The blockage of both GSH and TXN systems with combined use of BSO and SSA or AUR synergistically induces cancer cell death in established tumors [188]. Moreover, BSO effectively amplifies the growth inhibitory effect of As_2O_3 on cell lines derived from breast, prostate, colon, lung, bladder, cervix, and kidney cancers. The synergistic treatment of As_2O_3 and BSO inhibits the H_2O_2-scavenging enzymes, including catalase, glutathione, and glutathione peroxidase, to aggravate the cytotoxicity to prostate cancer cells with less harm on normal cells. The effectiveness of this combination treatment is further verified with an orthotopic model of prostate cancer metastasis, indicating an application potential of synergistic therapy using pro-oxidants and antioxidants in cancer treatment [189].

Recently, dual stimuli-responsive drugs that harbor the ability of both weakening antioxidant defense and enhancing ROS production are developed. Cinnamaldehyde, a major component of the cinnamon tree used as a food additive, has been reported to trigger the generation of mitochondrial ROS and suppress the growth of human cancer cells with minimal effect on normal cells. However, the clinical applications of cinnamaldehyde in tumor treatment have been restricted by its lower drug efficacy and poor bioavailability. To overcome these drawbacks, QCA [4-(1,3,2-dioxaborinan-2-yl)benzyl ((5-methyl-2-styryl-1,3-dioxan-5-yl) methyl) carbonate] is designed by coupling a glutathione-scavenging quinone methide-generating moiety to ROS-generating cinnamaldehyde. The novel hybrid anticancer prodrug, QCA, can be activated by ROS to release QM and cinnamaldehyde in tumor environments, which magnifies oxidative stress, resulting in the preferential killing of cancer cells [10]. Aminoferrocene-based prodrugs, after activation, not only release QMs that inhibit the cellular antioxidant defense via alkylating glutathione but also form catalysts (iron or ferrocenium ions) that promote the catalytic generation of hydroxyl radicals. This most potent prodrug is capable of selectively killing cancer cells, such as human promyelocytic leukemia (HL-60) and human glioblastoma-astrocytoma (U373), with minimal cytotoxicity to representative non-malignant fibroblasts [190]. Therefore, these dual stimuli-responsive drugs can exacerbate cancer cell death via amplifying oxidative stress.

10.4 Conclusion and Perspective

A broad spectrum of old drugs, including some traditional medicine, antiviral drugs, anti-inflammatory drugs, and nature metabolites, are able to modulate the redox homeostasis, thus hold the potential for the treatment and prevention of cancer. However, multiple concerns have to be addressed before these drug repurposing projects can be largely applied in the clinic. Firstly, the regulation of redox signaling within cells is extremely complex and largely undefined. Thus, further investigation of underlying mechanisms is a prerequisite. Secondly, it is necessary to deliver an excessive amount of ROS into the tumor, e.g., millimolar concentration; otherwise, adverse effect, namely, tumor progression, may occur. Thirdly, the clinical applications of oxidative medicine have also been restrained by its poor bioavailability (a short half-life in plasma) owing to the rapid oxidation of active group and lower drug efficacy than common anticancer drugs. Additionally, a relatively high dose of drugs, tumor-targeted delivery, and time-lagged infusion of drugs should be developed to further raise the intratumor ROS level and minimize the side effects. The full potential of stress medicine will be developed along with advanced drug repurposing strategy, and the drug repurposing of oxidative stress medicine will be of great importance in devising additional options in cancer treatment.

Acknowledgments This work was supported by grants from the National Natural Science Foundation of China (81821002, 81790251 and 81972665), National Key Research and Development Project (2020YFA0509400, 2020YFC2002705), and Guangdong Basic and Applied Basic Research Foundation (2019B030302012).

Competing Interests The authors declare no competing interests.

References

1. Hanahan D, Weinberg RA. Hallmarks of cancer: the next generation. Cell. 2011;144 (5):646–74.
2. Cairns RA, Harris IS, Mak TW. Regulation of cancer cell metabolism. Nat Rev Cancer. 2011;11(2):85–95.
3. Schieber M, Chandel NS. ROS function in redox signaling and oxidative stress. Curr Biol. 2014;24(10):R453–62.
4. Kim HS, et al. SIRT3 is a mitochondria-localized tumor suppressor required for maintenance of mitochondrial integrity and metabolism during stress. Cancer Cell. 2010;17(1):41–52.
5. Grivennikov SI, Greten FR, Karin M. Immunity, inflammation, and cancer. Cell. 2010;140 (6):883–99.
6. Liu X, et al. NADPH oxidase 1-dependent ROS is crucial for TLR4 signaling to promote tumor metastasis of non-small cell lung cancer. Tumour Biol. 2015;36(3):1493–502.
7. Peiris-Pages M, et al. Metastasis and oxidative stress: are antioxidants a metabolic driver of progression? Cell Metab. 2015;22(6):956–8.
8. Klaunig JE. Oxidative stress and cancer. Curr Pharm Des. 2018;24(40):4771–8.
9. Sanchez M, et al. Cross talk between eIF2alpha and eEF2 phosphorylation pathways optimizes translational arrest in response to oxidative stress. iScience. 2019;20:466–80.
10. Noh J, et al. Amplification of oxidative stress by a dual stimuli-responsive hybrid drug enhances cancer cell death. Nat Commun. 2015;6:6907.
11. Perillo B, et al. ROS in cancer therapy: the bright side of the moon. Exp Mol Med. 2020;52 (2):192–203.
12. Pessetto ZY, et al. Drug repurposing for gastrointestinal stromal tumor. Mol Cancer Ther. 2013;12(7):1299–309.
13. Bertolini F, Sukhatme VP, Bouche G. Drug repurposing in oncology--patient and health systems opportunities. Nat Rev Clin Oncol. 2015;12(12):732–42.
14. Pushpakom S, et al. Drug repurposing: progress, challenges and recommendations. Nat Rev Drug Discov. 2019;18(1):41–58.
15. Xue H, et al. Review of drug repositioning approaches and resources. Int J Biol Sci. 2018;14 (10):1232–44.

16. Corsello SM, et al. Discovering the anticancer potential of non-oncology drugs by systematic viability profiling. Nat Cancer. 2020;1(2):235–48.
17. Levine RL, et al. Methionine residues as endogenous antioxidants in proteins. Proc Natl Acad Sci U S A. 1996;93(26):15036–40.
18. Davies AM, Holt AG. Why antioxidant therapies have failed in clinical trials. J Theor Biol. 2018;457:1–5.
19. Tiganis T. Reactive oxygen species and insulin resistance: the good, the bad and the ugly. Trends Pharmacol Sci. 2011;32(2):82–9.
20. Holmstrom KM, Finkel T. Cellular mechanisms and physiological consequences of redox-dependent signalling. Nat Rev Mol Cell Biol. 2014;15(6):411–21.
21. Moldogazieva NT, et al. ROS and RNS signalling: adaptive redox switches through oxidative/nitrosative protein modifications. Free Radic Res. 2018;52(5):507–43.
22. Di Meo S, et al. Role of ROS and RNS sources in physiological and pathological conditions. Oxidative Med Cell Longev. 2016;2016:1245049.
23. Sies H, Berndt C, Jones DP. Oxidative stress. Annu Rev Biochem. 2017;86:715–48.
24. Tripathi DN, Walker CL. The peroxisome as a cell signaling organelle. Curr Opin Cell Biol. 2016;39:109–12.
25. Zhang J, et al. A tuberous sclerosis complex signalling node at the peroxisome regulates mTORC1 and autophagy in response to ROS. Nat Cell Biol. 2013;15(10):1186–96.
26. Schrader M, Fahimi HD. Peroxisomes and oxidative stress. Biochim Biophys Acta. 2006;1763(12):1755–66.
27. Laurindo FR, Araujo TL, Abrahão TB. Nox NADPH oxidases and the endoplasmic reticulum. Antioxid Redox Signal. 2014;20(17):2755–75.
28. Ramming T, et al. A PDI-catalyzed thiol-disulfide switch regulates the production of hydrogen peroxide by human Ero1. Free Radic Biol Med. 2015;83:361–72.
29. Yoboue ED, Sitia R, Simmen T. Redox crosstalk at endoplasmic reticulum (ER) membrane contact sites (MCS) uses toxic waste to deliver messages. Cell Death Dis. 2018;9(3):331.
30. Cao SS, Kaufman RJ. Endoplasmic reticulum stress and oxidative stress in cell fate decision and human disease. Antioxid Redox Signal. 2014;21(3):396–413.
31. Weydert CJ, Cullen JJ. Measurement of superoxide dismutase, catalase and glutathione peroxidase in cultured cells and tissue. Nat Protoc. 2010;5(1):51–66.
32. Tan BL, et al. Antioxidant and oxidative stress: a mutual interplay in age-related diseases. Front Pharmacol. 2018;9:1162.
33. McCord JM, Roy RS, Schaffer SW. Free radicals and myocardial ischemia. The role of xanthine oxidase. Adv Myocardiol. 1985;5:183–9.
34. Meister A. Glutathione metabolism and its selective modification. J Biol Chem. 1988;263(33):17205–8.
35. Meister A, Anderson ME. Glutathione. Annu Rev Biochem. 1983;52:711–60.
36. Mattill HA. Antioxidants. Annu Rev Biochem. 1947;16:177–92.
37. Veal EA, Day AM, Morgan BA. Hydrogen peroxide sensing and signaling. Mol Cell. 2007;26(1):1–14.
38. Alfonso-Prieto M, et al. The molecular mechanism of the catalase reaction. J Am Chem Soc. 2009;131(33):11751–61.
39. Lu J, Holmgren A. The thioredoxin antioxidant system. Free Radic Biol Med. 2014;66:75–87.
40. Lubos E, Loscalzo J, Handy DE. Glutathione peroxidase-1 in health and disease: from molecular mechanisms to therapeutic opportunities. Antioxid Redox Signal. 2011;15(7):1957–97.
41. Morris CM, et al. Brain iron homeostasis. J Inorg Biochem. 1992;47(3–4):257–65.
42. Bento I, et al. Ceruloplasmin revisited: structural and functional roles of various metal cation-binding sites. Acta Crystallogr D Biol Crystallogr. 2007;63(Pt 2):240–8.
43. Gomme PT, McCann KB, Bertolini J. Transferrin: structure, function and potential therapeutic actions. Drug Discov Today. 2005;10(4):267–73.

44. Fernandez-Marcos PJ, Nóbrega-Pereira S. NADPH: new oxygen for the ROS theory of aging. Oncotarget. 2016;7(32):50814–5.
45. Cheng X, Ku CH, Siow RC. Regulation of the Nrf2 antioxidant pathway by microRNAs: new players in micromanaging redox homeostasis. Free Radic Biol Med. 2013;64:4–11.
46. Kobayashi M, Yamamoto M. Molecular mechanisms activating the Nrf2-Keap1 pathway of antioxidant gene regulation. Antioxid Redox Signal. 2005;7(3–4):385–94.
47. Cho HY, Reddy SP, Kleeberger SR. Nrf2 defends the lung from oxidative stress. Antioxid Redox Signal. 2006;8(1–2):76–87.
48. Kim S, et al. Nrf2 activator via interference of Nrf2-Keap1 interaction has antioxidant and anti-inflammatory properties in Parkinson's disease animal model. Neuropharmacology. 2020;167:107989.
49. Vriend J, Reiter RJ. The Keap1-Nrf2-antioxidant response element pathway: a review of its regulation by melatonin and the proteasome. Mol Cell Endocrinol. 2015;401:213–20.
50. Waxman S, Anderson KC. History of the development of arsenic derivatives in cancer therapy. Oncologist. 2001;6(Suppl 2):3–10.
51. Shen ZY, et al. Arsenic trioxide induces apoptosis of oesophageal carcinoma in vitro. Int J Mol Med. 1999;4(1):33–7.
52. Shen ZX, et al. Use of arsenic trioxide (As2O3) in the treatment of acute promyelocytic leukemia (APL): II. Clinical efficacy and pharmacokinetics in relapsed patients. Blood. 1997;89(9):3354–60.
53. Chow SK, Chan JY, Fung KP. Inhibition of cell proliferation and the action mechanisms of arsenic trioxide (As2O3) on human breast cancer cells. J Cell Biochem. 2004;93(1):173–87.
54. Nakagawa Y, et al. Arsenic trioxide-induced apoptosis through oxidative stress in cells of colon cancer cell lines. Life Sci. 2002;70(19):2253–69.
55. Zheng J, et al. Arsenic trioxide induces apoptosis of HPV16 DNA-immortalized human cervical epithelial cells and selectively inhibits viral gene expression. Int J Cancer. 1999;82 (2):286–92.
56. Zhao XY, et al. Resveratrol protects against arsenic trioxide-induced cardiotoxicity in vitro and in vivo. Br J Pharmacol. 2008;154(1):105–13.
57. Yu M, et al. Resveratrol protects against arsenic trioxide-induced nephrotoxicity by facilitating arsenic metabolism and decreasing oxidative stress. Arch Toxicol. 2013;87(6):1025–35.
58. Thomas-Schoemann A, et al. Arsenic trioxide exerts antitumor activity through regulatory T cell depletion mediated by oxidative stress in a murine model of colon cancer. J Immunol. 2012;189(11):5171–7.
59. Griffin RJ, et al. Arsenic trioxide induces selective tumour vascular damage via oxidative stress and increases thermosensitivity of tumours. Int J Hyperth. 2003;19(6):575–89.
60. Jing Y, et al. Arsenic trioxide selectively induces acute promyelocytic leukemia cell apoptosis via a hydrogen peroxide-dependent pathway. Blood. 1999;94(6):2102–11.
61. Gartenhaus RB, et al. Arsenic trioxide cytotoxicity in steroid and chemotherapy-resistant myeloma cell lines: enhancement of apoptosis by manipulation of cellular redox state. Clin Cancer Res. 2002;8(2):566–72.
62. Akay C, et al. Antiretroviral drugs induce oxidative stress and neuronal damage in the central nervous system. J Neurovirol. 2014;20(1):39–53.
63. Wang X, et al. Human immunodeficiency virus protease inhibitor ritonavir inhibits cholesterol efflux from human macrophage-derived foam cells. Am J Pathol. 2007;171(1):304–14.
64. Deng W, et al. HIV protease inhibitors elicit volume-sensitive Cl- current in cardiac myocytes via mitochondrial ROS. J Mol Cell Cardiol. 2010;49(5):746–52.
65. Soprano M, et al. Oxidative stress mediates the antiproliferative effects of nelfinavir in breast cancer cells. PLoS One. 2016;11(6):e0155970.
66. Xiang T, et al. Nelfinavir, an HIV protease inhibitor, induces apoptosis and cell cycle arrest in human cervical cancer cells via the ROS-dependent mitochondrial pathway. Cancer Lett. 2015;364(1):79–88.

67. Kushchayeva Y, et al. The HIV protease inhibitor nelfinavir down-regulates RET signaling and induces apoptosis in medullary thyroid cancer cells. J Clin Endocrinol Metab. 2014;99(5):E734–45.
68. Gratton R, et al. Lopinavir/ritonavir treatment induces oxidative stress and caspaseindependent apoptosis in human glioblastoma U-87 MG cell line. Curr HIV Res. 2018;16(2):106–12.
69. Lü JM, et al. Ginsenoside Rb1 blocks ritonavir-induced oxidative stress and eNOS downregulation through activation of estrogen receptor-beta and upregulation of SOD in human endothelial cells. Int J Mol Sci. 2019;20(2):294.
70. Waibel S, et al. Ritonavir-induced suicidal death of human erythrocytes. Basic Clin Pharmacol Toxicol. 2016;119(1):51–7.
71. Arend C, Brandmann M, Dringen R. The antiretroviral protease inhibitor ritonavir accelerates glutathione export from cultured primary astrocytes. Neurochem Res. 2013;38(4):732–41.
72. Larder BA, Darby G, Richman DD. HIV with reduced sensitivity to zidovudine (AZT) isolated during prolonged therapy. Science. 1989;243(4899):1731–4.
73. Larder BA, Kemp SD. Multiple mutations in HIV-1 reverse transcriptase confer high-level resistance to zidovudine (AZT). Science. 1989;246(4934):1155–8.
74. Wheeler S, et al. Zidovudine-induced diaphragmatic contractile dysfunction: impact of an antioxidant diet. Respirology. 2005;10(2):171–6.
75. Papparella I, et al. Vitamin C prevents zidovudine-induced NAD(P)H oxidase activation and hypertension in the rat. Cardiovasc Res. 2007;73(2):432–8.
76. Banerjee A, et al. Zidovudine (AZT) and hepatic lipid accumulation: implication of inflammation, oxidative and endoplasmic reticulum stress mediators. PLoS One. 2013;8(10):e76850.
77. Mattson DM, et al. Cisplatin combined with zidovudine enhances cytotoxicity and oxidative stress in human head and neck cancer cells via a thiol-dependent mechanism. Free Radic Biol Med. 2009;46(2):232–7.
78. Apostolova N, et al. Enhanced oxidative stress and increased mitochondrial mass during efavirenz-induced apoptosis in human hepatic cells. Br J Pharmacol. 2010;160(8):2069–84.
79. Bertrand L, Toborek M. Dysregulation of endoplasmic reticulum stress and Autophagic responses by the antiretroviral drug Efavirenz. Mol Pharmacol. 2015;88(2):304–15.
80. Hecht M, et al. Cytotoxic effect of Efavirenz in BxPC-3 pancreatic cancer cells is based on oxidative stress and is synergistic with ionizing radiation. Oncol Lett. 2018;15(2):1728–36.
81. Apostolova N, et al. ER stress in human hepatic cells treated with Efavirenz: mitochondria again. J Hepatol. 2013;59(4):780–9.
82. Weiß M, et al. Efavirenz causes oxidative stress, endoplasmic reticulum stress, and autophagy in endothelial cells. Cardiovasc Toxicol. 2016;16(1):90–9.
83. Lin CC, Yin MC. Vitamins B depletion, lower iron status and decreased antioxidative defense in patients with chronic hepatitis C treated by pegylated interferon alfa and ribavirin. Clin Nutr. 2009;28(1):34–8.
84. Noshy MM, Hussien NA, El-Ghor AA. Evaluation of the role of the antioxidant silymarin in modulating the in vivo genotoxicity of the antiviral drug ribavirin in mice. Mutat Res. 2013;752(1–2):14–20.
85. Sharma S, Baksi R, Agarwal M. Repositioning of anti-viral drugs as therapy for cervical cancer. Pharmacol Rep. 2016;68(5):983–9.
86. Hileman EO, et al. Intrinsic oxidative stress in cancer cells: a biochemical basis for therapeutic selectivity. Cancer Chemother Pharmacol. 2004;53(3):209–19.
87. Huang P, et al. Superoxide dismutase as a target for the selective killing of cancer cells. Nature. 2000;407(6802):390–5.
88. LaVallee TM, et al. 2-Methoxyestradiol inhibits proliferation and induces apoptosis independently of estrogen receptors alpha and beta. Cancer Res. 2002;62(13):3691–7.
89. Golab J, et al. Antitumor effects of photodynamic therapy are potentiated by 2-methoxyestradiol. A superoxide dismutase inhibitor. J Biol Chem. 2003;278(1):407–14.
90. Florczak U, et al. 2-Methoxyestradiol-induced radiosensitization is independent of SOD but depends on inhibition of Akt and DNA-PKcs activities. Radiother Oncol. 2009;92(3):334–8.

91. Zhang Q, et al. Involvement of reactive oxygen species in 2-methoxyestradiol-induced apoptosis in human neuroblastoma cells. Cancer Lett. 2011;313(2):201–10.
92. Ding L, et al. Underlying mechanism of 2-methoxyestradiol-induced apoptosis and growth arrest in SKOV3 human ovarian cancer cells. Eur Rev Med Pharmacol Sci. 2015;19 (11):2084–90.
93. Zhou J, Du Y. Acquisition of resistance of pancreatic cancer cells to 2-methoxyestradiol is associated with the upregulation of manganese superoxide dismutase. Mol Cancer Res. 2012;10(6):768–77.
94. Gao N, et al. 2-Methoxyestradiol-induced apoptosis in human leukemia cells proceeds through a reactive oxygen species and Akt-dependent process. Oncogene. 2005;24(23):3797–809.
95. Chen Y, et al. Oxidative stress induces autophagic cell death independent of apoptosis in transformed and cancer cells. Cell Death Differ. 2008;15(1):171–82.
96. Loven DP, Leeper DB, Oberley LW. Superoxide dismutase levels in Chinese hamster ovary cells and ovarian carcinoma cells after hyperthermia or exposure to cycloheximide. Cancer Res. 1985;45(7):3029–33.
97. Thews O, et al. Possible protective effects of alpha-tocopherol on enhanced induction of reactive oxygen species by 2-methoxyestradiol in tumors. Adv Exp Med Biol. 2005;566:349–55.
98. Lu J, Holmgren A. Thioredoxin system in cell death progression. Antioxid Redox Signal. 2012;17(12):1738–47.
99. Sachweh MC, et al. Redox effects and cytotoxic profiles of MJ25 and auranofin towards malignant melanoma cells. Oncotarget. 2015;6(18):16488–506.
100. Gallegos A, et al. Transfection with human thioredoxin increases cell proliferation and a dominant-negative mutant thioredoxin reverses the transformed phenotype of human breast cancer cells. Cancer Res. 1996;56(24):5765–70.
101. Bhatia M, et al. The thioredoxin system in breast cancer cell invasion and migration. Redox Biol. 2016;8:68–78.
102. Wangpaichitr M, et al. The relationship of thioredoxin-1 and cisplatin resistance: its impact on ROS and oxidative metabolism in lung cancer cells. Mol Cancer Ther. 2012;11(3):604–15.
103. Zhang J, et al. Targeting the thioredoxin system for cancer therapy. Trends Pharmacol Sci. 2017;38(9):794–808.
104. Stafford WC, et al. Irreversible inhibition of cytosolic thioredoxin reductase 1 as a mechanistic basis for anticancer therapy. Sci Transl Med. 2018;10(428):eaaf7444.
105. McKeage MJ. Gold opens mitochondrial pathways to apoptosis. Br J Pharmacol. 2002;136 (8):1081–2.
106. Roder C, Thomson MJ. Auranofin: repurposing an old drug for a golden new age. Drugs R D. 2015;15(1):13–20.
107. Fan C, et al. Enhancement of auranofin-induced lung cancer cell apoptosis by selenocystine, a natural inhibitor of TrxR1 in vitro and in vivo. Cell Death Dis. 2014;5(4):e1191.
108. Fiskus W, et al. Auranofin induces lethal oxidative and endoplasmic reticulum stress and exerts potent preclinical activity against chronic lymphocytic leukemia. Cancer Res. 2014;74 (9):2520–32.
109. Zou P, et al. Auranofin induces apoptosis by ROS-mediated ER stress and mitochondrial dysfunction and displayed synergistic lethality with piperlongumine in gastric cancer. Oncotarget. 2015;6(34):36505–21.
110. Marzano C, et al. Inhibition of thioredoxin reductase by auranofin induces apoptosis in cisplatin-resistant human ovarian cancer cells. Free Radic Biol Med. 2007;42(6):872–81.
111. Hou GX, et al. Elimination of stem-like cancer cell side-population by auranofin through modulation of ROS and glycolysis. Cell Death Dis. 2018;9(2):89.
112. Griffith OW. Biologic and pharmacologic regulation of mammalian glutathione synthesis. Free Radic Biol Med. 1999;27(9–10):922–35.
113. Meister A. Selective modification of glutathione metabolism. Science. 1983;220(4596):472–7.

114. Yang Y, et al. γ-glutamylcysteine exhibits anti-inflammatory effects by increasing cellular glutathione level. Redox Biol. 2019;20:157–66.
115. Doxsee DW, et al. Sulfasalazine-induced cystine starvation: potential use for prostate cancer therapy. Prostate. 2007;67(2):162–71.
116. Estrela JM, Ortega A, Obrador E. Glutathione in cancer biology and therapy. Crit Rev Clin Lab Sci. 2006;43(2):143–81.
117. Schnelldorfer T, et al. Glutathione depletion causes cell growth inhibition and enhanced apoptosis in pancreatic cancer cells. Cancer. 2000;89(7):1440–7.
118. Lo M, et al. The xc- cystine/glutamate antiporter: a mediator of pancreatic cancer growth with a role in drug resistance. Br J Cancer. 2008;99(3):464–72.
119. Ma MZ, et al. Xc- inhibitor sulfasalazine sensitizes colorectal cancer to cisplatin by a GSH-dependent mechanism. Cancer Lett. 2015;368(1):88–96.
120. Lim JKM, et al. Cystine/glutamate antiporter xCT (SLC7A11) facilitates oncogenic RAS transformation by preserving intracellular redox balance. Proc Natl Acad Sci U S A. 2019;116 (19):9433–42.
121. Wahl C, et al. Sulfasalazine: a potent and specific inhibitor of nuclear factor kappa B. J Clin Invest. 1998;101(5):1163–74.
122. Gout PW, et al. Sulfasalazine, a potent suppressor of lymphoma growth by inhibition of the x (c)- cystine transporter: a new action for an old drug. Leukemia. 2001;15(10):1633–40.
123. Cildag S, Senturk T. Sulfasalazine-related hypersensitivity reactions in patients with rheumatic diseases. J Clin Rheumatol. 2017;23(2):77–9.
124. Lo M, et al. Potential use of the anti-inflammatory drug, sulfasalazine, for targeted therapy of pancreatic cancer. Curr Oncol. 2010;17(3):9–16.
125. Narang VS, et al. Suppression of cystine uptake by sulfasalazine inhibits proliferation of human mammary carcinoma cells. Anticancer Res. 2003;23(6c):4571–9.
126. Okazaki S, et al. Glutaminolysis-related genes determine sensitivity to xCT-targeted therapy in head and neck squamous cell carcinoma. Cancer Sci. 2019;110(11):3453–63.
127. Zheng Z, et al. The X(c)(-) inhibitor sulfasalazine improves the anti-cancer effect of pharmacological vitamin C in prostate cancer cells via a glutathione-dependent mechanism. Cell Oncol (Dordr). 2020;43(1):95–106.
128. Hamilton D, et al. A novel missense mutation in the gamma-glutamylcysteine synthetase catalytic subunit gene causes both decreased enzymatic activity and glutathione production. Blood. 2003;102(2):725–30.
129. Drew R, Miners JO. The effects of buthionine sulphoximine (BSO) on glutathione depletion and xenobiotic biotransformation. Biochem Pharmacol. 1984;33(19):2989–94.
130. Dusre L, et al. Potentiation of doxorubicin cytotoxicity by buthionine sulfoximine in multidrug-resistant human breast tumor cells. Cancer Res. 1989;49(3):511–5.
131. Lin LC, et al. γ-Glutamylcysteine synthetase (γ-GCS) as a target for overcoming chemo- and radio-resistance of human hepatocellular carcinoma cells. Life Sci. 2018;198:25–31.
132. Kramer RA, et al. Chemosensitization of L-phenylalanine mustard by the thiol-modulating agent buthionine sulfoximine. Cancer Res. 1987;47(6):1593–7.
133. Hedley DW, et al. Antileukemic action of buthionine sulfoximine: evidence for an intrinsic death mechanism based on oxidative stress. Leukemia. 1998;12(10):1545–52.
134. Reliene R, Schiestl RH. Glutathione depletion by buthionine sulfoximine induces DNA deletions in mice. Carcinogenesis. 2006;27(2):240–4.
135. O'Dwyer PJ, et al. Phase I trial of buthionine sulfoximine in combination with melphalan in patients with cancer. J Clin Oncol. 1996;14(1):249–56.
136. Kito M, et al. Arsenic trioxide-induced apoptosis and its enhancement by buthionine sulfoximine in hepatocellular carcinoma cell lines. Biochem Biophys Res Commun. 2002;291(4):861–7.
137. Anderson CP, et al. Depletion of glutathione by buthionine sulfoxine is cytotoxic for human neuroblastoma cell lines via apoptosis. Exp Cell Res. 1999;246(1):183–92.

138. Rigas B, Kashfi K. Nitric-oxide-donating NSAIDs as agents for cancer prevention. Trends Mol Med. 2004;10(7):324–30.
139. Williams JL, et al. Nitric oxide-releasing nonsteroidal anti-inflammatory drugs (NSAIDs) alter the kinetics of human colon cancer cell lines more effectively than traditional NSAIDs: implications for colon cancer chemoprevention. Cancer Res. 2001;61(8):3285–9.
140. Lavagna C, et al. Antiproliferative effects of nitrosulindac on human colon adenocarcinoma cell lines. Biochem Biophys Res Commun. 2001;284(3):808–16.
141. Kashfi K, et al. Nitric oxide-donating nonsteroidal anti-inflammatory drugs inhibit the growth of various cultured human cancer cells: evidence of a tissue type-independent effect. J Pharmacol Exp Ther. 2002;303(3):1273–82.
142. Yeh RK, et al. NO-donating nonsteroidal antiinflammatory drugs (NSAIDs) inhibit colon cancer cell growth more potently than traditional NSAIDs: a general pharmacological property? Biochem Pharmacol. 2004;67(12):2197–205.
143. Hulsman N, et al. Chemical insights in the concept of hybrid drugs: the antitumor effect of nitric oxide-donating aspirin involves a quinone methide but not nitric oxide nor aspirin. J Med Chem. 2007;50(10):2424–31.
144. Bak AW, et al. Cyclooxygenase-independent chemoprevention with an aspirin derivative in a rat model of colonic adenocarcinoma. Life Sci. 1998;62(23):367–73.
145. Williams JL, et al. Growth inhibition of human colon cancer cells by nitric oxide (NO)-donating aspirin is associated with cyclooxygenase-2 induction and beta-catenin/T-cell factor signaling, nuclear factor-kappaB, and NO synthase 2 inhibition: implications for chemoprevention. Cancer Res. 2003;63(22):7613–8.
146. Yoo D, et al. Glutathione-depleting pro-oxidant as a selective anticancer therapeutic agent. ACS Omega. 2019;4(6):10070–7.
147. Shelton MD, Chock PB, Mieyal JJ. Glutaredoxin: role in reversible protein s-glutathionylation and regulation of redox signal transduction and protein translocation. Antioxid Redox Signal. 2005;7(3–4):348–66.
148. Wadley AJ, Aldred S, Coles SJ. An unexplored role for Peroxiredoxin in exercise-induced redox signalling? Redox Biol. 2016;8:51–8.
149. Trzeciecka A, et al. Dimeric peroxiredoxins are druggable targets in human Burkitt lymphoma. Oncotarget. 2016;7(2):1717–31.
150. O'Leary PC, et al. Peroxiredoxin-1 protects estrogen receptor α from oxidative stress-induced suppression and is a protein biomarker of favorable prognosis in breast cancer. Breast Cancer Res. 2014;16(4):R79.
151. Lehtonen ST, et al. Peroxiredoxins, a novel protein family in lung cancer. Int J Cancer. 2004;111(4):514–21.
152. Perkins A, et al. Peroxiredoxins: guardians against oxidative stress and modulators of peroxide signaling. Trends Biochem Sci. 2015;40(8):435–45.
153. Zhang H, et al. Glutaredoxin 2 reduces both thioredoxin 2 and thioredoxin 1 and protects cells from apoptosis induced by auranofin and 4-hydroxynonenal. Antioxid Redox Signal. 2014;21(5):669–81.
154. Zhang Y, et al. Total synthesis of griseusins and elucidation of the griseusin mechanism of action. Chem Sci. 2019;10(32):7641–8.
155. Ye Q, et al. Frenolicin B targets peroxiredoxin 1 and glutaredoxin 3 to trigger ROS/4E-BP1-mediated antitumor effects. Cell Chem Biol. 2019;26(3):366–377.e12.
156. Liu CX, et al. Adenanthin targets peroxiredoxin I and II to induce differentiation of leukemic cells. Nat Chem Biol. 2012;8(5):486–93.
157. Hou JK, et al. Adenanthin targets peroxiredoxin I/II to kill hepatocellular carcinoma cells. Cell Death Dis. 2014;5(9):e1400.
158. de la Vega MR, Chapman E, Zhang DD. NRF2 and the hallmarks of cancer. Cancer Cell. 2018;34(1):21–43.
159. Cuadrado A, et al. Therapeutic targeting of the NRF2 and KEAP1 partnership in chronic diseases. Nat Rev Drug Discov. 2019;18(4):295–317.

160. Best SA, et al. Synergy between the KEAP1/NRF2 and PI3K pathways drives non-small-cell lung cancer with an altered immune microenvironment. Cell Metab. 2018;27(4):935–943.e4.
161. Menegon S, Columbano A, Giordano S. The dual roles of NRF2 in cancer. Trends Mol Med. 2016;22(7):578–93.
162. Fox DB, et al. NRF2 activation promotes the recurrence of dormant tumour cells through regulation of redox and nucleotide metabolism. Nat Metab. 2020;2(4):318–34.
163. Kansanen E, et al. The Keap1-Nrf2 pathway: mechanisms of activation and dysregulation in cancer. Redox Biol. 2013;1(1):45–9.
164. Bailly C. Potential use of edaravone to reduce specific side effects of chemo-, radio- and immuno-therapy of cancers. Int Immunopharmacol. 2019;77:105967.
165. Jangra A, et al. Edaravone alleviates cisplatin-induced neurobehavioral deficits via modulation of oxidative stress and inflammatory mediators in the rat hippocampus. Eur J Pharmacol. 2016;791:51–61.
166. Kokura S, et al. The radical scavenger edaravone enhances the anti-tumor effects of CPT-11 in murine colon cancer by increasing apoptosis via inhibition of NF-kappaB. Cancer Lett. 2005;229(2):223–33.
167. Tang Y, et al. Effect of edaravone on radiation-induced brain necrosis in patients with nasopharyngeal carcinoma after radiotherapy: a randomized controlled trial. J Neuro-Oncol. 2014;120(2):441–7.
168. Li H, et al. Edaravone ameliorates experimental autoimmune thyroiditis in rats through HO-1-dependent STAT3/PI3K/Akt pathway. Am J Transl Res. 2018;10(7):2037–46.
169. Mengíbar JL, et al. Simultaneous onset of type 1 diabetes mellitus and silent thyroiditis under durvalumab treatment. Endocrinol Diabetes Metab Case Rep. 2019;2019(1):45.
170. Ferrari SM, et al. Thyroid disorders induced by checkpoint inhibitors. Rev Endocr Metab Disord. 2018;19(4):325–33.
171. Stehlin-Gaon C, et al. All-trans retinoic acid is a ligand for the orphan nuclear receptor RORβ. Nat Struct Mol Biol. 2003;10(10):820–5.
172. Degos L, Wang ZY. All trans retinoic acid in acute promyelocytic leukemia. Oncogene. 2001;20(49):7140–5.
173. Chatterjee A, Chatterji U. All-trans retinoic acid ameliorates arsenic-induced oxidative stress and apoptosis in the rat uterus by modulating MAPK signaling proteins. J Cell Biochem. 2017;118(11):3796–809.
174. Khafaga AF, El-Sayed YS. All-trans-retinoic acid ameliorates doxorubicin-induced cardiotoxicity: in vivo potential involvement of oxidative stress, inflammation, and apoptosis via caspase-3 and p53 down-expression. Naunyn Schmiedeberg's Arch Pharmacol. 2018;391 (1):59–70.
175. Ramya D, et al. Chemoprotective effect of all-trans retinoic acid (ATRA) on oxidative stress and lung metastasis induced by benzo(a)pyrene. Immunopharmacol Immunotoxicol. 2012;34 (2):317–25.
176. Carlström KE, et al. Therapeutic efficacy of dimethyl fumarate in relapsing-remitting multiple sclerosis associates with ROS pathway in monocytes. Nat Commun. 2019;10(1):3081.
177. Nicolay JP, et al. Dimethyl fumarate restores apoptosis sensitivity and inhibits tumor growth and metastasis in CTCL by targeting NF-κB. Blood. 2016;128(6):805–15.
178. Xie X, et al. Dimethyl fumarate induces necroptosis in colon cancer cells through GSH depletion/ROS increase/MAPKs activation pathway. Br J Pharmacol. 2015;172(15):3929–43.
179. Saidu NEB, et al. Dimethyl fumarate, a two-edged drug: current status and future directions. Med Res Rev. 2019;39(5):1923–52.
180. Bennett Saidu NE, et al. Dimethyl fumarate is highly cytotoxic in KRAS mutated cancer cells but spares non-tumorigenic cells. Oncotarget. 2018;9(10):9088–99.
181. Chau LY. Heme oxygenase-1: emerging target of cancer therapy. J Biomed Sci. 2015;22 (1):22.
182. Loboda A, Jozkowicz A, Dulak J. HO-1/CO system in tumor growth, angiogenesis and metabolism - targeting HO-1 as an anti-tumor therapy. Vasc Pharmacol. 2015;74:11–22.

183. Di Biase S, et al. Fasting-mimicking diet reduces HO-1 to promote T cell-mediated tumor cytotoxicity. Cancer Cell. 2016;30(1):136–46.
184. Labbé RF, Vreman HJ, Stevenson DK. Zinc protoporphyrin: a metabolite with a mission. Clin Chem. 1999;45(12):2060–72.
185. Hjortsø MD, Andersen MH. The expression, function and targeting of haem oxygenase-1 in cancer. Curr Cancer Drug Targets. 2014;14(4):337–47.
186. Regehly M, et al. Water-soluble polymer conjugates of ZnPP for photodynamic tumor therapy. Bioconjug Chem. 2007;18(2):494–9.
187. Fang J, et al. Carbon monoxide, generated by heme oxygenase-1, mediates the enhanced permeability and retention effect in solid tumors. Cancer Sci. 2012;103(3):535–41.
188. Harris IS, et al. Glutathione and thioredoxin antioxidant pathways synergize to drive cancer initiation and progression. Cancer Cell. 2015;27(2):211–22.
189. Maeda H, et al. Effective treatment of advanced solid tumors by the combination of arsenic trioxide and L-buthionine-sulfoximine. Cell Death Differ. 2004;11(7):737–46.
190. Schikora M, et al. Activity of aminoferrocene-based prodrugs against prostate cancer. Bioorg Med Chem Lett. 2015;25(17):3447–50.

Redox Regulation of Metabolic Enzymes in Cancer

11

Jingwen Jiang and Canhua Huang

Abstract

Metabolic network has been intensely studied and considered as fundamental biological phenomena that participate in numerous healthy and pathological processes, especially in tumorigenesis and cancer development. One hallmark of cancer cells is adaptive metabolic reprogramming which is highlighted by reduced oxidative phosphorylation and enhanced aerobic glycolysis. The metabolic reprogramming is accompanied by accumulated oxidative stress which is another important hallmark of cancer. Simultaneously, redox signaling is emerging as a critical aspect that maintains proper cellular functions, and redox imbalance is thought to be a hallmark of various diseases. Growing body of evidence has suggested an intrinsic link between metabolic network and redox biology. Redox status can change the shift of metabolic flux through regulating the expression, function, or subcellular distribution of metabolic enzymes. The alteration of metabolic flux can in turn modulate the redox status, which controls a wide range of cellular processes. Therefore, uncovering the intricate mechanisms underlying redox regulation of metabolic enzymes will lead to a deep understanding of tumorigenesis and tumor progression and facilitate the translational medicine.

J. Jiang
State Key Laboratory of Biotherapy and Cancer Center, West China Hospital, and West China School of Basic Medical Sciences and Forensic Medicine, Sichuan University, and Collaborative Innovation Center for Biotherapy, Chengdu, People's Republic of China

C. Huang (✉)
State Key Laboratory of Biotherapy and Cancer Center, West China Hospital, and West China School of Basic Medical Sciences and Forensic Medicine, Sichuan University, and Collaborative Innovation Center for Biotherapy, Chengdu, People's Republic of China

School of Basic Medical Sciences, Chengdu University of Traditional Chinese Medicine, Chengdu, People's Republic of China
e-mail: hcanhua@scu.edu.cn

C. Huang, Y. Zhang (eds.), *Oxidative Stress*,
https://doi.org/10.1007/978-981-16-0522-2_11

Keywords

Redox · ROS · Metabolic enzymes · Cancer therapy

Metabolic abnormalities are widely identified in cancer cells and reported to be closely linked with the tumorigenesis and tumor development [1]. Cancer cells are prone to utilize glycolysis to fulfil the substance requirement for rapid proliferation even in the presence of oxygen [2]. This phenomenon is known as aerobic glycolysis or Warburg effect [3]. These metabolic alterations are associated with oncogenic mutations and activation of pro-proliferation signaling pathway, leading to malignant transformation of cancer [4]. Therefore, targeting metabolic abnormalities holds great potential for cancer therapy [5]. However, since metabolic enzymes are also utilized by normal cells, targeting these metabolic enzymes may lead to side effect and thus impedes its application. On account of this, new insight is urgently needed to improve our understanding in the regulatory mechanism of metabolic reprogramming. Emerging evidence suggests that oxidative stress induced by oncogenic mutation or metabolic reprogramming is involved in the regulation of metabolic reprogramming through modulating the expression, function, or subcellular localization of metabolic enzymes [6]. Intriguingly, some metabolic enzymes are redox sensors which could be oxidized under oxidative stress [7]. These oxidative modifications are occurred on the free cysteine residues of metabolic enzymes, leading to alteration of the function and subcellular localization of metabolic enzymes, thus influencing cell fate [8–10]. Although some questions have not been fully understood, this field holds the potential to be a research hot spot that leads toward translational medicine. In this review, we will discuss the bidirectional regulation of redox homeostasis and metabolic abnormalities, as well as the potential therapeutic strategies derived.

Oxidative stress is exerted by accumulated reactive oxygen species (ROS) which include non-radical molecules such as hydrogen peroxide (H_2O_2) and free radicals such as superoxide ($O_2^{\bullet -}$) and ($HO^{\bullet}$) [11]. These oxygen-containing chemical species are principally produced in the mitochondria, the endoplasmic reticulum (ER), and the peroxisomes [12, 13]. Mitochondrion is the major organelle that produces superoxide through the electron leakage from the electron transport chain (ETC) [14]. The ER is another ROS-producing compartment in which electrons transfer from the $FADH_2$ to oxygen and generate ROS [15]. The peroxisome is considered to play dual roles in regulating ROS production: one is to scavenge ROS by catalase (CAT) which decomposes H_2O_2 to H_2O and the other is to produce ROS through β-oxidation of fatty acids [16–18] (Fig. 11.1). Moreover, some mitochondrial metabolic enzymes including glycerol-3-phosphate dehydrogenase (GPDH), flavoprotein-ubiquinone oxidoreductase (FQR), 2-ketoglutarate dehydrogenase (KGDH), and pyruvate dehydrogenase (PDH) can directly promote the conversion of O_2 to superoxide [6, 19, 20]. In addition, other metabolic-associated enzymes such as NADPH oxidases (NOXs) and xanthine oxidases (XOs) also contribute to the generation of ROS [21]. Furthermore, glycolytic enzymes such as

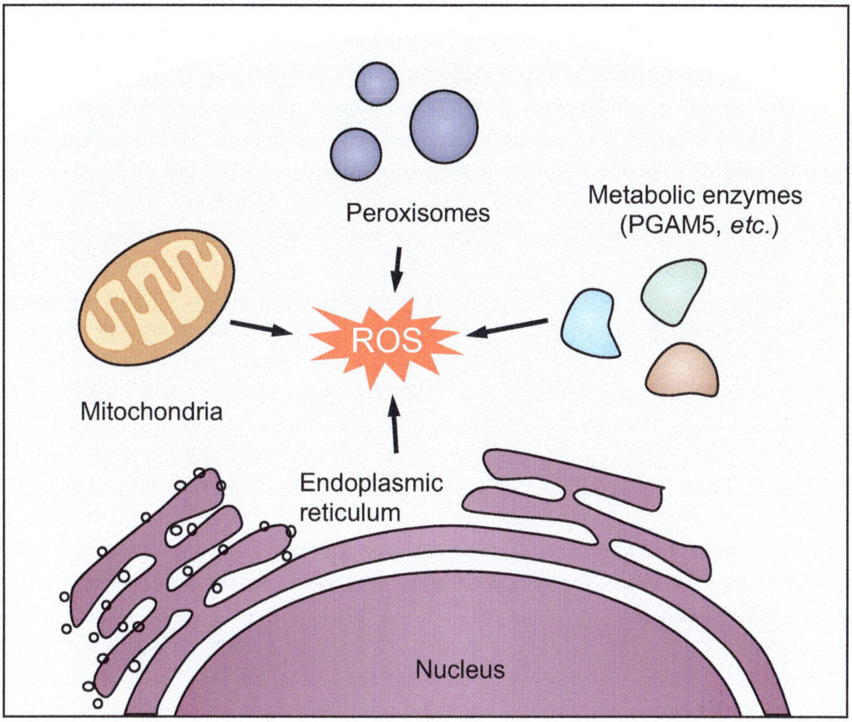

Fig. 11.1 The major sources of ROS production. Mitochondria, peroxisomes, and endoplasmic reticulum (ER) are major sources for ROS generation. In mitochondria, O_2 can be converted to superoxide, while in peroxisome, ROS can be produced through β-oxidation of fatty acids. ER favors the formation of protein disulfide bond and increases ROS levels. Besides, metabolic enzymes such as PGAM5 and GPDH can also increase ROS level

phosphoglycerate mutase family member 5 (PGAM5) are reported to influence the production of ROS [22] (Fig. 11.1). Taken together, these evidences suggest that these metabolic enzymes might be another leading cause of ROS generation.

To maintain cellular redox homeostasis, cells are developed with nimble antioxidant system. The transcription factor nuclear factor erythroid 2-related factor 2 (NRF2) is the master regulator of the transcription of numerous ROS-detoxifying enzymes such as glutathione peroxidase (GPX), thioredoxin (TXN), thioredoxin reductase 1 (TXNRD1), and peroxiredoxin 1 (PRDX1) [23]. Moreover, the metabolic byproducts including glutathione (GSH) and NADPH are important antioxidant metabolites and essential for redox homeostasis maintenance [24]. GSH is the most abundant non-enzymatic antioxidant molecule within the cell. GSH can be oxidized to GSSG, and the ratio of GSH/GSSG indicates the cellular reductive potential. GSH is regenerated from GSSG by glutathione reductase (GSR) with the consumption of NADPH [25, 26].

Oxidative stress has long been considered as a key pathological factor, especially in cancers [27]. Excessive ROS may oxidize and damage the intracellular targets such as DNA, protein, and lipid, which is conventionally regarded as the pivotal reasons for the occurrence of many diseases such as cancer and aging [28]. However, recent studies have revealed that moderate levels of ROS can serve as a second messenger, which is indispensable for physiological processes [29]. Appropriate level of ROS, together with antioxidant systems, coordinates reduction-oxidation (redox) signaling to maintain intracellular redox homeostasis and proper functions of organisms [30, 31]. Indeed, redox imbalance has been found to be involved in a variety of diseases, indicating a possible therapeutic strategy by redox resetting [32–34].

Redox regulation of protein function is under extensive study, although it's still far from resolved. Reduced form of thiol groups (-SH) in cysteine can undergo oxidative modification, thereby generating S-hydroxylated derivatives (-SOH), which may then produce disulfides (-s-s-). It allows the conformation changes and the protein-protein interaction, leading to function alterations. On the contrary, the oxidized form of protein can also be reduced by cellular antioxidant systems such as Trx reductase and Grx reductase [35, 36]. Therefore, the redox regulation can serve as molecular switches that control the function of some proteins. Importantly, a number of metabolism-related proteins were found to undergo redox modifications, suggesting that the redox regulation of some metabolic enzymes is critical for the metabolic alteration [6].

11.1 Redox Regulation of Metabolic Enzymes in Cancer Cells

11.1.1 AMPK

5'-AMP-activated protein kinase (AMPK) is a well-known energy sensor which is activated to regulate cellular metabolism under energy stress [37]. Our previous work has demonstrated that PRKAA, a catalytic subunit of AMPK, was activated by hepatitis B virus (HBV) infection-induced oxidative stress, indicating that AMPK is a potential target for redox modification [38]. Furthermore, emerging evidence has proven that AMPK is sensitive to ROS-mediated oxidative posttranslational modifications (oxPTMs). Zmijewski et al. reported that the kinase activity of AMPK can be stimulated under H_2O_2 exposure through S-glutathionylation at Cys299 and Cys304. In this circumstance, AMPK activation may result from S-glutathionylation-induced conformational change [39] (Fig. 11.2). Intriguingly, oxidation at Cys130 and Cys174 of AMPKα can disrupt the interaction between AMPK and AMPK kinases (AMPKK), thus attenuating the activity of AMPK. In contrast, thioredoxin-1 (Trx1) can cleave the disulfide bond between Cys130 and Cys174 to protect AMPK from oxidation-induced inactivation [40]. These studies imply that redox regulation of AMPK is especially sophisticated, which may be essential for the establishment of the circuit between cellular redox states and metabolism. Whether the subtle redox regulation of AMPK activity is correlated with cancer initiation and progression remains to be further explored.

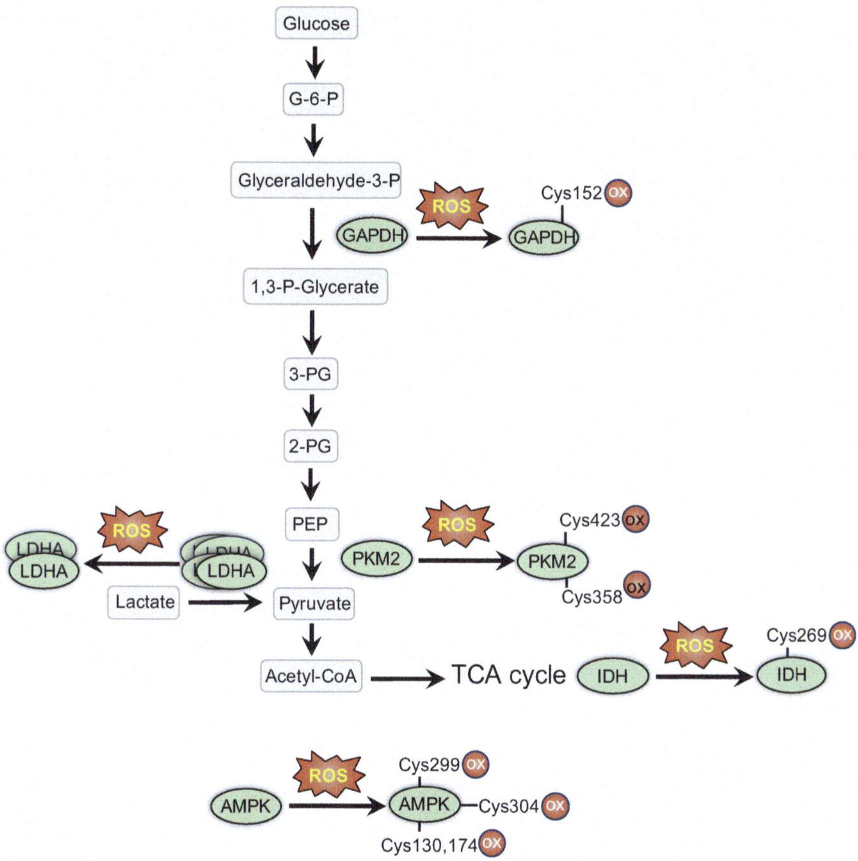

Fig. 11.2 The overview of redox modification in several metabolic enzymes. Under oxidative stress, several metabolic enzymes can undergo oxidative modification on their reduced thiol group in cysteine residues (GAPDH, PKM2, LDHA, IDH, AMPK, etc.). The oxidative modification of these metabolic enzymes may alter their metabolic activities and thus affect cell fate

11.1.2 GAPDH

It has been reported for a long time that GAPDH is one of the major targets in response to oxidative stress [41]. In 1994, Ravichandran and colleagues found that GAPDH can be S-thiolated with a decrease in enzymatic activity during the respiratory burst in blood monocytes [42]. Consistently, in yeast cells, S-thiolation of GAPDH may serve as an adaptive response during oxidative stress to protect cells from irreversible oxidation [43]. More in-depth studies revealed that Cys152 is an active site of GAPDH and can be oxidatively modified (Fig. 11.2). Oxidized GAPDH interacts with different proteins to exert moonlighting functions beyond metabolism [44]. Interestingly, Nakajima et al. found that the amyloid-like aggregation of GAPDH triggered by oxidative stress can lead to cell death, in which Cys152 plays a critical role. This study indicates the importance of GAPDH redox regulation in brain damage and some types of neuroblastoma [45]. Furthermore, in KRAS and BRAF mutant colorectal cancer cells, the S-glutathionylation level of GAPDH was demonstrated to be significantly higher during oxidative stress which resulted from high-dose vitamin C treatment. S-glutathionylation inhibits the activity of GAPDH and subsequent glycolysis, contributing to ATP depletion and cell death [46]. This study demonstrates that redox regulation of GAPDH plays an essential role in the homeostasis of cancer cell metabolism, thus determining cell fate. Taken together, GAPDH has been recognized as the critical mediator of oxidative stress responses. Redox regulation of GAPDH in cancer is now under wide investigation.

11.1.3 PKM2

The glycolytic enzyme PKM2 catalyzes the dephosphorylation reaction of phosphoenolpyruvate to pyruvate, which is accompanied by net ATP production [47]. PKM2 is upregulated in several types of cancers, and its oxidation form plays an important part in tumor progression. In human lung cancer cells, the enzymatic activity of PKM2 is inhibited through Cys358 oxidation, which rewires glycolysis flux to the oxidative pentose phosphate pathway (PPP) (Fig. 11.2). This diversion is especially critical for ROS detoxification through generating reducing potential, leading to lung cancer formation and progression [48]. This example illustrates that the redox regulation of PKM2 is necessary for cancer cell to withstand oxidative stress. In addition, oxidative stress can induce mitochondrial translocation of PKM2, where it phosphorylates Bcl2 to prevent the ubiquitin-mediated degradation of Bcl2. These events protect glioma cells from oxidative stress-induced apoptosis, which further highlights the importance of PKM2 in cancer cell adaption to high levels of ROS [49]. A recent study has reported that tetrameric PKM2 oxidized at its Cys423 can bind with and suppress the transcriptional activity of p53, thus inhibiting apoptosis. The high oxidation state of PKM2 is critical for its regulatory role in p53 activity [50] (Fig. 11.2). Thus, PKM2 is one of the typical enzymes sensitive to redox regulation, which is a key regulator in metabolic-redox circuits.

11.1.4 LDHA

LDHA catalyzes the conversion of pyruvate to L-lactate with the production of NADH in the last step of glycolysis. According to our recent work, LDHA translocates to the nucleus in response to human papilloma virus (HPV)-mediated ROS accumulation, which is accompanied by ROS-induced tetramer-to-dimer transition of LDHA (Fig. 11.2). Nuclear LDHA acquires a non-canonical enzymatic activity to produce α-hydroxybutyrate, which promotes the interaction between LDHA and DOT1L, and triggers histone H3K79 hypermethylation. This epigenetic alteration finally activates antioxidant responses and maintains cellular redox homeostasis. The non-canonical function of LDHA triggered by oxidative stress is crucial to HPV-induced cervical cancer progression [51]. This finding indicates that the redox status of LDHA is a potential biomarker for cervical cancer and that LDHA may have the potential to become a therapeutic target.

11.1.5 IDH

IDH catalyzes the conversion of isocitrate to α-ketoglutarate (α-KG) and carbon dioxide, which includes oxidation of isocitrate and the following decarboxylation. The point mutations of IDH have been widely documented in cancer, especially in gliomas and glioblastomas [52, 53]. Mutated IDH catalyzes the conversion of α-KG to 2-hydroxyglutarate (2-HG), which competitively suppresses α-KG-dependent enzymes, thus resulting in the global alterations of methylation levels [54]. It has been reported that IDH is a potential target of oxPTMs. Mitochondrial NADP(+)-dependent isocitrate dehydrogenase (IDPm) can be glutathionylated at Cys269 during oxidative stress, which leads to a decrease in IDPm activity (Fig. 11.2). The inactivated IDPm can also be reactivated through deglutathionylation, indicating that IDPm plays a crucial role in maintaining mitochondrial redox balance [55]. Moreover, a recent study on intracranial patient-derived xenografts of gliomas found that IDH mutations trigger metabolic landscape alteration, which affects cellular oxidative stress pathways [56]. This finding further highlights the regulatory role of IDH in redox hemostasis, in which oxPTMs of IDH may be involved. More interestingly, a competitive inhibitor of IDH, oxalomalate, has been proven to inhibit metastatic melanoma through ROS-dependent signaling pathways [57]. This finding suggests that IDH can act as a promising therapeutic target in clinical cancer management.

11.2 Targeting the Metabolic Abnormalities in Cancer Cell

As metabolism-derived cofactors are critical in regulating epigenetics, which is involved in tumorigenesis and cancer progression, it is feasible to target these metabolic cofactors or enzymes to interfere with chromatin modifications to benefit clinical outcomes. The most straightforward way to affect the chromatin modifications on the basis of metabolites adjustment is to deplete or replenish cellular metabolic cofactors such as SAM, α-KG, FAD, acetyl-CoA, NAD^+, and lactate. Multiple agents designed to inhibit the production or recycling of these metabolites are now under intensive investigations for cancer treatment. For example, 3-deazaneplanocin A (DZNep), which could inhibit the activity of SAH hydrolase and disrupt the recycling of SAM thus influencing the methylation of DNA and histones, has been found to induce apoptosis and reduce proliferation in AML cells and breast cancer cells, respectively [58, 59]. Moreover, quinoxaline compound 1-(2,3-di(thiophen-2-yl) quinoxalin-6-yl)-3-(2-methoxyethyl) urea could inhibit the activity of acyl-CoA synthetase short-chain family member 2 (ACSS2), leading to decreased acetate incorporation into both lipid and histones [60]. In addition, modifying the ratio between epigenetics-associated metabolites and their competitive metabolites (α-KG/2-HG, SAM/SAH, acetyl-CoA/CoA, etc.) could also influence the chromatin modifications. Moreover, biosynthesis of epigenetics-associated metabolites at different subcellular compartments could lead to specific alteration of chromatin modifications by changing the concentrations of local cofactors [61].

Glucose and glutamine are essential biomolecules which provide cancer cells with energy and intermediates to fulfil the requirement of rapid proliferation. Several compounds are developed for targeting the glucose or glutamine addiction in cancer cells. For example, 2-deoxy-D-glucose (2-DG), known as a glucose analog, could inhibit glycolysis and induce proliferation arrest in several cancer types [62]. Moreover, inhibition of LDHA could prevent the regeneration of NAD^+ which is essential for glycolysis and thus suppress tumor growth. Gossypol is a LDHA inhibitor which shows antitumor effect in several solid cancer types and is now under clinical trials [63, 64]. Metformin, known as a reference drug for type II diabetes treatment, could reduce cancer risk through the inhibition of mitochondrial complex I [65]. In addition, pyruvate dehydrogenase kinase (PDK) is overexpressed in several cancer types and promotes glycolytic phenotype [66]. PDK inhibitor dichloroacetate (DCA) can inhibit aerobic glycolysis and restrain tumor growth, indicating that targeting highly glycolytic tumors holds great potential for cancer therapy. Glutamine can be converted to α-KG to fuel TCA, during which glutaminase (GLS) and glutamate dehydrogenase (GDH) are required. Therefore, targeting GLS or GDH may benefit cancer therapy through inhibiting glutamine addiction in cancer cells. BPTES and CB-839 are two established GLS inhibitors for restricting the supply of α-KG to the TCA, and CB-839 is now under clinical studies in several cancers. Furthermore, aminooxyacetate (AOA) is an important aminotransferase inhibitor which exhibits potent antitumor effect in several cancer types including lung adenocarcinoma through targeting glutamine metabolism [67].

11.3 Modulating ROS for Cancer Therapy

The fact that some metabolic enzymes are regulated by cellular redox state highlights the promise of modulating the redox status for cancer therapy. For instance, as mentioned above, HPV16 E7 induces intracellular ROS accumulation, resulting in the translocation of LDHA in the nucleus where it gets a non-canonical function to generate α-HB and epigenetically regulate antioxidant responses [51]. Hence, ROS modulators that have been specifically developed to disable nuclear translocation of LDHA may be a promising choice to prevent and treat cervical cancer.

GSH, as discussed above, is an important free radical scavenger and detoxifying molecule in living cells [68]. GSH can maintain a reduced environment of subcellular compartments such as mitochondria, nucleus, and cytosol. Ten percent of synthesized GSH is found in mitochondria, where GSH can eliminate excessive ROS and protect cells from apoptosis. The high ratio of GSH/GSSG in the nucleus ensures the synthesis of deoxyribonucleotides and preserves the sulfhydryl groups of proteins for the biosynthesis of proper nucleic acids and DNA repair. In contrast, a low ratio of GSH/GSSG leads to an oxidized environment and enhanced addition reaction of disulfide bonds to the nascent proteins [69]. Moreover, reduced GSH can conjugate with many chemotherapeutic drugs such as cisplatin and doxorubicin, via a glutathione S-transferases (GST)-catalyzed reaction, resulting in increased clearance of these drugs [70, 71]. The efflux of chemotherapeutic drugs is often mediated by multidrug resistance protein 1 (MRP1), leading to cellular detoxification and apoptosis resistance [72]. Cancer cells are equipped with higher levels of GSH, which can be utilized for selectively targeting cancer cells. It has been reported that a doxorubicin-loaded glutathione-responsive cyclodextrin nanosponges (GSH-NS) could release doxorubicin in cancer cells which possess high GSH levels, resulting in higher antitumor efficacy and less cytotoxicity in normal cells [73]. Moreover, buthionine sulfoximine (BSO) is a potent inhibitor of GSH synthesis via inhibiting the activity of glutamate-cysteine ligase (GCL). BSO is now under intensive studies and has been used in clinic practice.

SAH, the byproduct of methyl transfer reaction, can be hydrolyzed to homocysteine and adenosine by SAH hydrolase (SAHH) [74]. Homocysteine is involved in the transsulfuration pathway to produce cysteine, a critical precursor and rate-limiting factor of GSH synthesis [75]. As GSH is the major antioxidant molecule to buffer ROS, the diversion of homocysteine from methionine regeneration into the cysteine production is important to maintain cellular redox homeostasis [76]. Conceivably, the direction (methionine/SAM regeneration or GSH biosynthesis) that homocysteine follows decides the fate of a cell by influencing the cellular redox states and epigenetic events [77]. Since several types of cancer cells often exhibit high ROS levels and DNA hypo-methylation, cancer cells are prone to utilize homocysteine to regenerate methionine during tumorigenesis. Therefore, targeting the recycle of homocysteine may be promising for tumor suppression.

Targeting the synthesis of GSH and NADPH or using pharmacological ROS inducers including As_2O_3, β-phenylethyl isothiocyanate (PEITC), and doxorubicin (Dox) constitutes the strategy of increasing ROS for anticancer therapy [78]. Importantly, scavenging ROS also exhibits favorable anticancer effect. For example, vitamin E is an important dietary antioxidant which reduces ROS levels through reacting with free radicals. Intriguingly, it has been reported that intake of vitamin E can reduce the risk of liver cancer, indicating that clearance of ROS by dietary supplementation of antioxidant holds great potential for preventing tumorigenesis [79]. However, this hypothesis is facing challenges since supplementation of β-carotene, or vitamin E, cannot reduce the risk of lung cancer [80]. The comprehensive understanding of involved regulatory mechanisms is urgently needed for facilitating the early intervention of tumorigenesis.

11.4 Conclusion

Metabolic network and redox homeostasis are closely intertwined with each other and constitute an orchestrated regulatory loop to participate in tumorigenesis and tumor development. Thus, targeting the vulnerabilities of metabolism and redox imbalance in cancer cells holds great potential for selectively killing malignant cells. However, our understanding regarding the complicated metabolic-redox loop still lacks, and more efforts should be paid to achieve effective approaches. More importantly, since the roles of ROS are multifaceted, it should be discreet to consider the application of ROS modulators (scavengers or inducers) in different cancer types. Furthermore, compounds targeting global metabolism, or redox homeostasis may elicit unexpected side effects on normal tissues, and more specific delivery system may be required for favorable therapeutic outcomes.

Acknowledgments This work was supported by grants from the National Natural Science Foundation of China (81821002, 81790251, and 82003098), National Key Research and Development Project (2020YFA0509400, 2020YFC2002705), and Guangdong Basic and Applied Basic Research Foundation (2019B030302012).

Disclosure Statement No competing financial interests exist.

References

1. Ward PS, Thompson CB. Metabolic reprogramming: a cancer hallmark even Warburg did not anticipate. Cancer Cell. 2012;21:297–308.
2. Pavlova NN, Thompson CB. The emerging hallmarks of cancer metabolism. Cell Metab. 2016;23:27–47.
3. Liberti MV, Locasale JW. The Warburg effect: how does it benefit cancer cells? Trends Biochem Sci. 2016;41:211–8.
4. Tarrado-Castellarnau M, de Atauri P, Cascante M. Oncogenic regulation of tumor metabolic reprogramming. Oncotarget. 2016;7:62726–53.
5. Vernieri C, Casola S, Foiani M, Pietrantonio F, de Braud F, Longo V. Targeting cancer metabolism: dietary and pharmacologic interventions. Cancer Discov. 2016;6:1315–33.
6. Wang K, Jiang J, Lei Y, Zhou S, Wei Y, Huang C. Targeting metabolic-redox circuits for cancer therapy. Trends Biochem Sci. 2019;44:401–14.
7. Bigelow DJ, Squier TC. Redox modulation of cellular signaling and metabolism through reversible oxidation of methionine sensors in calcium regulatory proteins. Biochim Biophys Acta. 1703;2005:121–34.
8. Klomsiri C, Karplus PA, Poole LB. Cysteine-based redox switches in enzymes. Antioxid Redox Signal. 2011;14:1065–77.
9. Paulsen CE, Carroll KS. Cysteine-mediated redox signaling: chemistry, biology, and tools for discovery. Chem Rev. 2013;113:4633–79.
10. Mailloux RJ, Jin X, Willmore WG. Redox regulation of mitochondrial function with emphasis on cysteine oxidation reactions. Redox Biol. 2014;2:123–39.
11. Hayes JD, Dinkova-Kostova AT, Tew KD. Oxidative stress in cancer. Cancer Cell. 2020;
12. Parvez S, Long MJC, Poganik JR, Aye Y. Redox signaling by reactive electrophiles and oxidants. Chem Rev. 2018;118:8798–888.
13. Lau AT, Wang Y, Chiu JF. Reactive oxygen species: current knowledge and applications in cancer research and therapeutic. J Cell Biochem. 2008;104:657–67.
14. O'Malley J, Kumar R, Inigo J, Yadava N, Chandra D. Mitochondrial stress response and cancer. Trends Cancer. 2020;
15. Lin Y, Jiang M, Chen W, Zhao T, Wei Y. Cancer and ER stress: mutual crosstalk between autophagy, oxidative stress and inflammatory response. Biomed Pharmacother. 2019;118:109249.

16. Sandalio LM, Rodriguez-Serrano M, Romero-Puertas MC, del Rio LA. Role of peroxisomes as a source of reactive oxygen species (ROS) signaling molecules. Subcell Biochem. 2013;69:231–55.
17. Bonekamp NA, Volkl A, Fahimi HD, Schrader M. Reactive oxygen species and peroxisomes: struggling for balance. Biofactors. 2009;35:346–55.
18. Schrader M, Fahimi HD. Peroxisomes and oxidative stress. Biochim Biophys Acta. 1763;2006:1755–66.
19. Holmstrom KM, Finkel T. Cellular mechanisms and physiological consequences of redox-dependent signalling. Nat Rev Mol Cell Biol. 2014;15:411–21.
20. Murphy MP. How mitochondria produce reactive oxygen species. Biochem J. 2009;417:1–13.
21. Magnani F, Mattevi A. Structure and mechanisms of ROS generation by NADPH oxidases. Curr Opin Struct Biol. 2019;59:91–7.
22. Schenk B, Fulda S. Reactive oxygen species regulate Smac mimetic/TNFalpha-induced necroptotic signaling and cell death. Oncogene. 2015;34:5796–806.
23. Lei XG, Zhu JH, Cheng WH, Bao Y, Ho YS, Reddi AR, et al. Paradoxical roles of antioxidant enzymes: basic mechanisms and health implications. Physiol Rev. 2016;96:307–64.
24. He L, He T, Farrar S, Ji L, Liu T, Ma X. Antioxidants maintain cellular redox homeostasis by elimination of reactive oxygen species. Cell Physiol Biochem. 2017;44:532–53.
25. Mironczuk-Chodakowska I, Witkowska AM, Zujko ME. Endogenous non-enzymatic antioxidants in the human body. Adv Med Sci. 2018;63:68–78.
26. Forman HJ, Zhang H, Rinna A. Glutathione: overview of its protective roles, measurement, and biosynthesis. Mol Asp Med. 2009;30:1–12.
27. Gorrini C, Harris IS, Mak TW. Modulation of oxidative stress as an anticancer strategy. Nat Rev Drug Discov. 2013;12:931–47.
28. Poprac P, Jomova K, Simunkova M, Kollar V, Rhodes CJ, Valko M. Targeting free radicals in oxidative stress-related human diseases. Trends Pharmacol Sci. 2017;38:592–607.
29. Wang J, Lu Q, Cai J, Wang Y, Lai X, Qiu Y, et al. Nestin regulates cellular redox homeostasis in lung cancer through the Keap1-Nrf2 feedback loop. Nat Commun. 2019;10:5043.
30. Schieber M, Chandel NS. ROS function in redox signaling and oxidative stress. Curr Biol. 2014;24:R453–62.
31. Dawane JS, Pandit VA. Understanding redox homeostasis and its role in cancer. J Clin Diagn Res. 2012;6:1796–802.
32. Liu H, Liu X, Zhang C, Zhu H, Xu Q, Bu Y, et al. Redox imbalance in the development of colorectal cancer. J Cancer. 2017;8:1586–97.
33. Pastore S, Korkina L. Redox imbalance in T cell-mediated skin diseases. Mediat Inflamm. 2010;2010:861949.
34. Banhegyi G, Mandl J, Csala M. Redox-based endoplasmic reticulum dysfunction in neurological diseases. J Neurochem. 2008;107:20–34.
35. Meyer Y, Buchanan BB, Vignols F, Reichheld JP. Thioredoxins and glutaredoxins: unifying elements in redox biology. Annu Rev Genet. 2009;43:335–67.
36. Berndt C, Lillig CH, Holmgren A. Thiol-based mechanisms of the thioredoxin and glutaredoxin systems: implications for diseases in the cardiovascular system. Am J Physiol Heart Circ Physiol. 2007;292:H1227–36.
37. Hardie DG. AMP-activated protein kinase: an energy sensor that regulates all aspects of cell function. Genes Dev. 2011;25:1895–908.
38. Xie N, Yuan K, Zhou L, Wang K, Chen HN, Lei Y, et al. PRKAA/AMPK restricts HBV replication through promotion of autophagic degradation. Autophagy. 2016;12:1507–20.
39. Zmijewski JW, Banerjee S, Bae H, Friggeri A, Lazarowski ER, Abraham E. Exposure to hydrogen peroxide induces oxidation and activation of AMP-activated protein kinase. J Biol Chem. 2010;285:33154–64.
40. Shao D, Oka S, Liu T, Zhai P, Ago T, Sciarretta S, et al. A redox-dependent mechanism for regulation of AMPK activation by Thioredoxin1 during energy starvation. Cell Metab. 2014;19:232–45.

41. Nicholls C, Li H, Liu JP. GAPDH: a common enzyme with uncommon functions. Clin Exp Pharmacol Physiol. 2012;39:674–9.
42. Ravichandran V, Seres T, Moriguchi T, Thomas JA, Johnston RB Jr. S-thiolation of glyceraldehyde-3-phosphate dehydrogenase induced by the phagocytosis-associated respiratory burst in blood monocytes. J Biol Chem. 1994;269:25010–5.
43. Shenton D, Grant CM. Protein S-thiolation targets glycolysis and protein synthesis in response to oxidative stress in the yeast Saccharomyces cerevisiae. Biochem J. 2003;374:513–9.
44. Hwang NR, Yim SH, Kim YM, Jeong J, Song EJ, Lee Y, et al. Oxidative modifications of glyceraldehyde-3-phosphate dehydrogenase play a key role in its multiple cellular functions. Biochem J. 2009;423:253–64.
45. Nakajima H, Amano W, Kubo T, Fukuhara A, Ihara H, Azuma YT, et al. Glyceraldehyde-3-phosphate dehydrogenase aggregate formation participates in oxidative stress-induced cell death. J Biol Chem. 2009;284:34331–41.
46. Yun J, Mullarky E, Lu C, Bosch KN, Kavalier A, Rivera K, et al. Vitamin C selectively kills KRAS and BRAF mutant colorectal cancer cells by targeting GAPDH. Science. 2015;350:1391–6.
47. Mazurek S. Pyruvate kinase type M2: a key regulator of the metabolic budget system in tumor cells. Int J Biochem Cell Biol. 2011;43:969–80.
48. Anastasiou D, Poulogiannis G, Asara JM, Boxer MB, Jiang JK, Shen M, et al. Inhibition of pyruvate kinase M2 by reactive oxygen species contributes to cellular antioxidant responses. Science. 2011;334:1278–83.
49. Liang J, Cao R, Wang X, Zhang Y, Wang P, Gao H, et al. Mitochondrial PKM2 regulates oxidative stress-induced apoptosis by stabilizing Bcl2. Cell Res. 2017;27:329–51.
50. Saleme B, Gurtu V, Zhang Y, Kinnaird A, Boukouris AE, Gopal K, et al. Tissue-specific regulation of p53 by PKM2 is redox dependent and provides a therapeutic target for anthracycline-induced cardiotoxicity. Sci Transl Med. 2019;11
51. Liu Y, Guo JZ, Liu Y, Wang K, Ding W, Wang H, et al. Nuclear lactate dehydrogenase A senses ROS to produce alpha-hydroxybutyrate for HPV-induced cervical tumor growth. Nat Commun. 2018;9:4429.
52. Cancer Genome Atlas Research N, Brat DJ, Verhaak RG, Aldape KD, Yung WK, Salama SR, et al. Comprehensive, integrative genomic analysis of diffuse lower-grade gliomas. N Engl J Med. 2015;372:2481–98.
53. Bleeker FE, Lamba S, Leenstra S, Troost D, Hulsebos T, Vandertop WP, et al. IDH1 mutations at residue p.R132 (IDH1(R132)) occur frequently in high-grade gliomas but not in other solid tumors. Hum Mutat. 2009;30:7–11.
54. Xu W, Yang H, Liu Y, Yang Y, Wang P, Kim SH, et al. Oncometabolite 2-hydroxyglutarate is a competitive inhibitor of alpha-ketoglutarate-dependent dioxygenases. Cancer Cell. 2011;19:17–30.
55. Kil IS, Park JW. Regulation of mitochondrial NADP+-dependent isocitrate dehydrogenase activity by glutathionylation. J Biol Chem. 2005;280:10846–54.
56. Fack F, Tardito S, Hochart G, Oudin A, Zheng L, Fritah S, et al. Altered metabolic landscape in IDH-mutant gliomas affects phospholipid, energy, and oxidative stress pathways. EMBO Mol Med. 2017;9:1681–95.
57. Kim SH, Kim H, Lee JH, Park JW. Oxalomalate suppresses metastatic melanoma through IDH-targeted stress response to ROS. Free Radic Res. 2019;53:418–29.
58. Tan J, Yang X, Zhuang L, Jiang X, Chen W, Lee PL, et al. Pharmacologic disruption of Polycomb-repressive complex 2-mediated gene repression selectively induces apoptosis in cancer cells. Genes Dev. 2007;21:1050–63.
59. Conte M, Altucci L. Molecular pathways: the complexity of the epigenome in cancer and recent clinical advances. Clin Cancer Res. 2012;18:5526–34.
60. Comerford SA, Huang Z, Du X, Wang Y, Cai L, Witkiewicz AK, et al. Acetate dependence of tumors. Cell. 2014;159:1591–602.
61. Meier JL. Metabolic mechanisms of epigenetic regulation. ACS Chem Biol. 2013;8:2607–21.

62. Lee C, Raffaghello L, Brandhorst S, Safdie FM, Bianchi G, Martin-Montalvo A, et al. Fasting cycles retard growth of tumors and sensitize a range of cancer cell types to chemotherapy. Sci Trans Med. 2012;4:124ra27.
63. Van Poznak C, Seidman AD, Reidenberg MM, Moasser MM, Sklarin N, Van Zee K, et al. Oral gossypol in the treatment of patients with refractory metastatic breast cancer: a phase I/II clinical trial. Breast Cancer Res Treat. 2001;66:239–48.
64. Xie H, Yin J, Shah MH, Menefee ME, Bible KC, Reidy-Lagunes D, et al. A phase II study of the orally administered negative enantiomer of gossypol (AT-101), a BH3 mimetic, in patients with advanced adrenal cortical carcinoma. Investig New Drugs. 2019;37:755–62.
65. Kamarudin MNA, Sarker MMR, Zhou JR, Parhar I. Metformin in colorectal cancer: molecular mechanism, preclinical and clinical aspects. J Exp Clin Cancer Res. 2019;38:491.
66. Kim JW, Tchernyshyov I, Semenza GL, Dang CV. HIF-1-mediated expression of pyruvate dehydrogenase kinase: a metabolic switch required for cellular adaptation to hypoxia. Cell Metab. 2006;3:177–85.
67. Wang X, Min S, Liu H, Wu N, Liu X, Wang T, et al. Nf1 loss promotes Kras-driven lung adenocarcinoma and results in Psat1-mediated glutamate dependence. EMBO Mol Med. 2019;11
68. Benhar M, Shytaj IL, Stamler JS, Savarino A. Dual targeting of the thioredoxin and glutathione systems in cancer and HIV. J Clin Invest. 2016;126:1630–9.
69. Bansal A, Simon MC. Glutathione metabolism in cancer progression and treatment resistance. J Cell Biol. 2018;217:2291–8.
70. Cole SP. Targeting multidrug resistance protein 1 (MRP1, ABCC1): past, present, and future. Annu Rev Pharmacol Toxicol. 2014;54:95–117.
71. Cole SP, Deeley RG. Transport of glutathione and glutathione conjugates by MRP1. Trends Pharmacol Sci. 2006;27:438–46.
72. Wiese M, Stefan SM. The A-B-C of small-molecule ABC transport protein modulators: from inhibition to activation-a case study of multidrug resistance-associated protein 1 (ABCC1). Med Res Rev. 2019;39:2031–81.
73. Daga M, Ullio C, Argenziano M, Dianzani C, Cavalli R, Trotta F, et al. GSH-targeted nanosponges increase doxorubicin-induced toxicity "in vitro" and "in vivo" in cancer cells with high antioxidant defenses. Free Radic Biol Med. 2016;97:24–37.
74. Xiao Y, Xia J, Cheng J, Huang H, Zhou Y, Yang X, et al. Inhibition of S-Adenosylhomocysteine hydrolase induces endothelial dysfunction via epigenetic regulation of p66shc-mediated oxidative stress pathway. Circulation. 2019;139:2260–77.
75. Locasale JW. Serine, glycine and one-carbon units: cancer metabolism in full circle. Nat Rev Cancer. 2013;13:572–83.
76. Aldini G, Altomare A, Baron G, Vistoli G, Carini M, Borsani L, et al. N-Acetylcysteine as an antioxidant and disulphide breaking agent: the reasons why. Free Radic Res. 2018;52:751–62.
77. Mentch SJ, Mehrmohamadi M, Huang L, Liu X, Gupta D, Mattocks D, et al. Histone methylation dynamics and gene regulation occur through the sensing of one-carbon metabolism. Cell Metab. 2015;22:861–73.
78. Moloney JN, Cotter TG. ROS signalling in the biology of cancer. Semin Cell Dev Biol. 2018;80:50–64.
79. Zhang W, Shu XO, Li H, Yang G, Cai H, Ji BT, et al. Vitamin intake and liver cancer risk: a report from two cohort studies in China. J Natl Cancer Inst. 2012;104:1173–81.
80. Alpha-Tocopherol BCCPSG. The effect of vitamin E and beta carotene on the incidence of lung cancer and other cancers in male smokers. N Engl J Med. 1994;330:1029–35.

Stress and Inflammation

12

Ting Li and Yi Zhun Zhu

Abstract

Both stress and inflammatory responses play a key role in maintaining homeostasis in the body. Stress responds to threats while inflammation responds to injury and infection by some physiological mechanisms in our life. In physiological status, our body is in dynamic equilibrium. Once the homeostasis is disrupted by acute or chronic stressor, a stress response will be triggered to restore normalcy, indicating that stress can be thought of as a protective function in maintaining physiological condition. Of interest, inflammatory response will be triggered to eliminate or promote adaptations to the stressor if the stress response is insufficient to recover homeostasis state. Growing evidences indicate that stress hormones and many types of stress such as endoplasmic reticulum stress and oxidative stress may induce inflammatory responses in multiple tissues and organs if the homeostasis cannot be recovered. On the other hand, long-term stress-induced inflammation will cause a majority of diseases including cardiovascular disease, metabolic diseases, neurodegenerative disease, cancer, and arthritis, indicating that the relationship between stress and inflammatory is ambiguous and it's well worth illuminating. Since stress-induced inflammation gains more foci in research and clinical practice, clarifying the relationship between the stress and inflammation and exploring correlated signaling pathways may provide new strategies for the treatment of inflammatory diseases. Therefore,

T. Li
State Key Laboratory of Quality Research in Chinese Medicine/Macau Institute for Applied Research in Medicine and Health, Macau University of Science and Technology, Avenida Wai Long, Taipa, Macau
e-mail: tli@must.edu.mo

Y. Z. Zhu (✉)
School of Pharmacy, Macau University of Science and Technology, Avenida Wai Long, Taipa, Macau
e-mail: yzzhu@must.edu.mo

C. Huang, Y. Zhang (eds.), *Oxidative Stress*,
https://doi.org/10.1007/978-981-16-0522-2_12

in this chapter, we systemically summarized and discussed the correlation between the stress and inflammation and associated diseases.

Keywords

Stress · Inflammation · Homeostasis · Inflammatory diseases

12.1 Stress and Diseases

Stress is not only a feeling but also a physiological mechanism which responds to threats in our life. The stress response system is made up of psychological, affective, and neuroendocrine components that can be activated by a range of physical and psychosocial stressors. When we're stressed, our blood pressure increases and pulse rises, and we breathe faster. Stress is generally consist of two types, acute stress and chronic stress. Acute stress (also known as the fight-or-flight response) is an immediate and short-term body response to the presence of something that is terrifying, either mentally or physically.

Chronic stress is the activation of a stress mechanism that is exposed to different stressors for a long time. Exposure to chronic stress can cause long-term or permanent changes in emotional, physiological, and behavioral responses, thereby affecting the susceptibility and duration of the disease [1]. Chronic stress could destroy tissue structure and raise blood pressure and damage the heart by inducing cortisol and adrenaline [2]. Therefore, prolonged or repeated activation of the stressor-elicited endocrine response systems may interfere with their control of other physiological systems, leading to an increased risk of physical and mental illness, such as cardiovascular dysfunction, diabetes, cancer, autoimmune syndrome, and mental illnesses [3].

As mentioned, chronic stress may be related to changes in parts of the brain, which may extend to other functionally related areas and cause cognitive, emotional, and behavioral dysfunctions. The effects of stress on the regulation of immune and inflammatory processes may affect the occurrence of depression; cause infections and autoimmune disease; and arise specific viruses-induced cancers and coronary artery disease [3]. And stress-induced depression can further affect susceptibility to disease and subsequent recovery [4].

Glucocorticoids (GCs) released under stress cause the aging process and neuron loss [5]. A recent study has shown that exposure to stress can accelerate the erosion of DNA fragments which are called telomeres [6]. Shorter telomere length is related to age, as well as to the disease morbidity and mortality. It is well known that the stress of early years can have an influential direct negative impact on future health. Hence, telomere erosion is a potential mechanism linking childhood stress to health problems later in life [7]. In addition, psychological stress induces acute phase reactions related to infection and tissue damage and increases the levels of circulating cytokines and various biomarkers of inflammation [8]. Therefore, stress plays an important role in multiple diseases.

12.2 Inflammation and Diseases

Inflammation is part of the body's immune response to injury and infection. It sends a signal to the body to defend against foreign invaders such as viruses and bacteria, as well as repair the damaged tissues. The symptoms of inflammation depend on whether the response is acute or chronic.

Acute inflammation is a short-term response that starts rapidly with local effects and may become severe in a short space of time. Generally, stress is correlated with chronic inflammation. Chronic inflammation is a long-term and whole-body effect. Under certain circumstances, acute inflammation can turn into chronic inflammation. An overreactive immune response will happen when it's unable to eliminate some factors remaining in the system that cause acute inflammation. For example, pathogens produces stable and low levels of inflammation throughout the body. This low level of inflammation can trigger an immune response. However, due to mistaking targets, the white blood cells start attacking internal organs or other healthy tissues and cells that eventually leads to a variety of chronic inflammatory or autoimmune diseases and conditions, including certain cancers, rheumatoid arthritis, and atherosclerosis [9, 10].

Chronic inflammation has been linked to heart disease and stroke because plaques can form when inflammatory cells stay in the blood vessels for a long time. The human body considers plaque as a foreign substance so that it tries to clear the plaque with flowing blood. If the plaque becomes unstable and ruptures, it can form a clot that blocks blood flow to the heart or brain, causing a heart attack or stroke [11, 12].

During inflammation, a large number of macrophages are accumulated, inflammasome are formed, and a large amount of tumor necrosis factor-α (TNF-α) and IL-6 are produced, which lead to atherosclerosis [13]. In addition, inflammation is a crucial factor in the development of metabolic diseases. Recent work has shown that inflammasome and caspase-1 play key roles in the regulation of obesity, insulin resistance, and cardiovascular disease (CVD) [14].

Chronic inflammation can also be caused by chronic infections. Hepatitis B and C virus infections are related to liver cancer; HIV increases the risk of other viruses and very rare cancers such as aggressive B-cell non-Hodgkin lymphoma and invasive cervical cancer [15]. In addition to being associated with many infectious diseases, inflammation is closely related to non-communicable diseases [16]. Cancer is one of the diseases associated with chronic inflammation. The tumor microenvironment composed of inflammatory cells is an essential participant in tumorigenesis by promoting tumor cell proliferation, survival, and migration [17]. Cytokines produced during the inflammatory process may stimulate the growth of blood vessels that bring oxygen and nutrients to the tumor. The process may also generate molecules called free radicals that further damage the DNA as well and implicate at every stage of tumorigenesis [18, 19]. Other evidence demonstrates that the Ras cancer gene plays a role in inflammation, which illustrates the role of chronic inflammation in many cancers at various stages of their progression [20].

Another major category of diseases related to chronic inflammation are those involving progressive and irreversible damage to the central nervous system. The brain is considered to be an "immune-privileged" organ. It is not susceptible to inflammation or immune activation and is considered largely immune to systemic inflammation and immune responses. However, recent results suggest that inflammation in the central nervous system can cause many acute and chronic degenerative diseases and may even lead to certain mental illnesses. Inflammation of the central nervous system has a potential role in diseases such as acute brain injury, stroke, epilepsy, multiple sclerosis, motor neuron disease, dyskinesia, and Alzheimer's disease. And some mental illnesses have also recently appeared, such as depression, anxiety, and schizophrenia [5].

12.3 Stress and Inflammation

Although it is uncontroversial that both stress and inflammatory responses belong to protective responses that maintain homeostasis, the relationship between them is ambiguous and the connection is needed to illuminate. Under physiological status, the system (cells, tissues, and organs) is in a homeostatic condition when the values of regulated variables are within an acceptable dynamic range. Then, a stress response will be triggered while the homeostatic capacity is insufficient to maintain these values (e.g., due to external interventions). If the stress response is insufficient to recover homeostasis, an inflammatory response will be engaged. Both stress response and inflammation are induced to eliminate the stressor, to promote adaptations to the stressor, and ultimately to return the system to the homeostatic state [21].

Accumulating evidence indicates that many types of stress can affect the inflammatory response in multiple tissues or organs. For example, GCs are one of the major stress hormones released during stress responses that are well known for their immunosuppressive and anti-inflammatory properties. GCs could decrease the expression of pro-inflammatory cytokines including TNF-α and IL-6 and increase the expression of anti-inflammatory cytokines including IL-10 and TNF-β [22]. However, the opposite role of GCs has been revealed by other studies. The studies found that GCs could promote the expression and function of inflammasome NLRP3 and in turn upregulate the secretion of IL-1β and IL-18 upon ATP [23]. Above all, the innate immune system has been activated through stimulated with GCs in response to danger signals.

The function of the endoplasmic reticulum (ER) tends to be particularly vital in immune cells because these cells produce a very large amount of proteins. ER stress usually causes inflammation, which in turn restrains impairment of tissue and promotes the repairment of tissue. In ER stress induced-inflammation, many signaling pathways are activated. Nucleotide-binding oligomerization domain 1 (NOD1), as an intracellular pattern recognition receptor (PRR), is associated with the innate immune response because its stimulation induces activation of NF-κB and promotes induction of cytokines upon ER stress-induced unfolded protein response (UPR). Likewise, the closely related PRR NOD2 has also been linked to inflammation by responding to the ER stress [24].

It has been found that ER stress could drive neuroinflammation involving in astrocytes and microglia [25]. ER stress-induced activation of the STAT3/JAK1 axis induces the expression of IL-6. Furthermore, the activation of the STAT3 signaling pathway is linked to PERK, a vital component of the ER stress response, which is phosphorylated by JAK1. Inhibition of PERK expression restrains ER stress-induced activation of STAT3 and subsequent gene expression. In addition, ER-stressed astrocytes trigger activation of microglia, leading to the production of IL-6 and oncostatin M (OSM) via paracrine signaling. Then, these cytokines can synergize with ER stress in astrocytes to drive inflammation [25, 26].

Oxidative stress is another typical stress response and is also linked to the inflammatory response. Reactive oxygen species (ROS) are generated during glucose or free fatty acid oxidation by mitochondria and from metabolic processes elsewhere in the cell. In physiological status, cells exhibit a self-protective activity against oxidative damage made up of enzymatic and non-enzymatic components [27]. However, the production of ROS is increased, and disruption of the balance between oxidant and antioxidant factors results in a pro-oxidative condition [28]. This oxidative stress can then damage cellular structures and trigger an inflammatory response [29].

Collectively, the stress response may induce an inflammatory response when the cell has no capacity to recover homeostasis. Long-term stress-induced inflammation would cause a majority of diseases, such as CVD, metabolic diseases, neurodegenerative disease, cancer, and arthritis diseases.

12.3.1 Stress, Inflammation, and Cardiovascular Disease

Cardiovascular disease (CVD) is a class of diseases that involve the heart and blood vessels. The disease has been classified to be the leading cause of death all over the world. Accumulating evidence has indicated that chronic stress, like early life stress or adult stress, has been linked to rising coronary heart disease (CHD) risk [30]. People who experienced more family conflict in childhood or high intensive stress usually led to higher blood pressure and cholesterol, leading to CVD [31, 32]. Besides, obesity and diabetes mellitus, which are associated with oxidative stress and metabolic stress, are linked to CVD [33].

More and more evidence reveals that chronic stress probably induced chronic low-grade inflammation and further contributed to the early process, progression, and thrombotic complications of atherosclerosis. Exposure to $PM_{2.5}$ could trigger oxidative stress and simultaneously increase the expression of the anti-oxidative gene, nuclear factor (erythroid-derived 2)-like 2-related factor (Nrf2), to activate the downstream anti-oxidative pathway [34]. Furthermore, long-term exposure to $PM_{2.5}$ induces cardiac dysfunction, which even could enhance the cardiovascular events and atherosclerosis [34]. In the $PM_{2.5}$ exposure animal model, $Nrf2^{-/-}$ mice display more severe inflammation and activate IκBα/NF-κB signaling pathway [35].

Atherosclerosis, a kind of CVD, mainly results from endothelial injury that causes cholesterol-containing LDL particles depositing in the arterial wall, and then the inflammatory response could be triggered [36]. Both adaptive and innate immunity are activated to contribute to the initiation and progression of atherosclerosis. For example, monocytes could differentiate into macrophages in the intima of the arterial wall and then transform into foam cells in the lipid necrotic core of the atheroma [37, 38]. Inflammation could be activated by crystals of cholesterol and other DAMPs located in the atherosclerotic lesion. The activation of inflammasomes promotes the release of IL-1β, IL-18, and other pro-inflammatory cytokines that are chemotactic for other immune cells for driving atherosclerosis [39, 40]. Late stage of atherosclerosis is featured by enormous number of cells undergoing senescence and apoptosis. In line with the results, necrotic core could be formed, finally causing plaque fragility and rupture, thrombus formation, and acute vascular occlusion [41].

It has been reported that IL-6 and C-reactive protein (CRP), rather than other inflammatory cytokines, are considered as the biomarkers of systematic inflammation and atherosclerosis [42]. High-sensitivity CRP (hs-CRP) levels could be decreased by statins, indicating that inhibition of inflammation could be applied as a therapeutic strategy to treat CVD [42, 43]. Collectively, these studies illuminate the correlation among stress, inflammation, and CVD, and anti-inflammatory drugs may have a synergistic effect with conventional antihypertensive drugs on the prevention and treatment of stress-related CVD.

12.3.2 Stress, Inflammation, and Metabolic Diseases

Stress is thought to be involved in the pathogenesis of several metabolic diseases, such as obesity, insulin resistance, and type 2 diabetes [44–46]. Metabolic ER stress disorders can disrupt the balance of glucose and energy metabolism. During overnutrition, obesity factors (such as fatty acids and LPS) activate macrophages, thereby enhancing M1 polarization in the intracellular environment that inhibits M2 polarization. Such M1-M2 imbalances in turn severely limit the energy utilization capabilities of brown and beige fats and enhance insulin resistance, which ultimately promote the progression of metabolic diseases [45].

It is thought that metabolic diseases are mainly driven by inflammation [47]. The emergence of chronic inflammation in obese individuals can lead to impaired insulin action, which in turn leads to the development of metabolic abnormalities [48, 49]. During obesity, the total number of macrophages dramatically increases. This is largely due to the recruitment of M1 polarized macrophages, which exhibit a more inflammatory phenotype and secrete TNF-α cytokines. In addition, the anti-inflammatory cytokines IL-4 and IL-13 produced by eosinophils can promote the differentiation of macrophages into alternately activated M2 polarized cells. Congenital type 2 lymphoid cells (ILC2s) play a key role in maintaining metabolic homeostasis by producing IL-5 and IL-13 in eosinophils and adipose tissue. An increase in the number of macrophages and an increase in the ratio of M1 to M2 macrophages are signs of adipose tissue inflammation caused by obesity and are related to the development of insulin resistance and metabolic diseases [50].

12.3.3 Stress, Inflammation, and Neurodegenerative Diseases

Neurological diseases include Alzheimer's disease (AD), vascular dementia (VAD), Parkinson's disease (PD), and multiple sclerosis (MS). Increasing evidence indicates that the pathogenesis of these diseases is not only limited to the neuronal compartment but also included the interactions with immune mechanisms in the brain, which is called neuroinflammation. Neuroinflammation is mainly driven by microglia, macrophages, and astrocytes in the brain. These cells play a central role in many aspects of brain metabolism and physiology under non-pathological conditions, maintaining the blood-brain barrier (BBB) integrity [51]. One of the common characteristics of these multiple diseases is that their neuropathological features differ from microglia and astrocytes and trigger an innate immune response characterized by the release of inflammatory mediators, leading to the progression and severity of neurodegenerative disease degree [52].

In the central nervous system, chronic neurodegenerative diseases are featured with neuroinflammation mitochondrial dysfunction. ROS and nitrogen. species (RNS) are overexpressed and result in intensively oxidative stress under the two conditions and further result in neuronal damage and subsequently induce inflammation, causing to a feed-forward loop of neurodegeneration.

Repeated immune attacks result in plaque formation characterized by the formation of glial scars, leading to mitochondrial damage and dysfunction [53]. In mitochondrial injury and dysfunction, excessive production and release of ROS produced by oxidative stress have been proposed as a general pathological mechanism for all major chronic neurodegenerative diseases [54, 55]. NF-κB activation can induce NOX2 and iNOS to effectively produce ROS/RNS under pro-inflammatory conditions [56]. In addition, the activation of NF-κB promotes COX-2 and cPLA2 expression correlated to prostaglandin production, thereby further promoting oxidative stress. It has been proved that there is a intensive correlation between the production of ROS/RNS and the induction of pro-inflammatory cytokines, which exacerbates neurodegeneration [56, 57].

One of the important cytokines, TNF, can activate and recruit immune cells through TNF receptor 1 (TNFR1), indicating that it plays a key role in inflammation pathogenesis. Furthermore, TNFR1 could increase ROS production and RNS-producing enzymes via induction of oxidative stress. Hence then, TNF-induced inflammation and oxidative stress cooperatively facilitate neurodegeneration. Additionally, TNF also interacts with TNF receptor 2 (TNFR) 2 to generate a neuroprotective effect and promote tissue regeneration, suggesting that TNF plays a dual role in neurodegenerative diseases [58].

Once TLR-mediated signaling is activated, astrocytes and microglia release cytokines, chemokines, and ROS to promote inflammation and exacerbate neuron damage [59]. The continued existence of neuronal damage will lead to the sustained activation of microglia and astrocytes, which in turn produces a neuroinflammatory environment to promote neurodegeneration [54, 60]. Collectively, there is a close correlation among stress, inflammation, and neurodegenerative diseases, and elucidation of the correlation is helpful for developing the intervention to treat neurodegenerative diseases.

12.3.4 Stress, Inflammation, and Cancer

Accumulating evidence suggests that psychological stress mediate influence on cancer by producing the key stress-related mediators and their corresponding receptors, which involve in multi-fold pathways [61]. These stress-related mediators act as immunosuppressors or mitogens in the micro-environment of tumor [62]. The psychological stress mediates the effects on cancer cells via the stress mediators, receptors and cytokines to activate the pro-migratory and pro-proliferative signaling pathway. Eventually, the molecular clock of cancer cells will be reset [61].

Classic stress signal, β-adrenergic signaling activation is considered as the main cause of progression and invasion in pancreatic cancer, acute lymphoblastic leukemia, and breast cancer [63]. Furthermore, selective α2-adrenergic blockade mimics the accelerating effect of chronic stress on breast cancer progression [64].

More recently, stress-mediated immune modulation of cytokines including IL-6, TNF-α, and IL-1, has been suggested as indicators of cancer progression, metastasis, and recurrence. Epidemiological data indicate that over 25% of all cancers are related to chronic infections and other types of unresolved inflammation [65, 66]. Certain cancers develop in tissues that are severely damaged by chronic inflammation, such as lung cancer caused by asbestos or cigarette smoke, gastric (stomach) cancer caused by *H. pylori* infection, liver cancer caused by hepatitis B or hepatitis C infection, cervical cancer caused by human papillomavirus infection, and ovarian cancer caused by pelvic inflammatory disease [67]. Accumulating study has addressed the role of the cytokines, chemokines, transcription factors, and eicosanoid to explain the correlation between inflammation and carcinogenesis [68].

Additionally, these cytokines secreted by tumor cells help to drive and sustain pro-tumorigenic inflammatory loops in colon cancer and lung cancer [69]. IL-6 trans-signaling-dependent activation of STAT3 can drive cancer progression through the transcription of target genes, including JunB, cFos, C/EBPβ, and C/EBPδ [70]. Although the connection among stress, inflammation, and cancer onset and progression has been investigated, the underlying mechanism needs to be intensively determined, which will be helpful to understanding every step of carcinogenesis and in turn propose effective therapy for cancer.

12.3.5 Stress, Inflammation, and Arthritis

Rheumatoid arthritis (RA) is an autoimmune disease that not only impairs psychological function but also mediates psychological stress on the patients. Stress is now recognized as an important risk in the pathogenesis of RA by influencing hypothalamic-pituitary-adrenal (HPA) axis, the sympathetic nervous system and the immune system [71]. It has been reported that the affected individuals are characterized by an increased level of anxiety and depression, and stress management interventions produce important clinical benefits for the persons with RA [72, 73].

As mentioned, the stress response results in the release of neurotransmitters (norepinephrine), hormones (cortisol), and immune cells which serve to send an efferent message from the brain to the periphery [74]. It has been reported that stress vulnerability, immune and HPA axis activity aggravate the symptoms of RA [75].

Signalings in cellular stress responses, including the heat shock response, ER stress response, NF-κB, and Nrf2-ARE-HO-1 axis, involve the activation of adaptive and innate immune responses, leading to arthritis [76–78]. Accordingly, ER stress-induced inflammation, the correlated pathways and their constituent elements are attractive targets for RA drug development [79].

Oxidative stress is significantly involved in cartilage degradation in experimental arthritis. Activation of the Nrf2-ARE-HO-1 axis contributes to reducing the oxidative stress by decreasing ROS production. It has been reported that Nrf2 is activated in the joints of arthritic mice and of patients with RA. Nrf2-knockout mice had more oxidative damage and more severe cartilage injuries compared to their wild-type littermates. Remarkably, a number of spontaneously fractured bones have been found in Nrf2-knockout mice with antibody-induced arthritis (AIA) [80]. In coinciding with the study, compounds targeting Nrf2 signaling significantly ameliorated symptoms of collagen-induced arthritis (CIA) in mice [81, 82]. Wu et al. demonstrated for the first time that S-propargyl-cysteine (SPRC, also known as ZYZ-802), an endogenous H_2S modulator, exerted anti-inflammatory properties in RA by upregulating the Nrf2-antioxidant response element (ARE) signaling pathway (Fig. 12.1) [83]. Recently, another study from the same research group reported that suppressing JMJD3 expression by the transcription factor Sp-1 was likely responsible for the ability of Cystathionine-γ-lyase (CSE) to negatively modulate the inflammatory response (Fig. 12.2) and reduce the progression of RA [84].

These studies illuminate the importance of stress in arthritis and pave a new way to treat arthritis.

12.4 Conclusion

In conclusion, stress-induced inflammation has drawn much attention not only in research but also in clinical practice. New proteins and signaling pathways might be discovered, and the efficient blockage of the upregulation of these unwanted proteins could open a new avenue for the treatment of inflammatory diseases.

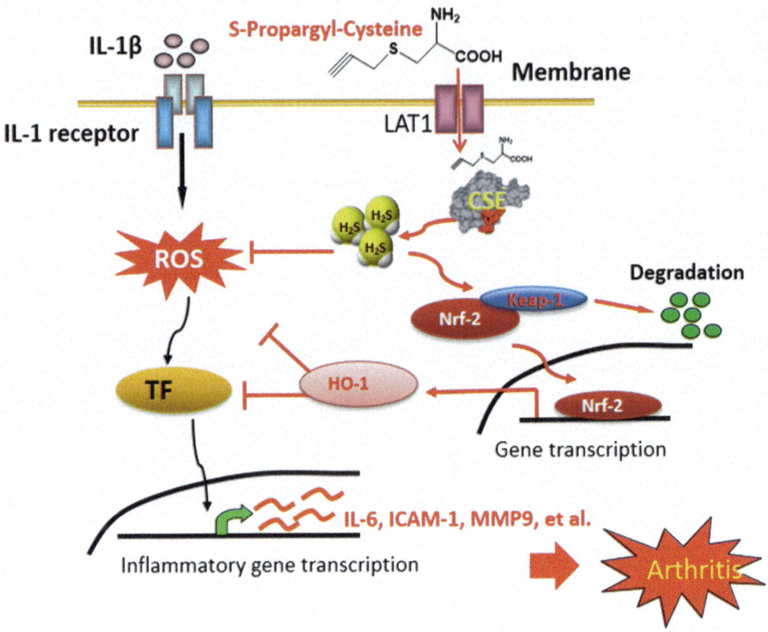

Fig. 12.1 S-propargyl-cysteine attenuates inflammatory response in rheumatoid arthritis by modulating the Nrf2-ARE signaling pathway

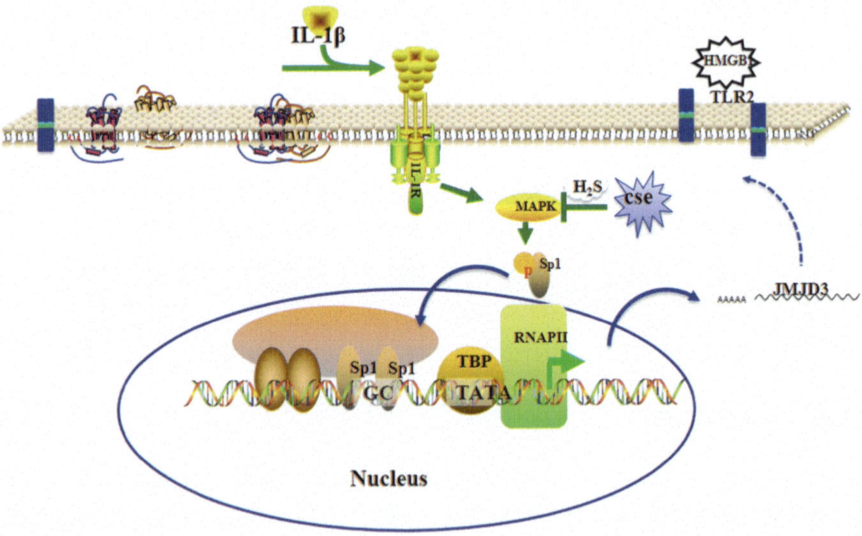

Fig. 12.2 Cystathionine-γ-lyase ameliorates the histone demethylase JMJD3-mediated autoimmune response in rheumatoid arthritis

References

1. Cohen S, Janicki-Deverts D, Miller GE. Psychological stress and disease. JAMA. 2007;298 (14):1685–7.
2. McEwen BS. Protective and damaging effects of stress mediators: central role of the brain. Dialogues Clin Neurosci. 2006;8(4):367–81.
3. Mariotti A. The effects of chronic stress on health: new insights into the molecular mechanisms of brain-body communication. Future Sci OA. 2015;1(3):FSO23.
4. Kiecolt-Glaser JK, McGuire L, Robles TF, Glaser R. Emotions, morbidity, and mortality: new perspectives from psychoneuroimmunology. Annu Rev Psychol. 2002;53:83–107.
5. Lucas SM, Rothwell NJ, Gibson RM. The role of inflammation in CNS injury and disease. Br J Pharmacol. 2006;147(Suppl 1):S232–40.
6. Brindley DN, Rolland Y. Possible connections between stress, diabetes, obesity, hypertension and altered lipoprotein metabolism that may result in atherosclerosis. Clin Sci (Lond). 1989;77 (5):453–61.
7. Shalev I. Early life stress and telomere length: investigating the connection and possible mechanisms: a critical survey of the evidence base, research methodology and basic biology. BioEssays. 2012;34(11):943–52.
8. Black PH. Stress and the inflammatory response: a review of neurogenic inflammation. Brain Behav Immun. 2002;16(6):622–53.
9. Liu YZ, Wang YX, Jiang CL. Inflammation: the common pathway of stress-related diseases. Front Hum Neurosci. 2017;11:316.
10. https://www.medicalnewstoday.com/articles/248423#common-treatments
11. Jin R, Yang G, Li G. Inflammatory mechanisms in ischemic stroke: role of inflammatory cells. J Leukoc Biol. 2010;87(5):779–89.
12. Lopez-Candales A, Hernandez Burgos PM, Hernandez-Suarez DF, Harris D. Linking chronic inflammation with cardiovascular disease: from normal aging to the metabolic syndrome. J Nat Sci. 2017;3(4).
13. Katsiari CG, Bogdanos DP, Sakkas LI, Manuscript NO: 41374 manuscript type: editorial inflammation and cardiovascular disease
14. Skeldon AM, Faraj M, Saleh M. Caspases and inflammasomes in metabolic inflammation. Immunol Cell Biol. 2014;92(4):304–13.
15. Shiels MS, Pfeiffer RM, Hall HI, Li J, Goedert JJ, Morton LM, Hartge P, Engels EA. Proportions of Kaposi sarcoma, selected non-Hodgkin lymphomas, and cervical cancer in the United States occurring in persons with AIDS, 1980-2007. JAMA. 2011;305(14):1450–9.
16. Hunter P. The inflammation theory of disease. The growing realization that chronic inflammation is crucial in many diseases opens new avenues for treatment. EMBO Rep. 2012;13 (11):968–70.
17. Coussens LM, Werb Z. Inflammation and cancer. Nature. 2002;420(6917):860–7.
18. Mantovani A, Allavena P, Sica A, Balkwill F. Cancer-related inflammation. Nature. 2008;454 (7203):436–44.
19. https://www.cancercenter.com/community/blog/2018/08/inflammation-linked-to-cancer-but-lifestyle-changes-may-help
20. Liu J, Yang G, Thompson-Lanza JA, Glassman A, Hayes K, Patterson A, Marquez RT, Auersperg N, Yu Y, Hahn WC, Mills GB, Bast RC Jr. A genetically defined model for human ovarian cancer. Cancer Res. 2004;64(5):1655–63.
21. Chovatiya R, Medzhitov R. Stress, inflammation, and defense of homeostasis. Mol Cell. 2014;54(2):281–8.
22. Sorrells SF, Caso JR, Munhoz CD, Sapolsky RM. The stressed CNS: when glucocorticoids aggravate inflammation. Neuron. 2009;64(1):33–9.
23. Busillo JM, Azzam KM, Cidlowski JA. Glucocorticoids sensitize the innate immune system through regulation of the NLRP3 inflammasome. J Biol Chem. 2011;286(44):38703–13.

24. Mendez JM, Kolora LD, Lemon JS, Dupree SL, Keestra-Gounder AM. Activation of the endoplasmic reticulum stress response impacts the NOD1 signaling pathway. Infect Immun. 2019;87(8).
25. Meares GP, Liu Y, Rajbhandari R, Qin H, Nozell SE, Mobley JA, Corbett JA, Benveniste EN. PERK-dependent activation of JAK1 and STAT3 contributes to endoplasmic reticulum stress-induced inflammation. Mol Cell Biol. 2014;34(20):3911–25.
26. Garfinkel BP, Hotamisligil GS. ER stress promotes inflammation through re-wIREd macrophages in obesity. Mol Cell. 2017;66(6):731–3.
27. Valdecantos MP, Perez-Matute P, Martinez JA. Obesity and oxidative stress: role of antioxidant supplementation. Rev de Invest Clin Org del Hosp Enfermedades de la Nutr. 2009;61 (2):127–39.
28. Hopps E, Noto D, Caimi G, Averna MR. A novel component of the metabolic syndrome: the oxidative stress. Nutr Metab Cardiovasc Dis. 2010;20(1):72–7.
29. Hotamisligil GS. Inflammation and metabolic disorders. Nature. 2006;444(7121):860–7.
30. Su S, Jimenez MP, Roberts CT, Loucks EB. The role of adverse childhood experiences in cardiovascular disease risk: a review with emphasis on plausible mechanisms. Curr Cardiol Rep. 2015;17(10):88.
31. Bleil ME, Adler NE, Appelhans BM, Gregorich SE, Sternfeld B, Cedars MI. Childhood adversity and pubertal timing: understanding the origins of adulthood cardiovascular risk. Biol Psychol. 2013;93(1):213–9.
32. Zhang Y, Vittinghoff E, Pletcher MJ, Allen NB, Zeki Al Hazzouri A, Yaffe K, Balte PP, Alonso A, Newman AB, Ives DG, Rana JS, Lloyd-Jones D, Vasan RS, Bibbins-Domingo K, Gooding HC, de Ferranti SD, Oelsner EC, Moran AE. Associations of blood pressure and cholesterol levels during young adulthood with later cardiovascular events. J Am Coll Cardiol. 2019;74(3):330–41.
33. Highlander P, Shaw GP. Current pharmacotherapeutic concepts for the treatment of cardiovascular disease in diabetics. Ther Adv Cardiovasc Dis. 2010;4(1):43–54.
34. Lawal AO. Air particulate matter induced oxidative stress and inflammation in cardiovascular disease and atherosclerosis: the role of Nrf2 and AhR-mediated pathways. Toxicol Lett. 2017;270:88–95.
35. Ge C, Hu L, Lou D, Li Q, Feng J, Wu Y, Tan J, Xu M. Nrf2 deficiency aggravates PM2.5-induced cardiomyopathy by enhancing oxidative stress, fibrosis and inflammation via RIPK3-regulated mitochondrial disorder. Aging. 2020;12(6):4836–65.
36. Libby P, Ridker PM, Hansson GK. Progress and challenges in translating the biology of atherosclerosis. Nature. 2011;473(7347):317–25.
37. Libby P, Ridker PM, Hansson GK, Leducq Transatlantic Network on Atherothrombosis. Inflammation in atherosclerosis: from pathophysiology to practice. J Am Coll Cardiol. 2009;54(23):2129–38.
38. Libby P, Okamoto Y, Rocha VZ, Folco E. Inflammation in atherosclerosis: transition from theory to practice. Circ J. 2010;74(2):213–20.
39. Duewell P, Kono H, Rayner KJ, Sirois CM, Vladimer G, Bauernfeind FG, Abela GS, Franchi L, Nunez G, Schnurr M, Espevik T, Lien E, Fitzgerald KA, Rock KL, Moore KJ, Wright SD, Hornung V, Latz E. NLRP3 inflammasomes are required for atherogenesis and activated by cholesterol crystals. Nature. 2010;464(7293):1357–61.
40. Warnatsch A, Ioannou M, Wang Q, Papayannopoulos V. Inflammation. Neutrophil extracellular traps license macrophages for cytokine production in atherosclerosis. Science. 2015;349 (6245):316–20.
41. Grootaert MO, Moulis M, Roth L, Martinet W, Vindis C, Bennett MR , De Meyer GR (2018) Vascular smooth muscle cell death, autophagy and senescence in atherosclerosis. Cardiovasc Res 114 (4): 622–634. 10.1093/cvr/cvy007
42. Nadrowski P, Chudek J, Skrzypek M, Puzianowska-Kuznicka M, Mossakowska M, Wiecek A, Zdrojewski T, Grodzicki T, Kozakiewicz K. Associations between cardiovascular disease risk factors and IL-6 and hsCRP levels in the elderly. Exp Gerontol. 2016;85:112–7.

43. von Kanel R, Bellingrath S, Kudielka BM. Association between burnout and circulating levels of pro- and anti-inflammatory cytokines in schoolteachers. J Psychosom Res. 2008;65(1):51–9.
44. Ozcan U, Cao Q, Yilmaz E, Lee AH, Iwakoshi NN, Ozdelen E, Tuncman G, Gorgun C, Glimcher LH, Hotamisligil GS. Endoplasmic reticulum stress links obesity, insulin action, and type 2 diabetes. Science. 2004;306(5695):457–61.
45. Shan B, Wang X, Wu Y, Xu C, Xia Z, Dai J, Shao M, Zhao F, He S, Yang L, Zhang M, Nan F, Li J, Liu J, Liu J, Jia W, Qiu Y, Song B, Han JJ, Rui L, Duan SZ, Liu Y. The metabolic ER stress sensor IRE1alpha suppresses alternative activation of macrophages and impairs energy expenditure in obesity. Nat Immunol. 2017;18(5):519–29.
46. Wang M, Kaufman RJ. Protein misfolding in the endoplasmic reticulum as a conduit to human disease. Nature. 2016;529(7586):326–35.
47. Kahn SE, Hull RL, Utzschneider KM. Mechanisms linking obesity to insulin resistance and type 2 diabetes. Nature. 2006;444(7121):840–6.
48. Grant RW, Dixit VD. Mechanisms of disease: inflammasome activation and the development of type 2 diabetes. Front Immunol. 2013;4:50.
49. Gregor MF, Hotamisligil GS. Inflammatory mechanisms in obesity. Annu Rev Immunol. 2011;29:415–45.
50. Lumeng CN, Bodzin JL, Saltiel AR. Obesity induces a phenotypic switch in adipose tissue macrophage polarization. J Clin Invest. 2007;117(1):175–84.
51. Sevenich L. Brain-resident microglia and blood-borne macrophages orchestrate central nervous system inflammation in neurodegenerative disorders and brain Cancer. Front Immunol. 2018;9:697.
52. Cervellati C, Trentini A, Pecorelli A, Valacchi G. Inflammation in neurological disorders: the thin boundary between brain and periphery. Antioxid Redox Signal. 2020;33:191–210.
53. Lassmann H, van Horssen J. The molecular basis of neurodegeneration in multiple sclerosis. FEBS Lett. 2011;585(23):3715–23.
54. Amor S, Puentes F, Baker D, van der Valk P. Inflammation in neurodegenerative diseases. Immunology. 2010;129(2):154–69.
55. Yamasaki R, Lu H, Butovsky O, Ohno N, Rietsch AM, Cialic R, Wu PM, Doykan CE, Lin J, Cotleur AC, Kidd G, Zorlu MM, Sun N, Hu W, Liu L, Lee JC, Taylor SE, Uehlein L, Dixon D, Gu J, Floruta CM, Zhu M, Charo IF, Weiner HL, Ransohoff RM. Differential roles of microglia and monocytes in the inflamed central nervous system. J Exp Med. 2014;211(8):1533–49.
56. Morgan MJ, Liu ZG. Crosstalk of reactive oxygen species and NF-kappaB signaling. Cell Res. 2011;21(1):103–15.
57. Hsieh HL, Yang CM. Role of redox signaling in neuroinflammation and neurodegenerative diseases. Biomed Res Int. 2013;2013:484613.
58. Fischer R, Maier O. Interrelation of oxidative stress and inflammation in neurodegenerative disease: role of TNF. Oxidative Med Cell Longev. 2015;2015:610813.
59. van Noort JM, Bsibsi M. Toll-like receptors in the CNS: implications for neurodegeneration and repair. Prog Brain Res. 2009;175:139–48.
60. Zipp F, Aktas O. The brain as a target of inflammation: common pathways link inflammatory and neurodegenerative diseases. Trends Neurosci. 2006;29(9):518–27.
61. Yuan A, Wang S, Li Z, Huang C. Psychological aspect of cancer: From stressor to cancer progression. Exp Ther Med. 2010;1(1):13–8.
62. Jin Shin K, Jin Lee Y, Ryoul Yang Y, Park S, Suh P-G, Yung Follo M, Cocco L, Ho Ryu S. Molecular mechanisms underlying psychological stress and cancer. Curr Pharm Des. 2016;22(16):2389–402.
63. Qin J-F, Jin F-J, Li N, Guan H-T, Lan L, Ni H, Wang Y. Adrenergic receptor β2 activation by stress promotes breast cancer progression through macrophages M2 polarization in tumor microenvironment. BMB Rep. 2015;48(5):295–300.
64. Lamkin DM, Sung HY, Yang GS, David JM, Ma JC, Cole SW, Sloan EK. α2-adrenergic blockade mimics the enhancing effect of chronic stress on breast cancer progression. Psychoneuroendocrinology. 2015;51:262–70.

65. Thaker PH, Han LY, Kamat AA, Arevalo JM, Takahashi R, Lu C, Jennings NB, Armaiz-Pena G, Bankson JA, Ravoori M. Chronic stress promotes tumor growth and angiogenesis in a mouse model of ovarian carcinoma. Nat Med. 2006;12(8):939–44.
66. Wong M, Inserra A, Lewis M, Mastronardi CA, Leong L, Choo J, Kentish S, Xie P, Morrison M, Wesselingh S. Inflammasome signaling affects anxiety-and depressive-like behavior and gut microbiome composition. Mol Psychiatry. 2016;21(6):797–805.
67. Zhang M, Zhou S, Zhang L, Ye W, Wen Q. Role of cancer-related inflammation in esophageal cancer. Crit Rev Eukaryotic Gene Exp. 2013;23(1):27–35.
68. Vendramini-Costa DB, Carvalho JE. Molecular link mechanisms between inflammation and cancer. Curr Pharm Des. 2012;18(26):3831–52.
69. Gao SP, Mark KG, Leslie K, Pao W, Motoi N, Gerald WL, Travis WD, Bornmann W, Veach D, Clarkson B. Mutations in the EGFR kinase domain mediate STAT3 activation via IL-6 production in human lung adenocarcinomas. J Clin Invest. 2007;117(12):3846–56.
70. Thiem S, Pierce TP, Palmieri M, Putoczki TL, Buchert M, Preaudet A, Farid RO, Love C, Catimel B, Lei Z. mTORC1 inhibition restricts inflammation-associated gastrointestinal tumorigenesis in mice. J Clin Invest. 2013;123(2),767–81.
71. Cutolo M, Straub RH. Stress as a risk factor in the pathogenesis of rheumatoid arthritis. Neuroimmunomodulation. 2006;13(5–6):277–82.
72. Ziarko M, Siemiątkowska K, Sieński M, Samborski W, Samborska J, Mojs E. Mental health and rheumatoid arthritis: toward understanding the emotional status of people with chronic disease. Biomed Res Int. 2019;2019:1473925.
73. Rhee SH, Parker JC, Smarr KL, Petroski GF, Johnson JC, Hewett JE, Wright GE, Multon KD, Walker SE. Stress management in rheumatoid arthritis: what is the underlying mechanism? Arthritis Care Res. 2000;13(6):435–42.
74. Straub RH, Dhabhar FS, Bijlsma JW, Cutolo M. How psychological stress via hormones and nerve fibers may exacerbate rheumatoid arthritis. Arthritis Rheumatism. 2005;52(1):16–26.
75. Evers AW, Verhoeven EW, van Middendorp H, Sweep FC, Kraaimaat FW, Donders ART, Eijsbouts AE, van Laarhoven AI, de Brouwer SJ, Wirken L. Does stress affect the joints? Daily stressors, stress vulnerability, immune and HPA axis activity, and short-term disease and symptom fluctuations in rheumatoid arthritis. Ann Rheum Dis. 2014;73(9):1683–8.
76. Muralidharan S, Mandrekar P. Cellular stress response and innate immune signaling: integrating pathways in host defense and inflammation. J Leukoc Biol. 2013;94(6):1167–84.
77. Cerri C, Genovesi S, Allegra M, Pistillo F, Püntener U, Guglielmotti A, Perry VH, Bozzi Y, Caleo M. The chemokine CCL2 mediates the seizure-enhancing effects of systemic inflammation. J Neurosci. 2016;36(13):3777–88.
78. Mahmoud AM, Mohammed HM, Khadrawy SM, Galaly SR. Hesperidin protects against chemically induced hepatocarcinogenesis via modulation of Nrf2/ARE/HO-1, PPARγ and TGF-β1/Smad3 signaling, and amelioration of oxidative stress and inflammation. Chem Biol Interact. 2017;277:146–58.
79. Rahmati M, Moosavi MA, McDermott MF. ER stress: a therapeutic target in rheumatoid arthritis? Trends Pharmacol Sci. 2018;39(7):610–23.
80. Wruck CJ, Fragoulis A, Gurzynski A, Brandenburg L-O, Kan YW, Chan K, Hassenpflug J, Freitag-Wolf S, Varoga D, Lippross S. Role of oxidative stress in rheumatoid arthritis: insights from the Nrf2-knockout mice. Ann Rheum Dis. 2011;70(5):844–50.
81. Su X, Li T, Liu Z, Huang Q, Liao K, Ren R, Lu L, Qi X, Wang M, Chen J. Licochalcone A activates Keap1-Nrf2 signaling to suppress arthritis via phosphorylation of p62 at serine 349. Free Radic Biol Med. 2018;115:471–83.
82. Su X, Huang Q, Chen J, Wang M, Pan H, Wang R, Zhou H, Zhou Z, Liu J, Yang F. Calycosin suppresses expression of pro-inflammatory cytokines via the activation of p62/Nrf2-linked heme oxygenase 1 in rheumatoid arthritis synovial fibroblasts. Pharmacol Res. 2016;113:695–704.

83. Wu WJ, Jia WW, Liu XH, Pan LL, Zhang QY, Yang D, Shen XY, Liu L, Zhu YZ. S-propargyl-cysteine attenuates inflammatory response in rheumatoid arthritis by modulating the Nrf2-ARE signaling pathway. Redox Biol. 2016;10:157–67.
84. Wu WJ, Qin M, Jia WW, Huang Z, Li ZZ, Yang D, Huang MW, Xiao CX, Long F, Mao JC, Moore PK, Liu XH, Zhu YZ. Cystathionine-γ-lyase ameliorates the histone demethylase JMJD3-mediated autoimmune response in rheumatoid arthritis. Cell Mol Immunol. 2019;16 (8),694–705.

Oxidative Stress in Cell Signaling and Cell Fate Determination Under Glucose Starvation

13

Yi Ren, Charlie Cheong, and Han-Ming Shen

Abstract

Glucose is the most important energy and carbon source in cells. At the same time, glucose metabolism also controls the redox homeostasis. Numerous studies have suggested that glucose starvation is closely linked with the generation of ROS and oxidative stress, and a tight redox regulation is crucial for cell survival. On the one hand, glucose starvation leads to a decreased ATP level, which activates the AMP-activated protein kinase (AMPK) to maintain cell energy and redox balance. On the other hand, oxidative stress caused by glucose starvation activates the c-Jun N-terminal kinase (JNK) to initiate apoptosis if ROS are not properly eliminated. Redox imbalance is implicated in many diseases such as cancers, neurodegenerative diseases, and ischemia-reperfusion injuries. Therefore, the cell fate determination is of great importance for the prevention or the treatment of these diseases. This chapter will focus on (1) how glucose is metabolized in cells; (2) how glucose metabolism is linked with ROS and oxidative stress; and (3) the role of AMPK and JNK in the determination of cell fate under glucose starvation.

Y. Ren
Department of Physiology, Yong Loo Lin School of Medicine, National University of Singapore, Singapore, Singapore

C. Cheong
Faculty of Health Sciences, University of Macau, Macau, People's Republic of China

H.-M. Shen (✉)
Department of Physiology, Yong Loo Lin School of Medicine, National University of Singapore, Singapore, Singapore

Faculty of Health Sciences, University of Macau, Macau, People's Republic of China
e-mail: hmshen@um.edu.mo

C. Huang, Y. Zhang (eds.), *Oxidative Stress*,
https://doi.org/10.1007/978-981-16-0522-2_13

Keywords

Stress · Inflammation · Homeostasis · Inflammatory diseases

13.1 Introduction

Sufficient energy supply is essential for the normal functioning of the body. Glucose, fat, and protein are the three major sources of energy. However, not all tissues and organs can utilize fat or protein, for example, the brain and the red blood cells. Glucose, therefore, is the most critical and widely used energy and carbon source in almost all living organisms.

Shortage in glucose supply has been implicated in various diseases including hypoglycemia, neurodegenerative diseases, as well as ischemia-reperfusion injuries in the brain and the heart [1–3]. Oxidative stress induced by glucose starvation is regarded as the major cause of cell death in these diseases [4]. In addition, glucose starvation is also frequently observed in tumors due to limited angiogenesis and rapid consumption of glucose by cancer cells [5, 6]. Therefore, targeting glucose metabolism and oxidative stress may provide novel insights into the therapeutics of these diseases. In this chapter, we will discuss the intersection between glucose metabolism and oxidative stress and the roles of AMP-activated protein kinase (AMPK) and the c-Jun N-terminal kinase (JNK) in cell fate determination under glucose starvation.

13.2 Glucose Uptake and Metabolism

Glucose mainly comes from the breakdown of polysaccharides in daily food. As a polar molecule, glucose cannot pass through the plasma membrane via simple diffusion. Thus, cells need glucose transporters to facilitate glucose transportation. There are two ways in which glucose uptake takes place, facilitated diffusion and secondary active transport. Facilitated diffusion is achieved by the glucose transporters (GLUTs) family in many tissues and organs. For example, red blood cells and the endothelial cells of the blood-brain barrier take up glucose from blood through GLUT1 [7]. In contrast, in the intestine and the kidney, where intracellular glucose concentration is higher than that in the bloodstream, glucose uptake is achieved by the symporter sodium-glucose cotransporters (SGLTs), particularly SGLT1 and SGLT2 [8]. Once getting into the cell, glucose is mainly metabolized by glycolysis to generate pyruvate, which feeds the tricarboxylic acid (TCA) cycle and oxidative phosphorylation (OXPHOS) in the mitochondria when oxygen supply is sufficient to produce ATP. In contrast, when oxygen supply is insufficient, pyruvate is converted to lactate by lactate dehydrogenase (LDH). Meanwhile, other parallel metabolic pathways are also branched from glycolysis, including the pentose phosphate pathway (PPP), the hexosamine pathway, and the serine biosynthesis pathway.

13.2.1 Glycolysis

The canonical glycolysis process consists of both reversible and irreversible reactions, generating two molecules of pyruvate from each molecule of glucose (Fig. 13.1). The first step is catalyzed by hexokinases (HKs) or glucokinase (GCK) to transfer the phosphate group of ATP to glucose to produce glucose-6-phosphate (G6P). Notably, this initial reaction is probably the most important step in glucose metabolism, based on the following facts: first, it prevents glucose from being exported from cells; and second, the other pathways including the PPP, the hexosamine pathway, and the glycogenesis pathway all branch from G6P [5]. The second step of glycolysis is catalyzed by the reversible glucose-6-phosphate isomerase (GPI) to convert G6P to fructose-6-phosphate (F6P). The third step is catalyzed by the phosphofructokinase 1 (PFK1), which phosphorylates F6P to fructose-1,6-bisphosphate (F1,6BP). PFK1 activity can be allosterically activated by fructose-2,6-bisphosphate (F2,6BP). The conversion between F2,6BP and F6P is mediated by 6-phosphofructo-2-kinase/fructose-2,6-bisphosphatase (PFKFB or PFK2/FBPase2), a protein family that possesses both kinase and phosphatase domains. In addition, 3–5% of F6P enters the hexosamine pathway [9]. The fourth step is the hydrolysis of F1,6BP into glyceraldehyde-3-phosphate (GAP) and dihydroxyacetone phosphate (DHAP) catalyzed by aldolase (ALDO). In the fifth step, DHAP can be isomerized to GAP by triosephosphate isomerase (TPI) for the subsequent reactions of glycolysis. Alternatively, glycerol-3-phosphate (G3P) is synthesized from DHAP by glycerol-3-phosphate dehydrogenase (GPDH) in a reduced nicotinamide adenine dinucleotide (NADH)-dependent manner. The resulting G3P serves as the starting material for the de novo synthesis of glycerolipids, such as triglycerides. So far, the first five steps of glycolysis as described above, known as the preparatory phase, consume two molecules of ATP to convert one molecule of glucose into two molecules of GAP. In contrast, the following steps of glycolysis generate two molecules of NADH and four molecules of ATP from each glucose molecule, thus regarded as the pay-off phase. The sixth step is catalyzed by glyceraldehyde-3-phosphate dehydrogenase (GAPDH), which converts GAP into 1,3-bisphosphoglycerate (1,3BPG) and reduces NAD^+ to NADH. One phosphate group from 1,3BPG is then transferred to ADP to produce ATP and 3-phosphoglycerate (3PG), catalyzed by phosphoglycerate kinase 1 (PGK1). The oxidation of 3PG is the initial step of serine biosynthesis. In glycolysis, 3PG is subsequently converted to 2-phosphoglycerate (2PG) by phosphoglycerate mutase 1 (PGAM1) and then to phosphoenolpyruvate (PEP) by enolase (ENO). The tenth and the final committed step of glycolysis is catalyzed by pyruvate kinases (PKs), including PKM1, PKM2, PKR, and PKL, generating pyruvate and ATP from PEP. A net gain of two molecules of NADH and two molecules of ATP is produced per glucose hereto through glycolysis.

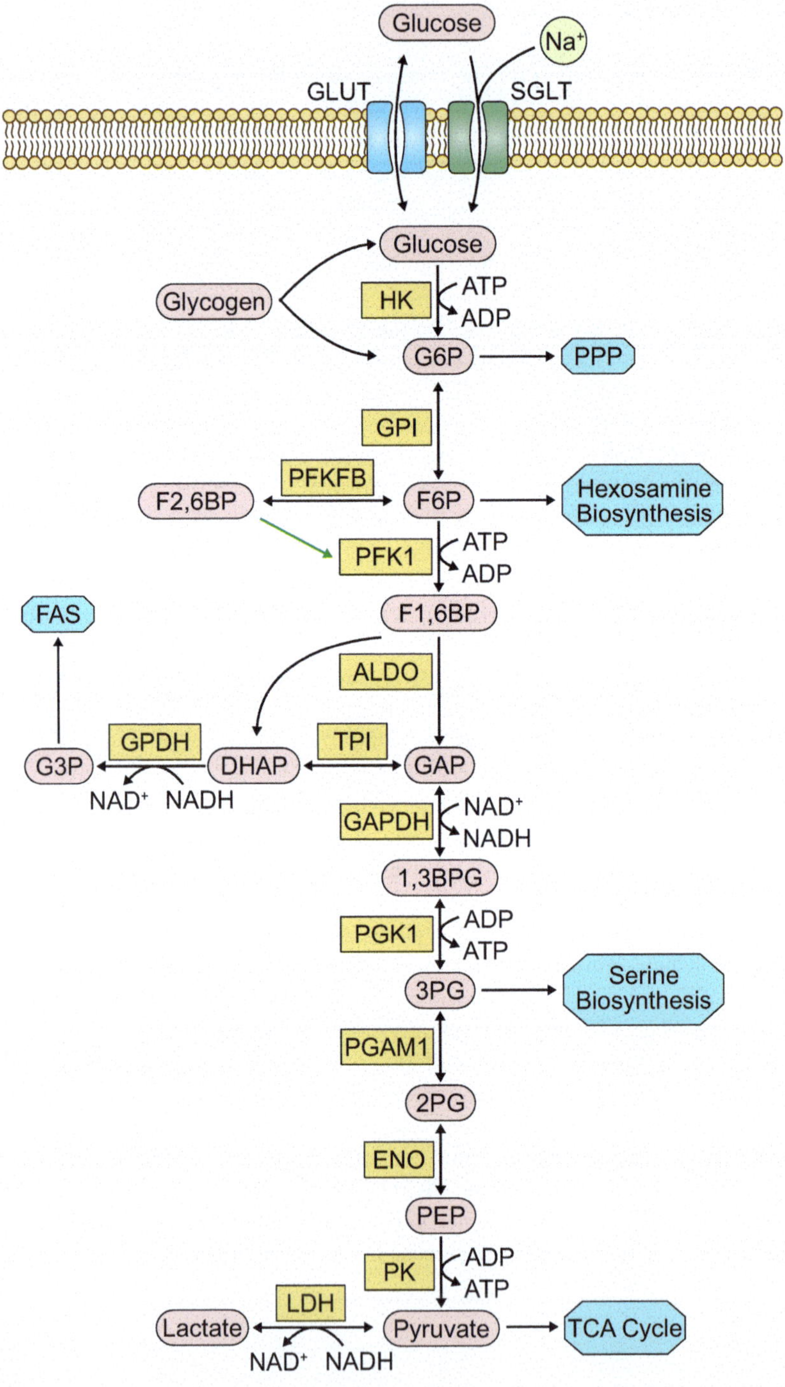

Fig. 13.1 Glucose uptake and glycolysis. Enzymes that catalyze the metabolic reactions are shown in rectangles. Metabolic pathways are shown in octagons. The green arrow indicates the positive regulation. Black arrows indicate the direction of the reaction. *1,3BPG* 1,3-bisphosphoglycerate,

13.2.2 The Pentose Phosphate Pathway

The PPP consists of two distinct phases: an irreversible oxidative phase and a reversible non-oxidative phase (Fig. 13.2a). The oxidative phase of PPP utilizes G6P to generate ribulose-5-phosphate (Ru5P) and two molecules of reduced nicotinamide adenine dinucleotide phosphate (NADPH), which are used in reductive biological reactions, such as fatty acid synthesis (FAS) and reduction of oxidized glutathione disulfide (GSSG) to reduced glutathione (GSH) [10]. The entire oxidative phase is composed of a series of dehydrogenation, hydrolysis, and oxidative decarboxylation catalyzed by glucose 6-phosphate dehydrogenase (G6PDH), 6-phosphogluconolactonase (6PGL), and 6-phosphogluconate dehydrogenase (6PGDH), respectively. Ru5P can be directly isomerized to ribose-5-phosphate (R5P) by ribose-5-phosphate isomerase (RPI), serving as an essential building block of ribonucleotides. Alternatively, Ru5P, as well as other metabolites such as F6P, F1,6BP, and GAP, can enter the non-oxidative PPP and be converted to R5P through a series of reversible reactions involving transaldolase, transketolase, and epimerase.

13.2.3 The Hexosamine Pathway

The hexosamine pathway, which generates UDP-N-acetylglucosamine (UDP-GlcNAc) as a sugar donor for protein glycosylation, includes the following four reactions (Fig. 13.2b) [11]: First, 3–5% of F6P generated by GPI is converted to glucosamine-6-phosphate (GlcN-6P) by glutamine:fructose-6-phosphate aminotransferase (GFAT), accompanied by the conversion from glutamine to glutamate. Second, GlcN-6P is converted to N-acetylglucosamine-6-phosphate (GlcNAc-6P) by glucosamine-6-phosphate N-acetyltransferase (GNPNAT), using acetyl-coenzyme A (acetyl-CoA). Then, GlcNAc-6P undergoes interconversion to N-acetylglucosamine-1-phosphate (GlcNAc-1P) catalyzed by phosphoacetylglucosamine mutase (PGM3), which is further converted to the final product UDP-GlcNAc by UDP-N-acetylhexosamine pyrophosphorylase (UAP1), utilizing UTP in the fourth step.

Fig. 13.1 (continued) 2PG 2-phosphoglycerate, *3PG* 3-phosphoglycerate, *ALDO* aldolase, *DHAP* dihydroxyacetone-phosphate, *ENO* enolase, *F1,6BP* fructose-1,6-bisphosphate, *F2,6BP* fructose-2,6-bisphosphate, *F6P* fructose-6-phosphate, *FAS* fatty acid synthesis, *G3P* glycerol-3-phosphate, *G6P* glucose-6-phosphate, *GAP* glyceraldehyde-3-phosphate, *GAPDH* glyceraldehyde-3-phosphate dehydrogenase, *GLUT* glucose transporter, *GPDH* glycerol-3-phosphate dehydrogenase, *GPI* glucose-6-phosphate isomerase, *HK* hexokinase, *LDH* lactate dehydrogenase, *NAD*$^+$ oxidized nicotinamide adenine dinucleotide, *NADH* reduced nicotinamide adenine dinucleotide, *PEP* phosphoenolpyruvate, *PFK1* phosphofructokinase 1, *PFKFB* 6-phosphofructo 2-kinase/fructose-2,6-bisphosphatase, *PGAM1* phosphoglycerate mutase 1, *PGK1* phosphoglycerate kinase 1, *PK* pyruvate kinase, *PPP* pentose phosphate pathway, *SGLT* sodium-glucose cotransporter, *TCA* tricarboxylic acid, *TPI* triosephosphate isomerase

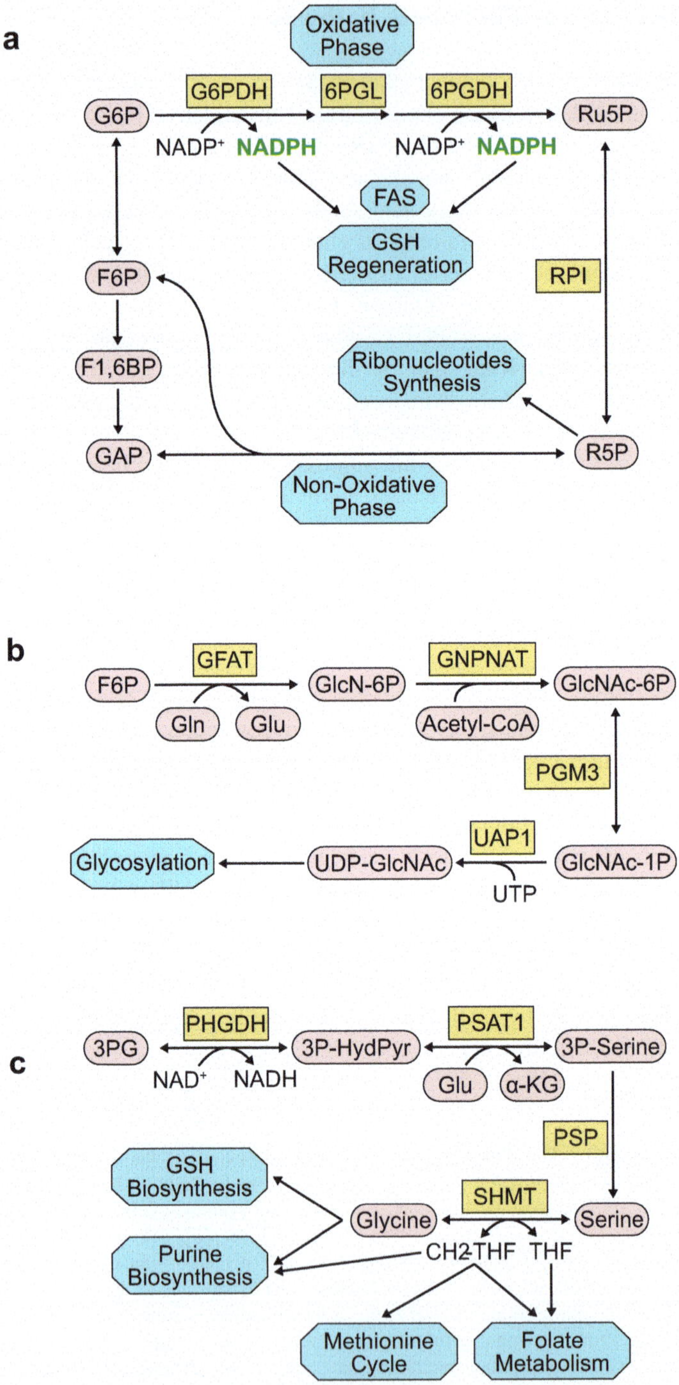

Fig. 13.2 Branching pathways from glycolysis. (**a**) the pentose phosphate pathway, (**b**) the hexosamine pathway, (**c**) the serine biosynthesis pathway. Enzymes that catalyze the metabolic reactions are shown in rectangles. Metabolic pathways are shown in octagons. NADPH generation

13.2.4 The Serine Biosynthesis Pathway

As mentioned in Sect. 13.4.1, besides the subsequent steps of glycolysis, 3PG also participates in the serine biosynthesis pathway, which extends to the biosynthesis of GSH, the one-carbon metabolism, as well as the biosynthesis of purines (Fig. 13.2c) [12]. Phosphoglycerate dehydrogenase (PHGDH) catalyzes the first reaction in the serine biosynthesis pathway by oxidizing 3PG to 3-phosphohydroxypyruvate (3P-HydPyr) using NAD^+. Second, 3P-HydPyr is converted to 3-phosphoserine (3P-serine) by phosphoserine aminotransferase (PSAT1), coupled with the generation of α-ketoglutarate (α-KG) from glutamate by transamination. The third and last dephosphorylation step is catalyzed by phosphoserine phosphatase (PSP), producing serine that can be further converted to glycine by serine hydroxymethyltransferase 1 (SHMT1) and SHMT2 in the cytosol and mitochondria, respectively [13]. The resulting glycine can be involved in the biosynthesis of GSH and purines. Notably, this process also produces 5,10-methylenetetrahydrofolate (CH_2-THF) from tetrahydrofolate (THF), an essential intermediate for the one-carbon metabolism, the methionine cycle, folate metabolism, and purine biosynthesis [12].

13.2.5 The TCA Cycle and OXPHOS

As we mentioned earlier, when oxygen supply is sufficient, pyruvate enters mitochondria for the TCA cycle and OXPHOS to produce a large amount of ATP. The TCA cycle, also known as the citric acid cycle or Krebs cycle, consists of a series of reactions that release the energy through the oxidation of acetyl-CoA (Fig. 13.3). Source of acetyl-CoA includes pyruvate from glycolysis, β-oxidation of fatty acids (FAO), and amino acids. Pyruvate passes mitochondrial membranes and enters the matrix via the mitochondrial pyruvate carrier (MPC), where it is

Fig. 13.2 (continued) is highlighted in green and bold. Black arrows indicate the direction of the reaction. *3P-HydPyr* 3-phosphohydroxypyruvate, *3P-serine* 3-phosphoserine, *3PG* 3-phosphoglycerate, *6PGDH* 6-phosphogluconate dehydrogenase, *6PGL* 6-phosphogluconolactonase, *α-KG* α-ketoglutarate, *acetyl-CoA* acetyl-coenzyme A, *CH_2-THF* 5,10-methylenetetrahydrofolate, *F1,6BP* fructose-1,6-bisphosphate, *F6P* fructose-6-phosphate, *FAS* fatty acid synthesis, *G6P* glucose-6-phosphate, *G6PDH* glucose-6-phosphate dehydrogenase, *GAP* glyceraldehyde-3-phosphate, *GFAT* glutamine:fructose-6-phosphate aminotransferase, *GlcN-6P* glucosamine-6-phosphate, *GlcNAc-1P* N-acetylglucosamine-1-phosphate, *GlcNAc-6P* N-acetylglucosamine-6-phosphate, *Gln* glutamine, *Glu* glutamate, *GNPNAT* glucosamine-6-phosphate N-acetyltransferase, *GSH* reduced glutathione, *GPI* glucose-6-phosphate isomerase, *NAD^+* oxidized nicotinamide adenine dinucleotide, *NADH* reduced nicotinamide adenine dinucleotide, *$NADP^+$* oxidized nicotinamide adenine dinucleotide phosphate, *NADPH* reduced nicotinamide adenine dinucleotide phosphate, *PGM3* phosphoacetylglucosamine mutase, *PHGDH* phosphoglycerate dehydrogenase, *PSAT1* phosphoserine aminotransferase, *PSP* phosphoserine phosphatase, *R5P* ribose-5-phosphate, *RPI* ribose-5-phosphate isomerase, *Ru5P* ribulose-5-phosphate, *SHMT* serine hydroxymethyltransferase, *THF* tetrahydrofolate, *UAP1* *N*-acetylhexosamine pyrophosphorylase, *UDP-GlcNAc* UDP-N-acetylglucosamine

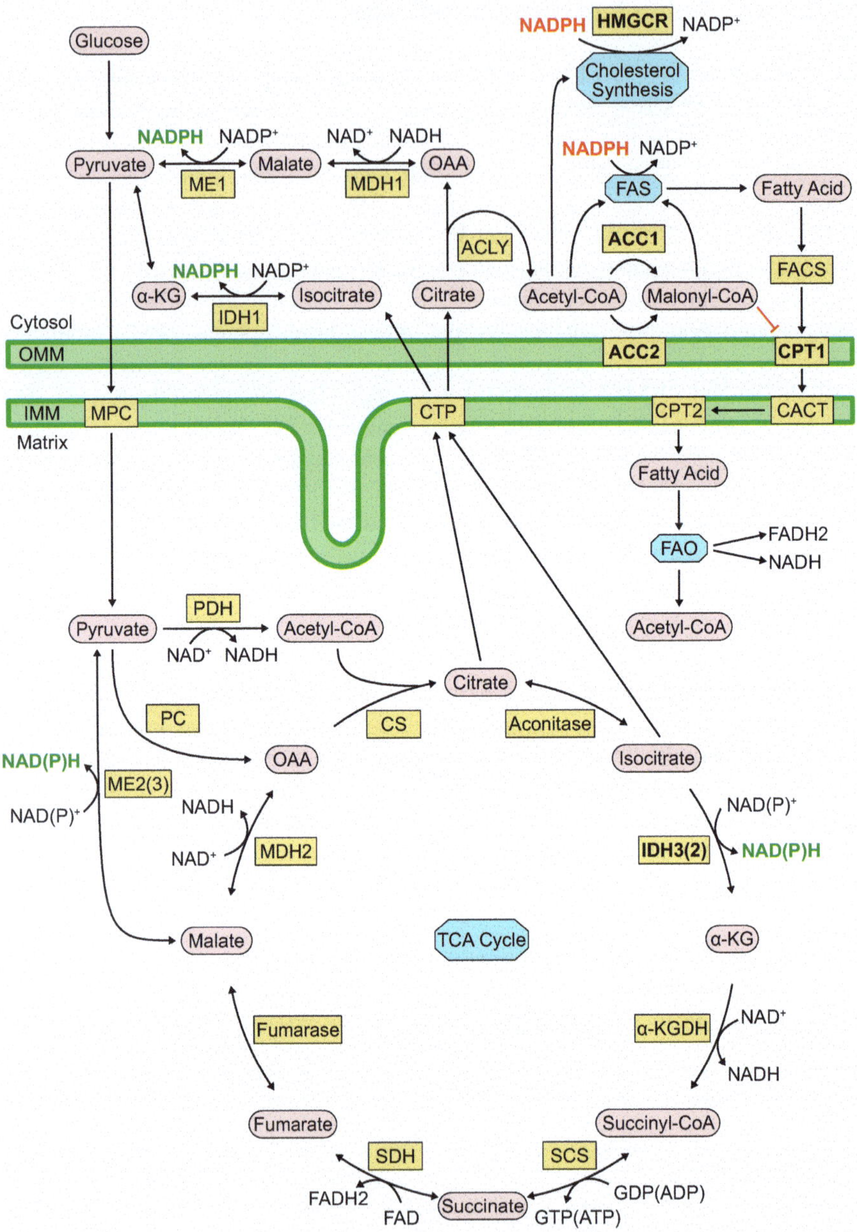

Fig. 13.3 The TCA cycle and its related metabolic pathways. Enzymes that catalyze the metabolic reactions or facilitate metabolite transport are shown in rectangles. The rate-limiting enzymes are highlighted in bold. Metabolic pathways are shown in octagons. NADPH generation is highlighted in green and bold. NADPH consumption is highlighted in red and bold. Black arrows indicate the direction of the reaction or the metabolite transport. The red blunt end arrow indicates the negative regulation. Note that ME2 and IDH3 are NAD^+-dependent, whereas ME1, ME3, IDH1, and IDH2 are $NADP^+$-dependent. *α-KG* α-ketoglutarate, *α-KGDH* α-ketoglutarate dehydrogenase, *ACC1/2* acetyl-CoA carboxylase 1/2, *acetyl-CoA* acetyl-coenzyme A, *ACLY* ATP citrate lyase, *CACT*

oxidized by pyruvate dehydrogenase (PDH) complex into acetyl-CoA. This process requires NAD^+ and CoA and generates NADH, carbon dioxide (CO_2), as well as acetyl-CoA. The TCA cycle contains ten basic reactions, starting with the extension of oxaloacetate to citrate by adding acetyl-CoA, mediated by citrate synthase (CS). After each cycle, one molecule of GTP or ATP, one molecule of ubiquinol (QH_2), two molecules of CO_2, and three molecules of NADH are generated. The product of the tenth reaction, oxaloacetate, enters the next TCA cycle, with the continuously supplied acetyl-CoA. Notably, two intermediates in the TCA cycle can be used to regenerate NADPH, i.e., isocitrate and malate. When TCA cycle is promoted by feeding more acetyl-CoA, the levels of isocitrate and malate are increased; thus the conversions from malate to pyruvate by the malic enzyme (ME), and from isocitrate to α-KG by isocitrate dehydrogenase (IDH), will generate more NADPH [14, 15].

OXPHOS is the last step that produces ATP using the NADH, reduced flavin adenine dinucleotide ($FADH_2$), and succinate generated from the TCA cycle (Fig. 13.4). During OXPHOS, electrons from NADH and $FADH_2$ are transferred to oxygen through a series of redox reactions, conducted by the electron transport chain (ETC) on the inner mitochondrial membrane (IMM). The energy stored in the double bond of oxygen is thus released as the form of proton gradient between the intermembrane space (IMS) and the matrix, which drives the ATP synthesis by the ATP synthase. Unlike glycolysis, which produces only two molecules of ATP per glucose molecule, OXPHOS produces 30 to 36 molecules of ATP, with the variation due to the electron leakage during ETC [16]. Of note, electrons leaked from the ETC complex I and III can be accepted by oxygen, forming superoxide ($O_2^{\bullet}$) or peroxide (O_2^{2-}) anions, which will be discussed later in detail.

13.3 Reactive Oxygen Species (ROS) and Oxidative Stress

ROS refer to a class of chemically reactive molecules containing oxygen, which mainly include superoxide, peroxide, and hydroxyl radical (${}^{\bullet}OH$) [17, 18]. ROS play critical roles in cell signaling and the determination of cell fate under different conditions. Low to moderate ROS levels control various signaling pathways,

Fig. 13.3 (continued) carnitine-acylcarnitine translocase, *CPT1/2* carnitine palmitoyltransferase 1/2, *CS* citrate synthase, *CTP* citrate transport protein, *FACS* fatty-acyl-CoA synthase, *FAD* oxidized flavin adenine dinucleotide, *$FADH_2$* reduced flavin adenine dinucleotide, *FAO* fatty acid oxidation, *FAS* fatty acid synthesis, *HMGCR* 3-hydroxy-3-methyl-glutaryl-coenzyme A reductase, *IDH1/2/3* isocitrate dehydrogenase 1/2/3, *IMM* inner mitochondrial membrane, *Malonyl-CoA* malonyl-coenzyme A, *MDH1/2* malate dehydrogenase 1/2, *ME1/2/3* malic enzyme 1/2/3, *MPC* mitochondrial pyruvate carrier, *NAD^+* oxidized nicotinamide adenine dinucleotide, *NADH* reduced nicotinamide adenine dinucleotide, *$NADP^+$* oxidized nicotinamide adenine dinucleotide phosphate, *NADPH* reduced nicotinamide adenine dinucleotide phosphate, *OAA* oxaloacetate, *OMM* outer mitochondrial membrane, *PC* pyruvate carboxylase, *PDH* pyruvate dehydrogenase, *SCS* succinyl-CoA synthetase, *SDH* succinate dehydrogenase, *succinyl-CoA* succinyl-coenzyme A, *TCA* tricarboxylic acid

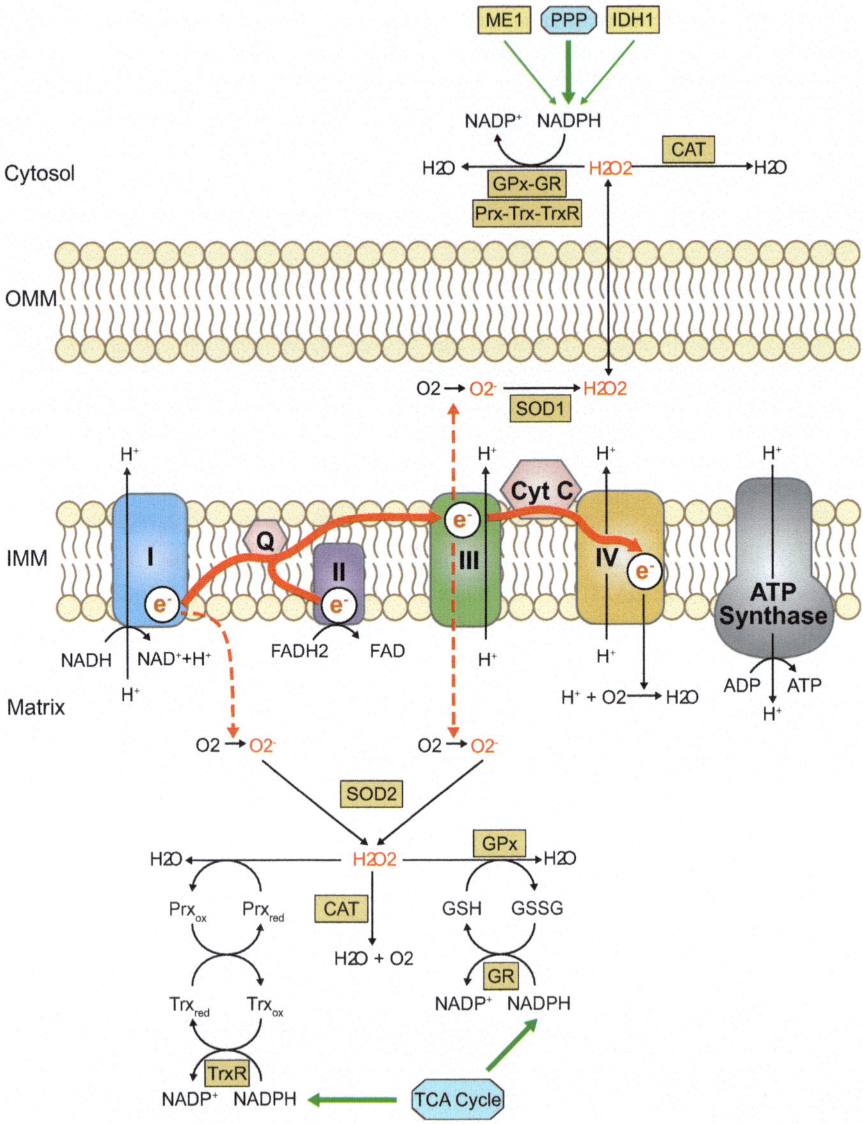

Fig. 13.4 OXPHOS, the generation and elimination of ROS. Enzymes that catalyze the metabolic reactions are shown in rectangles. Metabolic pathways are shown in octagons. Black arrows indicate the direction of the reaction or the metabolite transport. The generation of NADPH is shown in green arrows, with the thickness indicating the relative contribution. The four complexes (I to IV), the coenzyme Q (Q) and the cytochrome C (Cyt C) in the electron transport chain (ETC), and the ATP synthase are highlighted in bold. The transport direction of electrons (e^-) in the ETC is shown in the red solid arrow, whereas the leakage of e^- to O_2 is shown in red dash arrows. *CAT* catalase, *FAD* oxidized flavin adenine dinucleotide, *FADH*$_2$ reduced flavin adenine dinucleotide, *GPx* glutathione peroxidase, *GR* glutathione reductase, *IDH1* isocitrate dehydrogenase 1, *IMM* inner mitochondrial membrane, *ME1* malic enzyme 1, *NAD*$^+$ oxidized nicotinamide adenine dinucleotide, *NADH* reduced nicotinamide adenine dinucleotide, *NADP*$^+$ oxidized nicotinamide adenine dinucleotide phosphate, *NADPH* reduced nicotinamide adenine dinucleotide phosphate,

including cell growth, cell differentiation, and angiogenesis, whereas excessive ROS can cause oxidative stress, leading to irreversible oxidative modifications of protein and DNA and inducing cell death [19–21]. More importantly, oxidative stress is implicated in many diseases, such as cancer, neurological disease, cardiovascular disease, rheumatoid arthritis, etc. [20, 22, 23]. Therefore, the tight control of ROS generation and elimination by the antioxidant system is important to maintain intracellular redox homeostasis.

13.3.1 ROS

In cells, superoxide can be generated by electron leakage in OXPHOS in the mitochondria and by NADPH oxidase (NOX) complex at the plasma membrane. The mitochondria are the primary source of intracellular superoxide [24]. The ETC complexes I and III have been considered as the primary sites of superoxide generation (Fig. 13.4) [25]. Leaked electrons are accepted by oxygen, thus partially reducing oxygen and generating superoxide. Superoxide can be rapidly converted to H_2O_2 and oxygen by superoxide dismutase 1 (SOD1) and SOD2, deficiency of which is associated with accelerated aging and neurological diseases such as amyotrophic lateral sclerosis [26–28]. NOX, using NADPH as a reductant, produces superoxide which plays a role in immune response to kill the invading bacteria or fungi [29–31]. Interestingly, while some studies reported that glucose-derived NADPH supports NOX activity, others have shown that NOX-derived superoxide stimulates glucose uptake through GLUTs [32–35], indicating an intersection between glucose metabolism and NOX.

H_2O_2 is mainly generated by SOD, but it can also be formed spontaneously in peroxisomes. Compared to superoxide, H_2O_2 is more associated with cell signaling regulation, by mediating the oxidation of cysteine residues of proteins. Normally, cysteine residues exist as thiolate anions (Cys-S^-), which can be oxidized by H_2O_2 to the reversible sulfenic form (Cys-SOH) or further oxidized to the irreversible sulfinic (SO_2H) or sulfonic (SO_3H) forms. Sulfenic oxidation causes conformational changes of proteins and regulates their functions, whereas sulfinic and sulfonic oxidations can cause permanent damage to the protein and disrupt signaling pathways [10, 36, 37]. In the presence of ferrous (Fe^{2+}) or cuprous (Cu^+) ions, H_2O_2 can be converted to extremely reactive $^{\bullet}OH$ through the Fenton reaction, accompanied by the generation of ferric (Fe^{3+}) or cupric (Cu^{2+}) ions. $^{\bullet}OH$ mediates the most severe effects or ROS in cells, as it can indiscriminately oxidize and damage lipids, proteins, and DNA [38, 39].

Although H_2O_2 is less reactive than superoxide, it is capable of penetrating cellular membranes, thus causing damages in other organelles and associated with

Fig. 13.4 (continued) OMM outer mitochondrial membrane, *PPP* pentose phosphate pathway, *Prx* peroxiredoxin, *SOD1/2* superoxide dismutase 1/2, *TCA* tricarboxylic acid, *Trx* thioredoxin, *TrxR* thioredoxin reductase

carcinogenesis [40–42]. To balance this harmful effect, cells employ several dedicated enzymes to prevent the accumulation of H_2O_2, including catalases, glutathione peroxidase (GPx), thioredoxin (Trx), glutaredoxin (Grx), and peroxiredoxin (Prx) [43] (Fig. 13.4).

13.3.2 The Antioxidant System

To prevent the emergence of oxidative stress and maintain redox homeostasis, cells implement the antioxidant system to balance the production and effects of ROS. The antioxidant system consists of enzymatic antioxidants, such as the aforementioned SOD; catalase; GPx, Trx, Grx, and Prx; and non-enzymatic antioxidant, including GSH, vitamin C, and vitamin E (Fig. 13.4) [44, 45].

Although the dismutation of superoxide can occur spontaneously, SOD accelerates the reaction rate by 10^4 times [46]. As almost 100% of superoxide generated from the complex I is released into the mitochondria matrix, it is therefore mainly eliminated by the mitochondrial matrix localized SOD2 [47, 48]. In contrast, the complex III-derived superoxide is released into both the IMS and the mitochondrial matrix, with part of the IMS superoxide being eliminated by SOD1 [48, 49]. Superoxide may also be scavenged by vitamin E or oxidized back to oxygen by transition metal ions, such as Fe^{3+} [50, 51]. Notably, the redox reaction between superoxide and Fe^{3+} generates Fe^{2+}, which in turn can be involved in the Fenton reaction with H_2O_2, therefore boosting the production of $^{\bullet}OH$ and emergence of oxidative stress. Besides, the IMS superoxide can also be scavenged by the cytochrome c (Cyt c)-Cyt c oxidase system, with the participation of iron ions [52, 53]. This system is of particular importance as it not only regenerates oxygen from superoxide but also provides the proton gradient to ATP production [54, 55]. Although superoxide is not membrane permeable, it can be released from the mitochondria to the cytosol through the inner membrane anion channel (IMAC) and the voltage-dependent anion channels (VDACs) on the outer mitochondrial membrane (OMM) [56]. The cytosolic superoxide is also eliminated by SOD1.

Catalase is mainly localized in the peroxisomes to detoxify the product of SOD, H_2O_2. Catalase is highly efficient in breaking down H_2O_2 into water and oxygen, therefore preventing the accumulation of H_2O_2 [57]. As a critical antioxidant enzyme, catalase mutations or deficiencies have been associated with many diseases including diabetes and cancer [58].

Similar to catalase, GPx is also able to decompose H2O2. Eight isoforms of GPx have been identified in mammalian cells, which are localized to different subcellular compartments [59]. However, the process consumes GSH and converts it to GSSG, thus requiring the glutathione reductase (GR) to reduce GSSG back to GSH to sustain the reaction. The reduction of GSSG requests NADPH as the reducing agent. Therefore, the GSH/GSSG NADPH/$NADP^+$ cycling system is of paramount importance in maintaining redox homeostasis.

Trx, Grx, and Prx are three ubiquitous families of antioxidant enzymes. The basic mechanism of how Prx works is that a conserved cysteine on Prx can be oxidized to the sulfenic form by the peroxides, for example, H2O2, therefore eliminating peroxides [60]. Depending on the existence of another semi-conserved cysteine, the oxidized Prx is reduced by different mechanisms, with 2-Cys Prx reduced by Trx, Trx-like proteins or GSH, and 1-Cys Prx reduced by vitamin C or GSH [61]. Trx, which has two conserved vicinal cysteines as the CXXC motif, reduces the oxidized cysteine residues and cleaves the disulfide bonds on other oxidized proteins, for example, Prx [62]. The oxidized Trx is then reduced by thioredoxin reductase (TrxR) at the cost of NADPH [63]. The Prx-Trx-TrxR system protects proteins from being overoxidized by H2O2, thus maintaining the normal conformation and functions of them. Like Trx, Grx functions in a similar way by reducing the oxidized cysteine residues on the substrates [64]. However, the reduction of oxidized Grx is mediated by GSH in a non-enzymatical way [65].

Besides, the antioxidant system is also involving several signaling pathways, including Kelch-like ECH-associated protein 1 (KEAP1)-nuclear factor erythroid 2-related factor 2 (NRF2), liver kinase B1 (LKB1)-AMPK, as well as JNK, which determine the cell fate under oxidative stress conditions [10, 66–68]. Details of these signaling pathways will be discussed later.

13.3.3 Oxidative Stress Under Glucose Starvation

Many lines of evidence have suggested that glucose starvation is a strong inducer of ROS generation, especially H_2O_2 and mitochondrial superoxide [69–72]. The increase of H_2O_2 in cells under glucose starvation has been widely reported. On the one hand, as introduced earlier, GSH and NADPH are the two most important reducing agents that support the antioxidant system. The decrease of GSH and NADPH levels compromises the H2O2-detoxifying ability of cells. Thus, a sufficient glucose supply is required not only for NADPH generation through the PPP and the TCA cycle but also for GSH biosynthesis through the serine-glycine synthesis pathway. The importance of the PPP in maintaining NADPH level and redox homeostasis has been emphasized by the fact that the only the non-metabolizable glucose analog 2-deoxyglucose (2DG), but not 5-thioglucose (5TG), is able to suppress ROS generation and cell death induced by glucose starvation [71, 73]. The reason behind this phenomenon is that despite blocking glycolysis, 2DG still can enter the PPP in the form of 2-deoxy-D-glucose 6-phosphate (2DG6P), thereby supporting NADPH generation, leading to suppression of ROS and prevention of cell death.

On the other hand, the increased H_2O_2 level results from more superoxide generated by the mitochondria or NOX during glucose starvation. As mentioned previously, superoxide is generated in the mitochondria by leaked electron reduction of oxygen. It is estimated that in vitro, 1–2% of oxygen is converted to superoxide [25, 74]. However, the value under physiological conditions in vivo remains hard to estimate so far [75]. Despite being observed frequently, the mechanisms linking increased mitochondrial superoxide with glucose starvation remain to be elucidated [70, 76, 77]. Currently, there are two proposed models in which superoxide generation is enhanced in the mitochondria under glucose starvation. The first model is when there is an increased ratio of NADH/NAD^+ in the mitochondria matrix [78, 79]. The second model is when there is no ATP production; thus the proton gradient and the ratio of reduced coenzyme Q ($CoQH_2$)/CoQ are high [80–82]. However, based on the fact that glucose drives glycolysis and the TCA cycle to generate NADH, when cells undergo glucose starvation, the NADH level is expected to be decreased, which is also consistent with previous reports [83, 84]. On the contrary, glucose starvation disrupts the progression of glucose metabolism, leading to a disturbance of the metabolic intermediate levels, which causes a mismatch in the rates of electron donation and transfer in the ETC. Moreover, as ATP production is also halted by glucose starvation, the proton gradient that drives ATP synthesis keeps increasing. Therefore, glucose starvation may induce mitochondrial superoxide generation in a way that fits the second model. Other than mitochondria, NOX-derived superoxide has also been described in cells under glucose starvation [70, 85]. It is also reported that in cardiomyocytes, glucose starvation upregulates NOX4 in the endoplasmic reticulum (ER), which generates ROS and elicits autophagy for cell survival [86]. However, the mechanism underlying NOX activation under glucose starvation remains elusive, and the controversy that glucose supports NOX activity by providing NADPH also needs to be addressed [33, 70, 85]. Taken together, although there has been a solid correlation between glucose starvation and superoxide generation, the exact mechanism of which requires more studies to elucidate.

13.4 AMPK and JNK in Glucose Starvation-Induced Oxidative Stress

One of the most studied signaling pathways under glucose starvation is the LKB1-AMPK pathway. AMPK is a critical energy sensor and metabolism modulator which is evolutionarily conserved and ubiquitously expressed in almost all eukaryotes [87, 88]. AMPK responds to increased AMP and ADP levels when cells are facing scarce energy conditions, by activating catabolic pathways and inhibiting anabolic pathways, thus playing a pro-survival role [89]. Several studies have also reported that ROS, especially mitochondria superoxide, can activate AMPK, which maintains metabolic and redox homeostasis [90, 91]. On the contrary, JNK, which belongs to the mitogen-activated protein kinase (MAPK) family, is readily activated by ROS and involved in the determination of cell fate under oxidative stress [67, 92, 93].

13.4.1 The Role of AMPK in Redox Homeostasis and Cell Survival Under Glucose Starvation

When cells face energy shortage, especially glucose starvation or other ATP-depleting conditions, AMPK is activated by LKB1 through phosphorylation at the T172 residue on the AMPKα subunit [94–97]. Additionally, AMPK can also be activated by the calcium/calmodulin-dependent protein kinase kinase-β (CaMKKβ) in response to increased cytosolic calcium levels [98–100]. Recently, more sophisticated studies have shed lights on the spatiotemporal basis for hierarchical AMPK activation from different subcellular compartments during energy stress, in which ALDO, Axin, and lysosomes are involved [101–104].

The activated AMPK modulates cell metabolism in two ways, i.e., activating the catabolic pathways to regenerate the nutrients and inhibiting anabolic pathways to reduce ATP consumption (Fig. 13.5) [89]. Among these effects, the activation of FAO and the inhibition of FAS are directly linked with redox homeostasis. On the one hand, FAS synthesizes long-chain fatty acids using acetyl-CoA and NADPH. On the other hand, FAO can boost the TCA cycle by generating acetyl-CoA, which increases the levels of malate isocitrate, thus maintaining the NADPH level through ME and IDH as introduced in Sect. 13.2.5. Mechanistically, the activated AMPK phosphorylates and inhibits acetyl-CoA carboxylase 1 (ACC1) and ACC2 in the cytosol and on the mitochondria, respectively, thus inhibiting the conversion from acetyl-CoA to malonyl-CoA [105–107]. Malonyl-CoA is the precursor of FAS and a potent inhibitor of the rate-limiting enzyme in FAO, carnitine palmitoyltransferase 1 (CPT1) (Fig. 13.3) [108]. Therefore, the regulation of fatty acid metabolism by AMPK maintains the levels of both ATP and NADPH, thus protecting cells against glucose starvation-induced oxidative stress and cell death [71]. Besides ACC, AMPK directly phosphorylates and inhibits 3-hydroxy-3-methyl-glutaryl-coenzyme A reductase (HMGCR), which consumes NADPH for cholesterol synthesis [105, 109]. AMPK also downregulates lipogenesis through the inhibition of transcription factors, such as sterol regulatory element binding protein-1 (SREBP1) and carbohydrate response element-binding protein (ChREBP) (Fig. 13.5), which control the expression of genes encoding ACC, stearoyl-CoA desaturase-1 (SCD1), ATP citrate lyase (ACLY), and fatty acid synthase (FASN) [110, 111]. Although the studies were not performed under glucose starvation, they used mitochondria inhibitors and adenosine analogues, which partially mimicked the effect of glucose starvation.

In addition to the regulation of fatty acid metabolism, AMPK also upregulates gene expression of the antioxidant system in response to glucose starvation and oxidative stress (Fig. 13.5). First, AMPK activates NRF2, which is a master transcription factor that controls the expression of several antioxidant genes through binding to their antioxidant response element (ARE), including *NADPH dehydrogenase (quinone 1)* (*NQO1*) and *heme oxygenase 1* (*HO-1*) [112]. The activation of NRF2 by AMPK is achieved via several ways, including direct phosphorylation of NRF2 and enhanced autophagic degradation of KEAP1 [112–115]. KEAP1 mediates the interaction between NRF2 and the Cullin-3 E3 ubiquitin ligase complex, leading to NRF2 ubiquitination and proteasomal degradation [116, 117]. Second, AMPK can promote the expression of several antioxidant genes encoding catalase and SOD2, in a peroxisome proliferator-activated receptor (PPAR)-γ coactivator 1-α (PGC-1α)-dependent manner [91]. Interestingly, in cells that treated with mitochondria inhibitors, which induce severe energy stress and mitochondrial ROS, AMPK is required to phosphorylate unc-51-like kinase-1 (ULK1) and

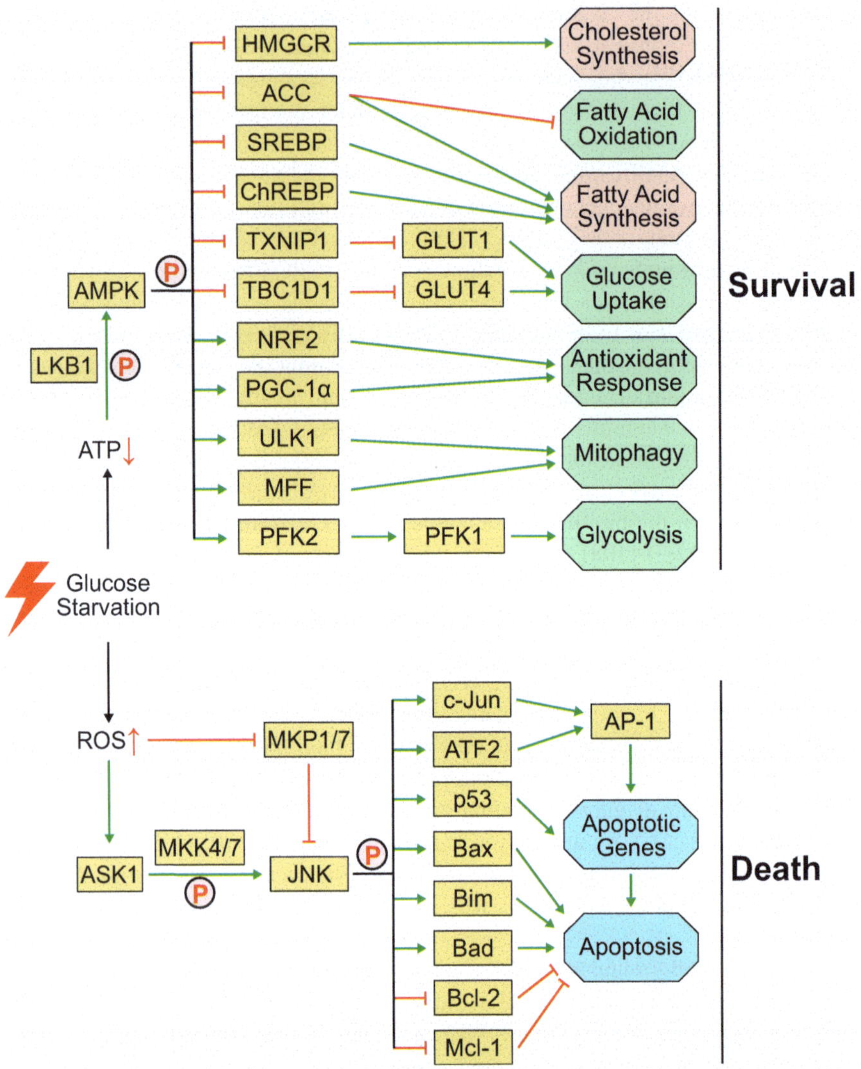

Fig. 13.5 Regulation of downstream substrates and biological pathways by AMPK and JNK in cells under glucose starvation. Proteins or protein complexes are shown in rectangles. P letters indicate protein phosphorylation. Biological pathways are shown in octagons. Pathways that contribute to redox homeostasis are highlighted in light green, pathways that consume NADPH are highlighted in light red, and pathways that promote cell death are highlighted in light blue. Green arrows indicate the positive regulation. Red blunt end arrows indicate the negative regulation. *ACC* acetyl-CoA carboxylase, *AMPK* AMP-activated protein kinase, *AP-1* activator protein 1, *ASK1* apoptosis signal-regulating kinase 1, *ATF2* activating transcription factor 2, *Bad* Bcl-2-associated agonist of cell death, *Bax* Bcl-2-associated X, *Bcl-2* B-cell lymphoma 2, *Bim* Bcl-2-interacting mediator of cell death, *ChREBP* carbohydrate response element-binding protein, *GLUT1/4* glucose transporter 1/4, *HMGCR* 3-hydroxy-3-methyl-glutaryl-coenzyme A reductase, *JNK* c-Jun N-terminal kinase, *LKB1* liver kinase B1, *Mcl-1* myeloid cell leukemia 1, *MFF* mitochondrial fission factor, *MKK4/7* mitogen-activated protein kinase kinase 4/7, *MKP1/7* mitogen-activated protein kinase phosphatase 1/7, *NRF2* nuclear factor erythroid 2-related factor 2, *PFK1/2* phosphofructokinase 1/2, *PGC-1α* peroxisome proliferator-activated receptor-γ coactivator 1-α,

mitochondrial fission factor (MFF) to remove damaged mitochondria via mitophagy [118–120].

At the same time, AMPKα subunit itself is subjected to cysteine oxidation. Under glucose starvation or exogenous H_2O_2 treatment, the Cys130 and Cys174 residues on the AMPKα subunit are oxidized, which interferes the interaction between LKB1 and AMPK, thus negatively regulating AMPK activity [121]. Interestingly, Trx1 cleaves the disulfide bonds and rescues AMPK oxidation, thus serving as an essential cofactor for AMPK activation during glucose starvation [121].

An additional way in which AMPK ameliorates oxidative stress is through stimulating glucose uptake by GLUT1 or GLUT4 [122–124], as well as promoting glycolysis through phosphorylation of PFK2 [125–127]. Notably, this adaptive response is more applicable in in vivo pathological conditions such as ischemia, where glucose supply is lowered but still available.

13.4.2 The Role of JNK in the Determination of Cell Fate

JNK, also known as stress-activated protein kinase (SAPK), has three isoforms, with JNK1 and JNK2 expressed in all tissues and JNK3 mainly found in the brain [128]. JNK was originally identified to phosphorylate and activate the transcription factor c-Jun [129]. Other targets of JNK include activating transcription factor 2 (ATF2), p53, c-Myc, and B-cell lymphoma 2 (Bcl-2) family proteins [130–133] (Fig. 13.5). JNK can be activated in response to growth factors, inflammatory cytokines, as well as various stress conditions, including oxidative stress and ultraviolet radiation [134]. The activated JNK, in turn, plays a critical role in the regulation of cell growth, proliferation, differentiation, and cell death [135]. As a member of the MAPK family, JNK is activated following the three-tiered kinase cascade. The direct upstream JNK kinases are MAPK kinase 4 (MKK4) and MKK7, which show synergistic activation of JNK [136–139]. Upstream of MKK4 and MKK7, several MAPK kinase kinases (MAP 3Ks) have been reported to initiate the JNK signaling pathway, including MAP 3 K1–4, apoptosis signal-regulating kinase 1 (ASK1), mixed-lineage kinase (MLK), and transforming growth factor-β-associated kinase 1 (TAK1) [140–142].

The activation of JNK under oxidative stress conditions has been well-established for a long time [135, 143]. ASK1 is largely involved in the ROS-induced JNK activation, forming a ROS-ASK1-MKK-JNK signaling axis [67]. Under normal conditions, reduced Trx binds to ASK1, thus leading to the ubiquitination and degradation of ASK1. In contrast, ROS cause Trx oxidation and dissociation with ASK1, therefore allowing ASK1 oligomerization, autophosphorylation, and subsequent activation [144–146]. It has been reported that Grx controls ASK1

Fig. 13.5 (continued) ROS reactive oxygen species, *SREBP* sterol regulatory element binding protein, *TXNIP1* thioredoxin-interacting protein 1, *ULK1* unc-51-like kinase-1

activity similarly to Trx [147, 148]. Additionally, different phosphatases have been described to regulate ASK1 activity in response to ROS. On the one hand, an okadaic acid-sensitive phosphatase (PP), presumably PP1 or PP2A, dephosphorylates ASK1 at Ser 967, leading to its dissociation with 14–3-3 and increased activity [149]. On the other hand, PP5 is able to dephosphorylate ASK1 in its kinase domain, thus inactivating the ASK1-JNK signaling pathway [150, 151]. In addition to ASK1, Src, Cas, and Gab1 have also been reported to link the oxidative stress with JNK activation, although the underlying mechanisms remain to be further investigated [152–154]. Independent of upstream kinases, JNK can be directly regulated by glutathione S-transferase Pi (GSTπ) or MAPK phosphatase 1 (MKP1) and MKP7. GSTπ monomer directly binds to JNK and suppresses its kinase activity, while ROS-induced oligomerization of GSTπ dissociates with JNK, thus alleviating JNK inactivation [155, 156]. MKPs are subjected to cysteine oxidation by ROS, which results in the inactivation of MKPs and can be reversed by Trx [157]. Therefore, oxidative stress inhibits MKP1 and MKP7, two negative regulators of JNK, leading to sustained JNK activation [157–159].

The exact role of JNK in cell fate determination is controversial [128, 160, 161]. However, it seems that JNK determines cell fate following a temporal pattern, with the early and transient activation supporting cell survival and the late and sustained activation mediating apoptosis [162, 163]. In oxidized low-density lipoprotein-treated macrophages, protocatechuic acid (PCA), which exhibits antioxidant activity, activates JNK and subsequently increases the NRF2 level in the nucleus, leading to the early elimination of ROS overproduction and apoptosis [164]. Similarly, in vestibular schwannomas where JNK is persistently active, inhibition of JNK by several different methods induces apoptosis, which also linked with ROS generation [165]. Recently, it has been reported that JNK-mediated autophagy is responsible for its anti-apoptotic function in cells treated with various drugs that induce oxidative stress [132, 166, 167]. However, these studies mainly used JNK inhibitors as a tool for JNK inactivation, and the detailed mechanisms of how JNK promotes autophagy remain largely elusive, although ER stress and JNK-mediated p62 expression and Bcl-2 phosphorylation are likely involved.

In contrast, in most cases, JNK exerts a pro-apoptotic function in response to oxidative stress (Fig. 13.5). Several reports have provided strong evidence that ASK1 knockout, selenoorganic antioxidant compounds, and direct inhibition of JNK by either JNK inhibitor or overexpression of the dominant-negative form of JNK can block ROS-induced apoptosis [168–170]. Two major pathways of which JNK causes apoptosis include the nuclear signaling pathway and the mitochondrial signaling pathway. First, activated JNK can translocate to the nucleus where it can phosphorylate c-Jun and ATF2, leading to formation and activation of the activator protein 1 (AP-1) transcription factor, therefore promoting the transcription of various pro-apoptotic genes including *tumor suppressor 53* (*TP53*), *Fas Ligand* (*FasL*), and *Bcl-2 homologous antagonist killer* (*Bak*) [171–173]. Moreover, JNK also phosphorylates p53, therefore promoting apoptosis by controlling the expression of both pro-apoptotic genes such as *Bcl-2-associated X* (*Bax*) and *p53 upregulated modulator of apoptosis* (*PUMA*), and anti-apoptotic genes, for example, *myeloid cell leukemia 1* (*Mcl-1*) [174–176]. Second, JNK is able to directly phosphorylate various proteins to promote apoptosis, including the inhibitory phosphorylation of anti-apoptotic proteins Bcl-2 and Mcl-1 [177, 178] and the activatory phosphorylation of pro-apoptotic proteins Bax and Bcl-2-interacting mediator of cell death (Bim) [179, 180].

Collectively, the exact role of JNK in determining cell fate is largely dependent on the cell type, the stimuli, and the duration of the treatment. Nonetheless, a sustained JNK activation in response to oxidative stress is closely associated with apoptosis undoubtedly. However, it is still worth to further explore the spatial-temporal activation of JNK in both transient and prolonged stress conditions. Furthermore, investigation of how cells determine to activate autophagy or apoptosis pathway under oxidative stress conditions may shed more light on the knowledge of cell fate determination and provide therapeutic targets for diseases such as cancer, ischemia, aging, and neurodegenerative diseases.

13.4.3 The Crosstalk Between AMPK and JNK Pathways Under Glucose Starvation

Despite the well-established link between oxidative stress and JNK, only limited studies have investigated the role of JNK under the glucose starvation and the crosstalk between AMPK and JNK pathways. It has been reported that glucose starvation induces JNK activation and cell death, in a well-known ROS-Grx/Trx-ASK1-MKK4-dependent manner [147, 148, 181]. Later, a crosstalk between the LKB1-AMPK axis and JNK was reported [163]. They found that in cells without LKB1, glucose starvation activates JNK, which, at the early stage, functions upstream of AMPK and promotes AMPK activity, thus inhibiting ROS generation and apoptosis. However, at the late stage, when AMPK activation cannot be sustained due to loss of LKB1, JNK promotes apoptosis [163]. In contrast, in cells with functional LKB1, AMPK is consistently activated by glucose starvation, thus causing a sustained inhibition of ROS and JNK, as well as apoptosis [163]. One major unanswered question in this study is whether JNK directly or indirectly phosphorylates AMPK. Thus, more evidence is needed to establish a positive regulation of AMPK by JNK. Since here we mentioned AMPK kinases, it is of note that TAK1, which activates JNK, has also been reported to be a direct kinase for AMPK activation, although the physiological regulations and functions remain elusive [182, 183].

Interestingly, several studies have also described both positive and negative roles of AMPK in JNK activation, although some of the studies were not performed under glucose starvation. Under hypoxia or interleukin-1β treatment, it has been reported that AMPK negatively regulates JNK activity, through the repression of ROS or the inhibition of interleukin-1 receptor-associated kinase-4 (IRAK4)-IκB kinase (IKK)-MKK4 signaling pathway, respectively [184, 185]. On the contrary, low glucose or the AMP mimetic 5-aminoimidazole-4-carboxamide riboside (AICAR) treatment triggers apoptosis in beta cells and hepatocytes in an AMPK-JNK-dependent way

[186, 187]. This positive regulation was also supported by a recent report in which in hepatocytes treated with lipopolysaccharide/D-galactosamine, the AMPK inhibitor compound C blocks both AMPK and JNK phosphorylation, as well as apoptosis [188].

In summary, most of the studies so far support an overall pro-cell survival of AMPK and an overall pro-cell death role of JNK under stress including glucose starvation. However, differences in cell types, tissues, stimuli, and the duration of treatments may affect the functions and the regulatory network of AMPK and JNK. Moreover, other factors should also be considered such as the status of LKB1 and TAK1. Given that both kinases are implicated in a variety of diseases, including cancer, diabetes, and inflammatory diseases, it is of interest and importance to further elucidate the interaction between two signaling pathways and explore the potential of developing therapeutics targeting the two pathways.

13.5 Perspectives

Based on the prevalent thoughts that AMPK plays a pro-survival role, while JNK plays a pro-death role in response to oxidative stress induced by glucose starvation, manipulation of the kinase activities of these two critical kinases provides the opportunity to control cell death or survival under pathological conditions. Given that cancer cells are usually addicted to enhanced glucose uptake and glycolysis, a phenomenon known as the "Warburg effect," targeting glucose metabolism, has been shown to be effective in selectively killing cancer cells. For example, in LKB1-deficient lung cancer cells, where AMPK activation is impaired, glucose starvation causes oxidative stress leading to cancer cell death [71]. Although complete glucose starvation is physiologically hard to achieve, mitochondrial inhibitors and glucose transporter inhibitors have been implemented to mimic the effects of glucose starvation [189, 190]. Moreover, calorie restriction and its mimetics have also been widely suggested as novel therapeutics for cancers [191–193]. In contrast, in diseases such as ischemia-reperfusion injuries where glucose supply is insufficient, inhibition of JNK reduces cell death and shows protective effects [194–196]. Similarly, in Alzheimer's disease, where oxidative stress is caused by amyloid-β, JNK has been shown to be involved in neuronal cell death [197, 198]. Therefore, JNK is a potential target for the treatment of diseases caused by oxidative stress-induced aberrant cell death. Lastly, given the importance of AMPK and JNK in determining cell fate, the crosstalk between two kinase pathways is worthy of investigation for the exploration of novel therapeutics for many diseases.

Acknowledgments We apologize to colleagues whose work could not be cited owing to the space limitations. This work was supported by research grants from Singapore National Medical Research Council (NMRC/CIRG/1490/2018), Singapore Ministry of Education (MOE) (MOE2018-T2-1-060), and University of Macau SRG2020-00002-FHS and CPG2020-00029-FHS to HMS.

References

1. Geronimo-Olvera C, Montiel T, Rincon-Heredia R, Castro-Obregon S, Massieu L. Autophagy fails to prevent glucose deprivation/glucose reintroduction-induced neuronal death due to calpain-mediated lysosomal dysfunction in cortical neurons. Cell Death Dis. 2017;8:e2911.
2. Lauretti E, Li JG, Di Meco A, Pratico D. Glucose deficit triggers tau pathology and synaptic dysfunction in a tauopathy mouse model. Transl Psychiatry. 2017;7:e1020.
3. Lauretti E, Pratico D. Glucose deprivation increases tau phosphorylation via P38 mitogen-activated protein kinase. Aging Cell. 2015;14:1067–74.
4. Robbins NM, Swanson RA. Opposing effects of glucose on stroke and reperfusion injury: acidosis, oxidative stress, and energy metabolism. Stroke. 2014;45:1881–6.
5. Hay N. Reprogramming glucose metabolism in cancer: can it be exploited for cancer therapy? Nat Rev Cancer. 2016;16:635–49.
6. Simons AL, Mattson DM, Dornfeld K, Spitz DR. Glucose deprivation-induced metabolic oxidative stress and cancer therapy. J Cancer Res Ther. 2009;5(Suppl 1):S2–6.
7. Deng D, Yan N. GLUT, SGLT, and SWEET: structural and mechanistic investigations of the glucose transporters. Protein Sci. 2016;25:546–58.
8. Wright EM, Loo DD, Hirayama BA. Biology of human sodium glucose transporters. Physiol Rev. 2011;91:733–94.
9. Akella NM, Ciraku L, Reginato MJ. Fueling the fire: emerging role of the hexosamine biosynthetic pathway in cancer. BMC Biol. 2019;17:52.
10. Ren Y, Shen HM. Critical role of AMPK in redox regulation under glucose starvation. Redox Biol. 2019;25:101154.
11. Chiaradonna F, Ricciardiello F, Palorini R. The nutrient-sensing hexosamine biosynthetic pathway as the hub of cancer metabolic rewiring. Cell. 2018;7:53.
12. Mattaini KR, Sullivan MR, Vander Heiden MG. The importance of serine metabolism in cancer. J Cell Biol. 2016;214:249–57.
13. Ducker GS, Ghergurovich JM, Mainolfi N, Suri V, Jeong SK, Hsin-Jung Li S, Friedman A, Manfredi MG, Gitai Z, Kim H, et al. Human SHMT inhibitors reveal defective glycine import as a targetable metabolic vulnerability of diffuse large B-cell lymphoma. Proc Natl Acad Sci U S A. 2017;114:11404–9.
14. Jo SH, Son MK, Koh HJ, Lee SM, Song IH, Kim YO, Lee YS, Jeong KS, Kim WB, Park JW, et al. Control of mitochondrial redox balance and cellular defense against oxidative damage by mitochondrial NADP+-dependent isocitrate dehydrogenase. J Biol Chem. 2001;276:16168–76.
15. Murai S, Ando A, Ebara S, Hirayama M, Satomi Y, Hara T. Inhibition of malic enzyme 1 disrupts cellular metabolism and leads to vulnerability in cancer cells in glucose-restricted conditions. Oncogenesis. 2017;6:e329.
16. Rich PR. The molecular machinery of Keilin's respiratory chain. Biochem Soc Trans. 2003;31:1095–105.
17. Circu ML, Aw TY. Reactive oxygen species, cellular redox systems, and apoptosis. Free Radic Biol Med. 2010;48:749–62.
18. Schieber M, Chandel NS. ROS function in redox signaling and oxidative stress. Curr Biol. 2014;24:R453–62.
19. Kumari S, Badana AK, G MM, G S, Malla R. Reactive oxygen species: a key constituent in cancer survival. Biomark Insights. 2018;13:1177271918755391.
20. Pizzino G, Irrera N, Cucinotta M, Pallio G, Mannino F, Arcoraci V, Squadrito F, Altavilla D, Bitto A. Oxidative stress: harms and benefits for human health. Oxidative Med Cell Longev. 2017;2017:8416763.
21. Snezhkina AV, Kudryavtseva AV, Kardymon OL, Savvateeva MV, Melnikova NV, Krasnov GS, Dmitriev AA. ROS generation and antioxidant defense systems in normal and malignant cells. Oxidative Med Cell Longev. 2019;2019:6175804.

22. Hussain T, Tan B, Yin Y, Blachier F, Tossou MC, Rahu N. Oxidative stress and inflammation: what polyphenols can do for us? Oxidative Med Cell Longev. 2016;2016:7432797.
23. Liguori I, Russo G, Curcio F, Bulli G, Aran L, Della-Morte D, Gargiulo G, Testa G, Cacciatore F, Bonaduce D, et al. Oxidative stress, aging, and diseases. Clin Interv Aging. 2018;13:757–72.
24. Boveris A, Cadenas E. Mitochondrial production of hydrogen peroxide regulation by nitric oxide and the role of ubisemiquinone. IUBMB Life. 2000;50:245–50.
25. Turrens JF. Mitochondrial formation of reactive oxygen species. J Physiol. 2003;552:335–44.
26. Brand MD. The sites and topology of mitochondrial superoxide production. Exp Gerontol. 2010;45:466–72.
27. Ogata T, Senoo T, Kawano S, Ikeda S. Mitochondrial superoxide dismutase deficiency accelerates chronological aging in the fission yeast Schizosaccharomyces pombe. Cell Biol Int. 2016;40:100–6.
28. Strange RW, Antonyuk S, Hough MA, Doucette PA, Rodriguez JA, Hart PJ, Hayward LJ, Valentine JS, Hasnain SS. The structure of holo and metal-deficient wild-type human Cu, Zn superoxide dismutase and its relevance to familial amyotrophic lateral sclerosis. J Mol Biol. 2003;328:877–91.
29. Lambeth JD. NOX enzymes and the biology of reactive oxygen. Nat Rev Immunol. 2004;4:181–9.
30. Slauch JM. How does the oxidative burst of macrophages kill bacteria? Still an open question. Mol Microbiol. 2011;80:580–3.
31. Warris A, Ballou ER. Oxidative responses and fungal infection biology. Semin Cell Dev Biol. 2019;89:34–46.
32. Brennan AM, Suh SW, Won SJ, Narasimhan P, Kauppinen TM, Lee H, Edling Y, Chan PH, Swanson RA. NADPH oxidase is the primary source of superoxide induced by NMDA receptor activation. Nat Neurosci. 2009;12:857–63.
33. Gupte SA, Levine RJ, Gupte RS, Young ME, Lionetti V, Labinskyy V, Floyd BC, Ojaimi C, Bellomo M, Wolin MS, et al. Glucose-6-phosphate dehydrogenase-derived NADPH fuels superoxide production in the failing heart. J Mol Cell Cardiol. 2006;41:340–9.
34. Henriquez-Olguin C, Knudsen JR, Raun SH, Li Z, Dalbram E, Treebak JT, Sylow L, Holmdahl R, Richter EA, Jaimovich E, et al. Cytosolic ROS production by NADPH oxidase 2 regulates muscle glucose uptake during exercise. Nat Commun. 2019;10:4623.
35. Prata C, Maraldi T, Fiorentini D, Zambonin L, Hakim G, Landi L. Nox-generated ROS modulate glucose uptake in a leukaemic cell line. Free Radic Res. 2008;42:405–14.
36. Finkel T. From sulfenylation to sulfhydration: what a thiolate needs to tolerate. Sci Signal. 2012;5:pe10.
37. Winterbourn CC, Hampton MB. Thiol chemistry and specificity in redox signaling. Free Radic Biol Med. 2008;45:549–61.
38. Pastor N, Weinstein H, Jamison E, Brenowitz M. A detailed interpretation of OH radical footprints in a TBP-DNA complex reveals the role of dynamics in the mechanism of sequence-specific binding. J Mol Biol. 2000;304:55–68.
39. Reiter RJ, Melchiorri D, Sewerynek E, Poeggeler B, Barlow-Walden L, Chuang J, Ortiz GG, Acuna-Castroviejo D. A review of the evidence supporting melatonin's role as an antioxidant. J Pineal Res. 1995;18:1–11.
40. Mates JM, Sanchez-Jimenez FM. Role of reactive oxygen species in apoptosis: implications for cancer therapy. Int J Biochem Cell Biol. 2000;32:157–70.
41. Ray G, Husain SA. Oxidants, antioxidants and carcinogenesis. Indian J Exp Biol. 2002;40:1213–32.
42. Valko M, Izakovic M, Mazur M, Rhodes CJ, Telser J. Role of oxygen radicals in DNA damage and cancer incidence. Mol Cell Biochem. 2004;266:37–56.
43. Quijano C, Trujillo M, Castro L, Trostchansky A. Interplay between oxidant species and energy metabolism. Redox Biol. 2016;8:28–42.

44. Ahmadinejad F, Geir Moller S, Hashemzadeh-Chaleshtori M, Bidkhori G, Jami MS. Molecular mechanisms behind free radical scavengers function against oxidative stress. Antioxidants (Basel). 2017;6:51.
45. Karihtala P, Soini Y. Reactive oxygen species and antioxidant mechanisms in human tissues and their relation to malignancies. APMIS. 2007;115:81–103.
46. Venditti P, Di Stefano L, Di Meo S. Mitochondrial metabolism of reactive oxygen species. Mitochondrion. 2013;13:71–82.
47. Fridovich I. Superoxide radical and superoxide dismutases. Annu Rev Biochem. 1995;64:97–112.
48. Muller FL, Liu Y, Van Remmen H. Complex III releases superoxide to both sides of the inner mitochondrial membrane. J Biol Chem. 2004;279:49064–73.
49. Okado-Matsumoto A, Fridovich I. Subcellular distribution of superoxide dismutases (SOD) in rat liver: Cu,Zn-SOD in mitochondria. J Biol Chem. 2001;276:38388–93.
50. Chow CK. Vitamin E regulation of mitochondrial superoxide generation. Biol Signals Recept. 2001;10:112–24.
51. Starke PE, Farber JL. Ferric iron and superoxide ions are required for the killing of cultured hepatocytes by hydrogen peroxide. Evidence for the participation of hydroxyl radicals formed by an iron-catalyzed Haber-Weiss reaction. J Biol Chem. 1985;260:10099–104.
52. Butler J, Jayson GG, Swallow AJ. The reaction between the superoxide anion radical and cytochrome c. Biochim Biophys Acta. 1975;408:215–22.
53. Korshunov SS, Krasnikov BF, Pereverzev MO, Skulachev VP. The antioxidant functions of cytochrome c. FEBS Lett. 1999;462:192–8.
54. Mailer K. Superoxide radical as electron donor for oxidative phosphorylation of ADP. Biochem Biophys Res Commun. 1990;170:59–64.
55. Pereverzev MO, Vygodina TV, Konstantinov AA, Skulachev VP. Cytochrome c, an ideal antioxidant. Biochem Soc Trans. 2003;31:1312–5.
56. Han D, Antunes F, Canali R, Rettori D, Cadenas E. Voltage-dependent anion channels control the release of the superoxide anion from mitochondria to cytosol. J Biol Chem. 2003;278:5557–63.
57. Aebi H. Catalase in vitro. Methods Enzymol. 1984;105:121–6.
58. Goth L, Rass P, Pay A. Catalase enzyme mutations and their association with diseases. Mol Diagn. 2004;8:141–9.
59. Kim Y, Jang HH. Role of cytosolic 2-Cys Prx1 and Prx2 in redox signaling. Antioxidants (Basel). 2019;8:169.
60. Claiborne A, Yeh JI, Mallett TC, Luba J, Crane EJ 3rd, Charrier V, Parsonage D. Protein-sulfenic acids: diverse roles for an unlikely player in enzyme catalysis and redox regulation. Biochemistry. 1999;38:15407–16.
61. Monteiro G, Horta BB, Pimenta DC, Augusto O, Netto LE. Reduction of 1-Cys peroxiredoxins by ascorbate changes the thiol-specific antioxidant paradigm, revealing another function of vitamin C. Proc Natl Acad Sci U S A. 2007;104:4886–91.
62. Nakamura H, Nakamura K, Yodoi J. Redox regulation of cellular activation. Annu Rev Immunol. 1997;15:351–69.
63. Mustacich D, Powis G. Thioredoxin reductase. Biochem J. 2000;346(Pt 1):1–8.
64. Foloppe N, Nilsson L. The glutaredoxin -C-P-Y-C- motif: influence of peripheral residues. Structure. 2004;12:289–300.
65. Fernandes AP, Holmgren A. Glutaredoxins: glutathione-dependent redox enzymes with functions far beyond a simple thioredoxin backup system. Antioxid Redox Signal. 2004;6:63–74.
66. Kansanen E, Kuosmanen SM, Leinonen H, Levonen AL. The Keap1-Nrf2 pathway: mechanisms of activation and dysregulation in cancer. Redox Biol. 2013;1:45–9.
67. Shen HM, Liu ZG. JNK signaling pathway is a key modulator in cell death mediated by reactive oxygen and nitrogen species. Free Radic Biol Med. 2006;40:928–39.

68. Son Y, Cheong YK, Kim NH, Chung HT, Kang DG, Pae HO. Mitogen-activated protein kinases and reactive oxygen species: how can ROS activate MAPK pathways? J Signal Transduct. 2011;2011:792639.
69. Ahmad IM, Aykin-Burns N, Sim JE, Walsh SA, Higashikubo R, Buettner GR, Venkataraman S, Mackey MA, Flanagan SW, Oberley LW, et al. Mitochondrial O2*- and H2O2 mediate glucose deprivation-induced stress in human cancer cells. J Biol Chem. 2005;280:4254–63.
70. Graham NA, Tahmasian M, Kohli B, Komisopoulou E, Zhu M, Vivanco I, Teitell MA, Wu H, Ribas A, Lo RS, et al. Glucose deprivation activates a metabolic and signaling amplification loop leading to cell death. Mol Syst Biol. 2012;8:589.
71. Jeon SM, Chandel NS, Hay N. AMPK regulates NADPH homeostasis to promote tumour cell survival during energy stress. Nature. 2012;485:661–5.
72. Li L, Chen Y, Gibson SB. Starvation-induced autophagy is regulated by mitochondrial reactive oxygen species leading to AMPK activation. Cell Signal. 2013;25:50–65.
73. Li XX, Wang ZJ, Zheng Y, Guan YF, Yang PB, Chen X, Peng C, He JP, Ai YL, Wu SF, et al. Nuclear receptor Nur77 facilitates melanoma cell survival under metabolic stress by protecting fatty acid oxidation. Mol Cell. 2018;69:480–92.
74. Boveris A, Chance B. The mitochondrial generation of hydrogen peroxide. General properties and effect of hyperbaric oxygen. Biochem J. 1973;134:707–16.
75. Murphy MP. How mitochondria produce reactive oxygen species. Biochem J. 2009;417:1–13.
76. Lee JY, Kapur M, Li M, Choi MC, Choi S, Kim HJ, Kim I, Lee E, Taylor JP, Yao TP. MFN1 deacetylation activates adaptive mitochondrial fusion and protects metabolically challenged mitochondria. J Cell Sci. 2014;127:4954–63.
77. Raut GK, Chakrabarti M, Pamarthy D, Bhadra MP. Glucose starvation-induced oxidative stress causes mitochondrial dysfunction and apoptosis via Prohibitin 1 upregulation in human breast cancer cells. Free Radic Biol Med. 2019;145:428–41.
78. Kudin AP, Bimpong-Buta NY, Vielhaber S, Elger CE, Kunz WS. Characterization of superoxide-producing sites in isolated brain mitochondria. J Biol Chem. 2004;279:4127–35.
79. Kussmaul L, Hirst J. The mechanism of superoxide production by NADH:ubiquinone oxidoreductase (complex I) from bovine heart mitochondria. Proc Natl Acad Sci U S A. 2006;103:7607–12.
80. Hansford RG, Hogue BA, Mildaziene V. Dependence of H2O2 formation by rat heart mitochondria on substrate availability and donor age. J Bioenerg Biomembr. 1997;29:89–95.
81. Korshunov SS, Skulachev VP, Starkov AA. High protonic potential actuates a mechanism of production of reactive oxygen species in mitochondria. FEBS Lett. 1997;416:15–8.
82. Lambert AJ, Brand MD. Superoxide production by NADH:ubiquinone oxidoreductase (complex I) depends on the pH gradient across the mitochondrial inner membrane. Biochem J. 2004;382:511–7.
83. Circu ML, Maloney RE, Aw TY. Low glucose stress decreases cellular NADH and mitochondrial ATP in colonic epithelial cancer cells: influence of mitochondrial substrates. Chem Biol Interact. 2017;264:16–24.
84. Luo X, Li R, Yan LJ. Roles of pyruvate, NADH, and mitochondrial complex I in redox balance and imbalance in beta cell function and dysfunction. J Diabetes Res. 2015;2015:512618.
85. Owada S, Shimoda Y, Tsuchihara K, Esumi H. Critical role of H2O2 generated by NOX4 during cellular response under glucose deprivation. PLoS One. 2013;8:e56628.
86. Sciarretta S, Zhai P, Shao D, Zablocki D, Nagarajan N, Terada LS, Volpe M, Sadoshima J. Activation of NADPH oxidase 4 in the endoplasmic reticulum promotes cardiomyocyte autophagy and survival during energy stress through the protein kinase RNA-activated-like endoplasmic reticulum kinase/eukaryotic initiation factor 2alpha/activating transcription factor 4 pathway. Circ Res. 2013;113:1253–64.
87. Miranda-Saavedra D, Gabaldon T, Barton GJ, Langsley G, Doerig C. The kinomes of apicomplexan parasites. Microbes Infect. 2012;14:796–810.

88. Miranda-Saavedra D, Stark MJ, Packer JC, Vivarès CP, Doerig C, Barton GJ. The complement of protein kinases of the microsporidium Encephalitozoon cuniculi in relation to those of Saccharomyces cerevisiae and Schizosaccharomyces pombe. BMC Genomics. 2007;8:309.
89. Lin SC, Hardie DG. AMPK: sensing glucose as well as cellular energy status. Cell Metab. 2018;27:299–313.
90. Hinchy EC, Gruszczyk AV, Willows R, Navaratnam N, Hall AR, Bates G, Bright TP, Krieg T, Carling D, Murphy MP. Mitochondria-derived ROS activate AMP-activated protein kinase (AMPK) indirectly. J Biol Chem. 2018;293:17208–17.
91. Rabinovitch RC, Samborska B, Faubert B, Ma EH, Gravel SP, Andrzejewski S, Raissi TC, Pause A, St-Pierre J, Jones RG. AMPK maintains cellular metabolic homeostasis through regulation of mitochondrial reactive oxygen species. Cell Rep. 2017;21:1–9.
92. Davis W Jr, Ronai Z, Tew KD. Cellular thiols and reactive oxygen species in drug-induced apoptosis. J Pharmacol Exp Ther. 2001;296:1–6.
93. Matsuzawa A, Ichijo H. Redox control of cell fate by MAP kinase: physiological roles of ASK1-MAP kinase pathway in stress signaling. Biochim Biophys Acta. 2008;1780:1325–36.
94. Hawley SA, Boudeau J, Reid JL, Mustard KJ, Udd L, Makela TP, Alessi DR, Hardie DG. Complexes between the LKB1 tumor suppressor, STRAD alpha/beta and MO25 alpha/beta are upstream kinases in the AMP-activated protein kinase cascade. J Biol. 2003;2:28.
95. Kato K, Ogura T, Kishimoto A, Minegishi Y, Nakajima N, Miyazaki M, Esumi H. Critical roles of AMP-activated protein kinase in constitutive tolerance of cancer cells to nutrient deprivation and tumor formation. Oncogene. 2002;21:6082–90.
96. Shaw RJ, Kosmatka M, Bardeesy N, Hurley RL, Witters LA, DePinho RA, Cantley LC. The tumor suppressor LKB1 kinase directly activates AMP-activated kinase and regulates apoptosis in response to energy stress. Proc Natl Acad Sci U S A. 2004;101:3329–35.
97. Woods A, Johnstone SR, Dickerson K, Leiper FC, Fryer LG, Neumann D, Schlattner U, Wallimann T, Carlson M, Carling D. LKB1 is the upstream kinase in the AMP-activated protein kinase cascade. Curr Biol. 2003;13:2004–8.
98. Hawley SA, Pan DA, Mustard KJ, Ross L, Bain J, Edelman AM, Frenguelli BG, Hardie DG. Calmodulin-dependent protein kinase kinase-beta is an alternative upstream kinase for AMP-activated protein kinase. Cell Metab. 2005;2:9–19.
99. Hurley RL, Anderson KA, Franzone JM, Kemp BE, Means AR, Witters LA. The Ca2+/calmodulin-dependent protein kinase kinases are AMP-activated protein kinase kinases. J Biol Chem. 2005;280:29060–6.
100. Woods A, Dickerson K, Heath R, Hong SP, Momcilovic M, Johnstone SR, Carlson M, Carling D. Ca2+/calmodulin-dependent protein kinase kinase-beta acts upstream of AMP-activated protein kinase in mammalian cells. Cell Metab. 2005;2:21–33.
101. Zhang CS, Hawley SA, Zong Y, Li M, Wang Z, Gray A, Ma T, Cui J, Feng JW, Zhu M, et al. Fructose-1,6-bisphosphate and aldolase mediate glucose sensing by AMPK. Nature. 2017;548:112–6.
102. Zhang CS, Jiang B, Li M, Zhu M, Peng Y, Zhang YL, Wu YQ, Li TY, Liang Y, Lu Z, et al. The lysosomal v-ATPase-ragulator complex is a common activator for AMPK and mTORC1, acting as a switch between catabolism and anabolism. Cell Metab. 2014;20:526–40.
103. Zhang YL, Guo H, Zhang CS, Lin SY, Yin Z, Peng Y, Luo H, Shi Y, Lian G, Zhang C, et al. AMP as a low-energy charge signal autonomously initiates assembly of AXIN-AMPK-LKB1 complex for AMPK activation. Cell Metab. 2013;18:546–55.
104. Zong Y, Zhang CS, Li M, Wang W, Wang Z, Hawley SA, Ma T, Feng JW, Tian X, Qi Q, et al. Hierarchical activation of compartmentalized pools of AMPK depends on severity of nutrient or energy stress. Cell Res. 2019;29:460–73.
105. Carling D, Zammit VA, Hardie DG. A common bicyclic protein kinase cascade inactivates the regulatory enzymes of fatty acid and cholesterol biosynthesis. FEBS Lett. 1987;223:217–22.
106. Carlson CA, Kim KH. Regulation of hepatic acetyl coenzyme a carboxylase by phosphorylation and dephosphorylation. J Biol Chem. 1973;248:378–80.

107. Fullerton MD, Galic S, Marcinko K, Sikkema S, Pulinilkunnil T, Chen ZP, O'Neill HM, Ford RJ, Palanivel R, O'Brien M, et al. Single phosphorylation sites in Acc1 and Acc2 regulate lipid homeostasis and the insulin-sensitizing effects of metformin. Nat Med. 2013;19:1649–54.
108. Hardie DG, Pan DA. Regulation of fatty acid synthesis and oxidation by the AMP-activated protein kinase. Biochem Soc Trans. 2002;30:1064–70.
109. Beg ZH, Allmann DW, Gibson DM. Modulation of 3-hydroxy-3-methylglutaryl coenzyme a reductase activity with cAMP and wth protein fractions of rat liver cytosol. Biochem Biophys Res Commun. 1973;54:1362–9.
110. Kawaguchi T, Osatomi K, Yamashita H, Kabashima T, Uyeda K. Mechanism for fatty acid "sparing" effect on glucose-induced transcription: regulation of carbohydrate-responsive element-binding protein by AMP-activated protein kinase. J Biol Chem. 2002;277:3829–35.
111. Li Y, Xu S, Mihaylova MM, Zheng B, Hou X, Jiang B, Park O, Luo Z, Lefai E, Shyy JY, et al. AMPK phosphorylates and inhibits SREBP activity to attenuate hepatic steatosis and atherosclerosis in diet-induced insulin-resistant mice. Cell Metab. 2011;13:376–88.
112. Endo H, Owada S, Inagaki Y, Shida Y, Tatemichi M. Glucose starvation induces LKB1-AMPK-mediated MMP-9 expression in cancer cells. Sci Rep. 2018;8:10122.
113. Joo MS, Kim WD, Lee KY, Kim JH, Koo JH, Kim SG. AMPK facilitates nuclear accumulation of Nrf2 by phosphorylating at serine 550. Mol Cell Biol. 2016;36:1931–42.
114. Komatsu M, Kurokawa H, Waguri S, Taguchi K, Kobayashi A, Ichimura Y, Sou YS, Ueno I, Sakamoto A, Tong KI, et al. The selective autophagy substrate p62 activates the stress responsive transcription factor Nrf2 through inactivation of Keap1. Nat Cell Biol. 2010;12:213–23.
115. Matzinger M, Fischhuber K, Poloske D, Mechtler K, Heiss EH. AMPK leads to phosphorylation of the transcription factor Nrf2, tuning transactivation of selected target genes. Redox Biol. 2020;29:101393.
116. McMahon M, Itoh K, Yamamoto M, Hayes JD. Keap1-dependent proteasomal degradation of transcription factor Nrf2 contributes to the negative regulation of antioxidant response element-driven gene expression. J Biol Chem. 2003;278:21592–600.
117. Zhang DD, Lo SC, Cross JV, Templeton DJ, Hannink M. Keap1 is a redox-regulated substrate adaptor protein for a Cul3-dependent ubiquitin ligase complex. Mol Cell Biol. 2004;24:10941–53.
118. Basit F, van Oppen LM, Schockel L, Bossenbroek HM, van Emst-de Vries SE, Hermeling JC, Grefte S, Kopitz C, Heroult M, Hgm Willems P, et al. Mitochondrial complex I inhibition triggers a mitophagy-dependent ROS increase leading to necroptosis and ferroptosis in melanoma cells. Cell Death Dis. 2017;8:e2716.
119. Egan DF, Shackelford DB, Mihaylova MM, Gelino S, Kohnz RA, Mair W, Vasquez DS, Joshi A, Gwinn DM, Taylor R, et al. Phosphorylation of ULK1 (hATG1) by AMP-activated protein kinase connects energy sensing to mitophagy. Science. 2011;331:456–61.
120. Schofield JH, Schafer ZT. Mitochondrial reactive oxygen species and mitophagy: a complex and nuanced relationship. Antioxid Redox Signal. 2020; https://doi.org/10.1089/ars.2020.8058.
121. Shao D, Oka S, Liu T, Zhai P, Ago T, Sciarretta S, Li H, Sadoshima J. A redox-dependent mechanism for regulation of AMPK activation by Thioredoxin1 during energy starvation. Cell Metab. 2014;19:232–45.
122. Gonzalez-Menendez P, Hevia D, Alonso-Arias R, Alvarez-Artime A, Rodriguez-Garcia A, Kinet S, Gonzalez-Pola I, Taylor N, Mayo JC, Sainz RM. GLUT1 protects prostate cancer cells from glucose deprivation-induced oxidative stress. Redox Biol. 2018;17:112–27.
123. Pehmoller C, Treebak JT, Birk JB, Chen S, Mackintosh C, Hardie DG, Richter EA, Wojtaszewski JF. Genetic disruption of AMPK signaling abolishes both contraction- and insulin-stimulated TBC1D1 phosphorylation and 14-3-3 binding in mouse skeletal muscle. Am J Physiol Endocrinol Metab. 2009;297:E665–75.

124. Wu N, Zheng B, Shaywitz A, Dagon Y, Tower C, Bellinger G, Shen CH, Wen J, Asara J, McGraw TE, et al. AMPK-dependent degradation of TXNIP upon energy stress leads to enhanced glucose uptake via GLUT1. Mol Cell. 2013;49:1167–75.
125. Chaube B, Malvi P, Singh SV, Mohammad N, Viollet B, Bhat MK. AMPK maintains energy homeostasis and survival in cancer cells via regulating p38/PGC-1alpha-mediated mitochondrial biogenesis. Cell Death Discov. 2015;1:15063.
126. Marsin AS, Bertrand L, Rider MH, Deprez J, Beauloye C, Vincent MF, Van den Berghe G, Carling D, Hue L. Phosphorylation and activation of heart PFK-2 by AMPK has a role in the stimulation of glycolysis during ischaemia. Curr Biol. 2000;10:1247–55.
127. Wu SB, Wei YH. AMPK-mediated increase of glycolysis as an adaptive response to oxidative stress in human cells: implication of the cell survival in mitochondrial diseases. Biochim Biophys Acta. 2012;1822:233–47.
128. Bode AM, Dong Z. The functional contrariety of JNK. Mol Carcinog. 2007;46:591–8.
129. Pulverer BJ, Kyriakis JM, Avruch J, Nikolakaki E, Woodgett JR. Phosphorylation of c-Jun mediated by MAP kinases. Nature. 1991;353:670–4.
130. Fuchs SY, Adler V, Pincus MR, Ronai Z. MEKK1/JNK signaling stabilizes and activates p53. Proc Natl Acad Sci U S A. 1998;95:10541–6.
131. Gupta S, Campbell D, Derijard B, Davis RJ. Transcription factor ATF2 regulation by the JNK signal transduction pathway. Science. 1995;267:389–93.
132. Wei Y, Pattingre S, Sinha S, Bassik M, Levine B. JNK1-mediated phosphorylation of Bcl-2 regulates starvation-induced autophagy. Mol Cell. 2008;30:678–88.
133. Yu C, Minemoto Y, Zhang J, Liu J, Tang F, Bui TN, Xiang J, Lin A. JNK suppresses apoptosis via phosphorylation of the proapoptotic Bcl-2 family protein BAD. Mol Cell. 2004;13:329–40.
134. Dhanasekaran DN, Reddy EP. JNK-signaling: a multiplexing hub in programmed cell death. Genes Cancer. 2017;8:682–94.
135. Weston CR, Davis RJ. The JNK signal transduction pathway. Curr Opin Cell Biol. 2007;19:142–9.
136. Derijard B, Raingeaud J, Barrett T, Wu IH, Han J, Ulevitch RJ, Davis RJ. Independent human MAP-kinase signal transduction pathways defined by MEK and MKK isoforms. Science. 1995;267:682–5.
137. Lin A, Minden A, Martinetto H, Claret FX, Lange-Carter C, Mercurio F, Johnson GL, Karin M. Identification of a dual specificity kinase that activates the Jun kinases and p38-Mpk2. Science. 1995;268:286–90.
138. Moriguchi T, Toyoshima F, Masuyama N, Hanafusa H, Gotoh Y, Nishida E. A novel SAPK/JNK kinase, MKK7, stimulated by TNFalpha and cellular stresses. EMBO J. 1997;16:7045–53.
139. Tournier C, Whitmarsh AJ, Cavanagh J, Barrett T, Davis RJ. Mitogen-activated protein kinase kinase 7 is an activator of the c-Jun NH2-terminal kinase. Proc Natl Acad Sci U S A. 1997;94:7337–42.
140. Ichijo H, Nishida E, Irie K, ten Dijke P, Saitoh M, Moriguchi T, Takagi M, Matsumoto K, Miyazono K, Gotoh Y. Induction of apoptosis by ASK1, a mammalian MAPKKK that activates SAPK/JNK and p38 signaling pathways. Science. 1997;275:90–4.
141. Rana A, Gallo K, Godowski P, Hirai S, Ohno S, Zon L, Kyriakis JM, Avruch J. The mixed lineage kinase SPRK phosphorylates and activates the stress-activated protein kinase activator, SEK-1. J Biol Chem. 1996;271:19025–8.
142. Shirakabe K, Yamaguchi K, Shibuya H, Irie K, Matsuda S, Moriguchi T, Gotoh Y, Matsumoto K, Nishida E. TAK1 mediates the ceramide signaling to stress-activated protein kinase/c-Jun N-terminal kinase. J Biol Chem. 1997;272:8141–4.
143. Weston CR, Davis RJ. The JNK signal transduction pathway. Curr Opin Genet Dev. 2002;12:14–21.

144. Gotoh Y, Cooper JA. Reactive oxygen species- and dimerization-induced activation of apoptosis signal-regulating kinase 1 in tumor necrosis factor-alpha signal transduction. J Biol Chem. 1998;273:17477–82.
145. Liu Y, Min W. Thioredoxin promotes ASK1 ubiquitination and degradation to inhibit ASK1-mediated apoptosis in a redox activity-independent manner. Circ Res. 2002;90:1259–66.
146. Saitoh M, Nishitoh H, Fujii M, Takeda K, Tobiume K, Sawada Y, Kawabata M, Miyazono K, Ichijo H. Mammalian thioredoxin is a direct inhibitor of apoptosis signal-regulating kinase (ASK) 1. EMBO J. 1998;17:2596–606.
147. Song JJ, Lee YJ. Differential role of glutaredoxin and thioredoxin in metabolic oxidative stress-induced activation of apoptosis signal-regulating kinase 1. Biochem J. 2003;373:845–53.
148. Song JJ, Rhee JG, Suntharalingam M, Walsh SA, Spitz DR, Lee YJ. Role of glutaredoxin in metabolic oxidative stress. Glutaredoxin as a sensor of oxidative stress mediated by H2O2. J Biol Chem. 2002;277:46566–75.
149. Goldman EH, Chen L, Fu H. Activation of apoptosis signal-regulating kinase 1 by reactive oxygen species through dephosphorylation at serine 967 and 14-3-3 dissociation. J Biol Chem. 2004;279:10442–9.
150. Morita K, Saitoh M, Tobiume K, Matsuura H, Enomoto S, Nishitoh H, Ichijo H. Negative feedback regulation of ASK1 by protein phosphatase 5 (PP5) in response to oxidative stress. EMBO J. 2001;20:6028–36.
151. Zhou G, Golden T, Aragon IV, Honkanen RE. Ser/Thr protein phosphatase 5 inactivates hypoxia-induced activation of an apoptosis signal-regulating kinase 1/MKK-4/JNK signaling cascade. J Biol Chem. 2004;279:46595–605.
152. Chen K, Vita JA, Berk BC, Keaney JF Jr. C-Jun N-terminal kinase activation by hydrogen peroxide in endothelial cells involves SRC-dependent epidermal growth factor receptor transactivation. J Biol Chem. 2001;276:16045–50.
153. Holgado-Madruga M, Wong AJ. Gab1 is an integrator of cell death versus cell survival signals in oxidative stress. Mol Cell Biol. 2003;23:4471–84.
154. Yoshizumi M, Abe J, Haendeler J, Huang Q, Berk BC. Src and Cas mediate JNK activation but not ERK1/2 and p38 kinases by reactive oxygen species. J Biol Chem. 2000;275:11706–12.
155. Adler V, Yin Z, Fuchs SY, Benezra M, Rosario L, Tew KD, Pincus MR, Sardana M, Henderson CJ, Wolf CR, et al. Regulation of JNK signaling by GSTp. EMBO J. 1999;18:1321–34.
156. Wang T, Arifoglu P, Ronai Z, Tew KD. Glutathione S-transferase P1-1 (GSTP1-1) inhibits c-Jun N-terminal kinase (JNK1) signaling through interaction with the C terminus. J Biol Chem. 2001;276:20999–1003.
157. Kamata H, Honda S, Maeda S, Chang L, Hirata H, Karin M. Reactive oxygen species promote TNFalpha-induced death and sustained JNK activation by inhibiting MAP kinase phosphatases. Cell. 2005;120:649–61.
158. Hou N, Torii S, Saito N, Hosaka M, Takeuchi T. Reactive oxygen species-mediated pancreatic beta-cell death is regulated by interactions between stress-activated protein kinases, p38 and c-Jun N-terminal kinase, and mitogen-activated protein kinase phosphatases. Endocrinology. 2008;149:1654–65.
159. Mizuno R, Oya M, Shiomi T, Marumo K, Okada Y, Murai M. Inhibition of MKP-1 expression potentiates JNK related apoptosis in renal cancer cells. J Urol. 2004;172:723–7.
160. Tafolla E, Wang S, Wong B, Leong J, Kapila YL. JNK1 and JNK2 oppositely regulate p53 in signaling linked to apoptosis triggered by an altered fibronectin matrix: JNK links FAK and p53. J Biol Chem. 2005;280:19992–9.
161. Tournier C, Hess P, Yang DD, Xu J, Turner TK, Nimnual A, Bar-Sagi D, Jones SN, Flavell RA, Davis RJ. Requirement of JNK for stress-induced activation of the cytochrome c-mediated death pathway. Science. 2000;288:870–4.

162. Ventura JJ, Hubner A, Zhang C, Flavell RA, Shokat KM, Davis RJ. Chemical genetic analysis of the time course of signal transduction by JNK. Mol Cell. 2006;21:701–10.
163. Yun H, Kim HS, Lee S, Kang I, Kim SS, Choe W, Ha J. AMP kinase signaling determines whether c-Jun N-terminal kinase promotes survival or apoptosis during glucose deprivation. Carcinogenesis. 2009;30:529–37.
164. Vari R, Scazzocchio B, Santangelo C, Filesi C, Galvano F, D'Archivio M, Masella R, Giovannini C. Protocatechuic acid prevents oxLDL-induced apoptosis by activating JNK/Nrf2 survival signals in macrophages. Oxidative Med Cell Longev. 2015;2015:351827.
165. Yue WY, Clark JJ, Fernando A, Domann F, Hansen MR. Contribution of persistent C-Jun N-terminal kinase activity to the survival of human vestibular schwannoma cells by suppression of accumulation of mitochondrial superoxides. Neuro-Oncology. 2011;13:961–73.
166. Huang XL, Zhang H, Yang XY, Dong XY, Xie XY, Yin HB, Gou X, Lin Y, He WY. Activation of a c-Jun N-terminal kinase-mediated autophagy pathway attenuates the anticancer activity of gemcitabine in human bladder cancer cells. Anti-Cancer Drugs. 2017;28:596–602.
167. Shen Y, Yang J, Zhao J, Xiao C, Xu C, Xiang Y. The switch from ER stress-induced apoptosis to autophagy via ROS-mediated JNK/p62 signals: a survival mechanism in methotrexate-resistant choriocarcinoma cells. Exp Cell Res. 2015;334:207–18.
168. Li Y, Arita Y, Koo HC, Davis JM, Kazzaz JA. Inhibition of c-Jun N-terminal kinase pathway improves cell viability in response to oxidant injury. Am J Respir Cell Mol Biol. 2003;29:779–83.
169. Tobiume K, Matsuzawa A, Takahashi T, Nishitoh H, Morita K, Takeda K, Minowa O, Miyazono K, Noda T, Ichijo H. ASK1 is required for sustained activations of JNK/p38 MAP kinases and apoptosis. EMBO Rep. 2001;2:222–8.
170. Yoshizumi M, Kogame T, Suzaki Y, Fujita Y, Kyaw M, Kirima K, Ishizawa K, Tsuchiya K, Kagami S, Tamaki T. Ebselen attenuates oxidative stress-induced apoptosis via the inhibition of the c-Jun N-terminal kinase and activator protein-1 signalling pathway in PC12 cells. Br J Pharmacol. 2002;136:1023–32.
171. Fan M, Chambers TC. Role of mitogen-activated protein kinases in the response of tumor cells to chemotherapy. Drug Resist Updat. 2001;4:253–67.
172. Saha MN, Jiang H, Yang Y, Zhu X, Wang X, Schimmer AD, Qiu L, Chang H. Targeting p53 via JNK pathway: a novel role of RITA for apoptotic signaling in multiple myeloma. PLoS One. 2012;7:e30215.
173. Shaulian E, Karin M. AP-1 as a regulator of cell life and death. Nat Cell Biol. 2002;4:E131–6.
174. Oleinik NV, Krupenko NI, Krupenko SA. Cooperation between JNK1 and JNK2 in activation of p53 apoptotic pathway. Oncogene. 2007;26:7222–30.
175. Shi Y, Nikulenkov F, Zawacka-Pankau J, Li H, Gabdoulline R, Xu J, Eriksson S, Hedstrom E, Issaeva N, Kel A, et al. ROS-dependent activation of JNK converts p53 into an efficient inhibitor of oncogenes leading to robust apoptosis. Cell Death Differ. 2014;21:612–23.
176. Yang H, Xie Y, Yang D, Ren D. Oxidative stress-induced apoptosis in granulosa cells involves JNK, p53 and Puma. Oncotarget. 2017;8:25310–22.
177. Inoshita S, Takeda K, Hatai T, Terada Y, Sano M, Hata J, Umezawa A, Ichijo H. Phosphorylation and inactivation of myeloid cell leukemia 1 by JNK in response to oxidative stress. J Biol Chem. 2002;277:43730–4.
178. Yamamoto K, Ichijo H, Korsmeyer SJ. BCL-2 is phosphorylated and inactivated by an ASK1/Jun N-terminal protein kinase pathway normally activated at G(2)/M. Mol Cell Biol. 1999;19:8469–78.
179. Putcha GV, Le S, Frank S, Besirli CG, Clark K, Chu B, Alix S, Youle RJ, LaMarche A, Maroney AC, et al. JNK-mediated BIM phosphorylation potentiates BAX-dependent apoptosis. Neuron. 2003;38:899–914.
180. Tsuruta F, Sunayama J, Mori Y, Hattori S, Shimizu S, Tsujimoto Y, Yoshioka K, Masuyama N, Gotoh Y. JNK promotes Bax translocation to mitochondria through phosphorylation of 14-3-3 proteins. EMBO J. 2004;23:1889–99.

181. Song JJ, Lee YJ. Cross-talk between JIP3 and JIP1 during glucose deprivation: SEK1-JNK2 and Akt1 act as mediators. J Biol Chem. 2005;280:26845–55.
182. Momcilovic M, Hong SP, Carlson M. Mammalian TAK1 activates Snf1 protein kinase in yeast and phosphorylates AMP-activated protein kinase in vitro. J Biol Chem. 2006;281:25336–43.
183. Xie M, Zhang D, Dyck JR, Li Y, Zhang H, Morishima M, Mann DL, Taffet GE, Baldini A, Khoury DS, et al. A pivotal role for endogenous TGF-beta-activated kinase-1 in the LKB1/AMP-activated protein kinase energy-sensor pathway. Proc Natl Acad Sci U S A. 2006;103:17378–83.
184. Chen X, Li X, Zhang W, He J, Xu B, Lei B, Wang Z, Cates C, Rousselle T, Li J. Activation of AMPK inhibits inflammatory response during hypoxia and reoxygenation through modulating JNK-mediated NF-kappaB pathway. Metabolism. 2018;83:256–70.
185. Mancini SJ, White AD, Bijland S, Rutherford C, Graham D, Richter EA, Viollet B, Touyz RM, Palmer TM, Salt IP. Activation of AMP-activated protein kinase rapidly suppresses multiple pro-inflammatory pathways in adipocytes including IL-1 receptor-associated kinase-4 phosphorylation. Mol Cell Endocrinol. 2017;440:44–56.
186. Kefas BA, Cai Y, Ling Z, Heimberg H, Hue L, Pipeleers D, Van de Casteele M. AMP-activated protein kinase can induce apoptosis of insulin-producing MIN6 cells through stimulation of c-Jun-N-terminal kinase. J Mol Endocrinol. 2003;30:151–61.
187. Meisse D, Van de Casteele M, Beauloye C, Hainault I, Kefas BA, Rider MH, Foufelle F, Hue L. Sustained activation of AMP-activated protein kinase induces c-Jun N-terminal kinase activation and apoptosis in liver cells. FEBS Lett. 2002;526:38–42.
188. Hu K, Gong X, Ai Q, Lin L, Dai J, Cai L, Jiang R, Ge P, Zhang L. Endogenous AMPK acts as a detrimental factor in fulminant hepatitis via potentiating JNK-dependent hepatocyte apoptosis. Cell Death Dis. 2017;8:e2637.
189. Scafoglio CR, Villegas B, Abdelhady G, Bailey ST, Liu J, Shirali AS, Wallace WD, Magyar CE, Grogan TR, Elashoff D, et al. Sodium-glucose transporter 2 is a diagnostic and therapeutic target for early-stage lung adenocarcinoma. Sci Transl Med. 2018;10:eaat5933.
190. Shackelford DB, Abt E, Gerken L, Vasquez DS, Seki A, Leblanc M, Wei L, Fishbein MC, Czernin J, Mischel PS, et al. LKB1 inactivation dictates therapeutic response of non-small cell lung cancer to the metabolism drug phenformin. Cancer Cell. 2013;23:143–58.
191. Hatzivassiliou G, Zhao F, Bauer DE, Andreadis C, Shaw AN, Dhanak D, Hingorani SR, Tuveson DA, Thompson CB. ATP citrate lyase inhibition can suppress tumor cell growth. Cancer Cell. 2005;8:311–21.
192. Iliopoulos D, Hirsch HA, Struhl K. Metformin decreases the dose of chemotherapy for prolonging tumor remission in mouse xenografts involving multiple cancer cell types. Cancer Res. 2011;71:3196–201.
193. Mukherjee P, Mulrooney TJ, Marsh J, Blair D, Chiles TC, Seyfried TN. Differential effects of energy stress on AMPK phosphorylation and apoptosis in experimental brain tumor and normal brain. Mol Cancer. 2008;7:37.
194. Borsello T, Clarke PG, Hirt L, Vercelli A, Repici M, Schorderet DF, Bogousslavsky J, Bonny C. A peptide inhibitor of c-Jun N-terminal kinase protects against excitotoxicity and cerebral ischemia. Nat Med. 2003;9:1180–6.
195. Kim BJ, Silverman SM, Liu Y, Wordinger RJ, Pang IH, Clark AF. In vitro and in vivo neuroprotective effects of cJun N-terminal kinase inhibitors on retinal ganglion cells. Mol Neurodegener. 2016;11:30.
196. Uehara T, Bennett B, Sakata ST, Satoh Y, Bilter GK, Westwick JK, Brenner DA. JNK mediates hepatic ischemia reperfusion injury. J Hepatol. 2005;42:850–9.
197. Bozyczko-Coyne D, O'Kane TM, Wu ZL, Dobrzanski P, Murthy S, Vaught JL, Scott RW. CEP-1347/KT-7515, an inhibitor of SAPK/JNK pathway activation, promotes survival and blocks multiple events associated with Abeta-induced cortical neuron apoptosis. J Neurochem. 2001;77:849–63.

198. Morishima Y, Gotoh Y, Zieg J, Barrett T, Takano H, Flavell R, Davis RJ, Shirasaki Y, Greenberg ME. Beta-amyloid induces neuronal apoptosis via a mechanism that involves the c-Jun N-terminal kinase pathway and the induction of Fas ligand. J Neurosci. 2001;21:7551–60.

The manufacturer's authorised representative in the EU is Springer Nature Customer Service Centre GmbH, Europaplatz 3, 69115 Heidelberg, Germany. If you have any concerns regarding our products, please contact ProductSafety@springernature.com

Printed and bound by CPI Group (UK) Ltd, Croydon, CR0 4YY
30/09/2026
02225240-0005